Wearable Communication Systems and Antennas (Second Edition)

Design, efficiency, and miniaturization techniques

Wearable Communication Systems and Antennas (Second Edition)

Design, efficiency, and miniaturization techniques

Albert Sabban
ORT Braude Engineering College, Karmiel, Israel

IOP Publishing, Bristol, UK

ISBN 978-0-7503-5224-6 (ebook)
ISBN 978-0-7503-5220-8 (print)
ISBN 978-0-7503-5221-5 (myPrint)
ISBN 978-0-7503-5223-9 (mobi)

DOI 10.1088/978-0-7503-5224-6

Version: 20220601

IOP ebooks

British Library Cataloguing-in-Publication Data: A catalogue record for this book is available from the British Library.

Published by IOP Publishing, wholly owned by The Institute of Physics, London

IOP Publishing, Temple Circus, Temple Way, Bristol, BS1 6HG, UK

US Office: IOP Publishing, Inc., 190 North Independence Mall West, Suite 601, Philadelphia, PA 19106, USA

The book is dedicated to the memory of my father, mother and sister.
David Sabban, Dolly Sabban and Aliza Sabban.

Contents

Preface

The book's main objective is to present efficient wearable system and compact sensors and antennas for communication and healthcare systems. The major application of wearable body area networks, BANs, and of wireless body area networks, WBANs, is to help physicians to monitor the health of theirs patients. Optimal wearable communication systems must be developed by using system engineering techniques and tools such functional analysis system techniques, and quality function deployment Pugh method. The design of optimal wearable medical systems must be customer driven and meet customer needs.

Chapter 1 presents theory and techniques to develop wearable devices for communication and medical applications. Chapter 2 presents wearable technology employed to develop communication, medical and sport sensors and devices. Chapter 3 presents radio frequency and microwave theory used to develop wearable communication devices. These chapters are written to help researchers, engineers and students to study electromagnetic and communication theory. There are several microwave theory and antenna books for microwave and communication scientists. However, there are few books that can help engineers and students to study and practice microwave, communication and electromagnetic engineering by using basic mathematics, integral and differential theory.

Chapter 4 presents microwave technologies for wearable communication systems. Microwave and MM wave technologies are discussed in chapter 4. Chapter 4 presents MIC, MMIC, MEMS and LTCC technologies and devices. Chapter 5 presents microwave devices and modules development for wearable communication and medical systems. Body area networks, and wearable communication systems concept are presented in chapter 6. Moreover, this chapter presents system engineering fundamentals and tools. QFD, FAST and Pugh method are presented.

There are several 3D full-wave electromagnetics software ADS, AWR, HFSS, and CST programs used to develop and evaluate RF modules, antennas and communication systems. Communication systems and radiating elements developed and evaluated in this book were designed by using the commercial software, ADS and HFSS. For all devices, modules and antenna developed in this book there was good agreement between simulated and measured results.

Chapter 7 presents wideband wearable antennas for communication and medical applications. Antenna theory and small antennas design are discussed in this chapter. Several wearable antennas, such as printed and loop antennas, are presented in this chapter. Compact wideband antennas for wearable communication, 5G, and medical applications are presented in chapter 8. Design, development and measurements of wearable antennas close to the human body are presented in chapter 9. Antenna electrical parameters near the human body are discussed in chapter 9. New compact RFID antennas for wireless communication systems are discussed in chapter 10. Chapter 11 presents novel wearable antennas for wireless communication and medical systems. Novel compact wearable meta-materials antennas for communication, 5G and healthcare applications are presented in sections 11.1–11.6. Meta-material antenna electrical performance close to the human body is discussed in section 11.4. Wearable fractal printed antennas are presented in sections 11.7–11.11. Chapter 12 presents active wearable printed antennas for communication and medical applications. Tunable printed antennas for communication and medical applications are presented in sections 12.1–12.8. Active wearable receiving antennas are presented in section 12.9. Active transmitting antenna are presented in section 12.10. Chapter 12 presents development considerations, simulation and measured results of compact wearable transceivers, BANs, and healthcare devices. Chapter 13 presents new wearable wide band passive and active slot and notch antennas for communication and healthcare systems.

Base station aperture antennas for radar and communication systems are presented in chapter 14. Printed antenna arrays for RF communication systems are presented in section 14.5. Integrated MIC outdoor module for MM wave radar and satellite communication applications is presented in section 14.6. The solid state power amplifier, SSPA, is presented in section 14.7. The integrated, MIC, automatic Ku band tracking system is presented in section 14.9.

Compact circular patch wearable metamaterials sensors for healthcare, IoT, and communication systems are discussed in chapter 15. New passive and active sensors are presented in chapter 15.

Green electronic and communication technologies are presented in chapter 16. The continuous huge growth of the cellular communication industry in the last century has depleted and ruined the Earth's resources. The electronics, communication, and high tech industries are facing continuous pressure from legislators to replace toxic and hazardous material from their systems with green materials. In 2022 it is very important to develop green wearable sensors and antennas. This chapter presents comprehensive information about green technologies.

Analysis and design of wearable RF, healthcare and IoT devices are discussed in chapter 17. Wearable systems and antennas measurements are presented in chapter 18. This chapter presents RF, electromagnetic engineering, wearable systems, radiators analysis and design process. Computer aided design, CAD, software is presented in this chapter. Ethics considerations for healthcare and communication wearable systems are presented in chapter 19. Several ethical dilemmas must be considered when we develop wearable healthcare and communication systems such as patient rights, intellectual property, employee exploitation and justice.

Each chapter in the book presents sufficient information to help students, scientists, electrical and system engineers to study and employ the information discussed in the book. In the first chapters the book presents basic communication, RF and antenna theory to help students and researchers with no background in communication, microwave and RF theory to study and employ the fundamental development considerations and theory of antennas, wearable antennas, printed antennas and compact antennas for wireless communication and healthcare systems.

Only one design and fabrication iteration was needed in development process of the devices and antenna presented in this book.

The book presents for the first time new information, topics, sensors, designs and devices. This book may serve students and design engineers as a reference book.

The book presents new design in the area of wearable systems and antennas, metamaterial antennas, fractal antennas and active receiving and transmitting antennas. The text contains sufficient scientific information and design examples to help engineers, students and researchers to study all topics discussed in the new book.

Several new wearable antennas and communication systems are discussed in this book. Development tradeoffs, simulated and electrical performance of the new wearable systems and antennas is presented in the book.

Acknowledgements

Acknowledgements to my family.

My wife—Mazal Sabban
My daughters—Dolly and Lilach
My son—David Sabban
Grandchildren—Nooa, Avigail, Ido, Shira, Efrat, Yael Hodaia, Tamara.

Acknowledgements to my engineering colleagues who have helped me through my 40 years of my engineering and research career.

Author biography

Albert Sabban

Dr Albert Sabban received the BSc degree and MSc degree Magna Cum Laude in electrical engineering from Tel Aviv University, Israel in 1976 and 1986, respectively. He received his PhD degree in electrical engineering from University of Colorado at Boulder, USA, in 1991. Dr Sabban's research interests are biomedical communication systems, system engineering, wearable systems, computer science and computer aided design CAD, microwave and antenna engineering. Dr Sabban was an RF and antenna specialist at biomedical hi-tech companies from 2007 to 2010 where he designed wearable compact antennas and systems for medical systems. In 1976 he joined RAFAEL in Israel where he worked as a senior researcher, group leader and project leader in the electromagnetic department until 2008. In 2008 he retired from RAFAEL. From 2008 to 2010 he worked as RF specialist and project leader at hi-tech biomedical companies. From 2010 to date he has been a senior lecturer and researcher in the electrical engineering department at several Colleges in Israel. Dr Sabban served as the president of the academic staff at Ort Braude College in Israel. Dr Sabban served as the vice president of the academic staff organization of lecturers in colleges in Israel, 2014 to 2017. He has published over 100 research papers and holds patents in the compact antennas area. Dr Sabban has written five books on low visibility antennas for communication and medical applications, a book on electromagnetics and microwave theory for graduate students, and a book on wide band microwaves technologies for communication and medical applications. He has also written five chapters in books on microstrip and wearable printed antennas for medical applications. He has edited five books on green technologies and wearable printed antennas for medical applications.

Academic and professional awards and grants

Ort Braude College best lecturer in education Award October 2015.

Member of the winning team of Rafael Best Project Prize 2007.

Member of the winning team of Rafael Best Project Prize 1995.

Rafael award for outstanding engineering achievements, December 1984.

Dr Albert Sabban developed and led the development of the RF systems and modules in the projects that won the Israel Defence Prize in 1984, 1986 and 2006.

IOP Publishing

Wearable Communication Systems and Antennas (Second Edition)

Design, efficiency, and miniaturization techniques
Albert Sabban

Chapter 1

Theory of wireless wearable communication systems

This chapter is an introduction to wireless wearable communication systems. Receiving and transmitting communication systems transmit and receive electromagnetic waves. Electrical performance of wireless wearable communication and healthcare systems are affected by the human body tissues. Human body tissues attenuate electromagnetic waves.

1.1 Wireless wearable communication systems: frequency range

Wireless wearable communication and healthcare systems operate at frequencies from 100 MHz to up to 5 GHz, to date. There are novel wearable communication devices and sensors in mm wave frequencies. At mm waves the component size is of the same order of the wavelength. The oscillating voltages and currents period time in the circuit is comparable with the propagation time from one point of the circuit to another. At RF frequencies voltages and currents vary as a function of time and location. Low circuit analysis based Ohm's and Kirchhoff's laws cannot provide accurate analysis and description of electromagnetic fields, voltages, currents and impedances along the length of transmission lines and RF components. Lumped elements are components whose length is less than a tenth of the wavelength. Kirchhoff's and Ohm's laws can analyze lumped elements. Distributed elements are components whose length is higher than a tenth of wavelength. Ohm's and Kirchhoff's laws cannot provide accurate analysis of distributed elements.

Communication systems and services are allocated in frequency bands, see [1–4]. The frequency spectrum is divided into frequency bands to avoid electromagnetic interference of transmitted electromagnetic waves. RF frequency bands are divided from KHz frequencies up to THz frequencies. The frequency bands have a basic band plan that provides the information for how they can be used to prevent

Table 1.1. Frequency spectrum and applications of wireless wearable communication systems.

Frequency band abbreviation	ITU	Frequency band name	Frequency/ wavelength ($\lambda 0$)	Communication systems
LF	5	Low frequency	30–300 kHz; 10 km–1 km	Healthcare, wearable RFID
MF	6	Medium frequency	300–3 GHz; 1 km–100 m	Cellular phones, LAN, healthcare, wearable RFID
HF	7	High frequency	3–30 MHz; 100 m–10 m	RF shortwave broadcasts, wearable RFID, mobile radio telephony
VHF	8	Very high frequency	30–300 MHz; 10 m–1 m	RFID, FM, television broadcasts, land mobile communication, weather radio
UHF	9	Ultra-high frequency	300–3 GHz, 1 m–100 mm	Wireless LAN, Mobile phones, ZigBee, bluetooth, GPS and two-way radios-land mobiles, wireless wearable communication systems
SHF	10	Super high frequency	3–30 GHz; 100 to 10 mm	DBS, wireless LAN, wireless wearable communication systems

electromagnetic interference and to set a protocol for the compatibility of the transmitting and receiving channels. The frequency spectrum and main applications of wireless wearable communication systems are listed in table 1.1. The International Telecommunication Union (ITU) bands are listed in table 1.2. In table 1.3 IEEE standard frequency bands for communication systems are listed.

1.2 Free space propagation

The fundamentals of wireless communication systems are presented in several papers and books [5–7]. Equation (1.1) presents the flux density at distance R of an isotropic source radiating P_t watts uniformly into free space. The area of the spherical shell at distance R from the source center is $4\pi R^2$. The variation of power with angle is $P(\theta)$. The gain at θ direction is $G(\theta)$, usually expressed in Decibels (dB) see equation (1.2). P_0 is the total power transmitted.

$$F = \frac{P_t}{4\pi R^2} \text{W/m}^{-2} \tag{1.1}$$

$$G(\theta) = \frac{P(\theta)}{P_0/4\pi}; G\text{ [dB]} = 10\text{ log10 } G \tag{1.2}$$

Table 1.2. The ITU bands for wireless communication systems and applications.

Band symbol	Frequency range	Wavelength range	Band #	Main applications
VLF	3–30 kHz	10–100 km	4	Healthcare, wearable RFID
LF	30–300 kHz	1–10 km	5	Healthcare, wearable RFID
MF	300–3000 kHz	100–1000 m	6	Healthcare, wearable RFID
HF	3–30 MHz	10–100 m	7	RF shortwave broadcasts, RFID, mobile radio telephony
VHF	30–300 MHz	1–10 m	8	RFID, FM, television broadcasts, land mobile communication, weather radio
UHF	300–3000 MHz	10–100 cm	9	wireless LAN, mobile phones, ZigBee, bluetooth, GPS, wireless wearable RF systems
SHF	3–30 GHz	1–10 cm	10	DBS, wireless LAN, wearable communication systems

Table 1.3. Frequency bands, IEEE standard, for wireless wearable communication systems.

Symbols	Frequency range	Main applications
L band	1–2 GHz	wireless LAN, mobile phones, ZigBee, bluetooth, GPS
S band	2–4 GHz	wireless LAN, mobile phones, ZigBee, bluetooth, GPS
C band	4–8 GHz	WLAN, radars, seekers
X band	8–12.4 GHz	Radars, seekers, communication systems
Ku band	12.4–18 GHz	Radars, seekers, communication systems
Ka band	18–40 GHz	Radars, seekers, communication systems

A radiator that radiates in all directions equally is an isotropic radiator. The amount of power that the transmitter should produce to radiate equally to all directions is called effective isotropic radiated power, EIRP. The antenna gain varies with angle causing the EIRP to vary as a function of angle. We now want to find the power density at the receiver. In a lossless medium power is conserved. The power radiated from a transmitter passes through a spherical shell, with $4\pi R^2$ area, on the receiver surface. The spherical spreading loss is $1/4\pi R^2$. The power flux density is written in equation (1.3) as a function of the transmitting antenna gain.

$$F = \frac{\text{EIRP}}{4\pi R^2} = \frac{P_t G_t}{4\pi R^2} \text{Wm}^{-2} \tag{1.3}$$

The received power by an antenna of area A_r is written in equation (1.4):

$$P_r = F \times A_r = \frac{P_t G_t A_r}{4\pi R^2} \tag{1.4}$$

The antenna's effective aperture area is A_e which is less than the physical antenna aperture area, where, $A_e = A_{\text{phy}} \times \eta$. η = the antenna aperture efficiency.

The antenna maximum gain G is related to the effective aperture area as written in equation (1.5), where A_e is the effective aperture area.

$$G = \text{Gain} = \frac{4\pi A_e}{\lambda^2} \tag{1.5}$$

Horns and reflectors are called aperture antennas. They have a physical effective area that can be calculated from their dimensions.

$$A_{\text{phy}} = \pi r^2 = \pi \frac{D^2}{4} \tag{1.6}$$

Using equations (1.5) and (1.6) the gain equation for aperture antenna is written in equations (1.7) to (1.10).

$$\text{Gain} = \frac{4\pi A_e}{\lambda^2} = \frac{4\pi A_{\text{phy}}}{\lambda^2} \times \eta \tag{1.7}$$

$$\text{Gain} = \left(\frac{\pi D}{\lambda}\right)^2 \times \eta \tag{1.8}$$

$$\begin{gathered} \text{Gain} \cong \eta\left(\frac{75\pi}{\theta_{3\text{dB}}}\right)^2 = \eta\frac{(75\pi)^2}{\theta_{3\text{dBH}}\,\theta_{3\text{dBE}}} \\ \text{where } \theta_{3\text{dB}} \cong \frac{75\lambda}{D} \end{gathered} \tag{1.9}$$

$\theta_{3\text{dB}}$—The antenna half power beamwidth. A typical aperture efficiency of 0.55 gives equation (1.10).

$$\text{Gain} \cong \frac{30000}{(\theta_{3\text{dB}})^2} = \frac{30000}{\theta_{3\text{dBH}}\,\theta_{3\text{dBE}}} \tag{1.10}$$

1.3 Electromagnetic transmission, Friis formula

Electromagnetic transmission is presented by the Friis transmission equation, equation (1.11).

$$P_r = P_t G_t G_r \left(\frac{\lambda}{4\pi R}\right)^2 \tag{1.11}$$

Propagation loss in free space is presented by (L_p). L_a, presents losses due to attenuation in the atmosphere. These losses should also be accounted for in the Friis transmission equation, where, $L_p = \left(\frac{4\pi R}{\lambda}\right)^2$. The received power may be given as: $P_r = \frac{P_t G_t G_r}{L_p}$.

L_{pol} represents losses due to the polarization mismatch. L_{ra} represents losses associated with the receiving antenna, and L_r represents losses with the receiver. Losses associated with the transmitting antenna are written as L_{ta}. These losses cannot be neglected in computation of the communication system budget. P_r, is written in equation (1.12). Transmitting antenna power is P_t. L_t is the loss between the power source and the antenna.

$$P_r = \frac{P_t G_t G_r}{L_p L_a L_{ta} L_{ra} L_{pol} L_o L_r} \tag{1.12}$$

$P_t = P_{out}/L_t$
EIRP $= P_t\, G_t$

$$\begin{aligned} P_r &= \frac{P_t G_t G_r}{L_p L_a L_{ta} L_{ra} L_{\text{pol}} L_{\text{other}} L_r} \\ &= \frac{\text{EIRP} \times G_r}{L_p L_a L_{ta} L_{ra} L_{\text{pol}} L_{\text{other}} L_r} \\ &= \frac{P_{\text{out}} G_t G_r}{L_t L_p L_a L_{ta} L_{ra} L_{\text{pol}} L_{\text{other}} L_r} \end{aligned} \tag{1.13}$$

where,

$$G = 10 \log\left(\frac{P_{\text{out}}}{P_{\text{in}}}\right) \text{dB Gain in dB.}$$

$$L = 10 \log\left(\frac{P_{\text{in}}}{P_{\text{out}}}\right) \text{dB Loss in dB.}$$

Equation (1.14) presents the channel gain.

$$\begin{aligned} &P_{\text{in}} = \frac{V_{\text{in}}^2}{R_{\text{in}}} \qquad P_{\text{out}} = \frac{V_{\text{out}}^2}{R_{\text{out}}} \\ &G = 10 \log\left(\frac{P_{\text{out}}}{P_{\text{in}}}\right) = 10 \log\left(\frac{\frac{V_{\text{out}}^2}{R_{\text{out}}}}{\frac{V_{\text{in}}^2}{R_{\text{in}}}}\right) \\ &G = 10 \log\left(\frac{V_{\text{out}}^2}{V_{\text{in}}^2}\right) + 10 \log\left(\frac{R_{\text{in}}}{R_{\text{out}}}\right) = 20 \log\left(\frac{V_{\text{out}}}{V_{\text{in}}}\right) + 10 \log\left(\frac{R_{\text{in}}}{R_{\text{out}}}\right) \end{aligned} \tag{1.14}$$

Basic logarithmic equations

Equations (1.15–1.18) present useful logarithmic operations.

$$\begin{aligned} 10 \log_{10}(A \times B) &= 10 \log_{10}(A) + 10 \log_{10}(B) \\ &= A\text{dB} + B\text{dB} \\ &= (A + B)\text{dB} \end{aligned} \tag{1.15}$$

$$\begin{aligned} 10 \log_{10}(A/B) &= 10 \log_{10}(A) - 10 \log_{10}(B) \\ &= A\text{dB} - B\text{dB} \\ &= (A - B)\text{dB} \end{aligned} \tag{1.16}$$

$$10\log_{10}(A^2) = 2 \times 10\ \log_{10}(A) = 20\log_{10}(A) = 2 \times (A \text{ in dB}) \tag{1.17}$$

$$10\log_{10}(\sqrt{A}) = \frac{10}{2}\log_{10}(A) = \frac{1}{2} \times (A \text{ in dB}) \tag{1.18}$$

Linear ratios versus logarithmic ratios are listed in table 1.4.

Equation (1.19) lists the received power P_r in dBm. The received power P_r is referred to as the 'carrier power', P_c.

$$P_r = \text{EIRP} - L_{ta} - L_p - L_a - L_{\text{pol}} - L_{ra} - L_{\text{other}} + G_r - L_r \tag{1.19}$$

The power flow per unit area W at distance d from a transmitter antenna with input power P_T and antenna gain G_T is written in equation (1.20). The surface area of the sphere of radius d is $4\pi d^2$.

$$W = \frac{P_r G_r}{4\pi d^2}\ (\text{Watts m}^{-2}) \tag{1.20}$$

The effective aperture and size or of the receiving antenna dictate the antenna gain, $G_{R,}$ and the received signal strength. For an antenna with effective area A, the received signal power may be calculated by using equation (1.21).

$$P_R = P_T G_T (A/(4\pi d^2)) \tag{1.21}$$

Table 1.4. Logarithmic values for a given linear values.

Linear values	Logarithmic values, dB	Linear values	Logarithmic values, dB
0.0005	−33.0	1.000	0.0
0.001	−30	2.000	3.0
0.005	−23.0	3.000	4.8
0.010	−20.0	4.000	6.0
0.05	−13	5.000	7.0
0.1	−10	6.000	7.8
0.200	−7.0	7.000	8.5
0.300	−5.2	8.000	9.0
0.400	−4.0	9.000	9.5
0.500	−3.0	10.000	10.0
0.600	−2.2	18.000	12.6
0.700	−1.5	100.000	20.0
0.800	−1.0	1000.000	30.0
0.900	−0.5	10000.000	40.0

where G_R, $G_R = 4\pi A/\lambda^2$, is the receiving antenna gain and λ the signal wavelength is $\lambda = c/f$.

1.4 Wearable communication channel budget

Communication channel parameters are listed in table 1.5.
Calculation of receiver power in dB.
$P_R\ (\text{dBm}) = P_T\ (\text{dBm}) + G_T\ (\text{dBi}) + G_R\ (\text{dBi}) + 10\ \log_{10}((\lambda/4\pi)^2) - 10\ \log_{10}(d^2)$.

Wireless communication system downlink
$P_R\ (\text{dBm}) = (P_T\ G_T\ G_R\ L)\ (\text{dBm}) - 40\ \text{dB} + 10\ \log_{10}(1/d^2)$
$P_R - (P_T + G_T + G_R + L) - 40\ \text{dB} = 10\ \log_{10}(1/d^2)$
$115 = 10\ \log_{10}\ (1/d^2)$. The distance d is: $d = 10^{((155\ -\ 40)/20)} = 562$ km.

Wireless communication system uplink: For standard 802.11
$d = 10^{((153\ -\ 40)/20)} = 446$ km.

- $P_R - P_T = -113.2$ dBm,

For 6 Mbps:
$d = 10^{(113.2\ -\ 40)/20} = 4500$ m.
$d = 10^{(113.2\ -\ 40\ -\ 3)/20} = 3235$ m with 3 dB gain margin.
$d = 10^{(113.2\ -\ 40\ -\ 3\ -\ 9)/20} = 1148$ m with 3 dB gain margin and neglecting antenna gains.

For 54 Mbps the link needs −85 dBm:
$d = 10^{(99.2\ -\ 40)/20} = 912$ m.
$d = 10^{(99.2\ -\ 40\ -\ 3)/20} = 646$ m with 3 dB gain margin.
$d = 10^{(99.2\ -\ 40\ -\ 3\ -\ 9)/20} = 230$ m with 3 dB gain margin and neglecting antenna gains.

Signal strength

- Measure signal strength in dBW = 10 log (power in watts)

  - dBm = 10 log (power in mW)

- For standard 802.11 can legally transmit at 30 dBm.
- Most 802.11 PCMCIA cards transmit at 10–20 dBm.
- Mobile phone base station: 20 W, but 60 users, so 0.3 W/user, antenna gain = 18 dBi.
- Mobile phone handset: 21 dBm.

Table 1.5. Communication channel parameters.

Parameter	Value	Wavelength cm	Propagation loss dB
Frequency GHz	2.4	$\lambda = c/f = 12.5$ cm	$10\ \log_{10}\ ((\lambda/4\pi)^2) = -40$
Frequency GHz	0.933	$\lambda = c/f = 32$ cm	$10\ \log_{10}\ ((\lambda/4\pi)^2) = -32$

1.5 Noise

Noise in a communication system is unwanted disturbance generated by several different effects such as temperature, atmospheric noise, flicker noise and several other types. Communication system dynamic range is defined as the system's capability to detect weak signals in the presence of large-amplitude signals. Noise limits the communication system ability to detect and analyze weak signals. There are several noise sources.

Noise sources

1. Thermal noise present in all electronics devices due to thermal agitation of electrons.
   N_t = Thermal noise = $kTBw(W)$
   Where k is Boltzmann's constant = 1.38×10^{-23}. Bw is bandwidth.
   Where T is temperature in Kelvin (C + 273). For example—
   For temperature = 293 °C, => −203 dB, −173 dBm Hz^{-1}.
   For temperature = 293 °C and 22 MHz => −130 dB, −100 dBm.
2. Random noise in resistors and transistors.
3. Noise generated by mixers.
4. Undesired cross-coupling noise from other transmitters and equipment.
5. Noise generated by power supply devices.
6. **Burst noise** consists of sudden step-like transitions between two or more discrete voltages or current levels, as high as several hundred microvolts, at random and unpredictable times.

Random noise

- Noise caused by external sources.
- Atmospheric noise. Also called static noise. Caused by electrical disturbances occurring in nature such as lightning.
- Interstellar noise.
- Cosmic noise. Stars generate noise called cosmic noise. These stars are too far away to individually affect communications systems, their large number leads to appreciable collective effects. Cosmic noise has been observed in a range from 8 MHz to 1.43 GHz.

Receiving channel (internal) Noise

- Thermal noise. Thermal noise is generated by the random thermal motion of charge carriers inside a conductor which happens regardless of any applied voltage.
- Flicker noise. $1/f$ noise which is negligible at high frequencies.
- Shot noise. Shot noise in electronic devices results from unavoidable random fluctuations of the electric current when the charge carriers transverse a gap.

Signal-to-noise ratio is written as SNR and is given in equation (1.22). SNR varies with frequency. SNR = signal power/noise power. Noise factor, F, is a measure of the degradation of SNR due to the noise added as we process the signal. F is given in equations (1.23) and (1.24).

$$\text{SNR} = \frac{S(f)}{N(f)} = \frac{\text{average} - \text{signal} - \text{power}}{\text{average} - \text{noise} - \text{power}} \tag{1.22}$$

SNR (dB) = 10 $\log_{10}$(signal power/noise power)

$$F = \frac{\text{available} - \text{output} - \text{noise} - \text{power}}{\text{available} - \text{output} - \text{noise} - \text{due} - \text{to} - \text{source}} \tag{1.23}$$

Noise figure = $NF = 10\log(F)$.

By using equation (1.24) the **noise figure** of a communication channel can be calculated.

$$F = F_1 + \frac{F_2 - 1}{G_1} + \frac{F_3 - 1}{G_1 G_2} + \ldots + \frac{F_n - 1}{G_1 G_2 \cdots G_{n-1}} \tag{1.24}$$

Signal strength is the transmitted power multiplied by a gain minus losses.

Loss sources in communication systems

- Distance between the receiving and transmitting channels.
- Rain and fog signal losses at high frequencies.
- Environmental media losses.
- Reflection losses.
- Multi-path losses. Multi-signal interference.
- The gain of the receiving and transmitting antennas is function of the relative angle between them. The received signal must have a strength that is larger than the receiver's sensitivity.

Sensitivity is defined as the minimum detectable input signal level for a given output SNR, also called the noise floor. A system SNR of 20 dB or better is a good SNR value.

1.6 Communication systems channel budget calculation

A link budget evaluation determines if there is enough power at the receiver to recover the information. Evaluation of the link budget determines if the received signal is larger than the receiver's sensitivity. The link budget must account for effective transmission power. Example of a transmitting link power budget is presented in table 1.6.

Transmitting channel main parameters

- Transmission power.
- Transmitting antenna gain.
- Transmission lines losses in cables and connectors.

Path losses are due to:

- Electromagnetic waves are attenuated as function of the distances from the transmitter.
- Reflections from the ground.
- Fading (self-interference).

Table 1.6. Transmitting link power budget for wireless wearable communication systems.

Component	Gain (dB)/loss (dB)	Power (dBm)	Remarks
Input power		−12	
Transmitter gain	42		
Power amplifier output power		30	
Filter losses	1	29	
Transmission line losses	1.5	27.5	
Matching losses	0.5	27	
Transmitted power		27	

Table 1.7. Receiving link power budget for wireless wearable communication systems.

Component	Gain (dB)/loss (dB)	Power (dBm)	Remarks
Input power		−20	
Receiver gain	25		
Transmission line losses	1	4	
Filter losses	1	3	
Matching losses	1	2	
Low noise amplifier (LNA) output power		2	

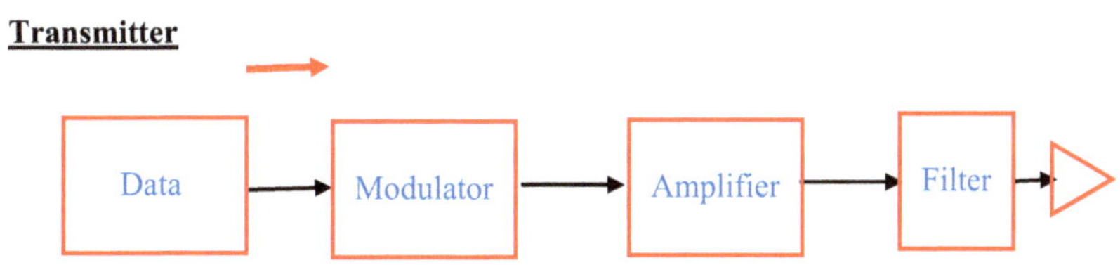

Figure 1.1. Transmitting channel block diagram for wireless wearable communication systems.

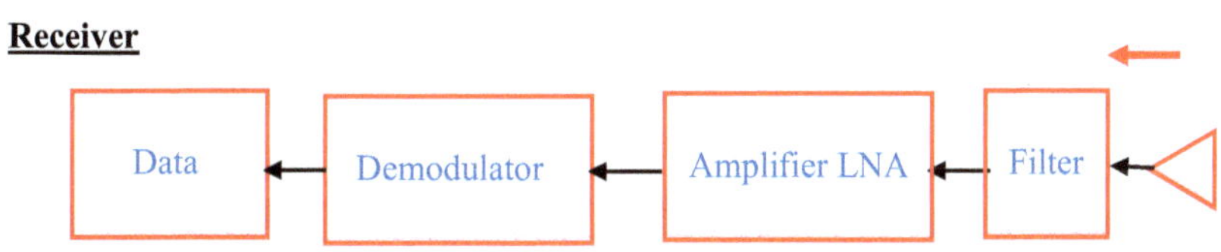

Figure 1.2. Receiving channel block diagram for wireless wearable communication systems.

Receiver

- Receiver sensitivity.
- Losses in cable and connectors.

The receiving channel power budget is listed in table 1.7. A transmitter block diagram for wireless communication systems is shown in figure 1.1. A receiver block diagram is shown in figure 1.2.

1.7 Communication system path loss

Electromagnetic waves are attenuated as function of the distance from the transmitter. Path losses are losses that are a direct result of the distance between the receiver and the transmitter.

There are many models used in the industry today to estimate the path loss and the most common are the free space and Hata models. Each model has its own requirements that need to be met in order to be utilized correctly. The free space path loss is the reference point that other models use.

Free space path loss

The free space model typically underestimates the path loss experienced for mobile communications, and predicts point-to-point fixed path loss.Free space path loss (dB) =

$$20\log_{10}f + 20\log_{10}d - 147.56$$

where F is the frequency in Hz, d is the distance in meters.

Hata model

The Hata model is used extensively in cellular communications. The basic model is for urban areas, with extensions for suburbs and rural areas.

The Hata model is valid for these ranges only:

- Distance 1–20 km.
- Base height 30–200 m.
- Mobile height 1–10 m.
- 150–1500 MHz.

The Hata formula for urban areas is:

$$L_H = 69.55 + 26.16\log_{10}f_c - 13.82\log_{10}h_b - \text{env}(h_m) + (44.9 - 6.55\log_{10}h_b)\log_{10}R$$

- h_b is the base station antenna's height in meters.
- h_m is the mobile antenna's height also measured in meters.
- R is the distance from the cell site to the mobile in km.
- f_c is the transmitting frequency in MHz.
- $\text{env}(h_m)$ is an adjustment factor for the type of environment and the height of the mobile. $\text{env}(h_m) = 0$ for urban environments with a mobile height of 1.5 m.

1.8 Receiver sensitivity

Sensitivity describes the weakest signal power level that the receiver is able to detect and decode. Sensitivity is determined by the lowest SNR at which the signal can be recovered. Different modulation and coding schemes have different minimum SNRs. Sensitivity is determined by adding the required SNR to the noise present at the receiver.

Noise sources

- Thermal noise.
- Noise introduced by the receiver's amplifier.

Thermal noise = $N = kTBw$ (watts)
- $k = 1.3803 \times 10^{-23}$ J K^{-1}.
- T = temperature in Kelvin.
- Bw = receiver's bandwidth.

N (dBm) = 10 $\log_{10}(kTBw) + 30$
Thermal noise is usually very small for reasonable bandwidths.
For $T = 290$ K (17 °C), $Bw = 1$ Hz
Thermal Noise = $1.38 \times 10^{-23} \times 290 \times 1 = 4.002 \times 10^{-21}$ W Hz^{-1} = 4.002×10^{-18} mW Hz^{-1}.
If we convert it to dBm, then,
4.002×10^{-18} mW Hz^{-1} = 10 $\log_{10}$ (4.002×10^{-18}) =6.0 − 180 = −174 dBm Hz^{-1}
For example thermal noise for −40 °C is −174.9 dBm Hz^{-1}. Thermal noise for 75 °C is −173.2 dBm Hz^{-1}.
For T = 290 K (17 °C), Bw = 1 MHz. Thermal Noise = −174 dBm+60 = −114 dBm.

Basic receiver's sensitivity calculation
Sensitivity (W) = $kTBw$ NF (linear) minimum SNR required (linear).
Sensitivity (dBm) = 10 $\log_{10}(kTBw\ 1000)$ + NF(dB) + minimum SNR required (dB).
Sensitivity (dBm) = 10 $\log_{10}(kTBw)$ + 30 + NF(dB) + minimum SNR required (dB).

Sensitivity decreases in communication systems when:
- The system bandwidth increases;
- The environmental temperature increases;
- The amplifier noise figure increases;
- Rain and snow increase considerably the path losses.

1.9 Definitions and characteristics of receiving channel

Figure 1.3 presents a receiving channel block diagram.
Basic definitions of receiving links
Communication system RF, IF and LO frequencies
The RF frequency is the frequency of the input signal of the receiving link that uses a mixer as a downconverter. The local oscillator provides a signal, LO frequency, to convert down the RF frequency. The product of the mixing process is a low frequency which is defined as the intermediate frequency (IF). The IF frequency is the difference between the RF frequency and the LO frequency.

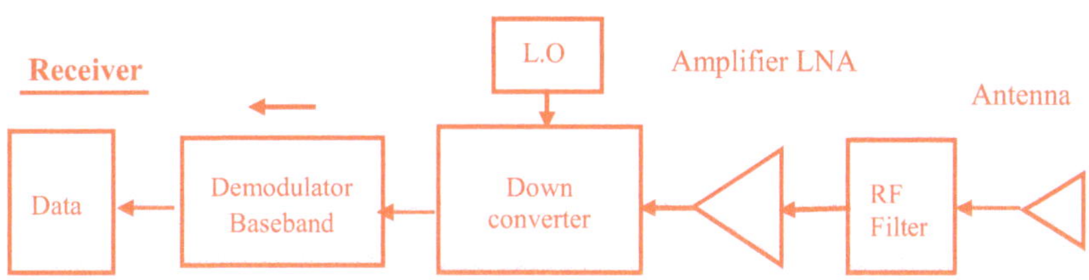

Figure 1.3. Receiving channel block diagram for wearable communication systems.

Communication system base band frequency
The base band frequency is the frequency at which the data is evaluated and analyzed.
Gain of the receiving channel
The gain of the receiving channel is the ratio of the input signal power to the output signal power.
Communication system pre-selector filter
Undesired RF signals are attenuated by using a pre-selector filter. The pre-selector filter of a cellular phone attenuates the high-power signals of transmitting radars that may saturate the receiving channel.
Communication system amplitude and phase matching
In communication systems that consist of several receiving links it is important that the magnitude and phase of the received signals match and track each other over the system frequency range. Amplitude and phase *matching* is achieved when the relative magnitude and phase of the received signals are almost equal.
Communication system instantaneous bandwidth and tuneable bandwidth
In a receiver with a fixed LO frequency, when the input frequency is swept to measure the system frequency response, we get the system instantaneous bandwidth. The resulting instantaneous bandwidth depends on the frequency responses of all the modules in the channel. The minimum detectable signal of the communication system is a function of the system instantaneous bandwidth. Tuneable bandwidth is achieved when we sweep the LO frequency to track the RF frequency. The tuneable bandwidth depends on the bandwidth of the LNA, pre-selector filter, and the mixer. Usually the instantaneous bandwidth is often lower than the tuneable bandwidth.
Receiving channel noise figure
Noise is an important parameter in the development of receiving channels that should detect extremely weak RF signals. The system noise figure defines the minimum detectable signal. The noise figure, NF, of a receiving channel is a measure of how much the receiving channel degrades the ratio of the input signal SNR. The noise contribution of each module in the signal path must be low enough that it will not significantly degrade the system SNR. If the LO signal has a high AM or FM noise, it may degrade the receiver noise figure. Far from the carrier signal, the AM and FM noise originate from the thermal noise. The effect of LO AM noise is reduced by using a balanced mixer.
1 dB compression point
The 1 dB compression point is the power level where the gain of the receiver is reduced by 1 dB due to compression.
Linearity
In a linear receiver a 1 dB increase in input signal power results in a 1 dB increase in IF output signal power.
Dynamic range
The dynamic range of a receiving channel is the difference between the maximum signal that will start to compress the receiver to the minimum detectable signal.
Signal-to-noise ratio (SNR)
SNR is the ratio between the signal and the noise in a communication system. The communication system SNR is a measure of how far a signal is above the noise floor.

Noise in communication systems. Noise factor, noise figure and noise temperature

- The noise factor is the ratio between the SNR at the input of a device to the SNR at the output of the device. $N_{in} = KT$. K is Boltzmann's constant. T is room temperature in Kelvin. KT is around −174 dBm Hz^{-1}. The noise factor of a device is specified as noise from a noise source at room temperature.
  - Noise factor = $F = (S_{in}/N_{in})/(S_{out}/N_{out})$
  - The signal level at the input is S_{in}
  - The noise level at the input is N_{in}
  - The signal level at the output is S_{out}
  - The noise level at the output is N_{out}

- Noise figure is the noise factor, expressed in decibels.

Noise figure = NF (decibels) = 10 log (F).
T = noise temperature = 290 (F − 1).
1 dB NF is about 75 K, and 3 dB is 288 K.

Equation (1.25) presents the noise factor contributions of each stage in a four-stage system.

$$F = F_1 + \frac{F_2 - 1}{G_1} + \frac{F_3 - 1}{G_1 G_2} + \frac{F_4 - 1}{G_1 G_2 G_3} \tag{1.25}$$

1.10 Basic features of radars

Mono-static radars: In mono-static radars the same antenna is used to transmit the RF signals and to receive the RF signals. Many radars are mono-static.

Bi-static radar: Bi-static radars have a transmitting antenna and a receiving antenna. RF signals are transmitted via the transmitting antenna. RF signals are received via the receiving antenna.

Doppler radars are used to measure the velocity of a target, using the Doppler shift. Missile and police radars are a good example of Doppler radars.

FMCW radar: FMCW radar can gather both range and velocity information of the target. FMCW means **F**requency **M**odulated/**C**ontinuous **W**ave. Frequency modulated/continuous wave implies that the radar signal is 'chirped', or the system frequency is varied in time. FMCW radars have extremely high resolutions, and relatively clutter-free displays.

Synthetic aperture radar (SAR): Satellite radar images are mostly using SAR radars. SAR is a radar that is used to create two-dimensional images or three-dimensional reconstructions of objects. SAR uses the motion of the radar antenna over a target region to provide finer spatial resolution than conventional stationary beam-scanning radars. SAR is typically mounted on a moving platform, such as an aircraft or spacecraft. SAR uses the moving platform to scan the radar in one or two dimensions. The distance the SAR travels over a target during the period when the target scene is illuminated creates the large synthetic antenna aperture. Typically, the larger the aperture, the higher the image resolution will be. SAR creates high-resolution images with comparatively small physical antennas.

1.11 Communication systems transmitters—definitions and features

Figure 1.4 presents a basic communication system transmitting channel block diagram.

High-power amplifiers, HPA, and medium power amplifiers are basic modules in transmitting channels. Amplifiers may be operated as class A, B, C, and D amplifiers.

Class A—Usually class A amplifiers have low efficiency 10% to 30%. Drain efficiencies of 50% are possible theoretically in class A. Class A amplifiers provides the best linearity. A Class A amplifier is biased at close to half of its saturated current. The output conducts during all 360° of phase of the input signal sine wave.

Class B—A class B amplifier can be very efficient, with efficiency of 80%–85% theoretically. The power amplifier is biased at a point where it draws nearly zero DC current. For a FET, this means that it is biased at pinch-off. During one half of the input signal sine wave it conducts, but not during the other half. The gain of a class B amplifier may be lower by six dB than the gain of a class A amplifier.

Class C—Usually class C amplifiers operate as high-power nonlinear amplifiers. Class C amplifiers operate when the amplifier is biased so that the output conducts for even less than 180° of the input signal. The amplifier gain decrease.

Power density is a measure of power level divided by the transistor's size. For FETs it is expressed in W mm^{-1}. GaN power transistors have more than 10 W mm^{-1} power density.

Saturated output power—PSAT is the output power where the P_{in}/P_{out} curve slope goes to zero.

Load pull—The process of varying the impedance seen by the *output* of an active device to other than 50 ohms in order to measure the gain of the device or other electrical parameter. In a power amplifier, a load pull power bench is used to explore large signal parameters such as saturated power, compression characteristics, efficiency and linearity as the output load is varied across the Smith chart.

Harmonic load pull—The process of varying the impedance at the output of an active device, with separate control of the impedances at harmonic frequencies such as F0, 2F0, 3F0, and so on.

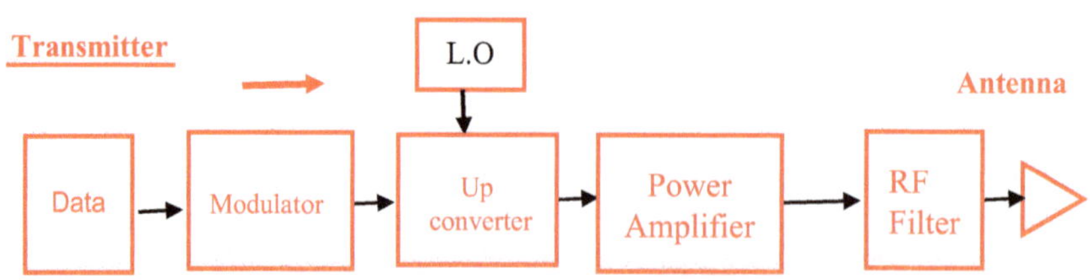

Figure 1.4. Basic transmitting channel block diagram for wearable communication systems.

Source pull—The process of varying the impedance seen by the input of an amplifier to other than 50 ohms in order to measure the electrical performance of the device. In the case of a low noise device, source pull is used in a noise parameter extraction setup to evaluate how the SNR and NF vary with the source impedance.

Temperature considerations for RF Amplifiers

The gain drop of FET amplifier with temperature is around 0.006 dB/stage/°C.

In an LNA, the first stage will dominate the module temperature effect. The noise figure of an LNA increases by +0.006 dB/°C. The amplifier noise figure increases with temperature.

RF systems power amplifiers

Power amplifiers transmit high power and they are used as transmitters in RF systems. Power amplifiers boost a small signal to a large signal. Solid state amplifiers and tube amplifiers are usually employed as power amplifiers. The output power range of power amplifiers is presented in table 1.8. The transmitted power range is from 0.1 W and up to 100 W at low frequencies. However, at high frequencies, 30–40 GHz, the transmitted power is lower than 3 W.

1.12 Introduction to wearable communication and IOT systems basics

Wireless communication systems play a major role in our lifestyle and each minute of our life. Almost every simple procedure in our life is accompanied and monitored by cellular phones and smartphones. Since 1980, cellular phone technology has progressed every year. Communication engineers developed five generations of cellular phones and smartphones. There was a great improvement from the first generation of cellular phones up to the fifth generation of cellular phones and smartphones. The sixth generation of cellular smartphones is in development stage. Communication theory and design are presented in several books and scientific papers, see [4–8]. Theory and design of wearable communication systems are presented in several books and scientific papers, see [4–14].

Generations of mobile networks

First generation (1G)

The first generation of cell phones began with 1G technology in 1980 and it is the first generation of wireless cellular technology. The first generation of mobile networks was based on analog radio systems. 1G supports voice calls only.

Table 1.8. RF HPAs output power capabilities.

Frequency range, GHz	Solid state amplifiers	Tube power sources
1–8	250 W (LDMOS) GaN	
8–12.4	60 W (GaN HEMT device)	3200 W (TWT)
12.4–18	8 W (GaAs PHEMT device)	1200 W (klystron)
18–26	5 W (GaAs PHEMT device)	

Mobile network users could make phone calls only. Users couldn't send or receive text messages. The maximum speed of 1G technology is 2.4 Kbps.

1G basic performance and features

- Mobile network speed—2.4 kbps;
- 1G technology could provide voice calls in one country;
- 1G technology employ analog signal;
- 1G technology had poor voice quality;
- 1G technology had poor battery life;
- 1G technology mobile network had large phone size;
- 1G technology had limited capacity;
- Poor handoff reliability;
- 1G technology had poor security;
- 1G technology provided very low level of spectrum efficiency.

Bits per second: The data rate of a communication system is normally measured in units of bits per second. Kbps is 1000 bits per second. Mbps is 1000 Kbps. Gbps is 1000 Mbps or one million Kbps.

GSM: GSM is a cell phone standard and means global system for mobile communications. GSM provides standard features like SMS, data networking, caller ID, phone call encryption, call forwarding, call waiting, and conferencing. GSM cell phone technology operates in the 1900 MHz frequency in the United States and the 900 MHz frequency in Europe and Asia. GSM phones can store the subscriber's information such as phone number and other data that proves that a user is in fact a subscriber to that carrier on a SIM card. *SIM is a subscriber identification module.* Cellular phones need a SIM card to communicate with the mobile network and to identify the phone owner. GSM provides standard features such as caller ID, phone call encryption, SMS, call forwarding, call waiting, data networking, and conferencing.

Second Generation (2G)

2G cellular phones employ GSM standard. GSM advanced cellular phones to digital communication technology. The 2G telephone technology introduced call and text encryption, along with data services such as SMS and MMS. The maximum speed of 2G with general packet radio service (GPRS) is 50 Kbps.

SMS: SMS means short message service, which is the formal name for the technology used for text messaging. The messages are sent through a cellular data network. It's a way to send only short text messages from one phone to another phone.

MMS: MMS, multimedia messaging service. MMS allows cellular phones and smartphone users to send each other SMS messages with audio files, images, photos, videos, and other attachments. Using MMS, the cell phone can send audio files, contact details, ringtones, photos, videos, and other data to any other phone with a text messaging plan.

EDGE: EDGE is a faster version of GSM with a speed and latency advancement in GSM technology. EDGE stands for enhanced data rates for GSM evolution. EDGE networks were developed to deliver multimedia applications such as streaming television, audio, and video to cellular phones at speeds up to 384 Kbps.

2G basic performance and features

- 2G has data speed up to 64 kbps;
- 2G employs digital signals and technology;
- 2G enables services such as text and audio messages, picture messages and multimedia messages;
- 2G provides better quality and capacity;
- 2G technology handle complex data such as videos, audio messages and photos;
- 2G technology require strong digital signals to help mobile phones work.

2.5G and 2.75G

A continuous growth in mobile technology led to the intermediate 2.5G and 2.75G generations. 2.5G and 2.75G introduced a new packet-switching technique that was more efficient than 2G technology. 2.75G provided a theoretical threefold speed increase.

2.5G and 2.75G performance and features

- These technologies provide phone calls;
- These technologies provide send/receive email messages;
- These technologies provide web browsing;
- The speed rate of 2.5G and 2.75G is 64–144 kbps;
- These technologies provide camera phones;
- In 2.5G and 2.75G it takes a time of 6–9 min to download a 3 min MP3 song.

Third generation (3G): 3G technology has high-speed access to data and voice services. A 3G network is a high-speed mobile broadband network with data speed rate of at least 144 Kbps. 3G networks can offer speeds of 3.1 Mbps.

3G basic performance and features

- This technology provides speed rate 2 Mbps;
- This technology provides increased bandwidth and data transfer rates to accommodate web-based applications and audio and video files;
- 3G technology provides faster communication;
- This technology provides send and receive large email messages;
- 3G technology provides high-speed web, more security and video conferencing;
- 3G technology provides 3D gaming;
- 3G technology provides large capacities and broadband capabilities;
- 3G technology provides TV streaming, mobile TV and phone calls;
- By using 3G technology we can download a three minute MP3 song;
- Disadvantage of this technology is the expensive fees for 3G licenses services;
- 3G technology has high bandwidth requirement;
- Disadvantage of this technology is the expensive price of 3G phones.

Fourth generation (4G): 4G technology provides mobile web access, HD mobile TV, video conferencing, 3D TV, gaming services, and other features that require high speed. The max speed of a 4G network for moving devices is 100 Mbps.

4G performance and features

- 4G technology may provide speed rate of 10 Mbps to 1 Gbps;
- 4G technology may provide high quality streaming video;
- 4G technology may provide a combination of Wi-Fi and Wi-Max;
- 4G technology may provide high security;
- 4G technology may support any kind of service at any time as per user requirements anywhere;
- 4G technology may provide expanded multimedia services;
- 4G technology provide low-cost per-bit prices.

Fifth generation (5G): 5G technology provides significantly faster data rates, much lower latency, higher connection density, and energy savings and several other improvements. The max speed of a 5G network will be up to 20 Gbps.

5G performance and features

- 5G technology is highly supportable to Wireless World Wide Web;
- 5G technology provides high speed, much lower latency, higher connection density, high capacity;
- 5G technology provides large broadcasting of data in Gbps;
- 5G technology supports multimedia newspapers, watching TV programs with HD clarity;
- 5G technology provides faster data transmission than that of the previous generation;
- 5G technology provides large phone memory, clarity in audio and video, dialing speed;
- 5G provides high-resolution and bi- directional large bandwidth sharing to cellular phone users.

1.13 Internet of things IoT basics

The internet of things, IoT, is a system of interrelated computing networks, personal sensors and devices, commodities, mechanical and digital devices, people, healthcare sensors, animals, and wireless communication systems that have a unique identifier (UID) and can transfer data over a network without requiring human-to-human or human-to-computer assistant. The IoT is an advanced 21st century emerging technology. It helps people in everyday life to get complete control over daily services and procedures and work smarter. IoT helps to automate, home smart devices, companies, business and healthcare centers. IoT provides companies with a real time inspection and observation of how their organization functions online. IoT enables medical organizations, business, and companies to automate processes and reduce everyday expenses. It may improve company services, cut down on waste and improve product delivery. IoT helps to manufacture and deliver products efficiently and with less expense.

IoT is relevant to all organizations, education, finance, healthcare, retail, research and development, manufacturing and marketing. Smart devices and cities help citizens to reduce waste, energy consumption, and daily life expenses. Sensors are used in agriculture and farming to help inspect and monitor crop and cattle yields and predict growth patterns.

IoT has emerged by converging new technologies such as computing technologies, wireless communication technologies, internet, and microelectromechanical systems. The convergence helped to combine between computing technologies, operational technology, and information technology, allowing unstructured machine-generated data to be monitored and evaluated for insights to drive significant improvements. IoT has emerged from machine-to-machine, M2M, communication, machines connecting to machines through a network without human assistance. M2M refers to connecting machines and devices to the cloud, monitoring the devices and collecting information. M2M provides the connectivity that enables IoT. IoT is a sensor network of thousands of smart sensors and devices that connect people, machines, organizations, systems and other applications to monitor, analyse, collect and share information.

The IoT is an extension of SCADA (supervisory control and data acquisition), which is a software application program to monitor process and collecting data in real time from remote locations to monitor devices, machines and conditions. SCADA systems include hardware and software devices. The devices collect and feed information into a computer that uses SCADA software, where the information is analyzed and processed in real time. New SCADA systems and software were developed into the first generation of IoT systems.

An IoT system consists of internet-enabled smart sensors and devices that use embedded processors, sensors and communication hardware to gather, send and analyze information they collect from their environments. IoT systems share the sensors information they gather by connecting to an IoT gateway or other IoT system where data is processed or sent to the cloud to be evaluated and analyzed. Sometimes, these systems communicate with other devices and process the data they receive from other devices. These devices do a great part of the process without human assistance, although users can interact with the devices, to turn them on and off, monitor them, or access the information. The networking, connectivity, and communication protocols employed with these internet-enabled devices depend on the specific IoT system deployed. An IOT system block diagram is shown in figure 1.5.

IOT benefits to companies and organizations

- Observe and monitor companies' and organizations' processes;
- Improve companies' and organizations' customer experience;
- Decrease companies' expenses, save time and money;
- IoT gives companies and organizations the tools to improve their strategies;
- Improve organizations' employee productivity, decrease work cost;
- Improve and adapt company models to maximize the company productivity;
- Optimize and improve business decisions, and generate more revenue.

IOT advantages to companies and organizations

- IoT provides access to information from anywhere at any time on any device;
- IoT improves online communication between connected electronic systems;

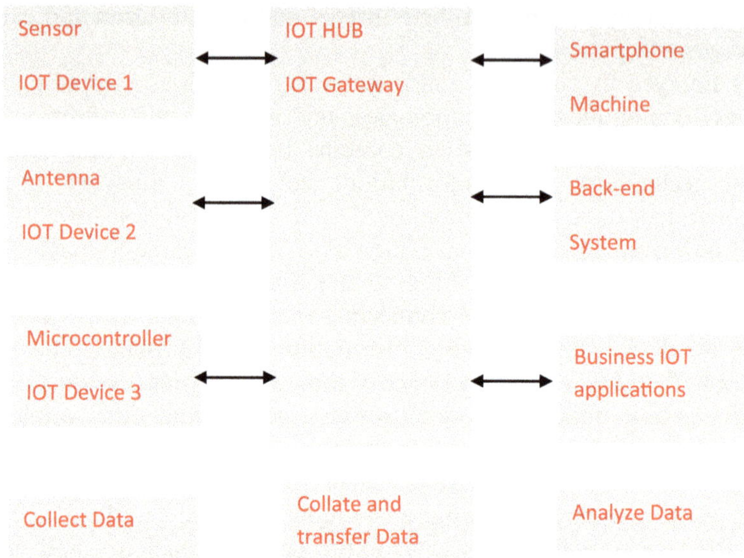

Figure 1.5. IOT system block diagram.

- IoT provides data packets transfer over connected networks and saves time and money;
- IoT automates tasks to improve the efficiency and quality of a companies' services and reduces human labour.

IOT disadvantages to companies and organizations

- As the number of connected systems increases and more data is shared between systems, the potential that hackers could steal confidential information increases;
- Enterprises may deal with huge numbers, up to millions, of IoT systems and gathering and monitoring the information from all those systems will be challenging;
- If there's a bug in the IoT system, probably every connected device will be corrupted;
- To date there's no compatibility international standard for IoT systems, it is almost impossible for systems from different manufacturers to communicate with each other.

1.14 Satellite communication transceiver

This section presents an example of a satellite communication transceiver. This section presents the development, electrical performance, and fabrication of an MIC RF module for satellite communication systems. Compact surface mount components were used to fabricate the RF head.

1.14.1 Introduction

Satellite communication, mobile communications, and the wireless communication industry are currently growing [8–15]. The great public demand for smartphones, cellular and cordless phones, and other smart devices has stimulated a huge interest in novel mobile services such as portable satellite communication terminals and smart watches. For example, the InMarsat-M system and VSAT system provides digital communications between the public switched terrestrial networks and mobile cellular users. Table 1.8 presents HPAs' output power capabilities.

Communication channels to and from mobile installations can be established through a satellite such as InMarsat geostationary satellite or other satellites and the associated ground station.

The RF communication system consists of a receiver and transmitter, synthesizers, RF controller, modem, and a DC supply and control unit. The RF system size is 300 × 200× 25 mm and weighs 1 kg. The transmitted power level is controlled by an automatic control unit to ensure low power consumption over all the frequency and temperature range. The transmitter may be operated in high-power mode to transmit 10 W or in low power mode to transmit 4 W. The RF system vent is set to 'on' and 'off' automatically by the RF controller. The gain of the receiving channel is 76 dB and is temperature-compensated by using a temperature sensor and a voltage-controlled attenuator.

1.14.2 Receiving channel development and specifications

Figure 1.6 presents the block diagram of the receiver.

The receiving link includes low noise amplifiers, filters, saw filter, active mixer, temperature sensor, voltage-controlled attenuator and IF amplifiers.

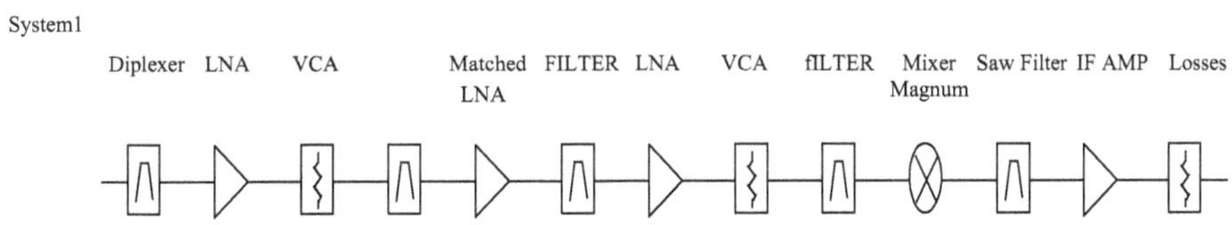

														Total
Gain (dB)	-1.00	20.00	-3.50	-2.00	14.00	-2.00	10.00	-3.50	-2.00	15.00	-8.00	42.50	-2.50	76.33*
NF (dB)	1.00	0.95	3.50	2.00	1.45	2.00	0.90	3.50	2.00	20.00	8.00	2.50	2.50	2.35
IP1dB (dBm)	20.00	-2.00	20.00	20.00	4.00	20.00	7.00	20.00	20.00	8.00	20.00	10.00	20.00	-59.50
NF+ (dB)	0.71	0.86	0.04	0.04	0.04	0.00	0.00	0.00	0.00	0.26	0.00	0.00	0.00	

Input Pwr (dBm)	-60.00	System Temp (K)	290.00

Modulation: FM

System BW (MHz)	0.03	MDS (dBm)	-126.86	Input IP3 (dBm)	N/A
S/N (dB, Actual)	66.18	S/N (dB, Req'd)	6.00	Output IP3 (dBm)	N/A
Srce Temp (K)	290.00	Sens. Loss (dB)	0.00	OIM3 (dBm)	N/A
Te Eff. (K)	208.30	Sensitivity (dBm)	-120.86	ORR3 (dB)	N/A
SFDR3 (dB)	N/A	G/T (dB/K)	-16.97	IRR3 (dB)	N/A

Figure 1.6. Block diagram and gain budget of the receiver.

The LNA, GaAS FET, has around 11 dB gain and 0.8 dB noise figure for frequencies from 1.525 to 1.559 GHz. The total gain of the receiving link is around 75 dB. The LNA price is $1.8.

Table 1.9 presents the receiver, power budget and noise figure computation results. The link gain is temperature-compensated by connecting the output port of a temperature sensor to the reference voltage port of a voltage-controlled attenuator. The variation of the sensor output voltage as a function of temperature varies the attenuation level of the attenuator. Gain stability as a function of temperature is around 1 dB. The receiver noise figure is less than 2.1 dB. The diplexer losses are 1 dB. The receiver rejects undesired signals with a power level lower than −20 dBm.

Receiving link requirements

Receiver frequency: 1.525–1.559 GHz.
LO frequency: 1.355–1.389 GHz.
IF frequency: 0.17 GHz.
Receiving link 1 dB bandwidth: 50 KHz.
Receiver Noise figure: 2.8 dB.
Input signals: −100– −130 dBm.
Output signals: 20–50 dBm.
Receiving sensitivity: $C/N = 41$ dB Hz^{-1} for −130 dBm input signal.

1.14.3 Development and fabrication of the receiver

The major goal in the receiver development was the low price target in the fabrication of two hundreds modules. Each module was selected to achieve the electrical specifications to meet the assigned low price target to the module. The size and low production cost of the receiver is achieved by using MIC technology.

Table 1.9. Power, gain, noise figure calculation of the receiving channel.

Module	Noise figure (dB)	Gain (dB)	P_{out} (dBm)
Diplexer	1	−1	−131
LNA	0.9	20	−111
Voltage-controlled attenuator	3.0	−3.0	−114.0
Band pass filter	2	−2	−116.0
Matched LNA	1.5	13.5	−102.5
Filter	2	−2	−104.5
LNA	0.9	10	−94.5
Voltage-controlled attenuator	3.0	−3.0	−97.5
Filter	2	−2	−99.5
Active Mixer	20	14.5	−85
Saw filter	8	−8	−93
IF amplifiers	2.5	42.0	−51
Transmission line losses	2.0	−2.0	−53
Total	2.30	77	−53

Tuning was not required in the fabrication process of the receiver. The electrical parameters, such as gain and noise figure, of the receiver was measured for each receiver.

1.14.4 Transmitter development

The block diagram of the transmitter is presented in figure 1.7. The transmitter includes low power amplifiers, pass band filters, voltage-controlled attenuator, active mixer, medium power and HPAs, high-power isolator, coupler, power detector, DC supply and control unit. The gain and power budget of the transmitter is listed in table 1.10.

Five stages of medium- and high-power amplifiers amplify the input signal from 0 to 40 dBm. The fifth stage is a compact high efficiency 10 W power amplifier. In the high-power mode the amplifier transmits 10 W. In the low power mode the amplifier transmits 4 W. The DC bias voltage of the HPA is automatically controlled by the RF controller to set the HPA to high or low mode. The output power level is measured via a −30 dB coupler and a power detector. The measured power level is transferred via an A/D converter to the RF controller to monitor the output power level of the transmitter by varying the attenuation of the voltage-controlled attenuator. The power monitoring process ensures high efficiency of the transmitter and low DC power consumption. In the standby mode the RF controller sets the transmitter bias voltages to ON and OFF. A temperature sensor is used to measure

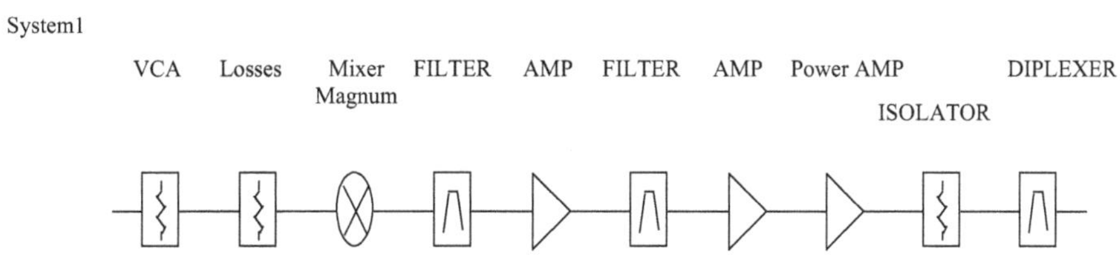

	VCA	Losses	Mixer Magnum	FILTER	AMP	FILTER	AMP	Power AMP	ISOLATOR	DIPLEXER	Total
Gain (dB)	-4.00	-0.50	2.00	-2.00	20.50	-2.00	16.00	11.50	-0.50	-1.00	40.00
NF (dB)	4.00	0.50	2.00	2.00	3.00	2.00	3.00	5.00	0.50	1.00	9.23
IP1dB (dBm)	20.00	20.00	10.00	20.00	10.00	20.00	10.00	20.00	20.00	20.00	-21.67
NF+ (dB)	0.87	0.16	0.95	0.58	1.77	0.01	0.02	0.00	0.00	0.00	

Input Pwr (dBm)	-60.00	System Temp (K)	290.00
Modulation: FM			

System BW (MHz)	0.03	MDS (dBm)	-119.98	Input IP3 (dBm)	N/A
S/N (dB, Actual)	59.98	S/N (dB, Req'd)	6.00	Output IP3 (dBm)	N/A
Srce Temp (K)	290.00	Sens. Loss (dB)	0.00	OIM3 (dBm)	N/A
Te Eff. (K)	2136.86	Sensitivity (dBm)	-113.98	ORR3 (dB)	N/A
SFDR3 (dB)	N/A	G/T (dB/K)	-23.85	IRR3 (dB)	N/A

Figure 1.7. Block diagram and gain budget of the transmitter.

Table 1.10. Transmitter channel gain and power budget.

Component	Gain/loss (dB)	P_{out} (dBm)
Input	0	0
Voltage-controlled attenuator	−3.5	−3.5
Transmission line losses	−0.5	−4.0
Active mixer	2	−2.0
Band pass filter	−2	−4.0
Low power amplifiers	20.0	16
Band pass filter	–2	14
Medium power amplifier	16	30
Power amplifier	11.5	41.5
Isolator	−0.5	41
Diplexer	−1	40
Total	40.0	40

the HPA temperature. The RF controller sets the transmitter vent to ON and OFF according to the measured transceiver temperature.

1.14.4.1 Transmitter reqirements

Transmitter frequency: 1.6265–1.6605 GHz.

I.F frequency range: 0.995–0.1335 GHz.

Transmitter LO frequency: 1.760 GHz.

Transmitter input power: 0 dBm.

Transmitter high mode output power: 38–40 dBm.

Transmitter low mode output power: 34–36 dBm.

Power consumption: 42 W.

The transmitter channel gain and power budget is listed in table 1.10.

1.14.4.2 RF head diplexer requirements

A very compact, lightweight and low-cost diplexer connects the receiver and transmitter to the transceiver antenna. The diplexer's simple structure ensures the diplexer low cost in production.

Transmitting filter

Filter pass band frequency: 1.6265–1.6605 GHz.

Filter pass band insertion loss: 0.7 dB.

Filter pass band $S_{11} < -13$ dB.

Filter rejection level > 54 dB at 1.525–1.559 GHz.

Receiving filter

Pass band frequency range: 1.525–1.559 GHz.

Pass band $S_{11} < -13$ dB.

Filter rejection level > 65 dB at 1.6265–1.6605 GHz.

Diplexer size: 8.6 × 3.6 × 2.5 cm.

1.14.5 Transmitter development and manufacturing

Figure 1.8 presents a photo of the transceiver prototype of the InMarsat-M ground terminal. A quick trimming procedure is required in the fabrication of the transmitter to achieve the required output power and efficiency. Around 500 transmitters were manufactured during the first production cycle. The major goal in the receiver development was the low price target in the fabrication of two hundred modules. Each module was selected to achieve the electrical specifications to meet the assigned low price target of the module. The size and low production cost of the transmitter and receiver is achieved by using MIC technology. The electrical parameters, such as gain and noise figure, of the transmitter was measured for each transmitter.

Figure 1.9 presents a photo of the transceiver modules for an InMarsat-M ground terminal. The transceiver is separated into five sections. Receiver and transmitter, diplexer, synthesizers, RF controller and a DC supply unit. The transmitter and receiver are separated by metallic fences and covers.

1.14.6 RF controller

The RF controller is based on an 87c51 microcontroller. The RF controller sets the transmitter to 'on' and 'off' by controlling the DC bias voltage switching unit. The RF controller monitors the transmitter output power level by varying the attenuation of the voltage-controlled attenuator in the transmitting module. The RF controller sets the transmitter to burst or single channel per carrier (SCPC) modes with high or low power levels, and produces the clock data and enables signals for the

Figure 1.8. Transceiver prototype for an InMarsat-M ground terminal.

Figure 1.9. Transceiver modules for an InMarsat-M ground terminal.

Rx and Tx synthesizers. The RF controller communicates with the transceiver controller via a full duplex serial bus. The communication is based on message transfer.

1.15 Conclusions

A compact and low-cost transceiver for InMarsat-M applications is presented in this chapter. The transceiver is part of a portable satellite communication ground terminal, 'Cary-phone', which supplies phone and fax services to users.

The RF controller automatically monitors the transmitter output power level to ensure low DC power consumption. A dc-to-dc converter supplies a controlled DC bias voltage to the power amplifier to a high-power level mode of 10 or 4 W in the case of the low power mode.

The total gain of the receiver is 76 dB with a gain stability of 1 dB as function of temperature. The receiver noise figure is less than 2.3 dB. The transceiver size is 300 × 200 × 25 mm.

Parts of this chapter have been reproduced from reference [8].

References

[1] ITU-R Recommendation V.431 2015 *Nomenclature of the Frequency and Wavelength Bands used in Telecommunications* (Geneva: Int. Telecommunication Union)

[2] IEEE Standard 521-2002 2003 *Standard Letter Designations for Radar-Frequency Bands* (Piscataway, NJ: IEEE)

[3] AFR 55-44/AR 105-86/OPNAVINST 3430.9A/MCO 3430.1, 27 October 1964 superseded by AFR 55-44/AR 105-86/OPNAVINST 3430.1A/MCO 3430.1A, 6 December 1978: Performing Electronic Countermeasures in the United States and Canada, Attachment 1, ECM Frequency Authorizations
[4] Belov L A, Smolskiy S M and Kochemasov V N 2012 *Handbook of RF, Microwave, and Millimeter-Wave Components* (Boston, MA: Artech House) pp 27–8
[5] Lee W C Y 1982 *Mobile Communication Engineering* (New York: McGraw Hill)
[6] Fujimoto K and James J R 1994 *Mobile Antenna Systems Handbook* (Boston, MA: Artech House)
[7] Skolnik M L 1980 *Introduction to Radar Systems* (New York: McGraw Hill)
[8] Sabban A 2016 *Wideband RF Technologies and Antenna in Microwave Frequencies* (New York: Wiley)
[9] Sabban A 2016 Ultra-wideband RF modules for communication systems *PARIPEX Indian J. Res.* **5** 91–5
[10] Sabban A 2015 Wideband RF modules and antennas at microwave and MM wave frequencies for communication applications *J. Mod. Commun. Technol. Res.* **3** 89–97
[11] Sabban A, Cabiri J and Carmeli E 2007 18 to 40 GHz Integrated compact switched filter bank module *ISSSE 2007 Conf. (Montreal, Canada)* 347–50
[12] Sabban A 2007 Applications of MM wave microstrip antenna arrays ISSSE 2007 Conf. (Montreal, Canada) 119–22
[13] Sabban A, Britebard A and Shemesh Y 1997 Development and fabrication of a compact integrated RF-head for InMarsat-M ground terminal 1997 *EuMW Conf. (Jerusalem, Israel)* 1186–91
[14] Madjar A, Behar D, Sabban A and Shapir I 1996 RF front end prototype for a millimeter wave (Ka/K) band VSAT satellite communication earth terminal 1996 *European Microwave Conf. (Prague)* 953–5
[15] Sabban A and Shemesh Y 1995 A compact low power consumption integrated RF-head for InMarsat-M ground terminal *EUMC (Bologna, Italy)* 81–2

Chapter 2

Wearable communication technology for medical and sport applications

Wearable technology has several applications in personal wireless communication and medical devices as presented in [1–8]. In the last few years the biomedical industry has experienced continuous growth. Several medical devices and systems have been developed to monitor patient health as presented in many books and papers [1–44]. Wearable technology provides a powerful new tool to medical and surgical rehabilitation services. Wireless body-area networks (WBANs) can record electrocardiograms, measure body temperature, blood pressure, heart rate, arterial blood pressure, electro-dermal activity and other healthcare parameters.

2.1 Wearable technology

Accessories that can comfortably be worn on the body are called wearable devices. Wearable technology is a multidisciplinary developing field. Knowledge in bioengineering, electrical engineering, software engineering and mechanical engineering is needed to design and develop wearable communication and medical systems. Wearable medical systems and sensors are used to measure and monitor physiological parameters of the human body. Biomedical systems in the vicinity of the human body may be wired or wireless. Many physiological parameters may be measured and analyzed using wearable medical systems and sensors, such as body temperature, heart rate, blood pressure and sweat rate. Wearable technology may provide scanning and sensing features that are not offered by mobile phones and laptop computers, and usually has communication capabilities. The user may have access to information in real time. Several wireless technologies are used to handle the data collection and processing by medical systems. The collected data may be stored or transmitted to a medical center for analysis. Wearable devices gather raw data that is fed to a database or a software application, and the subsequent analysis

doi:10.1088/978-0-7503-5224-6ch2

typically may result in a response that might alert a physician to contact a patient who is experiencing abnormal symptoms. However, a message may be also sent to a person who achieves a fitness goal.

Examples of wearable devices include headbands, smart wristbands, belts, watches, glasses, contact lenses, e-textiles and smart fabrics, jewelry, bracelets, and hearing aid devices.

Usually, wearable communication systems consist of a transmitting unit, receiving unit, data processing unit and wearable antennas.

Wearable technology may influence the fields of transportation, health and medicine, fitness, aging, disabilities, education, finance, gaming, entertainment and music. In the next decade wearable devices will be an important part of people's daily lives.

2.2 Wearable medical systems

One of the main goals of wearable medical systems is to increase the prevention of disease. By using wearable medical devices a person can be more in control and have increased awareness of their health. Sophisticated analysis of continuously measured medical data of a large number of medical centers' patients may result in better low-cost medical treatment.

Applications of wearable medical systems. Wearable medical devices may be used to:

- help monitor hospital activities,
- assist diabetes and asthma patients,
- help in solving sleep disorders and obesity problems,
- assist in solving cardiovascular diseases,
- assist epilepsy patients,
- help in the treatment of Alzheimer's disease patients,
- gather data for clinical research trials and academic research studies.

2.3 Physiological parameters measured by wearable medical systems

Several physiological parameters may be measured by using wearable medical systems and sensors. Some of the physiological data is presented in this chapter.

Measurements of human body temperature

The temperature of a healthy person ranges between 35 °C to 38 °C. Temperatures below or above this range may indicate that a person is sick, and temperatures above 40 °C may cause death. Human body temperature data may be transmitted to a medical center and if needed the doctor may contact the patient if they require further assistance.

Measurements of blood pressure

Blood pressure indicates the arterial pressure of the blood circulating in the human body.

Some of the causes of changes in blood pressure may be stress and obesity. The blood pressure of a healthy person is around 80 by 120, where the systole is 120 and the diastole is 80.

Ten percent changes above or below these values is a matter of concern and should be examined. Usually blood pressure and heart rate are measured in the same set of measurements. Blood pressure and heart rate data may be transmitted to a medical center and if needed the doctor may contact the patient if they require further assistance.

Measurements of heart rate

Measurement of heart rate is one of the most important tests needed to examine the health of a patient. A change in heart rate will change the blood pressure and the amount of blood delivered to all parts of the body. The heart rate of a healthy person in 72 times per minute. Changes in heart rate may cause several kinds of cardiovascular diseases. Traditionally, heart rate is measured by using a stethoscope. However, this is a manual test and is not as accurate. To measure and analyze heart rate a wearable medical device may be connected to a patient's chest.

Medical devices that measure heart rate may be wired or wireless.

Measurements of respiration rate

Measurements of respiration rate indicate if a person breathes normally and if they are healthy. Elderly and overweight people may have difficulty breathing normally. Wearable medical devices are used to measure a person's respiration rate. A wired medical device used to measure respiration rate may make the patient feel uneasy and this may cause an error in the data. It is better to use a wireless medical device for this purpose. The measured respiration rate may be transmitted to a medical center and if needed the doctor may contact the patient if they require further assistance.

Measurements of sweat rate

Glucose is the primary energy source of human beings. Glucose is usually supplied to the human body as a sugar that is a monosaccharide that provides energy to the human body. When a person does extensive physical activity glucose comes out of the skin as sweat. A wearable medical device may be used to monitor and measure the sweat rate of a person when extensive physical activity is done. A wearable medical device may be attached to a person's clothes in proximity to the skin to monitor and measure the sweat rate. This device may be also used to measure the sweat PH, which is important in the diagnosis of diseases. The water vapor evaporated from the skin is absorbed in the medical device to determine the sweat PH. If the amount of sweat coming out of the body is too high the body may dehydrate. Dehydration causes tiredness and fatigue. Measurements of sweat rate and PH may be used to monitor the physical activity of a person.

Measurements of human gait

The movement of human limbs is called human gait. Human gait describes the various ways in which a human can move. Different gait patterns are characterized by differences in limb movement patterns, overall velocity, forces, kinetic and potential energy cycles, and changes in contact with the ground. Walking, jogging, skipping and sprinting are defined as natural human gait. Gait analysis is a helpful and fundamental research tool to characterize human locomotion. Wearable devices may be attached to different parts of the human body, and the movement signal recorded by these devices can be used to analyze human gait. Temporal characteristics of gait are collected and estimated from wearable accelerometers and pressure sensors inside footwear.

In sport, gait analysis based on wearable sensors can be used for training and analysis and for improvement of an athlete's performance. Ambulatory gait analysis results may determine whether or not a particular treatment is appropriate for a patient. Motion analysis of human limbs during gait is applied in pre-operative planning for patients with cerebral palsy and can alter medical treatment decisions. Parkinson's disease is characterized by motor difficulties, such as gait difficulty, slowing of movement and limb rigidity. Gait analysis has been verified as one of the most reliable diagnostic signs of this disease. For patients with neurological problems, such as Parkinson's disease and stroke, ambulatory gait analysis is an important tool in their recovery process and can provide low-cost and convenient rehabilitation monitoring.

Gait analysis based on wearable devices may be applied in healthcare monitoring, such as in the detection of gait abnormalities, the assessment of recovery, fall risk estimation, and in sports training. In healthcare centers, gait information is used to detect walking behavior abnormalities that may predict health problems or the progression of neurodegenerative diseases. Falling over is the most common type of home accident among elderly people and is a major threat to their health and independence. Gait analysis using wearable devices can be used to analyze and predict the risk of falling among elderly patients.

Wearable devices tracking and monitoring doctors and patients inside hospitals
Each patient may have a wearable device attached to their body. This wearable device is connected to several sensors and each sensor has its own specific task to perform. For example, one sensor node may be detecting heart rate or body temperature, while another is detecting blood pressure. Doctors can also carry a wearable device, which allows other personnel to locate them within a hospital.

2.4 Wearable body-area networks (WBANs)

The main goal of WBANs is to provide continuous biofeedback data. WBANs can record electrocardiograms, measure body temperature, blood pressure, heart rate, arterial blood pressure, electro-dermal activity and other healthcare parameters in an efficient way. For example, accelerometers can be used to sense heart rate, movement or even muscular activity. body-area networks (BANs) include applications and communication devices using wearable and implantable wireless networks. A sensor network that senses health parameters is called a body sensor network (BSN). A WBAN is a special purpose wireless sensor network that incorporates different networks and wireless devices to enable remote monitoring in various environments.

A common application of WBANs is in medical centers where the conditions of a large number of patients are constantly being monitored (as presented in figures 2.1–2.3). Wireless monitoring of the physiological signals of a large number of patients is needed in order to deploy a complete wearable sensor network (WSN) in healthcare centers. Human health monitoring is emerging as a significant application of embedded sensor networks. A WBAN can monitor vital signs, providing real-time feedback to allow many patient diagnostic procedures using continuous monitoring of chronic conditions, or progress of recovery from an illness. Recent technological advances in wireless networking promise a new generation of wireless sensor networks suitable for human body wearable network systems.

Data acquisition in WBAN devices can be point-to-point or multipoint-to-point, depending on specific applications. Detection of the condition of an athlete's health would require point-to-point data sharing across various on-body sensors. Human body monitoring of vital signs requires the routing of data from several wearable sensors, multipoint-to-point, to a sink node, which in turn can relay the information wirelessly to an out-of-body computer. Data may be transferred in real-time or non-real-time. Human body monitoring applications require real-time data transfer. Monitoring an athlete's physiological data can be collected offline for processing and analysis purposes.

A typical WBAN consists of a number of compact low-power sensing devices, a control unit and wireless transceivers. The power supply for these components should be compact, lightweight and long lasting. WBANs consist of compact devices that take up little space, with fewer opportunities for redundancy. To improve the efficiency of WBANs it is important to minimize the number of nodes in the network. Adding more devices and path redundancy for solving node failure and

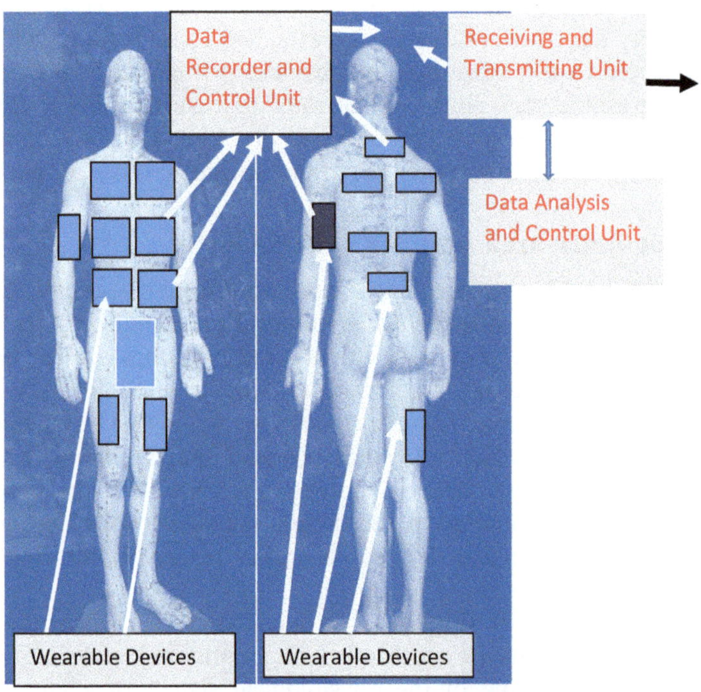

Figure 2.1. Wearable devices for various medical applications.

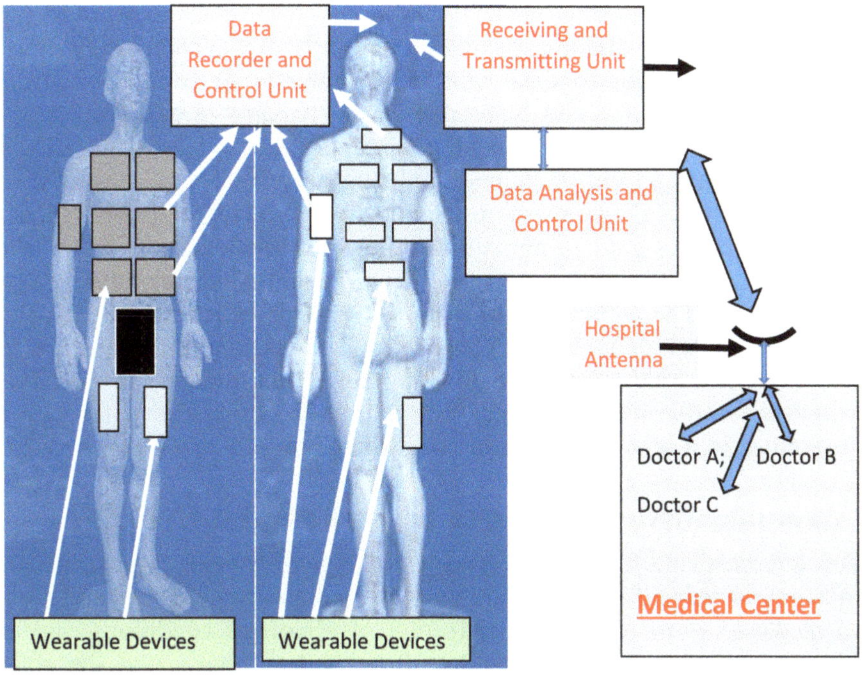

Figure 2.2. Wearable body-area network, WBAN, for various medical applications.

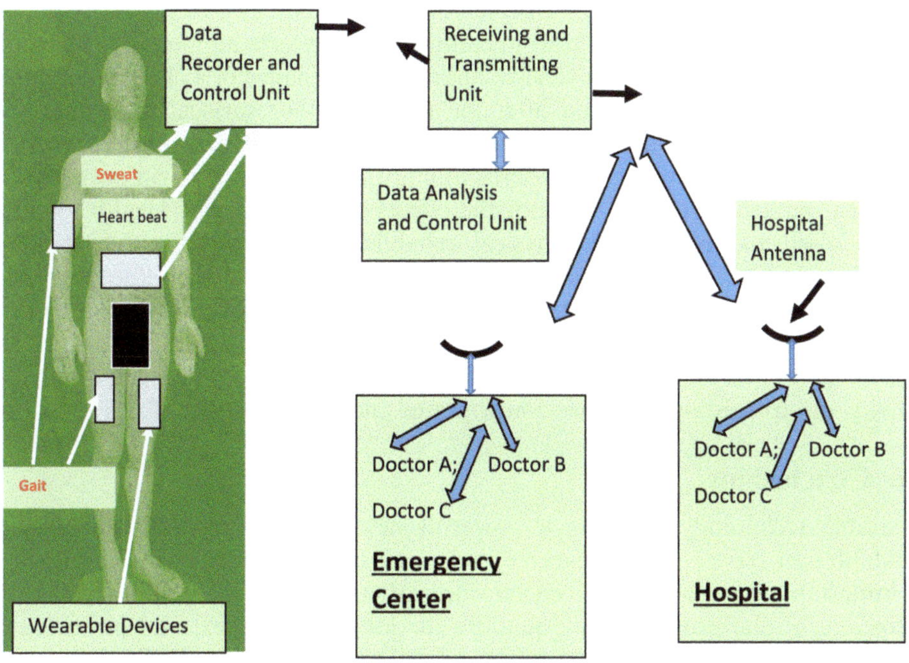

Figure 2.3. Wearable body-area network, WBAN, health monitoring system.

network problems cannot be a practical option in WBAN systems. WBANs receive and transmit a large amount of data constantly. Data processing must be hierarchical and efficient to deal with the asymmetry of several resources, to maintain system efficiency and to ensure the availability of data. WBANs in a medical area consist of wearable and implantable sensor nodes that can sense biological information from the human body and transmit it over a short distance wirelessly to a control device worn on the body or placed in an accessible location. The sensor electronics must be miniaturized, low-power and able to detect medical signals such as electrocardiograms, electroencephalography, pulse rate, pressure, and temperature. The gathered data from the control devices can then be transmitted to remote destinations in a WBAN for diagnostic and therapeutic purposes by including other wireless networks for long-range transmission.

A wireless control unit is used to collect information from sensors through wires and transmit it to a remote station for monitoring.

2.5 Wearable wireless body-area network (WWBAN)

Wireless communication systems offer a wide range of benefits to medical centers, patients, physicians and sports centers through continuous measuring and monitoring of medical information, early detection of abnormal conditions, supervised rehabilitation, and potential knowledge through data analysis of the collected information. Wearable health monitoring systems allow a person to closely follow changes in important health parameters, providing feedback for maintaining optimal health. If a WWBAN is part of a telemedicine system, it can be used to alert medical personnel when life-threatening events occur, as presented in figure 2.4. In addition, patients may benefit from continuous long-term monitoring as part of a diagnostic procedure. One can achieve optimal maintenance of a chronic condition, or monitor the recovery period after an acute event or surgical procedure. The collected medical data may be a very good indicator of the cardiac recovery of patients after heart surgery. Long-term monitoring can also confirm adherence to treatment guidelines or help monitor the effects of drug therapy. Health monitors can be used to monitor the physical rehabilitation of patients during stroke rehabilitation, or brain trauma rehabilitation, and after hip or knee surgeries. Many people are using WBAN devices such as wearable heart rate monitors, respiration rate monitors and pedometers for medical reasons or as part of a fitness regime. A WWBAN may be attached to a cotton shirt to measure respiratory activity, electrocardiograms, electromyograms and body posture.

2.6 Conclusions

Wearable technology provides a powerful new tool for medical and surgical rehabilitation services. The wearable body-area network (WBAN) is emerging as an important option for medical centers and patients. Wearable technology provides a convenient platform that may quantify the long-term context and physiological response of individuals, and support the development of individualized treatment systems with real-time feedback to help promote patients' health. Wearable medical

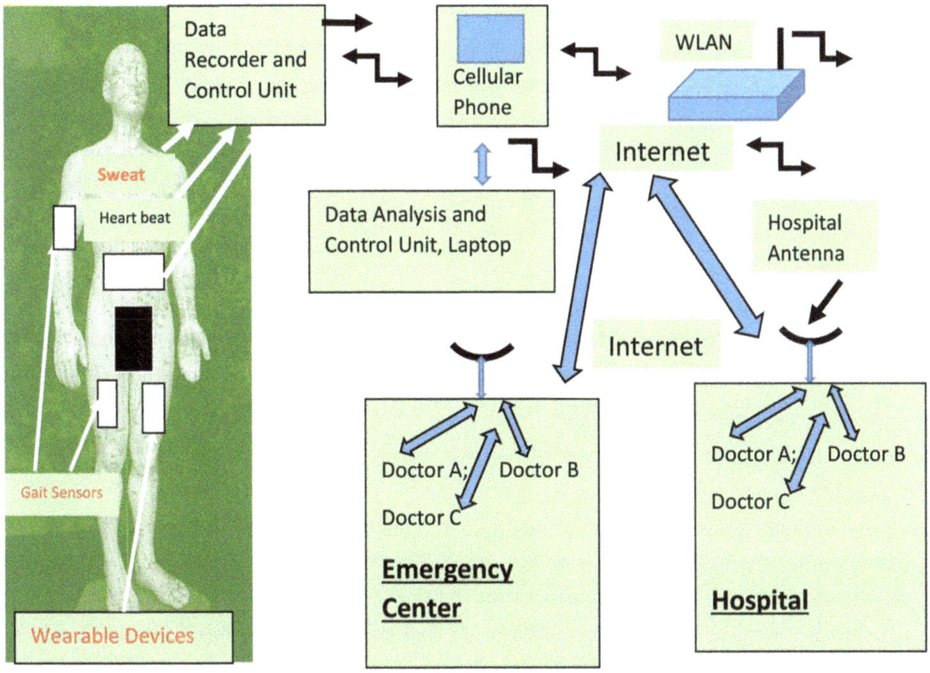

Figure 2.4. Wireless wearable body-area network (WWBAN) health monitoring system.

systems and sensors can measure body temperature, heart rate, blood pressure, sweat rate, perform gait analysis and other physiological parameters. Gait analysis is a useful tool both in clinical practice and biomechanical research. Gait analysis using wearable sensors provides quantitative and repeatable results over extended time periods with low cost and good portability, showing better prospects and making great progress in recent years. At present, commercialized wearable sensors have been adopted in various applications of gait analysis.

References

[1] Mukhopadhyay S C (ed) 2015 *Wearable Electronics Sensors* (Switzerland: Springer)

[2] Sabban A 2016 *Wideband RF Technologies and Antenna in Microwave Frequencies* (New York: Wiley)

[3] Sabban A 2015 *Low visibility Antennas for Communication Systems* (London: Taylor and Francis)

[4] Sabban A 2016 Small wearable meta materials antennas for medical systems *Appl. Comput. Electromagn. Soc.* **31**

[5] Sabban A 2011 Microstrip antenna arrays *Microstrip Antennas* ed N Nasimuddin (Rijeka: InTech) pp 361–84

[6] Sabban A 2013 New wideband printed antennas for medical applications *IEEE J. Trans. Antennas Propag.* **61** 84–91

[7] Sabban Albert 2012 *Dual polarized dipole wearable antenna U.S Patent number:* 8203497

[8] Bonfiglio A and De Rossi D (ed) 2011 *Wearable Monitoring Systems* (New York: Springer)

[9] Gao T, Greenspan D, Welsh M, Juang R R and Alm A 2005 *Vital signs monitoring and patient tracking over a wireless network Proc. of IEEE-EMBS 27th Annual Int. Conf. of the Engineering in Medicine and Biology (Shanghai, China)* 1–5 September pp 102–5

[10] Otto C A, Jovanov E and Milenkovic E A 2006 *WBAN-based system for health monitoring at home Proc. of IEEE/EMBS Int. Summer School, Medical Devices and Biosensors (Boston, MA, USA)* 4–6 September pp 20–3

[11] Zhang G H, Poon C C Y, Li Y and Zhang Y T 2009 *A biometric method to secure telemedicine systems Proc. of the 31st Annual Int. Conf. of the IEEE Engineering in Medicine and Biology Society (Minneapolis, MN, USA)* pp 701–4

[12] Bao S, Zhang Y and Shen L 2005 *Physiological signal based entity authentication for body area sensor networks and mobile healthcare systems Proc. of the 27th Annual Int. Conf. of the IEEE EMBS (Shanghai, China)* 1–4 September pp 2455–8

[13] Ikonen V and Kaasinen E 2008 *Ethical assessment in the design of ambient assisted living Proc. of Assisted Living Systems—Models, Architectures and Engineering Approaches (Schloss Dagstuhl, Germany)* pp 14–7

[14] Srinivasan V, Stankovic J and Whitehouse K 2008 *Protecting your daily in home activity information from a wireless snooping attack Proc. of the 10th Int. Conf. on Ubiquitous Computing (Seoul, Korea)* 21–24 September pp 202–11

[15] Casas R, Blasco M R, Robinet A, Delgado A R, Yarza A R, Mcginn J, Picking R and Grout V 2008 *User modelling in ambient intelligence for elderly and disabled people Proc. of the 11th Int. Conf. on Computers Helping People with Special Needs (Linz, Austria)* pp 114–22

[16] Jasemian Y 2008 *Elderly comfort and compliance to modern telemedicine system at home Proc. of the Second Int. Conf. on Pervasive Computing Technologies for Healthcare (Tampere, Finland)* 30 January–1 February pp 60–3

[17] Atallah L, Lo B, Yang G Z and Siegemund F 2008 *Wirelessly accessible sensor populations (WASP) for elderly care monitoring Proc. of the Second Int. Conf. on Pervasive Computing Technologies for Healthcare (Tampere, Finland)* 30 January–1 February pp 2–7

[18] Hori T, Nishida Y, Suehiro T and Hirai S 2000 *SELF-Network: Design and implementation of network for distributed embedded sensors Proc. of IEEE/RSJ Int. Conf. on Intelligent Robots and Systems (Takamatsu, Japan)* 30 October–5 November pp 1373–8

[19] Mori Y, Yamauchi M and Kaneko K 2000 *Design and implementation of the Vital Sign Box for home healthcare Proc. of IEEE EMBS Int. Conf. on Information Technology Applications in Biomedicine (Arlington, VA, USA)* pp 104–9

[20] Lauterbach C, Strasser M, Jung S and Weber W 2002 *Smart clothes self-powered by body heat Proc. of Avantex Symposium (Frankfurt, Germany)* pp 5259–63

[21] Marinkovic S and Popovici E 2009 *Network coding for efficient error recovery in wireless sensor networks for medical applications Proc. of Int. Conf. on Emerging Network Intelligence (Sliema, Malta)* 11–16 October pp 15–20

[22] Schoellhammer T, Osterweil E, Greenstein B, Wimbrow M and Estrin D 2004 *Lightweight temporal compression of microclimate datasets Proc. of the 29th Annual IEEE Int. Conf. on Local Computer Networks (Tampa, FL, USA)* 16–18 November pp 516–24

[23] Barth A T, Hanson M A, Powell H C Jr and Lach J 2009 *Tempo 3.1: A body area sensor network platform for continuous movement assessment Proc. of the 6th Int. Workshop on Wearable and Implantable Body Sensor Networks (Berkeley, CA, USA)* pp 71–6

[24] Gietzelt M, Wolf K H, Marschollek M and Haux R 2008 *Automatic self-calibration of body worn triaxial-accelerometers for application in healthcare Proc. of the Second Int. Conf. on Pervasive Computing Technologies for Healthcare (Tampere, Finland)* pp 177–80

[25] Gao T, Greenspan D, Welsh M, Juang R R and Alm A 2005 *Vital signs monitoring and patient tracking over a wireless network Proc. of the 27th Annual Int. Conf. of the IEEE EMBS (Shanghai, China)* 1–4 September pp 102–5

[26] Purwar A, Jeong D U and Chung W Y 2007 *Activity monitoring from realtime triaxial accelerometer data using sensor network Proc. of Int. Conf. on Control, Automation and Systems (Hong Kong)* 21–23 March pp 2402–6

[27] Baker C *et al* 2007 *Wireless sensor networks for home health care Proc. of the 21st Int. Conf. on Advanced Information Networking and Applications Workshops (Niagara Falls, Canada)* 21–23 May pp 832–7

[28] Schwiebert L, Gupta S K S and Weinmann J 2001 *Research challenges in wireless networks of biomedical sensors Proc. of the 7th Annual Int. Conf. on Mobile Computing and Networking (Rome, Italy)* 16–21 July pp 151–65

[29] Aziz O, Lo B, King R, Darzi A and Yang G Z 2006 *Pervasive body sensor network: An approach to monitoring the postoperative surgical patient Proc. of Int. Workshop on Wearable and implantable Body Sensor Networks (BSN 2006) (Cambridge, MA, USA)* pp 13–8

[30] Kahn J M, Katz R H and Pister K S J 1999 *Next century challenges: Mobile networking for smart dust Proc. of the ACM MobiCom'99 (Washington, DC, USA)* pp 271–8

[31] Noury N, Herve T, Rialle V, Virone G, Mercier E, Morey G, Moro A and Porcheron T 2000 *Monitoring behavior in home using a smart fall sensor Proc. of IEEE-EMBS Special Topic Conf.on Micro-technologies in Medicine and Biology (Lyon, France)* 12–14 October pp 607–10

[32] Kwon D Y and Gross M 2005 *Combining body sensors and visual sensors for motion training Proc. of the 2005 ACM SIGCHI Int. Conf. on Advances in Computer Entertainment Technology (Valencia, Spain)* 15–17 June pp 94–101

[33] Boulgouris N K, Hatzinakos D and Plataniotis K N 2005 Gait recognition: A challenging signal processing technology for biometric identification *IEEE Signal Process. Mag.* **22** 78–90

[34] Kimmeskamp S and Hennig E M 2001 Heel to toe motion characteristics in parkinson patients during free walking *Clin. Biomech.* **16** 806–12

[35] Turcot K, Aissaoui R, Boivin K, Pelletier M, Hagemeister N and de Guise J A 2008 New accelerometric method to discriminate between asymptomatic subjects and patients with medial knee osteoarthritis during 3-D gait *IEEE Trans. Biomed. Eng.* **55** 1415–22

[36] Furnée H 1997 Real-time motion capture systems *Three-Dimensional Analysis of Human Locomotion* ed P Allard, A Cappozzo, A Lundberg and C L Vaughan (Chichester, UK: Wiley) pp 85–108

[37] Bamberg S J M, Benbasat A Y, Scarborough D M, Krebs D E and Paradiso J A 2008 Gait analysis using a shoe-integrated wireless sensor system *IEEE Trans. Inf. Technol. Biomed.* **12** 413–23

[38] Choi J H, Cho J, Park J H, Eun J M and Kim M S 2008 *An efficient gait phase detection device based on magnetic sensor array Proc. of the 4th Kuala Lumpur Int. Conf. on Biomedical Engineering* **21** *(Kuala Lumpur, Malaysia)* 25–28 June pp 778–81

[39] Hidler J 2004 *Robotic-assessment of walking in individuals with gait disorders Proc. of the 26th Annual Int. Conf. of the IEEE Engineering in Medicine and Biology Society*7 *(San Francisco, CA, USA)* 1–5 September pp 4829–31

[40] Wahab Y and Bakar N A 2011 *Gait analysis measurement for sport application based on ultrasonic system Proc. of the 2011 IEEE 15th Int. Symp. on Consumer Electronics (Singapore)* 14–17 June pp 20–4

[41] De Silva B, Jan A N, Motani M and Chua K C 2008 *A real-time feedback utility with body sensor networks Proc. of the 5th Int. Workshop on Wearable and Implantable Body Sensor Networks (BSN 08) (Hong Kong)* 1–3 June pp 49–53

[42] Salarian A, Russmann H, Vingerhoets F J G, Dehollain C, Blanc Y, Burkhard P R and Aminian K 2004 Gait assessment in Parkinson's disease: Toward an ambulatory system for long-term monitoring *IEEE Trans. Biomed. Eng.* **51** 1434–43

[43] Atallah L, Jones G G, Ali R, Leong J J H, Lo B and Yang G Z 2011 *Observing recovery from knee-replacement surgery by using wearable sensors Proc. of the 2011 Int. Conf. on Body Sensor Networks (Dallas, TX, USA)* 23–25 May pp 29–34

[44] ElSayed M, Alsebai A, Salaheldin A, El Gayar N and ElHelw M 2010 *Ambient and wearable sensing for gait classification in pervasive healthcare environments Proc. of the 12th IEEE Int. Conf. on e-Health Networking Applications and Services (Healthcom) (Lyon, France)* 1–3 July pp 240–5

IOP Publishing

Wearable Communication Systems and Antennas
(Second Edition)
Design, efficiency, and miniaturization techniques
Albert Sabban

Chapter 3

Electromagnetic waves and transmission lines for wearable communication systems

This chapter provides the basic theory and a short introduction to wireless wearable communication systems. Transmitting and receiving information in microwave frequencies is based on the propagation of electromagnetic waves.

3.1 Electromagnetic spectrum

The electromagnetic spectrum corresponds to electromagnetic waves as presented in references [1–17]. Infrared light was discovered by Sir William Herschel in 1800. Johann Wilhelm Ritter discovered ultraviolet light in 1801. In 1867, James Clerk Maxwell predicted that there should be light with longer wavelengths than infrared light. In 1887, Heinrich Hertz demonstrated the existence of the waves predicted by Maxwell by producing radio waves in his laboratory. The frequency unit is named after Heinrich Hertz and is measured in hertz (Hz). X-rays were first observed and documented in 1895 by Wilhelm Conrad Röntgen in Germany. Gamma-rays were first observed in 1900 by Paul Villard when he was investigating radiation from radium.

At very low frequencies, 1 Hz to 1 MHz, the wavelength is of much higher order than the size of the circuit components used in electronic circuits. At very low frequencies voltage, current and impedance do not vary as a function of the device length. At very low frequencies relations between voltage, current and impedance are evaluated by using Kirchhoff's and Ohm's laws. Components with dimensions lower than a tenth of the wavelength are called lumped elements. At high frequencies the wavelength is of the same order of magnitude as the circuit devices used. At high frequencies, conventional circuit analysis based on Kirchhoff's and Ohm's laws could not analyze and describe the variation of fields, impedance, voltages and currents along the length of the components. Components with dimensions higher than a tenth of the wavelength are called distributed elements. Kirchhoff's and Ohm's laws may be applied to lumped elements, but they

doi:10.1088/978-0-7503-5224-6ch3

Table 3.1. Electromagnetic spectrum and applications.

Band name	Abbreviation	ITU	Frequency/λ_0	Applications
Tremendously low frequency	TLF		< 3 Hz > 100 000 km	Natural and artificial EM noise
Extremely low frequency	ELF		3–30 Hz 100 000 km–10 000 km	Communication with submarines
Super low frequency	SLF		30–300 Hz 10 000 km–1000 km	Communication with submarines
Ultra-low frequency	ULF		300–3000 Hz 1000 km–100 km	Submarine communication, communication within mines
Very low frequency	VLF	4	3–30 kHz 100 km–10 km	Navigation, time signals, submarine communication, wireless heart rate monitors, geophysics
Low frequency	LF	5	30–300 kHz 10 km–1 km	Navigation, clock time signals, AM longwave broadcasting (Europe and parts of Asia), RFID, amateur radio
Medium frequency	MF	6	300–3000 kHz 1 km–100 m	AM (medium-wave) broadcasts, amateur radio, avalanche beacons
High frequency	HF	7	3–30 MHz 100 m–10 m	Shortwave broadcasts, radio, amateur radio and aviation, communications, RFID, radar, near-vertical incidence sky wave (NVIS) radio communications, marine and mobile radio telephony
Very high frequency	VHF	8	30–300 MHz 10 m–1 m	FM, television broadcasts and line-of-sight ground-to-aircraft and aircraft-to-aircraft communications, land mobile and maritime mobile communications, amateur radio, weather radio
Ultra-high frequency	UHF	9	300–3000 MHz 1 m–100 mm	Television broadcasts, microwave ovens, radio astronomy, mobile phones, wireless LAN, Bluetooth, ZigBee, GPS and two-way radios such as land mobiles, FRS and GMRS radios
Super high frequency	SHF	10	3–30 GHz 100 mm–10 mm	Radio astronomy, wireless LAN, modern radars, communications satellites, satellite television broadcasting, DBS
Extremely high frequency	EHF	11	30–300 GHz 10 mm–1 mm	Radio astronomy, microwave radio relay, microwave remote sensing, directed-energy weapons, scanners
Terahertz or tremendously high frequency	THz or THF	12	300–3000 GHz 1 mm–100 μm	Terahertz imaging, ultrafast molecular dynamics, condensed-matter physics, terahertz time-domain spectroscopy, terahertz computing/communications

Table 3.2. IEEE standard radar frequency bands.

Microwave frequency bands—IEEE Standard	
Designation	Frequency range
L band	1–2 GHz
S band	2–4 GHz
C band	4–8 GHz
X band	8–12 GHz
Ku band	12–18 GHz
K band	18–26.5 GHz
Ka band	26.5–40 GHz
Q band	30–50 GHz
U band	40–60 GHz
V band	50–75 GHz
E band	60–90 GHz
W band	75–110 GHz
F band	90–140 GHz
D band	110–170 GHz

Table 3.3. The International Telecommunication Union bands and medical applications.

Band number	Symbols	Frequency range	Wavelength range	Medical and sport applications
4	VLF	3–30 kHz	10–100 km	Wireless (heart rate monitors)
5	LF	30–300 kHz	1–10 km	RFID tags
6	MF	300–3000 kHz	100–1000 m	Avalanche beacons
7	HF	3–30 MHz	10–100 m	RFID Tags
8	VHF	30–300 MHz	1–10 m	Mobile radio telephony
9	UHF	300–3000 MHz	10–100 cm	WBAN and WWBAN
10	SHF	3–30 GHz	1–10 cm	Wireless WBAN
11	EHF	30–300 GHz	1–10 mm	Microwave sensors
12	THF	300–3000 GHz	0.1–1 mm	Terahertz imaging

cannot be applied to distributed elements. To prevent interference and to provide efficient use of the radio spectrum, similar services are allocated in bands, see [11–14]. Bands are divided at wavelengths of 10^n meters, or frequencies of 3×10^n Hz. Each of these bands has a basic band plan, which dictates how it is to be used and shared, to avoid interference and to set protocol for the compatibility of transmitters and receivers. In table 3.1 the electromagnetic spectrum and applications are listed. In table 3.2 the IEEE standard for radar frequency bands is listed. In table 3.3 the International Telecommunication Union, ITU, bands are given. In table 3.4 radar frequency bands as defined by NATO for ECM systems are listed, see [13].

Table 3.4. Radar frequency bands as defined by NATO for ECM systems [13].

Band	Frequency range
A band	0–0.25 GHz
B band	0.25–0.5 GHz
C band	0.5–1.0 GHz
D band	1–2 GHz
E band	2–3 GHz
F band	3–4 GHz
G band	4–6 GHz
H band	6–8 GHz
I band	8–10 GHz
J band	10–20 GHz
K band	20–40 GHz
L band	40–60 GHz
M band	60–100 GHz

3.2 Basic electromagnetic wave definitions

Field—A field is a physical quantity that has a value for each point in space and time.
Wavelength—The wavelength is the distance between two sequential equivalent points. Wavelength, λ, is measured in meters.
Wave period—The period is the time, T, for one complete cycle of an oscillation of a wave. The period is measured in seconds.
Frequency—The frequency, f, is the number of periods per unit time (second) and is measured in hertz.
Phase velocity—The phase velocity, v, of a wave is the rate at which the phase of the wave propagates in space. Phase velocity measured in m s^{-1}.

$$\begin{aligned} \lambda &= V^*T \\ T &= 1/f \\ \lambda &= v/f \end{aligned} \tag{3.1}$$

Electromagnetic waves propagate in free space in the phase velocity of light.

$$v = 3 \times 10^8 \text{ ms}^{-1} \tag{3.2}$$

Wavenumber—A wavenumber, k, is the spatial frequency of the wave in radians per unit distance (per meter). $K = 2\pi/\lambda$.
Angular frequency—The angular frequency, ω, represents the frequency in radians per second. $\omega = v^*k = 2\pi f$.
Polarization—A wave is polarized if it oscillates in one direction or plane. The polarization of a transverse wave describes the direction of oscillation in the plane perpendicular to the direction of propagation.

Antennas—Antennas are used to efficiently radiate electromagnetic energy in certain desired directions. They match radio frequency systems to space. All antennas may be used to receive or radiate energy. They transmit or receive electromagnetic waves and convert electromagnetic radiation into electric current, or vice versa. Antennas transmit and receive electromagnetic radiation at radio frequencies, and are a necessary part of all communication links and radio equipment. They are used in systems such as radio and television broadcasting, point-to-point radio communication, wireless LAN, cell phones, radar, medical systems and spacecraft communication. They are most commonly employed in air or outer space, but can also be operated under water, on and inside the human body, or even through soil and rock at low frequencies for short distances.

3.3 Electromagnetic waves theory

Maxwell's equations

Maxwell's equations are presented in several books [1–10]. They describe how electric and magnetic fields are generated and altered by each other, and are a

Table 3.5. Symbols and abbreviations of physical parameters.

Dimensions	Parameter	Symbol
Wb m^{-1}	Magnetic potential	A
m s^{-2}	Acceleration	a
tesla	Magnetic field	B
F	Capacitance	C
V m^{-1}	Electric field displacement	D
V m^{-1}	Electric field	E
N	Force	F
A m^{-1}	Magnetic field strength	H
A	Current	I
A m^{-2}	Current density	J
H	Self-inductance	L
m	Length	l
H	Mutual-inductance	M
C	Charge	q
W m^{-2}	Poynting vector	**P**
W	Power	P
Ω	Resistance	R
m^3	Volume	V
m s^{-1}	Velocity	v
F m^{-1}	Dielectric constant	ε
	Relative dielectric constant	ε_r
H m^{-1}	Permeability	μ
Ω^{-1} m^{-1}	Conductivity	σ
Wb	Magnetic flux	ψ

Table 3.6. Symbols and abbreviations of electromagnetic fields.

Dimensions	Symbol	Parameter
$V\ m^{-1}$	D	Electric field displacement
$V\ m^{-1}$	E	Electric field
$A\ m^{-1}$	H	Magnetic field strength
$8.854 \times 10^{-12}\ Fm^{-1}$	ε_0	Dielectric constant in space
$\mu_0 = 4\pi \cdot 10^{-7}\ Hm^{-1}$	μ_0	Permeability in space
$C\ m^{-3}$	ρ_V	Volume charge density
Tesla, $Wb\ m^{-2}$	B	Magnetic field
$\Omega \cdot m^{-1}$	σ	Conductivity
Wb	ψ	Magnetic flux
M	δ_s	Skin depth

classical approximation to the more accurate and fundamental theory of quantum electrodynamics. Quantum deviations from Maxwell's equations are usually small. Inaccuracies occur when the particle nature of light is important or when electric fields are strong. Symbols and abbreviations of physical parameters and electromagnetic fields are listed in tables 3.5 and 3.6.

Gauss's law for electric fields

Gauss's law for electric fields states that the electric flux via any closed surface S is equal to the net charge q divided by the free space dielectric constant.

$$\begin{aligned} \oint_S E \cdot ds &= \frac{1}{\varepsilon_0}\int_V \rho dv = \frac{1}{\varepsilon_0} q \\ D &= \varepsilon E \\ \nabla \cdot D &= \rho_v \end{aligned} \tag{3.3}$$

Gauss's law for magnetic fields

Gauss's law for magnetic fields states that the magnetic flux via any closed surface S is equal to zero. There is no magnetic charge in Nature.

$$\begin{aligned} \oint_S B \cdot ds &= 0 \\ B &= \mu H \\ \psi_m &= \int_S B \cdot ds \\ \nabla \cdot B &= 0 \end{aligned} \tag{3.4}$$

Ampère's law

The original 'Ampère's law' stated that magnetic fields can be generated by electrical current. Ampère's law has been corrected by Maxwell, who stated that magnetic fields can be also generated by time variant electric fields. The corrected 'Ampère's law' shows that a changing magnetic field induces an electric field and also a time variant electric field induces a magnetic field.

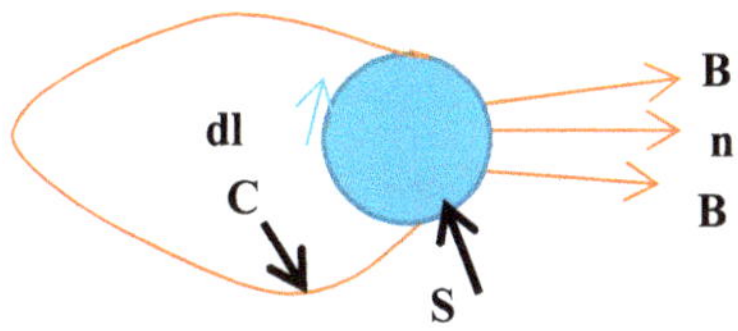

Figure 3.1. Faraday's law.

$$\oint_C \frac{B}{\mu_0} \cdot dl = \int_S J \cdot ds + \frac{d}{dt}\int_s \varepsilon_0 E \cdot ds$$

$$\nabla XH = J + \frac{\partial D}{\partial t} \tag{3.5}$$

$$J = \sigma E$$

$$i = \int_S J \cdot ds$$

Faraday's law

Faraday's law describes how a propagating time-varying magnetic field through a surface S creates an electric field, as shown in figure 3.1.

$$\oint_C E \cdot dl = -\frac{d}{dt}\int_s B \cdot ds$$

$$\nabla XE = -\frac{\partial B}{\partial t} \tag{3.6}$$

Wave equations

The variation of electromagnetic waves as a function of time may be written as $e^{j\omega t}$. The derivative as a function of time is $j\omega e^{j\omega t}$. Maxwell's equations may be written as:

$$\nabla XE = -j\omega\mu H$$

$$\nabla XH = (\sigma + j\omega\varepsilon)E \tag{3.7}$$

A ∇X(curl) operation on the electric field E results in:

$$\nabla X \nabla XE = -j\omega\mu \nabla XH \tag{3.8}$$

By substituting the expression of ∇XH in equation we get

$$\nabla X \nabla XE = -j\omega\mu(\sigma + j\omega\varepsilon)E \tag{3.9}$$

$$\nabla X \nabla XE = -\nabla^2 E + \nabla(\nabla \cdot E) \tag{3.10}$$

In free space there is no charge so $\nabla \cdot E = 0$. Equation (3.11) presents the wave equation for electric fields. Where γ is the complex propagation constant, α represents losses in the medium, and β represents the wave phase constant in radians per meter.

$$\nabla^2 E = j\omega\mu(\sigma + j\omega\varepsilon)E = \gamma^2 E$$

$$\gamma = \sqrt{j\omega\mu(\sigma + j\omega\varepsilon)} = \alpha + j\beta \tag{3.11}$$

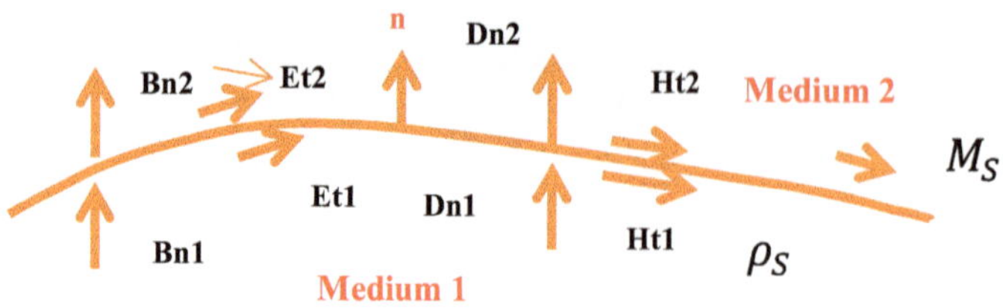

Figure 3.2. Fields between two media.

If we follow the same procedure on the magnetic field we will get the wave equation for magnetic field.

$$\nabla^2 H = j\omega\mu(\sigma + j\omega\varepsilon)H = \gamma^2 H \tag{3.12}$$

The law of conservation of energy imposes boundary conditions on the electric and magnetic fields. When an electromagnetic wave travels from medium 1 to medium 2 the electric and magnetic fields should be continuous as presented in figure 3.2.

General boundary conditions

$$n \cdot (D_2 - D_1) = \rho_S \tag{3.13}$$

$$n \cdot (B_2 - B_1) = 0 \tag{3.14}$$

$$(E_2 - E_1) \times n = M_S \tag{3.15}$$

$$n \times (H_2 - H_1) = J_S \tag{3.16}$$

Boundary conditions for a dielectric medium

$$n \cdot (D_2 - D_1) = 0 \tag{3.17}$$

$$n \cdot (B_2 - B_1) = 0 \tag{3.18}$$

$$(E_2 - E_1) \times n = 0 \tag{3.19}$$

$$n \times (H_2 - H_1) = 0 \tag{3.20}$$

Boundary conditions for a conductor

$$n \cdot (D_2 - D_1) = \rho_S \tag{3.21}$$

$$n \cdot (B_2 - B_1) = 0 \tag{3.22}$$

$$(E_2 - E_1) \times n = 0 \tag{3.23}$$

$$n \times (H_2 - H_1) = J_S \tag{3.24}$$

The condition for a good conductor is: $\sigma \gg \omega\epsilon$.

$$\gamma = \sqrt{j\omega\mu(\sigma + j\omega\varepsilon)} = \alpha + j\beta \approx (1 + j)\sqrt{\frac{\omega\mu\sigma}{2}} \tag{3.25}$$

$$\delta_s = \sqrt{\frac{2}{\omega\mu\sigma}} = \frac{1}{\alpha} \tag{3.26}$$

The conductor's skin depth is given as δ_s. The wave attenuation is α.

3.4 Wave propagation through the human body

The electrical properties of human body tissues should be considered in the design of wearable communication and medical systems. Conductivity and the dielectric constant of human body tissues are listed in table 3.7, see [15, 16]. The complex propagation constant γ, given in equation (3.11), can be calculated as a function of frequency by using the conductivity and dielectric constant values listed in table 3.7. The major issue in the design of wearable antennas is the interaction between RF transmission and the human body. Electrical properties of human body tissues should be considered in the design of wearable antennas. The attenuation, α, of RF

Table 3.7. Electrical properties of human body tissues.

Tissue	Property	434 MHz	800 MHz	1000 MHz
Prostate	σ	0.75	0.90	1.02
	ε	50.53	47.4	46.65
Stomach	σ	0.67	0.79	0.97
	ε	42.9	40.40	39.06
Colon, heart	σ	0.98	1.15	1.28
	ε	63.6	60.74	59.96
Kidney	σ	0.88	0.88	0.88
	ε	117.43	117.43	117.43
Nerve	σ	0.49	0.58	0.63
	ε	35.71	33.68	33.15
Fat	σ	0.045	0.056	0.06
	ε	5.02	4.58	4.52
Lung	σ	0.27	0.27	0.27
	ε	38.4	38.4	38.4

transmission through the human body is the real value of the propagation constant γ. Where, $\alpha = \text{Real}(\gamma)$. For example, stomach tissue attenuation is around 1.67 dB cm^{-1} at 500 MHz and 3.2 dB cm^{-1} at 1000 MHz. Blood attenuation is around 3.38 dB cm^{-1} at 500 MHz and 4.02 dB cm^{-1} at 1000 MHz. Attenuation of stomach tissues, skin and pancreas at frequencies from 150 MHz to 1000 MHz are listed in table 3.8. Attenuation of fat, small-intestine tissues and blood at frequencies from 150 MHz to 1000 MHz are listed in table 3.9.

3.5 Materials

Hard materials are presented in table 3.10. Alumina is the most popular hard substrate in MIC (microwave integrated circuits). Gallium arsenide is the most popular hard substrate in monolithic microwave integrated circuits (MIMIC) technology at microwave frequencies.

Soft materials are presented in table 3.11. Duroid is the most popular soft substrate in MIC and in the printed antennas industry. Dielectric losses in Duroid are significantly lower than dielectric losses in a FR-4 substrate. However, the cost of a FR-4 substrate is significantly lower than the cost of Duroid. Commercial MIC devices usually use the FR-4 substrate. Duroid is the most popular soft substrate used in the development of printed antennas with high efficiency at microwave frequencies.

Table 3.8. Attenuation of stomach tissues, skin and pancreas.

Frequency MHz	Attenuation stomach dB cm^{-1}	Attenuation skin dB cm^{-1}	Attenuation pancreas dB cm^{-1}
150	1.096 431 56	0.993 564 88	1.154 709 08
200	1.209 393 08	1.084 756 96	1.270 231 2
250	1.312 928 12	1.167 876 64	1.37479916
300	1.413 919 92	1.249 121 44	1.4755566
350	1.486 701 72	1.304 282 84	1.54858144
400	1.557 096 52	1.357 282 92	1.61897624
450	1.622 474 28	1.406 429 08	1.68601188
500	1.678 442 92	1.448 674 64	1.74763988
550	1.733 500 16	1.490 112 96	1.80851272
600	1.788 062 64	1.531 082 56	1.86901232
650	1.838 406 64	1.577 303 56	1.9296508
700	1.888 368 72	1.623 298 88	1.99015908
750	1.938 096 44	1.669 164	2.05065868
800	1.987 711 32	1.714 968 36	2.11124508
850	2.048 740 4	1.742 770 4	2.15668488
900	2.109 847 6	1.770 303 36	2.20203788
950	2.171 067 64	1.797 645 36	2.2473822
1000	2.232 426 56	1.824 813 76	2.292 769 92

Table 3.9. Attenuation of fat, small-intestine and blood.

Frequency MHz	Attenuation fat dB cm^{-1}	Attenuation small-intestine dB cm^{-1}	Attenuation blood dB cm^{-1}
150	0.162 767 36	1.452 606 68	2.301 875 24
200	0.199 694 684	1.862 892 92	2.559 002 88
250	0.224 987 336	2.082 835 44	2.769 484 2
300	0.247 448 572	2.212 931 28	2.951 512 48
350	0.269 270 092	2.294 575 36	3.082 302 72
400	0.291 230 492	2.348 269 84	3.193 562 96
450	0.303 265 312	2.385 012 28	3.291 751 12
500	0.315 080 528	2.411 052 28	3.382 023 12
550	0.327 376 616	2.430 052 8	3.464 023 08
600	0.340 818 464	2.444 288	3.539 599 84
650	0.354 259 444	2.455 198 76	3.607 191
700	0.367 729 068	2.463 731 2	3.670 589 72
750	0.371 782 628	2.470 501 6	3.730 611 92
800	0.375 756 332	2.475 978 68	3.787 908 6
850	0.379 671 88	2.480 448 88	3.849 128 64
900	0.383 546 632	2.484 155 24	3.908 543 24
950	0.388 768 52	2.487 254	3.966 456 2
1000	0.393 995 616	2.489 866 68	4.023 110 56

3.6 Transmission lines theory

Transmission lines are used to transfer electromagnetic energy from one point to another with minimum losses over a wide band of frequencies. There are three major types of transmission lines [1–10]. Transmission lines with a very small cross section compared to the wavelength of the dominant mode of propagation is the transverse electromagnetic mode (TEM). Closed rectangular and cylindrical conducting tubes on which the dominant modes of propagation are the transverse electric mode and transverse magnetic mode (TE and TM modes). Open boundary structures with cross sections greater than 0.1λ may support a surface wave mode of propagation.

For TEM modes Ez = Hz = 0. For TE modes Ez = 0. For TM modes Hz = 0.

Voltage and currents in transmission lines may be derived by using the transmission lines equations. Transmission lines equations may be derived by employing Maxwell's equations and the boundary conditions on the transmission line section shown in figure 3.3. Equation (3.32) is the first lossless transmission lines equation. l_e is the self-inductance per length.

$$\frac{\partial V}{\partial Z} = -\frac{\partial}{\partial t} l_e \ I = -l_e \frac{\partial I}{\partial t} \tag{3.27}$$

Table 3.10. Hard materials—ceramics.

Material	Symbol or formula	Dielectric constant	Dissipation factor (tan δ)	Coefficient of thermal expansion (ppm/°C)	Thermal cond. (W/m °C)	Mass density (gr/cc)
Alumina 99.5%	Al_2O_3	9.8	0.0001	8.2	35	3.97
Alumina 96%	Al_2O_3	9.0	0.0002	8.2	24	3.8
Aluminum nitride	AlN	8.9	0.0005	7.6	290	3.26
Beryllium oxide	BeO	6.7	0.003	6.05	250	
Gallium arsenide	GaAs	12.88	0.0004	6.86	46	5.32
Indium phosphide	InP	12.4				
Quartz		3.8	0.0001	0.6	5	
Sapphire		9.3, 11.5				
Silicon (high resistivity)	Si (HRS)	4		2.5	138	2.33
Silicon carbide	SiC	10.8	0.002	4.8	350	3.2

Table 3.11. Soft materials.

Manufacturer and material	Symbol or formula	Relative dielectric constant ε_R	Tolerance on dielectric constant	tan δ	Mass density (gr/cc)	Thermal conductivity (W/m °C)	Coefficient of thermal expansion (ppm/°C $x/y/z$)
Rogers Duroid 5870	PTFE/Random glass	2.33	0.02	0,0012	2.2	0.26	22/28/173
Rogers Duroid 5880	PTFE/Random glass	2.2	0.02	0.0012	2.2	0.26	31/48/237
Rogers Duroid 6002	PTFE/Random glass	2.94	0.04	0.0012	2.1	0.44	16/16/24
Rogers Duroid 6006	PTFE/Random glass	6	0.15	0.0027	2.7	0.48	38/42/24
Rogers Duroid 6010	PTFE/Random glass	10.2–10.8	0.25	0.0023	2.9	0.41	24/24/24
FR-4	Glass/epoxy	4.8		0.022		0.16	
Polyethylene		2.25					
Polyflon CuFlon	PTFE	2.1		0.00045			12.9
Polyflon PolyGuide	Polyolefin	2.32		0.0005			108
Polyflon Norclad	Thermoplastic	2.55		0.0011			53
Polyflon Clad Ultem	Thermoplastic	3.05		0.003	1.27		56
PTFE	PTFE	2.1		0.0002	2.1	0.2	
Rogers R/flex 3700	Thermally stable thermoplastic	2.0		0.002			8
Rogers RO3003	PTFE ceramic	3	0.04	0.0013	2.1	0.5	17/17/24
Rogers RO3006	PTFE ceramic	6.15	0.15	0.0025	2.6	0.61	17/17/24
Rogers RO3010	PTFE ceramic	10.2	0.3	0.0035	3	0.66	17/17/24
Rogers RO3203	PTFE ceramic	3.02					
Rogers RO3210	PTFE ceramic	10.2					
Rogers RO4003	Thermoset plastic ceramic glass	3.38	0.05	0.0027	1.79	0.64	11/14/46
Rogers RO4350B	Thermoset plastic ceramic glass	3.48	0.05	0.004	1.86	0.62	14/16/50
Rogers TMM 3	Ceramic/thermos	3.27	0.032	0.002	1.78	0.7	15/15/23
Rogers TMM 10	Ceramic/thermoset	9.2	0.23	0.0022	2.77	0.76	21/21/20

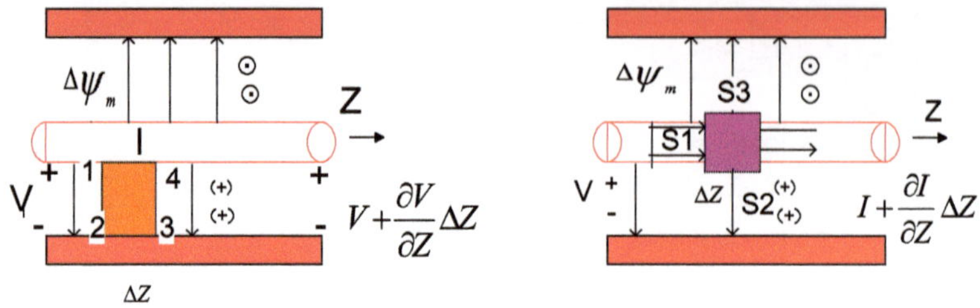

Figure 3.3. Transmission line geometry.

$$\frac{\partial I}{\partial Z} = -c\frac{\partial V}{\partial t} - gV \tag{3.28}$$

Equation (3.27) is the second lossless transmission lines equation: $c = \dfrac{\Delta C}{\Delta Z}$ Fm^{-1}.

$$g = \frac{\Delta G}{\Delta Z}\ \text{m}\Omega^{-1}.$$

Equation (3.28) is the first transmission lines equation with losses.

$$-\frac{\partial v}{\partial z} = Ri + L\frac{\partial i}{\partial t} \tag{3.29}$$

Equation (3.29) is the second transmission lines equation with losses.

$$-\frac{\partial i}{\partial z} = Gv + C\frac{\partial v}{\partial t} \tag{3.30}$$

Where $C = \dfrac{\Delta C(z)}{\Delta Z}$ Fm^{-1} $L = \dfrac{\Delta L(z)}{\Delta Z}$ Hm^{-1}$R = \dfrac{1}{G} = \dfrac{\Delta R(z)}{\Delta Z}$ Ωm^{-1}

By differentiating equation (3.28) with respect to z and by differentiating equation (3.29) with respect to t and adding the result we get:

$$-\frac{\partial^2 v}{\partial z^2} = RGv + (RC + LG)\frac{\partial v}{\partial t} + LC\frac{\partial^2 v}{\partial t^2} \tag{3.31}$$

By differentiating equation (3.35) with respect to z and by differentiating equation (3.31) with respect to t and adding the result we get:

$$-\frac{\partial^2 i}{\partial z^2} = RGi + (RC + LG)\frac{\partial i}{\partial t} + LC\frac{\partial^2 i}{\partial t^2} \tag{3.32}$$

Equations (3.31) and (3.32) are analog to the wave equations. The solution of these equations is a superposition of a forward, $+z$, and backward wave, $-z$.

$$\begin{aligned} V(z, t) &= V_+\left(t - \frac{z}{v}\right) + V_-\left(t + \frac{z}{v}\right) \\ I(z, t) &= Y_0\left\{V_+\left(t - \frac{z}{v}\right) - V_-\left(t + \frac{z}{v}\right)\right\} \end{aligned} \tag{3.33}$$

Y_0—is the characteristic admittance of the transmission line $Y_0 = \frac{1}{Z_0}$.

The variation of electromagnetic waves as a function of time may be written as $e^{j\omega t}$. The derivative as a function of time is $j\omega e^{j\omega t}$. By using these relations we may write phazor transmission lines equations.

$$\begin{aligned} \frac{dV}{dZ} &= -ZI \\ \frac{dI}{dZ} &= -YV \\ \frac{d^2V}{dz^2} &= \gamma^2 V \\ \frac{d^2I}{dz^2} &= \gamma^2 I \end{aligned} \tag{3.34}$$

Where:

$$\begin{aligned} Z &= R + j\omega L \quad \Omega\,\mathrm{m}^{-1} \\ Y &= G + j\omega C \quad \mathrm{m}\,\Omega^{-1} \\ \gamma &= \alpha + j\beta = \sqrt{ZY} \end{aligned} \tag{3.35}$$

The solution of the transmission lines equations in a harmonic steady state is:

$$\begin{aligned} v(z, t) &= \mathrm{Re}\; V(z)e^{j\omega t} \\ i(z, t) &= \mathrm{Re}\; I(z)e^{j\omega t} \end{aligned} \tag{3.36}$$

$$\begin{aligned} V(z) &= V_+e^{-\gamma z} + V_-\, e^{\gamma z} \\ I(z) &= I_+e^{-\gamma z} + I_-\, e^{\gamma z} \end{aligned} \tag{3.37}$$

For a lossless transmission line we may write:

$$\begin{aligned} \frac{dV}{dZ} &= -j\omega LI \\ \frac{dI}{dZ} &= - \; -j\omega CV \\ \frac{d^2V}{dz^2} &= -\omega^2 LCV \\ \frac{d^2I}{dz^2} &= -\omega^2 LCI \end{aligned} \tag{3.38}$$

The solution of the lossless transmission lines equations is:

$$\begin{aligned} V(z) &= e^{j\omega t}(V^+e^{-j\beta z} + V^-e^{j\beta z}) \\ I(z) &= Y_0(e^{j\omega t}(V^+e^{-j\beta z} - V^-e^{j\beta z})) \end{aligned} \tag{3.39}$$

$$v_p = \frac{\omega}{\beta} = \frac{1}{\sqrt{LC}} = \frac{1}{\sqrt{\mu\varepsilon}}$$

v_p—phase velocity.
Z_0—is the characteristic impedance of the transmission line.

$$Z_0 = \frac{V_+}{I_+} = \frac{V_-}{I_-} = \sqrt{\frac{(R + j\omega L)}{(G + j\omega C)}}$$
$$\text{FOR } R = 0 \quad G = 0 \tag{3.40}$$
$$Z_0 = \frac{V_+}{I_+} = \frac{V_-}{I_-} = \sqrt{\frac{L}{C}}$$

Waves in transmission lines.

A load Z_L is connected, at $z = 0$, to a transmission line with impedance Z_0. The voltage on the load is V_L. The current on the load is I_L (see figure 3.4).

$$V(0) = V_L = I(0) \cdot Z_L$$
$$I(0) = I_L \tag{3.41}$$

For $z = 0$ we can write:

$$V(0) = I(0) \cdot Z_L = V_+ + V_-$$
$$I(0) = Y_0(V_+ - V_-) \tag{3.42}$$

By substituting $I(0)$ in $V(0)$ we get:

$$Z_L = Z_0 \frac{V_+ + V_-}{(V_+ - V_-)}$$
$$Z_L = Z_0 \frac{1 + \frac{V_-}{V_+}}{1 - \frac{V_-}{V_+}} \tag{3.43}$$

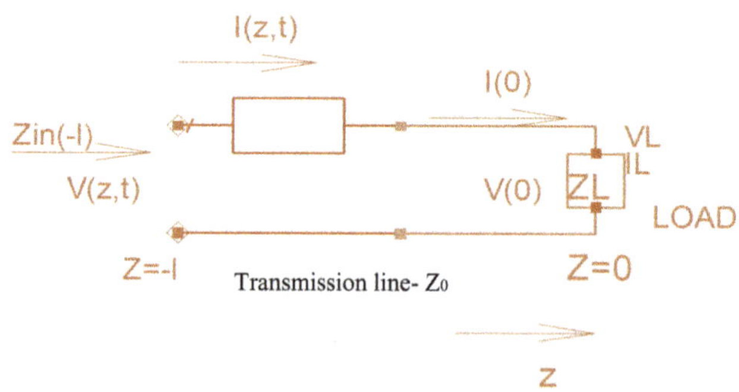

Figure 3.4. Transmission line with load.

The ratio $\frac{V_-}{V_+}$ is defined as a reflection coefficient. $\Gamma_L = \frac{V_-}{V_+}$.

$$Z_L = Z_0 \frac{1 + \Gamma_L}{1 - \Gamma_L} \tag{3.44}$$

$$\Gamma_L = \frac{\frac{Z_L}{Z_0} - 1}{\frac{Z_L}{Z_0} + 1} \tag{3.45}$$

The reflection coefficient as a function of z may be written as:

$$\Gamma(Z) = \frac{V_-}{V_+} = \frac{V_- e^{j\beta z}}{V_+ e^{-j\beta z}} = \Gamma_L e^{2j\beta z} \tag{3.46}$$

The input impedance as a function of z may be written as:

$$\begin{aligned} Z_{in}(z) &= \frac{V(z)}{I(z)} = \frac{V_+ e^{-j\beta z} + V_- e^{j\beta z}}{(V_+ e^{-j\beta z} - V_- e^{j\beta z}) Y_0} = \\ &= Z_0 \frac{1 + \Gamma_L e^{2j\beta z}}{1 - \Gamma_L e^{2j\beta z}} \end{aligned} \tag{3.47}$$

$$Z_{in}(-l) = Z_L \frac{\cos \beta l + j\frac{Z_0}{Z_L} \sin \beta l}{\cos \beta l + j\frac{Z_L}{Z_0} \sin \beta l} \tag{3.48}$$

The voltage and current as function of z may be written as:

$$\begin{aligned} V(z) &= V_+ e^{-j\beta z}(1 + \Gamma(z)) \\ I(z) &= Y_0 V_+ e^{-j\beta z}(1 - \Gamma(z)) \end{aligned} \tag{3.49}$$

The ratio between the maximum and minimum voltage along a transmission line is called a voltage standing wave ratio S, VSWR.

$$S = \left| \frac{V(z)\max}{V(z)\min} \right| = \frac{1 + |\Gamma(z)|}{1 - |\Gamma(z)|} \tag{3.50}$$

3.7 Matching techniques

Usually, in communication systems, the load impedance is not the same as the impedance of the commercial transmission lines. We will get maximum power transfer from source to load if the impedances are matched [1–10]. A perfect impedance match corresponds to a VSWR 1:1. A reflection coefficient magnitude of zero is a perfect match, a value of one is a perfect reflection. The reflection coefficient (Γ) of a short circuit has a value of −1 (1 at an angle of 180°). The reflection coefficient of an open circuit is one at an angle of 0°. The return loss of a load is merely the magnitude of the reflection coefficient expressed in decibels. The correct equation for a return loss is: Return loss = −20 × log [mag(Γ)].

For a maximum voltage V_m in a transmission line the maximum power will be:

$$P_{\max} = \frac{V_m^2}{Z_0} \tag{3.51}$$

For an unmatched transmission line the maximum power will be:

$$\begin{aligned} P_{\max} &= \frac{(1 - |\Gamma|^2)V_+^2}{Z_0} \\ V_{\max} &= (1 + |\Gamma|)V_+ \\ V_+^2 &= \frac{V_{\max}^2}{(1 + |\Gamma|)^2} \end{aligned} \tag{3.52}$$

$$P_{\max} = \frac{1 - |\Gamma|}{1 + |\Gamma|} \cdot \frac{V_{\max}^2}{Z_0} = \frac{V_{\max}^2}{VSWR \cdot Z_0} \tag{3.53}$$

A 2:1 VSWR will result in half of maximum power transferred to the load. The reflected power may cause damage to the source.

Equation (3.47) indicates that there is a one-to-one correspondence between the reflection coefficient and input impedance. A movement of distance z along a transmission line corresponds to a change of $e^{-2j\beta z}$, which represents a rotation via an angle of $2\beta z$. In the reflection coefficient plane we may represent any normalized impedance by contours of constant resistance, r, and contours of constant reactance, x. The corresponding impedance moves on a constant radius circle via an angle of $2\beta z$ to a new impedance value. Those relations may be presented by a graphical aid called a Smith chart. Those relations may be presented by the following set of equations:

$$\begin{aligned} \Gamma_L &= \frac{\dfrac{Z_L}{Z_0} - 1}{\dfrac{Z_L}{Z_0} + 1} \\ z(l) &= \frac{Z_L}{Z_0} = r + jx \\ \Gamma_L &= \frac{r + jx - 1}{r + jx + 1} = p + jq \end{aligned} \tag{3.54}$$

$$
\begin{aligned}
\frac{Z(z)}{Z_0} &= \frac{1 + \Gamma(z)}{1 - \Gamma(z)} = r + jx \\
\Gamma(z) &= u + jv \\
\frac{1 + u + jv}{1 - u - jv} &= r + jx
\end{aligned}
\tag{3.55}
$$

$$
\begin{aligned}
\left(u - \frac{r}{1 + r}\right)^2 + v^2 &= \frac{1}{(1 + r)^2} \\
(u - 1)^2 + \left(v - \frac{1}{x}\right)^2 &= \frac{1}{x^2}
\end{aligned}
\tag{3.56}
$$

Equation (3.56) presents two families of circles in the reflection coefficient plane. The first family are contours of constant resistance, r, and the second family are contours of constant reactance, x, see figures 3.5a and 3.5b. The center of the Smith chart is $r = 1$. Moving away from the load corresponds to moving around the chart in a clockwise direction. Moving away from the generator toward the load corresponds to moving around the chart in a counterclockwise direction (figure 3.5c).

A complete revolution around the chart in a clockwise direction corresponds to a movement of half-wavelength away from the load. The Smith chart may be employed to calculate the reflection coefficient and the input impedance for a given transmission line and load impedance. If we are at a matched impedance condition at the center of the Smith chart, any length of transmission line with impedance Z_0 does nothing to the input match. But if the reflection coefficient of the network (S_{11}) is at some non-ideal impedance, adding a transmission line between the network and the reference plane rotates the observed reflection coefficient clockwise about the center of the Smith chart. Further, the rotation occurs at a fixed radius (and VSWR magnitude) if the transmission line has the same characteristic impedance as the source impedance Z_0. Adding a quarter-wavelength, means a 180° phase rotation. Adding one quarter-wavelength from a short circuit moves us 180° to the right side of the chart, to an open circuit.

Smith chart guidelines

The Smith chart contains almost all of the possible complex impedances within one circle.

- The Smith chart's horizontal center line represents resistance/conductance.
- Zero resistance is located on the left end of the horizontal center line.
- Infinite resistance is located on the right end of the horizontal center line.
- Horizontal center line—resistive/conductive horizontal scale of the chart.

  Impedances in the Smith chart are normalized to the characteristic impedance of the transmission line and are independent of the characteristic impedance of the transmission.
- The center of the line and also of the chart is 1.0 point, where $R = Z_0$ or

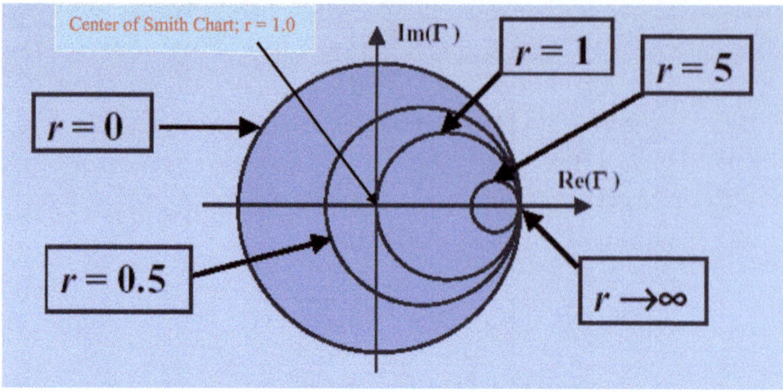

(a)

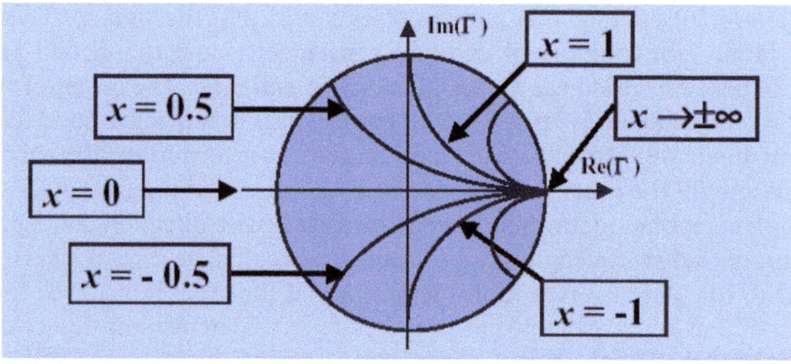

(b)

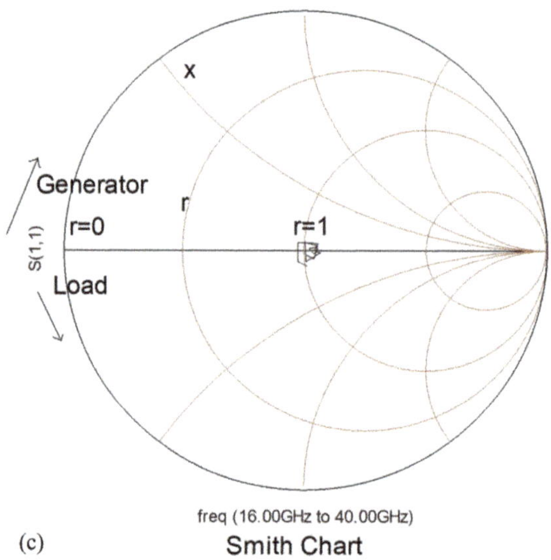

(c)

Figure 3.5. (a) r circles, (b) x circles and (c) Smith chart.

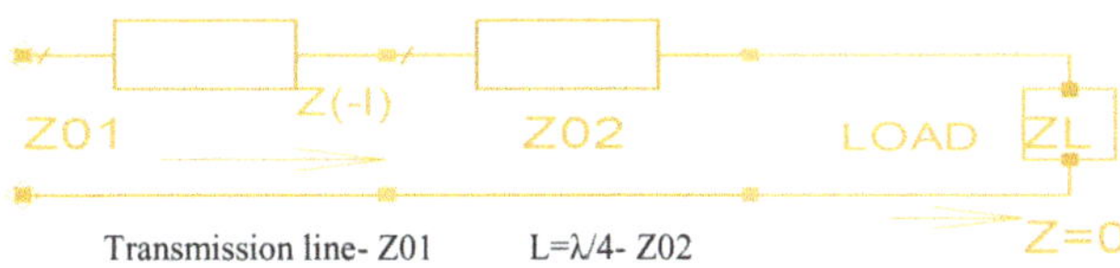

Figure 3.6. Quarter-wave transformer.

$G = Y_0$.
At point $r = 1.0$, $Z = Z_0$, and no reflection will occur.

3.7.1 Quarter-wave transformers

A quarter-wave transformer may be used to match a device with impedance ZL to a system with impedance Z_0, as shown in figure 3.6. A quarter-wave transformer is a matching network with a bandwidth somewhat inversely proportional to the relative mismatch we are trying to match. For a single-stage quarter-wave transformer, the correct transformer impedance is the geometric mean between the impedances of the load and the source. If we substitute to equation (3.47) $l = \frac{\lambda}{4}$, $\beta = \frac{2\pi}{\lambda}$ we get:

$$\bar{Z}(-l) = \frac{\frac{Z_L}{Z_{02}} \cos \beta\frac{\lambda}{4} + j \sin \beta\frac{\lambda}{4}}{\cos \beta\frac{\lambda}{4} + j\frac{Z_L}{Z_{02}} \sin \beta\frac{\lambda}{4}} = \frac{Z_{02}}{Z_L} \tag{3.57}$$

$$\bar{Z}(-l) = \frac{Z_{01}}{Z_{02}} = \frac{Z_{02}}{Z_L}$$

We will achieve matching when;

$$Z_{02} = \sqrt{Z_L Z_{01}} \tag{3.58}$$

For complex Z_L values Z_{02} also will be of complex impedance. However, standard transmission lines have real impedance values. To match a complex $Z_L = R + jX$ we transform Z_L to a real impedance $Z_{L1} - jX$ to Z_L. Connecting a capacitor $-jX$ to Z_L is not practical at high frequencies. A capacitor at high frequencies has parasitic inductance and resistance. A practical method to transform Z_L to a real impedance Z_{L1} is to add a transmission line with impedance Z_0 and length l to get a real value Z_{L1}.

3.7.2 Wideband matching multi-section transformers

Multi-section quarter-wave transformers are employed for wideband applications. Responses such as, Chebyshev (equi-ripple) and maximally flat are possible for multi-section transformers. Each section brings us to intermediate impedance. In four section transformers from 25 ohms to 50 ohms intermediate impedances are chosen by using an arithmetic series. For an arithmetic series the steps are equal, $\Delta Z = 6.25\ \Omega$, so the impedances are 31.25 Ω, 37.5 Ω, 43.75 Ω. Solving for the transformers yields $Z_1 = 27.951$, $Z_2 = 34.233$, $Z_3 = 40.505$ and $Z_4 = 46.771\ \Omega$. A second solution to multi-section transformers involves a geometric series from impedance Z_L to impedance Z_S. Here the impedance from one section to the next adjacent section is a constant ratio.

3.7.3 Single stub matching

A device with admittance Y_L can be a match to a system with admittance Y_0 by using a shunt or series single stub. At a distance l from the load we can get a normalized admittance $\overline{Y}in = 1 + j\overline{B}$. By solving equation (3.58) we can calculate l.

$$\overline{Y}(l) = \frac{1 + j\dfrac{Z_L}{Z_0}\tan\beta l}{\dfrac{Z_L}{Z_0} + j\tan\beta l} = 1 + j\overline{B} \tag{3.59}$$

At this location we can add a shunt stub with normalized input susceptance, $-j\overline{B}$ to yield $\overline{Y}in = 1$ as presented in equation (3.60). $\overline{Y}in = 1$ represents a matched load. The stub can be a short circuited line or open circuited line. The susceptance $\overline{B}$ is given in equation (3.60).

$$\begin{aligned} \overline{Y}in &= 1 + j\overline{B} \\ \overline{Y}_{1n} &= -j\overline{B} \\ \overline{B} &= ctg\beta l_1 \end{aligned} \tag{3.60}$$

The length $l1$ of the short circuited line may be calculated by solving equation (3.61) (figure 3.7).

$$\mathrm{Im}\left\{\frac{1 + j\dfrac{Z_L}{Z_0}\tan\beta l}{\dfrac{Z_L}{Z_0} + j\tan\beta l}\right\} = ctg\beta l_1 \tag{3.61}$$

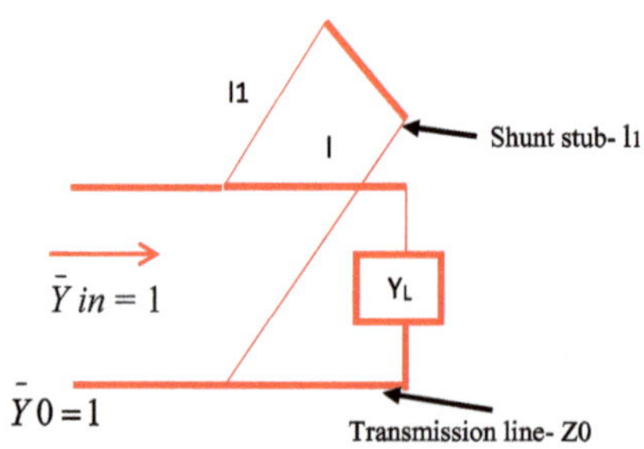

Figure 3.7. Single stub matching.

3.8 Coaxial transmission line

A coax transmission line consists of two round conductors in which one completely surrounds the other, with the two separated by a continuous solid dielectric [1–9]. The desired propagation mode is TEM. The major advantage of a coax over a microstrip line is that the transmission line does not radiate. The disadvantages are that coax lines are more expensive. Coax lines are usually employed up to 18 GHz, and are extremely expensive at frequencies higher than 18 GHz. To obtain a good performance at higher frequencies, small diameter cables are required to stay below the cutoff frequency. Maxwell's laws are employed to compute the electric and magnetic fields in the coax transmission line. A cross section of the coaxial transmission line is shown in figure 3.8.

$$\oint_S E \cdot ds = \frac{1}{\varepsilon_0}\int_V \rho dv = \frac{1}{\varepsilon_0} q$$
$$E_r = \frac{\rho_L}{2\pi r \varepsilon} \tag{3.62}$$

$$V = \int_a^b E \cdot dr = \int_a^b \frac{\rho_L}{2\pi r \varepsilon} \cdot dr = \frac{\rho_L}{2\pi\varepsilon} \ln \frac{b}{a} \tag{3.63}$$

Ampère's law is employed to calculate the magnetic field.

$$\oint H \cdot dl = 2\pi r H_\varphi = I$$
$$H_\varphi = \frac{I}{2\pi r} \tag{3.64}$$

$$V = \int_a^b E_r \cdot dr = -\int_a^b \eta H_\varphi \cdot dr = \frac{I\eta}{2\pi} \ln \frac{b}{a}$$
$$Z_0 = \frac{V}{I} = \frac{\eta}{2\pi} \ln \frac{b}{a} \qquad \eta = \sqrt{\mu/\varepsilon} \tag{3.65}$$

The power flow in the coaxial transmission line may be calculated by calculating the Poynting vector.

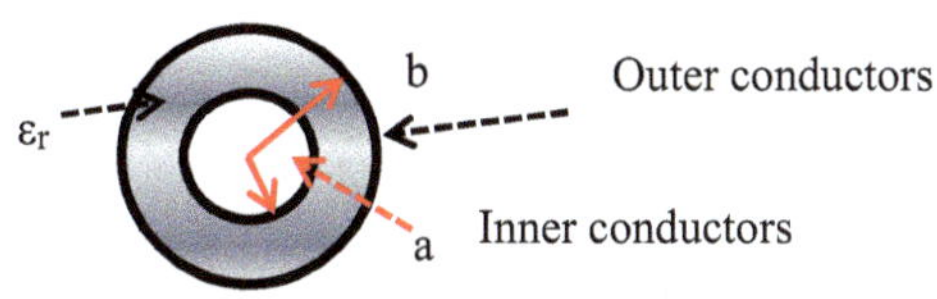

Figure 3.8. Coaxial transmission line.

Table 3.12. Industry-standard coax cables.

Cable type	Outer diameter (inches)	$2b$ (inches)	$2a$ (inches)	Z_0 (ohms)	f_c (GHz)
RG-8A	0.405	0.285	0.089	50	14.0
RG-58A	0.195	0.116	0.031	50	35.3
RG-174	0.100	0.060	0.019	50	65.6
RG-196	0.080	0.034	0.012	50	112
RG-214	0.360	0.285	0.087	50	13.9
RG-223	0.216	0.113	0.037	50	34.6
SR-085	0.085	0.066	0.0201	50	60.2
SR-141	0.141	0.1175	0.0359	50	33.8
SR-250	0.250	0.210	0.0641	50	18.9

$$
\begin{aligned}
P &= (E \times H) \cdot n = EH \\
P &= \frac{VI}{2\pi r^2 \ln(b/a)} \\
W &= \int_s P \cdot ds = \int_a^b \frac{VI}{2\pi r^2 \ln(b/a)} 2\pi r dr = VI
\end{aligned}
\tag{3.66}
$$

Table 3.12 presents several industry-standard coax cables. The cables' dimensions, impedance and cutoff frequency are given in table 3.12. RG cables are flexible cables. SR cables are semi-rigid cables.

Cutoff frequency and wavelength of coax cables

fc—cutoff frequency.

The criterion for the cutoff frequency is that the circumference at the midpoint inside the dielectric must be less than a wavelength. Therefore, the cutoff wavelength for the TE01 mode is: $\lambda_c = \pi(a + b)\sqrt{\mu_r \varepsilon_r}$.

3.9 Microstrip line

A microstrip is a planar printed transmission line, and is the most popular RF transmission line over the last 20 years [3–10]. Microstrip transmission lines consist of a conductive strip of width 'W' and thickness 't' and a wider ground plane, separated by a dielectric layer of thickness 'H'. In practice, a microstrip line is usually made by etching circuitry on a substrate that has a ground plane on the opposite face. A cross section of the microstrip line is shown in figure 3.9. The major advantage of a microstrip over a stripline is that all components can be mounted on top of the board. The disadvantages are that when high isolation is required such as in a filter or switch, some external shielding is needed. Microstrip circuits may radiate, causing unintended circuit response. A microstrip is dispersive; signals of different frequencies travel at slightly different speeds. Other microstrip line configurations are the offset stripline and suspended air microstrip line. For a microstrip line not all of the fields are constrained to the same dielectric. At the line

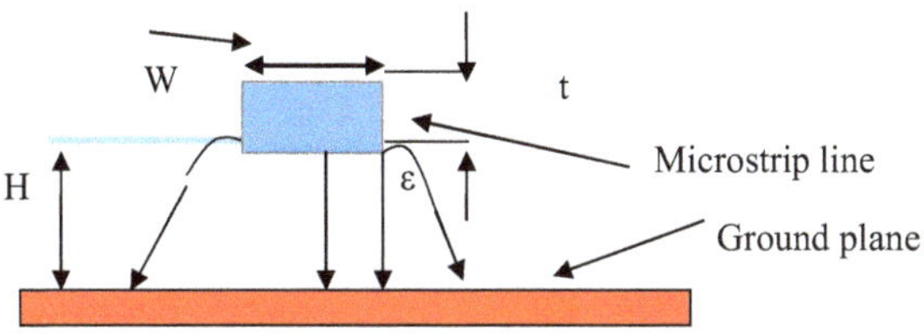

Figure 3.9. Microstrip line cross section.

edges the fields pass via air and the dielectric substrate. The effective dielectric constant should be calculated.

3.9.1 Effective dielectric constant

A part of the fields in the microstrip line structure exists in air and the other part of the fields exists in the dielectric substrate. The effective dielectric constant is somewhat less than the substrate's dielectric constant.

The effective dielectric constant of the microstrip line is calculated by:

For $\left(\frac{W}{H}\right) < 1$

$$\varepsilon_e = \frac{\varepsilon_r + 1}{2} + \frac{\varepsilon_r - 1}{2}\left[\left(1 + 12\left(\frac{H}{W}\right)\right)^{-0.5} + 0.04\left(1 - \left(\frac{W}{H}\right)\right)^2\right] \tag{3.67a}$$

For $\left(\frac{W}{H}\right) \geqslant 1$

$$\varepsilon_e = \frac{\varepsilon_r + 1}{2} + \frac{\varepsilon_r - 1}{2}\left[\left(1 + 12\left(\frac{H}{W}\right)\right)^{-0.5}\right] \tag{3.67b}$$

This calculation ignores strip thickness and frequency dispersion, but their effects are negligible.

3.9.2 Characteristic impedance

The characteristic impedance Z_0 is a function of the ratio of the height to the width W/H of the transmission line, and also has separate solutions depending on the value of W/H. The characteristic impedance Z_0 of a microstrip is calculated by:

For $\left(\frac{W}{H}\right) < 1$

$$Z_0 = \frac{60}{\sqrt{\varepsilon_e}}\ln\left[8\left(\frac{H}{W}\right)+0.25\left(\frac{H}{W}\right)\right]\Omega \quad (3.68a)$$

For $\left(\frac{W}{H}\right) \geqslant 1$

$$Z_0 = \frac{120\pi}{\sqrt{\varepsilon_e}\left[\left(\frac{H}{W}\right)+1.393 + 0.66_{*}\ln\left(\frac{H}{W}+1.444\right)\right]}\Omega \quad (3.68b)$$

We can calculate Z_0 by using equation (3.68) for a given $(\frac{W}{H})$. However, to calculate $(\frac{W}{H})$, for a given Z_0 we first should calculate ε_e. However, to calculate ε_e we should know $(\frac{W}{H})$. We first assume that $\varepsilon_e = \varepsilon_r$ and compute $(\frac{W}{H})$, for this value of $(\frac{W}{H})$ we compute ε_e. Then we compute a new value of $(\frac{W}{H})$. Two to three iterations are needed to calculate accurate values of $(\frac{W}{H})$ and ε_e. We may calculate $(\frac{W}{H})$ with around 10% accuracy by using equation (3.69). Table 3.13 presents examples of microstrip line parameters.

$$\frac{W}{H} = 8\frac{\sqrt{\left[e^{\frac{Z_0}{42.4}\sqrt{(\varepsilon_r+1)}}-1\right]\left[\frac{7+\frac{4}{\varepsilon r}}{11}\right]+\left[\frac{1+\frac{1}{\varepsilon r}}{0.81}\right]}}{\left[e^{\frac{Z_0}{42.4}\sqrt{(\varepsilon_r+1)}}-1\right]} \quad (3.69)$$

3.9.3 Higher order transmission modes in a microstrip line

In order to prevent higher order transmission modes we should limit the thickness of the microstrip substrate to 10% of a wavelength. The cutoff frequency of the higher order transmission mode is given as: $f_c = \frac{c}{4H\sqrt{\varepsilon - 1}}$.

Examples: Higher order modes will not propagate in microstrip lines printed on alumina substrate 15 mil thick up to 18 GHz. Higher order modes will not propagate in microstrip lines printed on GaAs substrate 4 mil thick up to 80 GHz. Higher order

Table 3.13. Examples of microstrip line parameters.

Substrate	W/H	Impedance Ω
Alumina ($\varepsilon_r = 9.8$)	0.95	50
GaAs ($\varepsilon_r = 12.9$)	0.75	50
$\varepsilon_r = 2.2$	3	50

modes will not propagate in microstrip lines printed on quartz substrate 5 mil thick up to 120 GHz.

Losses in a microstrip line

Losses in a microstrip line are due to conductor loss, radiation loss and dielectric loss.

3.9.4 Conductor loss

Conductor loss may be calculated by using equation (3.70).

$$\begin{aligned} \alpha_c &= 8.686 \log(R_S/(2WZ_0)) \quad \text{dB/length} \\ R_S &= \sqrt{\pi f \mu \rho} \qquad \text{skin resistance} \end{aligned} \tag{3.70}$$

Conductor losses may also be calculated by defining an equivalent loss tangent δc, given by $\delta c{=}\delta s/h$, where $\delta s = \sqrt{2/\omega\mu\sigma}$. Where σ is the strip conductivity, h is the substrate height, and μ is the free space permeability.

3.9.5 Dielectric loss

Dielectric loss may be calculated by using equation (3.71).

$$\begin{aligned} \alpha_d &= 27.3 \frac{\varepsilon_r}{\sqrt{\varepsilon_{eff}}} \frac{\varepsilon_{eff} - 1}{\varepsilon_r - 1} \frac{tg\delta}{\lambda_0} \ \text{dB cm}^{-1} \\ tg\delta &= \text{dielectric loss coefficent} \end{aligned} \tag{3.71}$$

Losses in microstrip lines are presented in tables 3.14–3.16 for several microstrip line structures. For example, the total loss of a microstrip line presented in table 3.15 at 40 GHz is 0.5 dB cm^{-1}, and the total loss of a microstrip line presented in table 3.16 at 40 GHz is 1.42 dB cm^{-1}. We may conclude that losses in microstrip lines limit the applications of microstrip technology at mm wave frequencies.

3.10 Waveguides

Waveguides are low loss transmission lines.

Waveguides may be rectangular or circular. A rectangular waveguide structure is presented in figure 3.10. The structure of a waveguide is uniform in the z direction. Fields in waveguides are evaluated by solving the Helmholtz equation. A wave

Table 3.14. Microstrip line losses for alumina substrate 10 mil thick.[a]

Frequency (GHz)	Loss tangent loss (dB cm^{-1})	Metal loss (dB cm^{-1})	Total loss (dB cm^{-1})
10	−0.005	−0.12	−0.124
20	−0.009	−0.175	−0.184
30	−0.014	−0.22	−0.23
40	−0.02	−0.25	−0.27

[a] Line parameters. Alumina, H = 254 μm (10 mils), W = 247 μm, ε_r = 9.9, Tan δ = 0.0002, 3 μm gold, conductivity 3.5 × 107 mhos m^{-1}.

Table 3.15. Microstrip line losses for alumina substrate 5 mil thick.[a]

Frequency (GHz)	Loss tangent loss (dB cm^{-1})	Metal loss (dB cm^{-1})	Total loss (dB cm^{-1})
10	−0.004	−0.23	−0.23
20	−0.009	−0.333	−0.34
30	−0.013	−0.415	−0.43
40	−0.018	−0.483	−0.5

[a] Alumina, $H = 127$ μm (5 mils), $W = 120$ μm, $\varepsilon_r = 9.9$, Tan $\delta = 0.0002$, 3 μm gold, conductivity 3.5 × 107 mhos m^{-1}.

Table 3.16. Microstrip line losses for GaAs substrate 2 mil thick.[a]

Frequency (GHz)	Tangent loss (dB cm^{-1})	Metal loss (dB cm^{-1})	Total loss (dB cm^{-1})
10	−0.010	−0.66	−0.67
20	−0.02	−0.96	−0.98
30	−0.03	−1.19	−1.22
40	−0.04	−1.38	−1.42

[a] GaAs, $H = 50$ μm (2 mils), $W = 34$ μm, $\varepsilon_r = 12.88$, Tan $\delta = 0.0004$, 3 μm gold, conductivity 3.5 × 107 mhos m^{-1}.

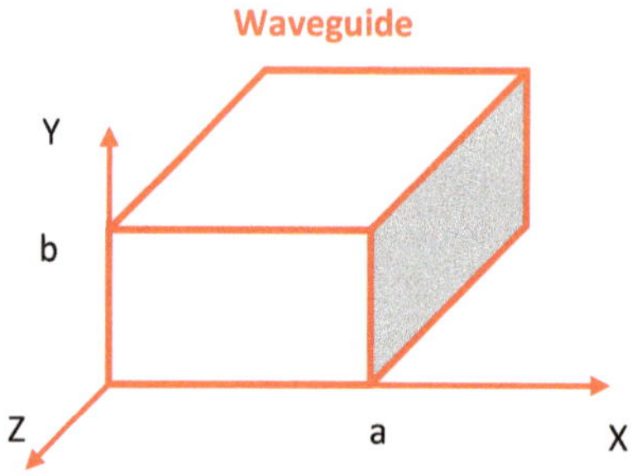

Figure 3.10. Rectangular waveguide structure.

equation is given in equation (3.72), and a wave equation in a rectangular coordinate system is given in equation (3.73).

$$\begin{aligned} \nabla^2 E &= \omega^2 \mu\varepsilon E = -k^2 E \\ \nabla^2 H &= \omega^2 \mu\varepsilon H = -k^2 H \\ k &= \omega\sqrt{\mu\varepsilon} = \frac{\omega}{v} = \frac{2\pi}{\lambda} \end{aligned} \tag{3.72}$$

$$\begin{aligned} \nabla^2 E &= \frac{\partial^2 E_i}{\partial x^2} + \frac{\partial^2 E_i}{\partial y^2} + \frac{\partial^2 E_i}{\partial z^2} + k^2 E_i = 0 \qquad i = x, y, z \\ \nabla^2 H &= \frac{\partial^2 H_i}{\partial x^2} + \frac{\partial^2 H_i}{\partial y^2} + \frac{\partial^2 H_i}{\partial z^2} + k^2 H_i = 0 \end{aligned} \tag{3.73}$$

The wave equation solution may be written as: $E = f(z)g(x, y)$. A field variation in the z direction may be written as $e^{-j\beta z}$. The derivative of this expression in the direction may be written as $-j\beta e^{j\beta z}$. Maxwell's equation may be presented as written in equations (3.74) and (3.75). A field may be represented as a superposition of waves in transverse and longitudinal directions.

$$\begin{aligned}
&E(x, y, z) = e(x, y)e^{-j\beta z} + e_z(x, y)e^{-j\beta z}\\
&H(x, y, z) = h(x, y)e^{-j\beta z} + h_z(x, y)e^{-j\beta z}\\
&\qquad \nabla XE = (\nabla_t - j\beta a_z) \times (e + e_z)e^{-j\beta z} = -j\omega\mu(h + h_z)e^{-j\beta z}\\
&\nabla_t \times e - j\beta a_z \times e + \nabla_t \times e_z - j\beta a_z \times e_z = -j\omega\mu(h + h_z)\\
&\qquad a_z \times e_z = 0\\
&\qquad \nabla_t \times e = -j\omega\mu h_z\\
&-j\beta a_z \times e + \nabla_t \times e_z = -a_z \times \nabla_t e_z - j\beta a_z \times e = -j\omega\mu h
\end{aligned} \tag{3.74}$$

$$\begin{aligned}
\nabla_t \times h &= -j\omega\varepsilon e_z\\
a_z \times \nabla_t h_z + j\beta a_z \times h &= -j\omega\varepsilon e
\end{aligned} \tag{3.75}$$

$$\begin{aligned}
&\nabla \cdot \mu H = (\nabla_t - j\beta a_z) \cdot (h + h_z)\mu e^{-j\beta z} = 0\\
&\nabla_t \cdot h - j\beta a_z \cdot h_z = 0\\
&\nabla \cdot \varepsilon E = 0 \quad \nabla_t \cdot e - j\beta a_z \cdot e_z = 0
\end{aligned} \tag{3.76}$$

Waves may be characterized as TEM, TE or TM waves. In TEM waves $e_z = h_z = 0$. In TE waves $e_z = 0$. In TM waves $h_z = 0$. TE—transverse electric, TM—transverse magnetic mode.

3.10.1 TE waves

In TE waves $e_z = 0$. h_z is as in the solution of equation (3.77). The solution to equation (3.77) may be written as $h_z = f(x)g(y)$.

$$\nabla_t^2 h_z = \frac{\partial^2 h_z}{\partial x^2} + \frac{\partial^2 h_z}{\partial y^2} + k_c^2 h_z = 0 \tag{3.77}$$

By applying $h_z = f(x)g(y)$ to equation (3.77) and dividing by fg we get equation (3.78).

$$\frac{f''}{f} + \frac{g''}{g} + k_c^2 = 0 \tag{3.78}$$

f is a function that varies in the x direction and g is a function that varies in the y direction. The sum of f and g may be equal to zero only if they are equal to a constant. These facts are written in equation (3.79).

$$\frac{f''}{f} = -k_x^2; \quad \frac{g''}{g} = -k_y^2$$
$$k_x^2 + k_y^2 = k_c^2 \tag{3.79}$$

The solutions for f and g are given in equation (3.80). A_1, $A_2 B_1$,B_2 are derived by applying h_z boundary conditions to equation (3.78).

$$f = A_1 \cos k_x x + A_2 \sin k_x x$$
$$g = B_1 \cos k_y y + B_2 \sin k_y y \tag{3.80}$$

h_z boundary conditions are written in equation (3.81).

$$\frac{\partial h_z}{\partial x} = 0@ \quad x = 0, a$$
$$\frac{\partial h_z}{\partial y} = 0@ \quad y = 0, b \tag{3.81}$$

By applying h_z boundary conditions to equation (3.80) we get the relations written in equation (3.82).

$$-k_x A_1 \sin k_x x + k_x A_2 \cos k_x x = 0$$
$$-k_y B_1 \sin k_y y + k_y B_2 \cos k_y y = 0$$
$$A_2 = 0 \quad k_x a = 0 \quad k_x = \frac{n\pi}{a} \quad n = 0, 1, 2$$
$$B_2 = 0 \quad k_y b = 0 \quad k_y = \frac{m\pi}{b} \quad m = 0, 1, 2 \tag{3.82}$$

The solution for h_z is given in equation (3.83).

$$h_z = A_{nm} \cos \frac{n\pi x}{a} \cos \frac{m\pi y}{b}$$
$$n = m \neq 0 \; n = 0, 1, 2 \; m = 0, 1, 2$$
$$k_{c,\,nm} = \left[\left(\frac{n\pi}{a}\right)^2 + \left(\frac{m\pi}{b}\right)^2\right]^{1/2} \tag{3.83}$$

Both n and m cannot be zero. The wavenumber at cutoff is $k_{c,\,nm}$ and depends on the waveguide dimensions. The propagation constant γ_{nm} is given in equation (3.84).

$$\gamma_{nm} = j\beta_{nm} = j(k_0^2 - k_c^2)^{1/2}$$
$$= j\left[\left(\frac{2\pi}{\lambda_0}\right)^2 - \left(\frac{n\pi}{a}\right)^2 - \left(\frac{m\pi}{b}\right)^2\right]^{1/2} \tag{3.84}$$

For $k_0 > k_{c,\,nm}$, β is real and the wave will propagate. For $k_0 < k_{c,\,nm}$, β is imaginary and the wave will decay rapidly. Frequencies that define propagating and decaying waves are called cutoff frequencies. We may calculate cutoff frequencies by using equation (3.85).

$$f_{c,\,nm} = \frac{c}{2\pi}k_{c,\,nm} = \frac{c}{2\pi}\left[\left(\frac{n\pi}{a}\right)^2 + \left(\frac{m\pi}{b}\right)^2\right]^{1/2} \tag{3.85}$$

For $a = 2$ the cutoff wavelength is computed by using equation (3.86).

$$\begin{aligned} \lambda_{c,\,nm} &= \frac{2ab}{[n^2b^2 + m^2a^2]^{1/2}} = \frac{2a}{[n^2 + 4m^2]^{1/2}} \\ \lambda_{c,\,10} &= 2a\lambda_{c,\,01} = a\lambda_{c,\,11} = 2a/\sqrt{5} \\ &\frac{c}{2a}\langle f_{01}\rangle\frac{c}{a} \end{aligned} \tag{3.86}$$

For $\frac{c}{2a}\langle f_{01}\rangle\frac{c}{a}$ TE_{10} is dominant.

By using equations (3.74)–(3.76) we can derive the electromagnetic fields that propagate in the waveguide as given in equation (3.87).

$$\begin{aligned} H_z &= A_{nm}\cos\frac{n\pi x}{a}\cos\frac{m\pi y}{b}e^{\pm j\beta_{nm}z} \\ H_x &= \pm j\frac{n\pi\beta_{nm}}{ak_{c,\,nm}^2}A_{nm}\sin\frac{n\pi x}{a}\cos\frac{m\pi y}{b}e^{\pm j\beta_{nm}z} \\ H_y &= \pm j\frac{m\pi\beta_{nm}}{bk_{c,\,nm}^2}A_{nm}\cos\frac{n\pi x}{a}\sin\frac{m\pi y}{b}e^{\pm j\beta_{nm}z} \\ E_X &= Z_{h,\,nm}j\frac{m\pi\beta_{nm}}{bk_{c,\,nm}^2}A_{nm}\cos\frac{n\pi x}{a}\sin\frac{m\pi y}{b}e^{\pm j\beta_{nm}z} \\ E_Y &= -jZ_{h,\,nm}\frac{n\pi\beta_{nm}}{ak_{c,\,nm}^2}A_{nm}\sin\frac{n\pi x}{a}\cos\frac{m\pi y}{b}e^{\pm j\beta_{nm}z} \end{aligned} \tag{3.87}$$

The impedance of the nm modes is given as: $Z_{h,\,nm} = \frac{e_x}{h_y} = \frac{k_0}{\beta_{nm}}\sqrt{\frac{\mu_0}{\varepsilon_0}}$.

The power of the nm mode is computed by using the Poynting vector calculation as shown in equation (3.88).

$$\begin{aligned} P_{nm} &= 0.5\,\mathrm{Re}\int_0^a\int_0^b E\times H^{*}{\cdot}a_z\,dxdy \\ &= 0.5\,\mathrm{Re}\,Z_{h,\,nm}\int_0^a\int_0^b (H_yH_Y^{*} + H_xH_x^{*})dxdy \\ \int_0^a\int_0^b &\cos^2\frac{n\pi x}{a}\sin^2\frac{m\pi y}{b}dxdy = \frac{ab}{4}\quad n\neq 0\ \ m\neq 0 \\ &or\ \frac{ab}{2}\ nor\ m = 0 \\ P_{nm} &= \frac{|A_{nm}|^2}{2\varepsilon_{0n}\varepsilon_{0m}}\left(\frac{\beta_{nm}}{k_{c,\,nm}}\right)^2 Z_{h,\,nm}\quad \varepsilon_{0n} = 1, n = 0\ \ \varepsilon_{0n} = 2, n{>}0 \end{aligned} \tag{3.88}$$

The TE mode with the lowest cutoff frequency in the rectangular waveguide is TE_{10}.

TE10 fields in the rectangular waveguide are shown in figure 3.11.

3.10.2 TM waves

In TM waves $h_z = 0$. e_z is as in the solution of equation (3.89). The solution to equation (3.18) may be written as $e_z = f(x)g(y).e_z$ should be zero at the metallic walls. e_z boundary conditions are written in equation (3.90). The solution for e_z is given in equation (3.96).

$$\nabla_t^2 e_z = \frac{\partial^2 e_z}{\partial x^2} + \frac{\partial^2 e_z}{\partial y^2} + k_c^2 e_z = 0 \tag{3.89}$$

$$\begin{aligned} e_z &= 0@\ x = 0,\ a \\ e_z &= 0@\ y = 0,\ b \end{aligned} \tag{3.90}$$

$$\begin{aligned} e_z &= A_{nm} \sin \frac{n\pi x}{a} \sin \frac{m\pi y}{b} \\ n &= m \neq 0 \ \ n = 0,\ 1,\ 2 \ \ m = 0,\ 1,\ 2 \\ k_{c,\,nm} &= \left[\left(\frac{n\pi}{a}\right)^2 + \left(\frac{m\pi}{b}\right)^2\right]^{1/2} \end{aligned} \tag{3.91}$$

The first propagating TM mode is TM_{11}, $n = m = 1$. By using equations (3.74)–(3.76) and (3.89) we can derive the electromagnetic fields that propagate in the waveguide as given in equation (3.92).

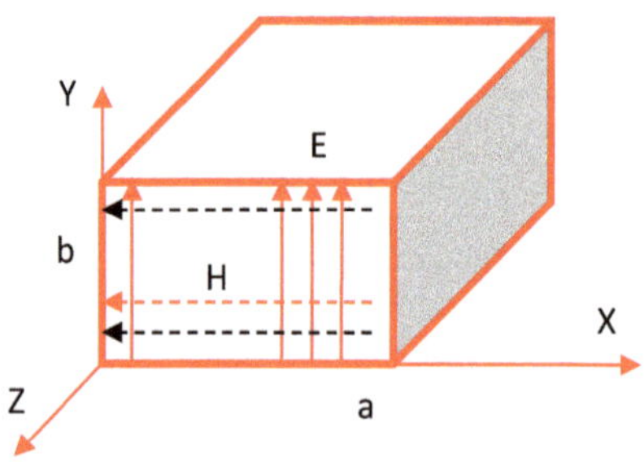

Figure 3.11. TE_{10} mode.

$$
\begin{aligned}
E_z &= \sin\frac{n\pi x}{a}\sin\frac{m\pi y}{b}e^{\pm j\beta_{nm}z} \\
E_x &= -j\frac{n\pi\beta_{nm}}{ak_{c,\,nm}^2}\cos\frac{n\pi x}{a}\sin\frac{m\pi y}{b}e^{\pm j\beta_{nm}z} \\
E_y &= -j\frac{m\pi\beta_{nm}}{bk_{c,\,nm}^2}A_{nm}\sin\frac{n\pi x}{a}\cos\frac{m\pi y}{b}e^{\pm j\beta_{nm}z} \\
H_X &= \frac{-E_y}{Z_{e,\,nm}} \\
H_Y &= \frac{E_x}{Z_{e,\,nm}}
\end{aligned}
\tag{3.92}
$$

The impedance of the nm modes: $Z_{e,\,nm} = \frac{\beta_{nm}}{k_0}\sqrt{\frac{\mu_0}{\varepsilon_0}}$.

The TM mode with the lowest cutoff frequency in the rectangular waveguide is TM_{11}.

TM_{11} fields in the rectangular waveguide are shown in figure 3.12.

3.11 Circular waveguide

A circular waveguide is used to transmit electromagnetic waves in a circular polarization. At high frequencies attenuation of several modes in the circular waveguide is lower than in the rectangular waveguide. A circular waveguide's structure is uniform in the *z* direction. The fields in waveguides are evaluated by solving the Helmholtz equation in a cylindrical coordinate system. A circular waveguide in a cylindrical coordinate system is presented in figure 3.13. A wave equation is given in equation (3.93), and a wave equation in a cylindrical coordinate is given in equation (3.94).

$$
\begin{aligned}
\nabla^2 E &= \omega^2\mu\varepsilon E = -k^2 E \\
\nabla^2 H &= \omega^2\mu\varepsilon H = -k^2 H \\
k &= \omega\sqrt{\mu\varepsilon} = \frac{\omega}{v} = \frac{2\pi}{\lambda}
\end{aligned}
\tag{3.93}
$$

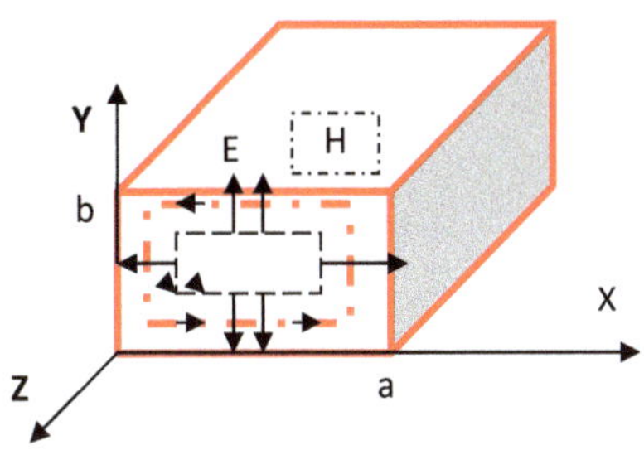

Figure 3.12. TM_{11} mode.

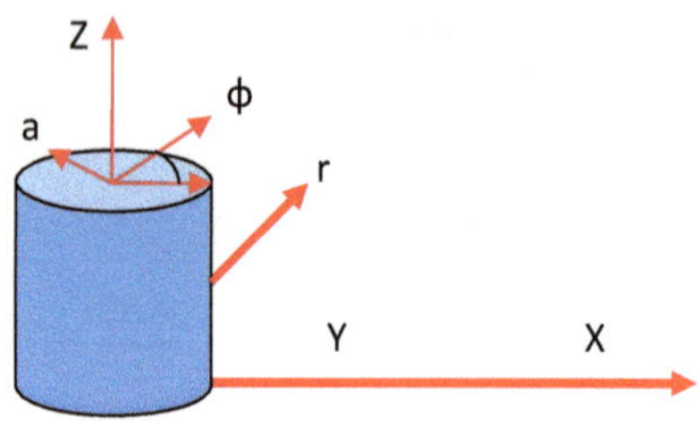

Figure 3.13. Circular waveguide structure.

$$\nabla^2 E = \frac{1}{r}\frac{\partial}{\partial r}\left(r\frac{\partial E_i}{\partial r}\right) + \frac{1}{r^2}\frac{\partial^2 E_i}{\partial \varphi^2} + \frac{\partial^2 E_i}{\partial z^2} - \gamma^2 E_i = 0 \qquad i = r, \varphi, z$$
$$\nabla^2 H = \frac{1}{r}\frac{\partial}{\partial r}\left(r\frac{\partial H_i}{\partial r}\right) + \frac{1}{r^2}\frac{\partial^2 H_i}{\partial \varphi^2} + \frac{\partial^2 H_i}{\partial z^2} - \gamma^2 H_i = 0 \tag{3.94}$$

The solution to equation (3.94) may be written as $E = f(r)g(\varphi)h(z)$. By applying E to equation (3.94) and dividing by $f(r)g(\varphi)h(z)$ we get equation (3.95).

$$\nabla^2 E = \frac{1}{rf}\frac{\partial}{\partial r}\left(r\frac{\partial f}{\partial r}\right) + \frac{1}{r^2 g}\frac{\partial^2 g}{\partial \varphi^2} + \frac{\partial^2 h}{h\partial z^2} - \gamma^2 = 0 \tag{3.95}$$

f is a function that varies in the r direction, g is a function that varies in the φ direction and h is a function that varies in the z direction. The sum of f, g and h may be equal to zero only if they are equal to a constant. The solution for h is written in equation (3.96). The propagation constant is γg.

$$\frac{\partial^2 h}{h\partial z^2} - \gamma_g^2 = 0$$
$$h = Ae^{-\gamma_g z} + Be^{\gamma_g z} \tag{3.96}$$

$$\frac{r}{f}\frac{\partial}{\partial r}\left(r\frac{\partial f}{\partial r}\right) + \frac{1}{g}\frac{\partial^2 g}{\partial \varphi^2} - (\gamma^2 - \gamma_g^2) = 0 \tag{3.97}$$

The solution for g is written in equation (3.98).

$$\frac{\partial^2 g}{g\partial \varphi^2} + n^2 = 0$$
$$g = A_n \sin n\varphi + B_n \cos n\varphi \tag{3.98}$$

$$r\frac{\partial}{\partial r}\left(r\frac{\partial f}{\partial r}\right) + ((k_c r)^2 - n^2)f = 0$$
$$k_c^2 + \gamma^2 = \gamma_g^2 \tag{3.99}$$

Equation (3.99) is a Bessel equation. The solution of this equation is written in equation (3.100). $J_n(k_c r)$ is a Bessel equation with order n and represents a standing wave. The wave varies a cosine function in the circular waveguide. $N_n(k_c r)$ is a Bessel equation with order n and represents a standing wave. The wave varies a sine function in the circular waveguide.

$$f = C_n J_n(k_c r) + D_n N_n(k_c r) \tag{3.100}$$

The general solution for the electric fields in the circular waveguide is given in equation (3.101). For $r = 0$ $N_n(k_c r)$ goes to infinity, so $D_n = 0$.

$$E(r, \varphi, z) = (C_n J_n(k_c r) + D_n N_n(k_c r))(A_n \sin n\varphi + B_n \cos n\varphi)e^{\pm\gamma_g z} \tag{3.101}$$

$$A_n \sin n\varphi + B_n \cos n\varphi = \sqrt{A_n^2 + B_n^2} \cos\left(n\varphi + \tan^{-1}\left(\frac{A_n}{B_n}\right)\right) = F_n \cos n\varphi \tag{3.102}$$

The general solution for the electric fields in the circular waveguide is given in equation (3.103).

$$\begin{aligned} E(r, \varphi, z) &= E_0(J_n(k_c r))(\cos n\varphi)e^{\pm\gamma_g z} \quad \text{If } \alpha = 0 \\ E(r, \varphi, z) &= E_0(J_n(k_c r))(\cos n\varphi)e^{\pm\beta_g z} \\ \beta_g &= \pm\sqrt{\omega^2\mu\varepsilon - k_c^2} \end{aligned} \tag{3.103}$$

3.11.1 TE waves in a circular waveguide

In TE waves $e_z = 0$. H_z is as in the solution of equation (3.104). The solution to equation (3.105) may be written as given in equation (3.105).

$$\nabla^2 H_z = \gamma^2 H_z \tag{3.104}$$

$$H_z = H_{0z}(J_n(k_c r))(\cos n\varphi)e^{\pm j\beta_g z} \tag{3.105}$$

The electric and magnetic fields are solutions of Maxwell's equations as written in equations (3.106) and (3.107).

$$\nabla XE = -j\omega\mu H \tag{3.106}$$

$$\nabla \times H = j\omega\varepsilon E \tag{3.107}$$

Field variation in the z direction may be written as $e^{-j\beta z}$. The derivative of this expression in the same direction may be written as $-j\beta e^{j\beta z}$. The electric and magnetic field components are solutions of equations (3.108) and (3.109).

$$E_r = -\frac{j\omega\mu}{k_c^2}\frac{1}{r}\left(\frac{\partial H_z}{\partial \varphi}\right)$$
$$E_\varphi = \frac{j\omega\mu}{k_c^2}\left(\frac{\partial H_z}{\partial r}\right) \tag{3.108}$$

$$H_\varphi = -\frac{-j\beta_g}{k_c^2}\frac{1}{r}\left(\frac{\partial H_z}{\partial \varphi}\right)$$
$$H_r = \frac{-j\beta_g}{k_c^2}\left(\frac{\partial H_z}{\partial r}\right) \tag{3.109}$$

H_z, H_r and E_φ boundary conditions are written in equation (3.110).

$$\frac{\partial H_z}{\partial r} = 0 \ @ \quad r = a$$
$$H_r = 0 \ @ \quad r = a \tag{3.110}$$
$$E_\varphi = 0 \ @ \quad r = a$$

By applying the boundary conditions to equation (3.105) we get the relations written in equation (3.111). The solutions of equation (3.111) are listed in table 3.17.

$$\frac{\partial H_z}{\partial r} \mid r = a = H_{0z}(J_n'(k_c a))(\cos n\varphi)e^{-\beta_g z} = 0$$
$$J_n'(k_c a) = 0 \tag{3.111}$$

The wavenumber at cutoff is $k_{c,\,np}$. $k_{c,\,np}$ depends on the waveguide dimensions. The propagation constant $\gamma_{g,\,np}$ is given in equation (3.112).

$$\gamma_{g,\,np} = j\beta_{g,\,np} = j(k_0^2 - k_c^2)^{1/2}$$
$$= j\left[\left(\frac{2\pi}{\lambda_0}\right)^2 - \left(\frac{X_{np}'}{a}\right)^2\right]^{1/2} \tag{3.112}$$

Table 3.17. Circular waveguide TE modes.

p	$(n = 0)\ X_{np}'$	$(n = 1)\ X_{np}'$	$(n = 2)\ X_{np}'$	$(n = 3)\ X_{np}'$	$(n = 4)\ X_{np}'$	$(n = 5)\ X_{np}'$
1	3.832	1.841	3.054	4.201	5.317	6.416
2	7.016	5.331	6.706	8.015	9.282	10.52
3	10.173	8.536	9.969	11.346	12.682	13.987
4	13.324	11.706	13.170			

For $k_0 > k_{c,\,nm}$, β is real and the wave will propagate. For $k_0 < k_{c,\,nm}$, β is imaginary and the wave will decay rapidly. Frequencies that define propagating and decaying waves are called cutoff frequencies. We may calculate the cutoff frequencies by using equation (3.113).

$$f_{c,\,nm} = \frac{cX'_{np}}{2\pi a} \tag{3.113}$$

We get the field components by solving equations (3.108) and (3.109). The field components are written in equation (3.114).

$$\begin{aligned}
H_z &= H_{0z}\left(J_n\left(\frac{X'_{np}r}{a}\right)\right)(\cos n\varphi)e^{-j\beta_g z} \\
H_\varphi &= \frac{E_{0r}}{Z_g}\left(J_n\left(\frac{X'_{np}r}{a}\right)\right)(\sin n\varphi)e^{-j\beta_g z} \\
H_r &= \frac{E_{0\varphi}}{Z_g}\left(J'_n\left(\frac{X'_{np}r}{a}\right)\right)(\cos n\varphi)e^{-j\beta_g z} \\
E_\varphi &= E_{0\varphi}\left(J'_n\left(\frac{X'_{np}r}{a}\right)\right)(\cos \varphi)e^{-j\beta_g z} \\
E_r &= E_{0r}\left(J_n\left(\frac{X'_{np}r}{a}\right)\right)(\sin \varphi)e^{-j\beta_g z}
\end{aligned} \tag{3.114}$$

The impedance of the *np* modes is written in equation (3.115).

$$Z_{g,\,np} = \frac{E_r}{H_\varphi} = \frac{\omega\mu}{\beta_{g,\,np}} = \frac{\eta}{\sqrt{1 - \left(\frac{f_c}{f}\right)^2}} \tag{3.115}$$

3.11.2 TM waves in a circular waveguide

In TM waves $h_z = 0$. e_z is as in the solution of equation (3.116). The solution to equation (3.116) is written in equation (3.117).

$$\nabla^2 E_z = \gamma^2 E_z \tag{3.116}$$

$$E_z = E_{0z}(J_n(k_c r))(\cos n\varphi)e^{\pm j\beta_g z} \tag{3.117}$$

The electric and magnetic fields are the solutions of Maxwell's equations as written in equations (3.118) and (3.119).

$$\nabla XE = -j\omega\mu H \tag{3.118}$$

$$\nabla \times H = j\omega\varepsilon E \tag{3.119}$$

Field variation in the z direction may be written as $e^{-j\beta z}$. The derivative of this expression in the same direction may be written as $-j\beta e^{j\beta z}$. The electric and magnetic fields components are solutions of equations (3.120) and (3.121).

$$\begin{aligned} H_r &= \frac{j\omega\varepsilon}{k_c^2}\frac{1}{r}\left(\frac{\partial H_z}{\partial \varphi}\right) \\ H_\varphi &= \frac{-j\omega\varepsilon}{k_c^2}\left(\frac{\partial H_z}{\partial r}\right) \end{aligned} \tag{3.120}$$

$$\begin{aligned} E_\varphi &= -\frac{-j\beta_g}{k_c^2}\frac{1}{r}\left(\frac{\partial E_z}{\partial \varphi}\right) \\ E_r &= \frac{-j\beta_g}{k_c^2}\left(\frac{\partial E_z}{\partial r}\right) \end{aligned} \tag{3.121}$$

The E_r boundary condition is written in equation (3.122).

$$E_z = 0 \;@\; r = a \tag{3.122}$$

By applying the boundary conditions to equation (3.122) we get the relations written in equation (3.123). The solutions of equation (3.123) are listed in table 3.18.

$$\begin{aligned} &E_z\,|(r = a) = H_{0z}(J_n(k_c a))(\cos n\varphi)e^{-j\beta_g z} = 0 \\ &J_n(k_c a) = 0 \end{aligned} \tag{3.123}$$

The wavenumber at cutoff is $k_{c,\,np}$. $k_{c,\,np}$ depends on the waveguide dimensions. The propagation constant $\gamma_{g,\,np}$ is given in equation (3.124).

$$\begin{aligned} \gamma_{g,\,np} &= j\beta_{g,\,np} = j(k_0^2 - k_c^2)^{1/2} \\ &= j\left[\left(\frac{2\pi}{\lambda_0}\right)^2 - \left(\frac{X_{np}}{a}\right)^2\right]^{1/2} \end{aligned} \tag{3.124}$$

Table 3.18. Circular waveguide TM modes.

p	$n = 0, X_{np}$	$n = 1, X_{np}$	$n = 2, X_{np}$	$n = 3, X_{np}$	$n = 4, X_{np}$	$n = 5, X_{np}$
1	2.405	3.832	5.136	6.38	7.588	8.771
2	5.52	7.106	8.417	9.761	11.065	12.339
3	8.645	10.173	11.62	13.015	14.372	
4	11.792	13.324	14.796			

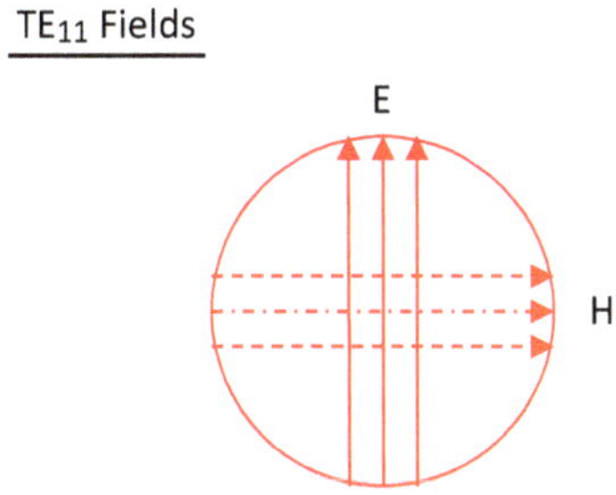

Figure 3.14. TE_{11} fields in circular waveguide.

For $k_0 > k_{c,\, nm}$, β is real and the wave will propagate. For $k_0 < k_{c,\, nm}$, β is imaginary and the wave will decay rapidly. Frequencies that define propagating and decaying waves are called cutoff frequencies. We may calculate the cutoff frequencies by using equation (3.125).

$$f_{c,\, nm} = \frac{cX_{np}}{2\pi a} \tag{3.125}$$

We get the field components by solving equations (3.120) and (3.121). The field components are written in equation (3.126).

$$\begin{aligned}
E_z &= E_{0z}\left(J_n\left(\frac{X_{np}r}{a}\right)\right)(\cos n\varphi)e^{-j\beta_g z} \\
H_\varphi &= \frac{E_{0r}}{Z_g}\left(J_n'\left(\frac{X_{np}r}{a}\right)\right)(\cos n\varphi)e^{-j\beta_g z} \\
H_r &= \frac{E_{0\varphi}}{Z_g}\left(J_n\left(\frac{X_{np}r}{a}\right)\right)(\sin n\varphi)e^{-j\beta_g z} \\
E_\varphi &= E_{0\varphi}\left(J_n\left(\frac{X_{np}r}{a}\right)\right)(\sin \varphi)e^{-j\beta_g z} \\
E_r &= E_{0r}\left(J_n'\left(\frac{X_{np}r}{a}\right)\right)(\cos \varphi)e^{-j\beta_g z}
\end{aligned} \tag{3.126}$$

The impedance of the *np* modes is written in equation (3.127).

$$Z_{g,\, np} = \frac{E_r}{H_\varphi} = \frac{\beta_{g,\, np}}{\omega\varepsilon} = \eta\sqrt{1 - \left(\frac{f_c}{f}\right)^2} \tag{3.127}$$

The mode with the lowest cutoff frequency in a circular waveguide is TE_{11}. TE_{11} fields in a circular waveguide are shown in figure 3.14.

References

[1] Ramo W and Duzer V 1994 *Fields and Waves in Communication Electronics* 3rd edn (New York: Wiley)

[2] Collin R E 1966 *Foundation for Microwave Engineering* (New York: McGraw-Hill)
[3] Balanis C A 1996 *Antenna Theory: Analysis and Design* 2nd edn (New York: Wiley)
[4] Sabban A 2015 *Low Visibility Antennas for Communication Systems* (London: Taylor and Francis)
[5] Sabban A 2016 *Wideband RF Technologies and Antenna in Microwave Frequencies* (New York: Wiley)
[6] Godara L C (ed) 2002 *Handbook of Antennas in Wireless Communications* (Boca Raton, FL: CRC Press)
[7] Kraus J D and Marhefka R J 2002 *Antennas for all Applications* 3rd edn (New York: McGraw-Hill)
[8] Ulaby F T 2005 *Electromagnetics for Engineers* (USA: Pearson)
[9] James J R, Hall P S and Wood C 1981 *Microstrip Antenna Theory and Design* (London: Peter Perigrinus)
[10] Sabban A 2014 *RF Engineering, Microwave and Antennas* (Israel: Saar publication)
[11] ITU-R Recommendation V.431 2015 *Nomenclature of the Frequency and Wavelength Bands Used in Telecommunications* (Geneva: International Telecommunication Union)
[12] IEEE Standard 521-2002 2003 *Standard Letter Designations for Radar-Frequency Bands* (Piscataway, NJ: IEEE) E-ISBN: 0-7381-3356-6
[13] AFR 55-44/AR 105-86/OPNAVINST 3430.9A/MCO 3430.1, 27 October 1964 superseded by AFR 55-44/AR 105-86/OPNAVINST 3430.1A/MCO 3430.1A, 6 December 1978: Performing Electronic Countermeasures in the United States and Canada, Attachment 1, ECM Frequency Authorizations.
[14] Belov L A, Smolskiy S M and Kochemasov V N 2012 *Handbook of RF, Microwave, and Millimeter-Wave Components* (Boston, MA: Artech House) pp 27–28
[15] Chirwa L C, Hammond P A, Roy S and Cumming D R S 2003 Electromagnetic radiation from ingested sources in the human intestine between 150 MHz and 1.2 GHz *IEEE Trans. Biomed. Eng.* **50** 484–92
[16] Werber D, Schwentner A and Biebl E M 2006 Investigation of RF transmission properties of human tissues *Adv. Radio Sci.* **4** 357–60
[17] Sabban A 2017 *Novel Wearable Antennas for Communication and Medical Systems* (London: Taylor and Francis)

Chapter 4

Microwave technologies for wearable communication systems

4.1 Introduction

Compact wearable communication systems may be designed only by employing modern microwave technologies such as monolithic microwave integrated circuit (MMIC), micro-electro-mechanical systems (MEMS) and low temperature co-fired ceramic (LTCC) technologies. The communication industry in microwave and mm wave frequencies is currently growing. Radio frequency modules such as front end, filters, power amplifiers, printed antennas, passive components and limiters are important modules in wearable communication devices, see [1–20]. The electrical performance of the modules determines if the system will meet the required specifications. Moreover, in several cases the module's performance limits the system's performance. Minimization of the size and weight of the RF modules is achieved by employing MMIC and microwave integrated circuit (MIC) technology. However, integration of MIC and MMIC components and modules raise several technical challenges. Design parameters that may be neglected for modular communication systems cannot be ignored in the design of wideband integrated RF modules. Powerful RF design software, such as ADS (advanced design software) and HFSS (high frequency structure simulator), are required to achieve accurate design of RF modules in microwave frequencies. Accurate design of mm wave RF modules is crucial. It is impossible to tune mm wave RF modules in the fabrication process.

Microwave technologies for wearable communication systems

- MIC,
- MMIC,
- MEMS,
- LTCC.

doi:10.1088/978-0-7503-5224-6ch4

4.2 MIC—microwave integrated circuit

Traditional microwave systems consists of connectorized components (such as amplifiers, filters and mixers) connected by coaxial cables. These modules have large dimensions and suffer from high losses and weight. Dimension and losses may be minimized by using microwave integrated circuits, MIC technology. There are three types of circuits: HMIC, standard MIC and miniature HMIC. HMIC is a hybrid microwave integrated circuit. Solid state and passive elements are bonded to the dielectric substrate. The passive elements are fabricated by using thick or thin film technology. Standard MICs use a single level metallization for conductors and transmission lines. Miniature HMICs use a multilevel process in which passive elements such as capacitors and resistors are batch-deposited on the substrate. Semiconductor devices such as amplifiers and diodes are bonded on the substrate. Figure 4.1 shows a MIC transceiver. Figure 4.2 presents the layout of the MIC receiving link. The receiving channel consists of a low noise amplifier, filters, dielectric resonant oscillators and a diode mixer.

4.3 Low noise K band compact receiving channel for a satellite communication ground terminal

In section 4.3 an example of a MIC receiving channel is presented.

4.3.1 Introduction

An increasing demand for wide bandwidth in communication links makes the K/Ka band attractive for future commercial systems. The frequency allocations for the Ka/K band VSAT system are 17.7–21.2 GHz for the receiving channel and 27.5–31 GHz for the transmitting channel. The communications industry is continually growing. In particular, very small aperture terminal (VSAT) networks have gained

Figure 4.1. A MIC transceiver.

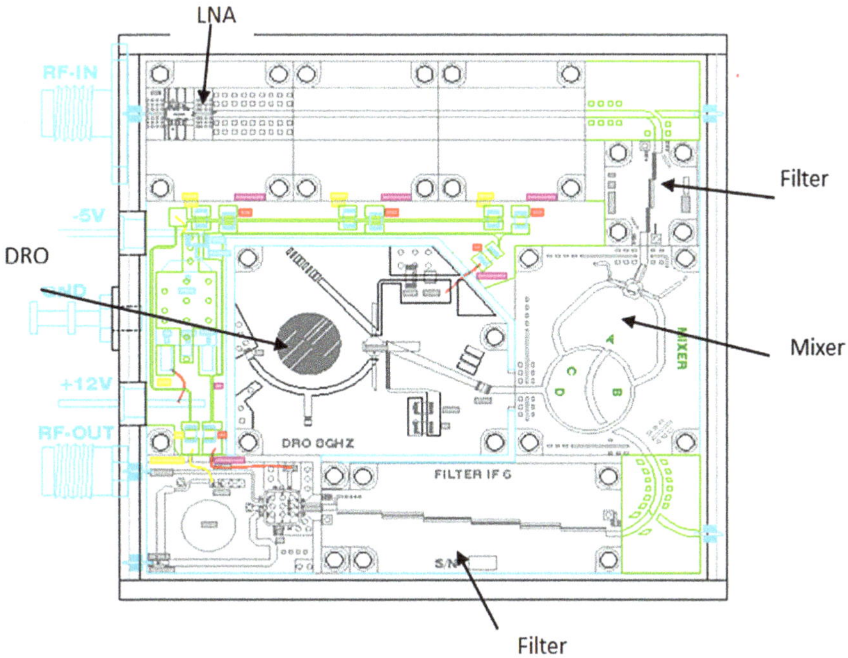

Figure 4.2. Layout of a MIC receiving link.

wide usage for business and private applications. Private organizations and banks are using VSAT networks to communicate between their various sites. VSAT applications cover a wide range such as telephony, message distribution, lottery, credit card approval and inventory management. Commercial VSAT systems operate in the C band and Ku band, however, there are many advantages to developing wideband K/Ka band communication systems. However, only some commercial low-cost power amplifiers, low noise amplifiers, mixers and DROs are published in commercial catalogs. Moreover, development of low-cost RF components is crucial for the K/Ka band satellite communication industry. This section describes the design and performance of a compact and low-cost K band receiving channel.

4.3.2 Receiving channel design

The major objectives in the design of the receiving channel were electrical specifications and cost.

Receiving channel specifications

The receiving channel specification is listed in table 4.1.

4.3.3 Description of the receiving channel

A block diagram of the receiving channel is shown in figure 4.3.

Table 4.1. Receiving channel specification.

Parameter	Specification
RF frequency range	18.8–19.3 GHz, 19.7–20.2 GHz
IF frequency range	0.95–1.45 GHz
Gain	50 dB
Noise figure	2 dB
Input VSWR	2:1
Output VSWR	2:1
Spurious level	−40 dBc
Frequency stability versus temperature	±2 MHz
Supply voltage	±5 V
Connectors	K-connectors
Operating temperature	−40 °C to 60 °C
Storage temperature	−50 °C to 80 °C
Humidity	95%

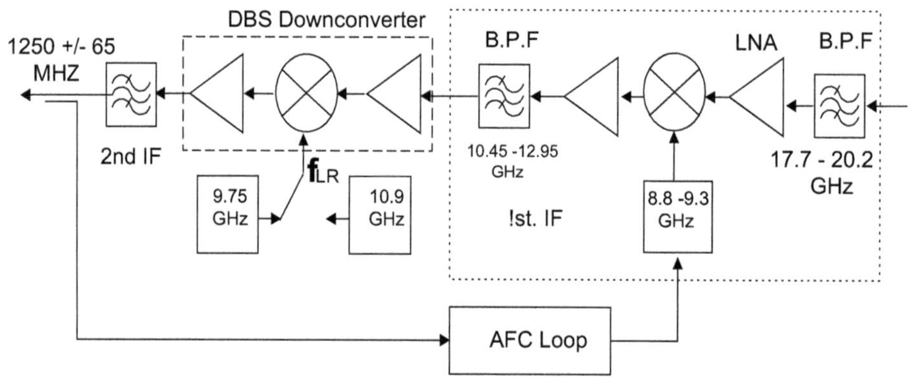

Figure 4.3. Block diagram of the receiving channel.

The receiving channel consists of a RF side-coupled band pass filter, low noise amplifier, mixer, DRO, IF filter, MMIC down-converter block. Noise figure, gain, and IP3 budget are given in table 4.2.

A receiving channel with improved NF (0.95 dB) and gain (77 dB) is obtained by adding a gain block after the low noise amplifier (LNA). However, from a cost consideration we decided to realize the first configuration shown in table 4.2. The major objectives in the design of the receiving channel were specifications and cost.

4.3.4 Development of the receiving channel

A MIC and MMIC LNA were developed for K band receiving links. The dimensions of the MMIC LNA are much smaller than the MIC LNA. However, the NF of the MIC LNA is around 1.2–1.5 dB and the NF of the MMIC LNA is around 1.7–2 dB. Components of the receiving channel were printed on 10 mil thick RT-5880 Duroid substrate. Drawings of the MMIC LNA, and of the receiving channel are shown in figures 4.4 and 4.5. A photograph of the receiving channel with a MMIC LNA is shown in figure 4.6.

4.3.5 Measured test results of the receiving channel

The measured test results of the receiving channel are summarized in table 4.3.

Table 4.2. Noise figure, gain, and IP3 budget.

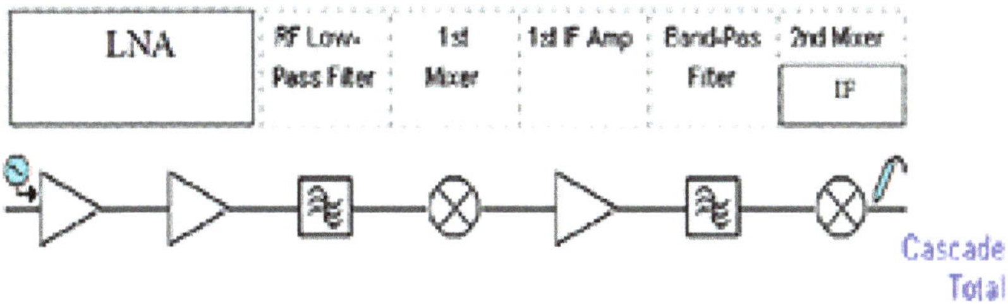

	LNA		RF Low-Pass Filter	1st Mixer	1st IF Amp	Band-Pass Filter	2nd Mixer	Cascade Total
NF (dB)	0.8	0.8	0.5	10	2.5	6.5	6.5	1.66
Gain (dB)	10	10	-0.5	-10	16	-6.5	37	56.00
OIP3 (dBm)	14	14	90	8	18	100	[illegible]	8.00

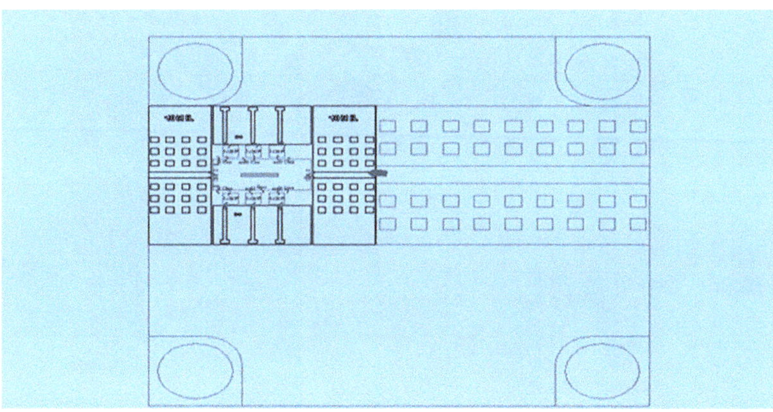

Figure 4.4. MMIC LNA on carrier.

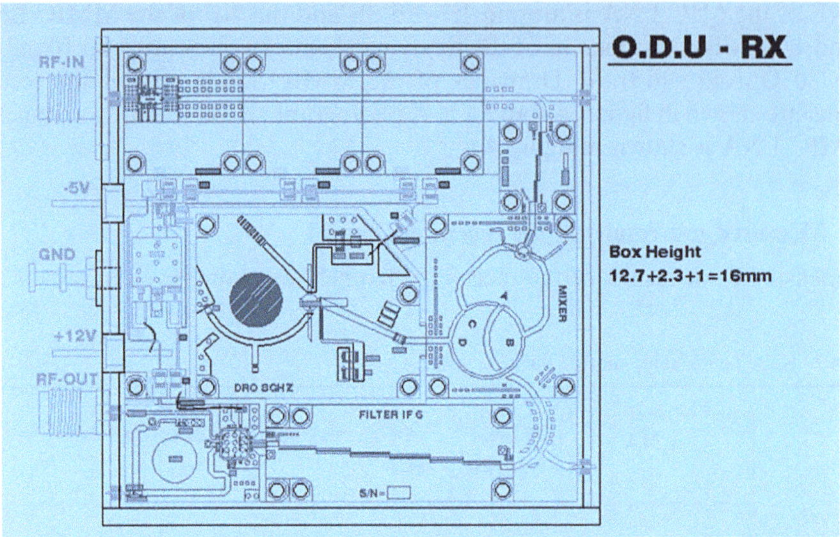

Figure 4.5. ODU receiving channel.

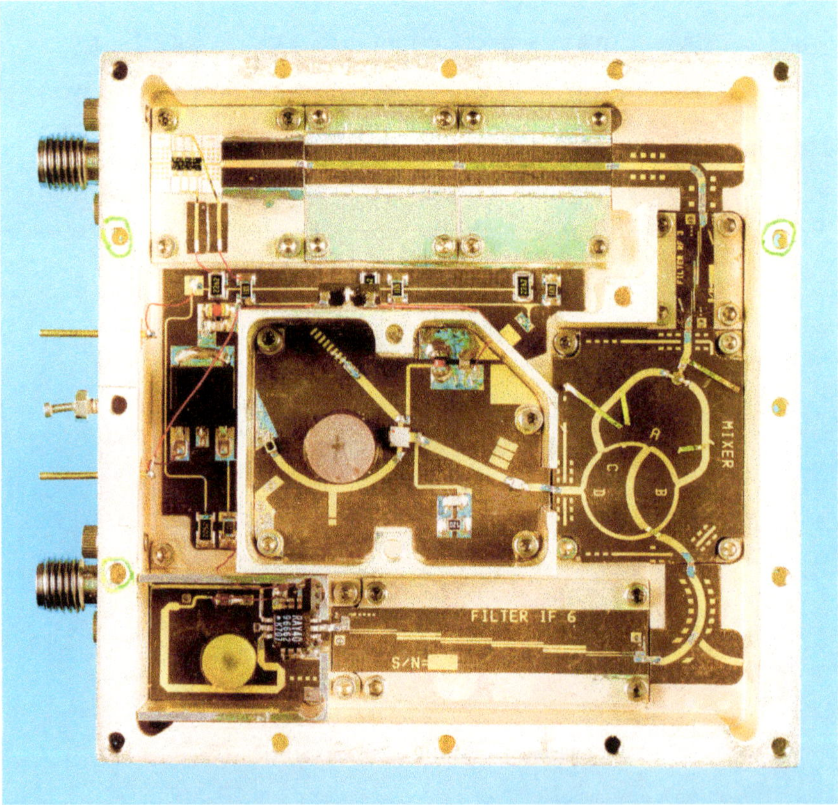

Figure 4.6. Photo of the ODU receiving channel.

4.4 MMICs—monolithic microwave integrated circuits

Monolithic microwave integrated circuits are circuits in which active and passive elements are formed on the same dielectric substrate, as presented in figure 4.7, by using a deposition scheme such as epitaxy, ion implantation, sputtering, evaporation and diffusion. In figure 4.7 the MMIC chip consists of passive elements such as resistors, capacitors, inductors and a field effect transistor (FET).

4.4.1 MMIC technology features

Accurate design is crucial in the design of MMICs. Accurate design may be achieved by using 3D electromagnetic software.

Materials employed in the design of MMICs are GaAs, InP, GaN and SiGe.

Table 4.3. Receiving channel measured test results.

Parameter	Measured results
RF frequency range	18.8–19.3 GHz
IF frequency range	0.95–2 GHz
Gain	50 dB
Noise figure	2 dB
Input VSWR	2:1
Output VSWR	2:1
Spurious level	−40 dBc
Frequency stability versus temperature	±2 MHz

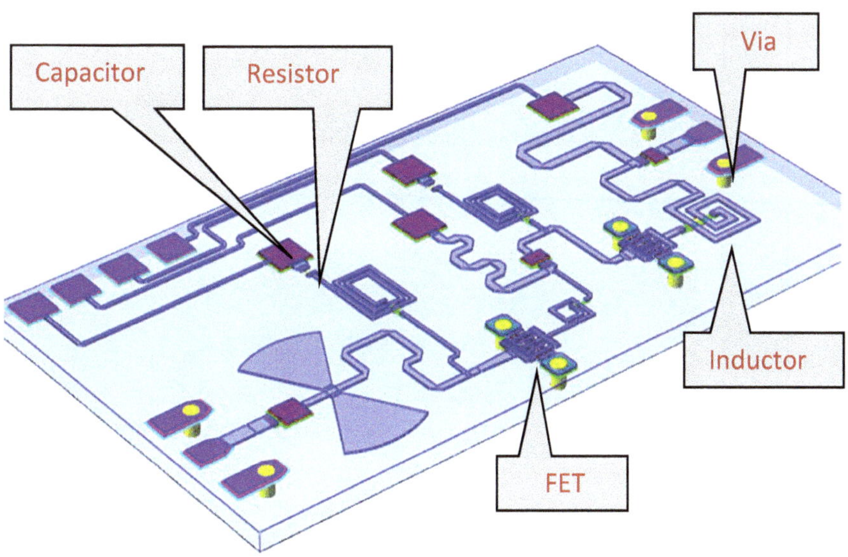

Figure 4.7. MMIC basic components.

Large statistic scattering of all the electrical parameters can cause sensitivity of the design.

FAB runs are expensive, around $200 000 per run. The miniaturization of the components yields lower cost MMICs. Figure 4.8 presents a MMIC design flow.

The designer's goal is to comply with customer specifications in one design iteration. MMIC components cannot be tuned.

- 0.25 micron GaAs PHEMT for power applications to the Ku band.
- 0.15 micron GaAs PHEMT for applications to the high Ka band.
- GaAs PIN process for low loss power switching applications.
- Future new process—InP HBT, SiGe, GaN, RFCMOS, RF MEMS.

Figure 4.9 presents a GaAs WAFER layout. The wafer size may be 3′, 5′ or 6′.

4.4.2 MMIC components

Several microwave components are fabricated by using MMIC technology.

- Mixers—balanced, star, sub-harmonic.
- Amplifiers, LNA, general, power amplifiers, wideband power amplifiers, distributed TWA.
- Switches—PIN, PHEMT, T/R matrix.
- Frequency multipliers—active, passive.
- Modulators—QPSK, QAM (PIN, PHEMT).
- Multifunction—RX chip, TX chip, switched Amp chip, LO chain.

FET—field effect transistor.

BJT—bipolar junction transistor.

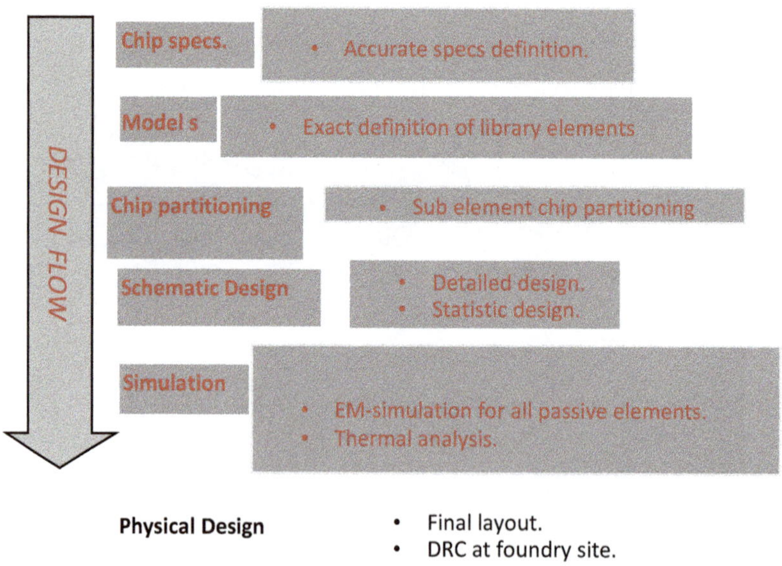

Figure 4.8. MMIC design flow.

HEMT—high electron mobility transistor.
PHEMT—pseudo-morphic HEMT.
MHEMT—metamorphic HEMT.
D-HBT—double hetero-structure bipolar transistor.
CMOS—complementary metal-oxide semiconductor.
Table 4.4 presents types of devices fabricated by using MMIC technology.

4.4.3 Advantages of GaAs versus silicon

MMICs are originally fabricated by using gallium arsenide (GaAs), a III–V compound semiconductor. They are dimensionally small (from around 1 mm^2 to 10 mm^2) and can be mass produced. GaAs has some electronic properties that are better than those of silicon. It has a higher saturated electron velocity and higher electron mobility, allowing transistors made from GaAs to function at frequencies higher than 250 GHz. Unlike silicon junctions, GaAs devices are relatively insensitive to heat due to their higher band gap. Also, GaAs devices tend to have

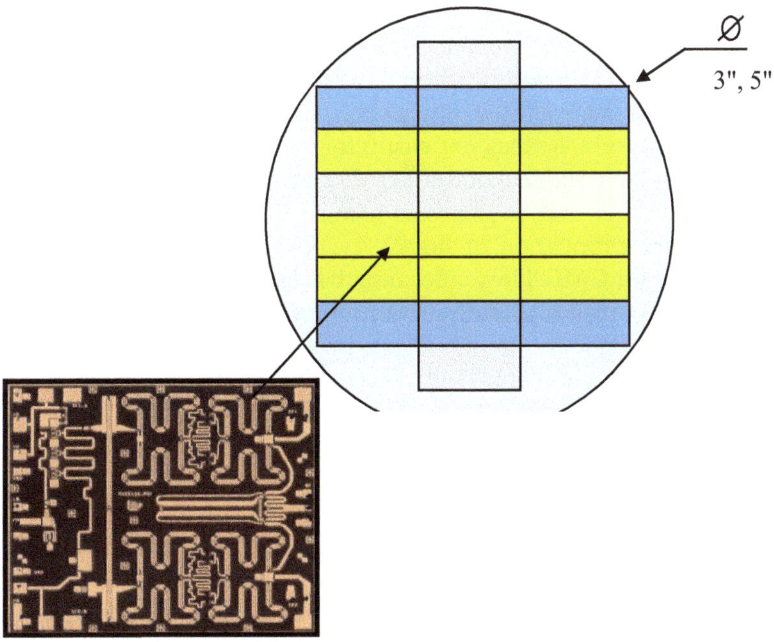

Figure 4.9. GaAs wafer layout.

Table 4.4. MMIC technology.

Material	FET	BJT	Diode
III–V-based	PHEMT GaAs HEMT InPMHEMT GaAs HEMT GaN	HBT GaAsD-HBT InP	Schottky GaAs
Silicon	CMOS	HBT SiGe	

less noise than silicon devices especially at high frequencies, which is a result of the higher carrier mobility and lower parasitic resistive device. These properties recommend GaAs circuitry for use in mobile phones, satellite communications, microwave point-to-point links, and higher frequency radar systems. It is used in the fabrication of Gunn diodes to generate microwaves. GaAs has a direct band gap, which means that it can be used to emit light efficiently. Silicon has an indirect band gap and so is very poor at emitting light. Nonetheless, recent advances may make silicon LEDs and lasers possible. Due to its lower band gap, Si LEDs cannot emit visible light and work in the IR range while GaAs LEDs function in visible red light. As a wide direct band gap material and its resulting resistance to radiation damage, GaAs is an excellent material for space electronics and optical windows in high power applications.

Silicon has three major advantages over GaAs for the integrated circuit manufacturer. First, silicon is a cheap material. In addition, a Si crystal has an extremely stable structure mechanically and it can be grown to very large diameter boules and be processed with very high yields. It is also a decent thermal conductor, thus enabling very dense packing of transistors. All of these factors are very attractive for the design and manufacturing of very large ICs. The second major advantage of Si is the existence of silicon dioxide—one of the best insulators. Silicon dioxide can easily be incorporated onto silicon circuits, and such layers are adherent to the underlying Si. GaAs does not easily form such a stable adherent insulating layer and does not have stable oxide either. The third, and perhaps most important, advantage of silicon is that it possesses a much higher hole mobility. This high mobility allows the fabrication of higher speed P-channel field effect transistors, which are required for CMOS logic. Because they lack a fast CMOS structure, GaAs logic circuits have a much higher power consumption, which has made them unable to compete with silicon logic circuits. The primary advantage of Si technology is its lower fabrication cost compared with GaAs. Silicon wafer diameters are larger. Typically 8′ or 12′ compared with 4′ or 6′ for GaAs. The cost is much lower than GaAs wafers, contributing to a less expensive Si IC.

Other III–V technologies, such as indium phosphide (InP), offer better performance than GaAs in terms of gain, higher cutoff frequency, and low noise. However, they are more expensive due to smaller wafer sizes and increased material fragility.

Silicon germanium (SiGe) is a Si-based compound semiconductor technology offering higher speed transistors than conventional Si devices but with similar cost advantages.

Gallium nitride (GaN) is also an option for MMICs. Because GaN transistors can operate at much higher temperatures and work at much higher voltages than GaAs transistors, they make ideal power amplifiers at microwave frequencies. In table 4.5 properties of the materials used in MMIC technology are compared.

4.4.4 Semiconductor technology

The cutoff frequency of Si CMOS MMIC devices is lower than 200 GHz. Si CMOS MMIC devices are usually low power and low-cost devices. The cutoff frequency of

SiGe MMIC devices is lower than 200 GHz, and they are used as medium power, high gain devices. The cutoff frequency of InP HBT devices is lower than 400 GHz, and they are used as medium power, high gain devices. The cutoff frequency of InP HEMT devices is lower than 600 GHz, and they are used as medium power, high gain devices. In table 4.6 the properties of MMIC technologies are compared. Figure 4.10 presents a 0.15 micron PHEMT on a GaAs substrate.

4.4.5 MMIC fabrication process

The MMIC fabrication process consists of several controlled processes in a semiconductor FAB. The process is listed in the next paragraph. In figure 4.11 a MESFET cross section on GaAs substrate is shown.

MMIC fabrication process list

Wafer fabrication—Preparing the wafer for fabrication.
Wet cleans—Wafer cleaning by a wet process.
Ion implantation—Dopants are embedded to create regions of increased or decreased conductivity. Selectively implant impurities. Create p or n type semiconductor regions.
Dry etching—Selectively remove materials.
Wet etching—Selectively remove materials chemical process.

Table 4.5. Comparison of material properties.

Property	Si	Si or Sapphire	GaAs	InP
Dielectric constant	11.7	11.6	12.9	14
Resistivity Ω cm^{-1}	10^3–10^5	$>10^{14}$	10^7–10^9	10^7
Mobility (cm^2 V^{-1} s^{-1})	700	700	4300	3000
Density (gr cm^{-3})	2.3	3.9	5.3	4.8
Saturation velocity (cm s^{-1})	9×10^6	9×10^6	1.3×10^7	1.9×10^7

Table 4.6. Summary of semiconductor technology.

	Si CMOS	SiGe HBT	InP HBT	InP HEMT	GaN HEMT
Cutoff frequency	>200 GHz	>200 GHz	>400 GHz	>600 GHz	>200 GHz
Published MMICs	170 GHz	245 GHz	325 GHz	670 GHz	200 GHz
Output power	Low	Medium	Medium	Medium	High
Gain	Low	High	High	Low	Low
RF noise	High	High	High	Low	Low
Yield	High	High	Medium	Low	Low
Mixed signal	Yes	Yes	Yes	No	No
$1/f$ noise	High	Low	Low	High	High
Breakdown voltage	−1 V	−2 V	−4 V	−2 V	>20 V

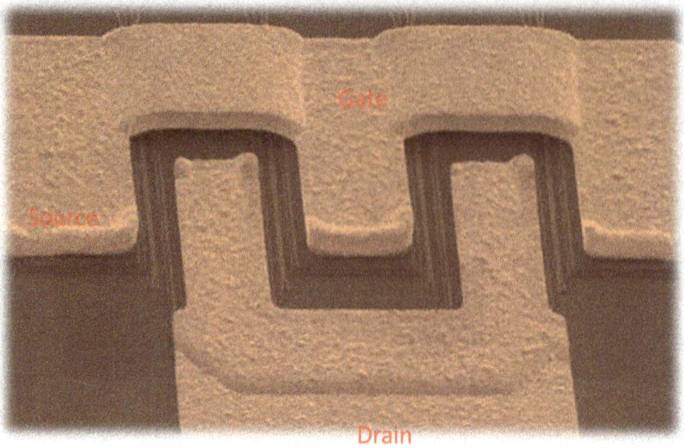

Figure 4.10. 0.15 micron PHEMT on GaAs substrate.

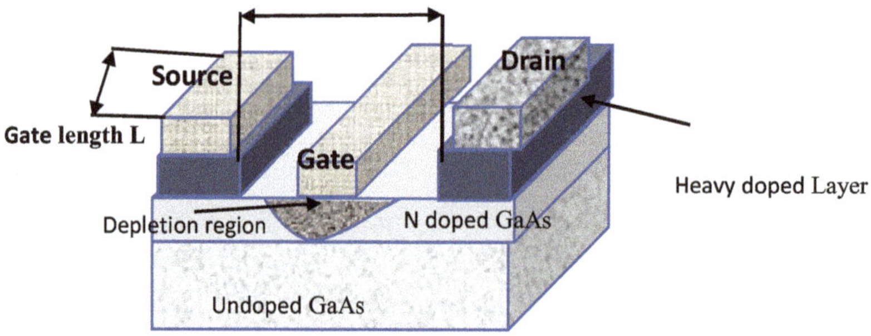

Figure 4.11. MESFET cross section on GaAs substrate.

Plasma etching—Selectively remove materials.
Thermal treatment—High temperature process to remove stress.
Rapid thermal anneal—High temperature process to remove stress.
Furnace anneal—After ion implantation, thermal annealing is required. Furnace annealing may take minutes and causes too much diffusion of dopants for some applications.
Oxidation—Substrate oxidation, for example: Dry oxidation—$Si + O_2 \rightarrow SiO_2$.
Wet oxidation $Si + 2H_2O \rightarrow SiO_2 + 2H_2$.
Chemical vapor deposition CVD—Chemical vapor deposited on the wafer. Pattern defined by the photo-resist.
Physical vapor deposition PVD—Vapor produced by evaporation or sputtering deposited on the wafer. Pattern defined by the photo-resist.

Molecular beam epitaxy MBE—A beam of atoms or molecules produced in high vacuum. Selectively grow layers of materials. Pattern defined by the photo-resist.
Electroplating—Electromechanical process used to add metal.
Chemical mechanical polish, CMP.
Wafer testing—Electrical test of the wafer.
Wafer back-grinding.
Die preparation.
Wafer mounting.
Die cutting.
Lithography—Lithography is the process of transferring a pattern onto the wafer by selectively exposing and developing the photo-resist. Photolithography consists of four steps, the order depends on whether we are etching or lifting off the unwanted material.
Contact lithography—A glass plate is used that contains the pattern for the entire wafer. It is literally led against the wafer during exposure of the photo-resist. In this case the entire wafer is patterned in one shot.
Electron-beam lithography is a form of direct-write lithography. Using E-beam lithography you can write directly onto the wafer without a mask. Because an electron beam is used, rather than light, much smaller features can be resolved.
Exposure can be done with light, UV light, or electron beam, depending on the accuracy needed. An E beam provides a much higher resolution than light, because the particles are bigger (greater momentum), the wavelength is shorter.

Etching versus lift-off removal processes
There are two principal means of removing material, etching and lift-off.
The steps for an etch-off process are:

1. Deposit material.
2. Deposit photo-resist.
3. Pattern (expose and develop).
4. Remove material where it is not wanted by etching.

Etching can be isotropic (etching wherever we can find the material we want to etch) or anisotropic (directional, etching only where the mask allows). Etches can be dry (reactive ion etching or RIE) or wet (chemical). Etches can be very selective (only etching what we intend to etch) or non-selective (attacking a mask to the substrate).

In a lift-off process, the photo-resist *forms a mold*, into which the desired material is deposited. The desired features are completed when photo-resist B under the unwanted areas is dissolved, and the unwanted material is 'lifted off'.

Lift-off process is:

- Deposit photo-resist.
- Pattern.
- Deposit material conductor or insulator.
- Remove material where it is not wanted by *lifting off*.

In figure 4.12 a MESFET cross section on GaAs substrate is shown. In figure 4.13 a MMIC resistor cross section is shown.

In figure 4.14 a MMIC capacitor cross section is shown.

Figure 4.15 presents the ion implantation process. Figure 4.16 presents the ion etch process. Figure 4.17 presents the wet etch process.

4.4.6 Generation of microwave signals in microwave and mm wave

Microwave signals can be generated by solid state devices and vacuum tube based devices. Solid state microwave devices are based on semiconductors such as silicon

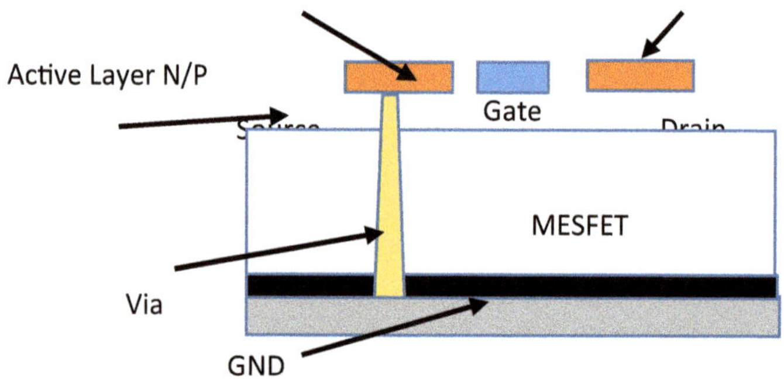

Figure 4.12. MESFET cross section.

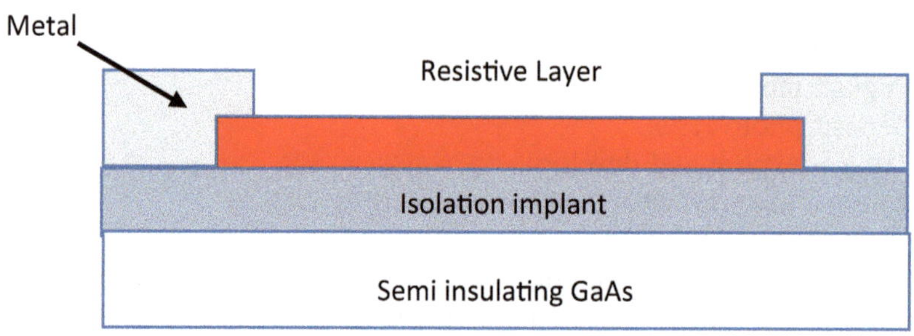

Figure 4.13. Resistor cross section.

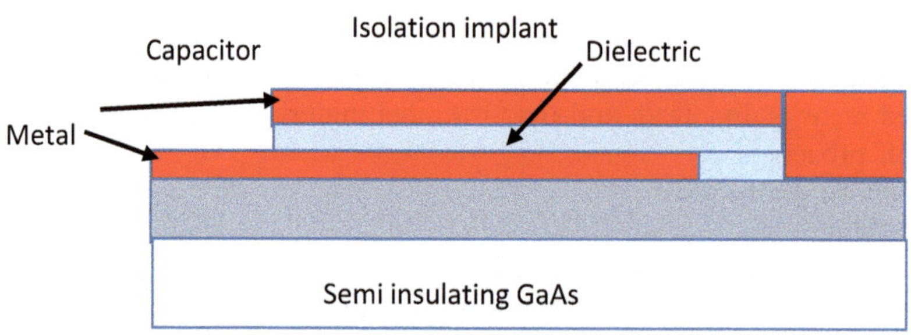

Figure 4.14. Capacitor cross section.

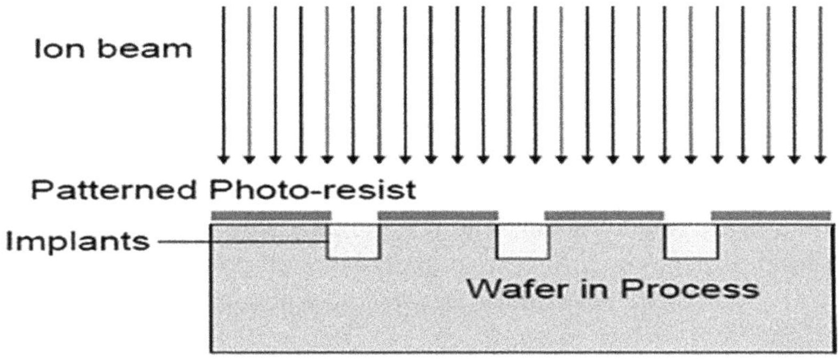

Figure 4.15. Ion implantation.

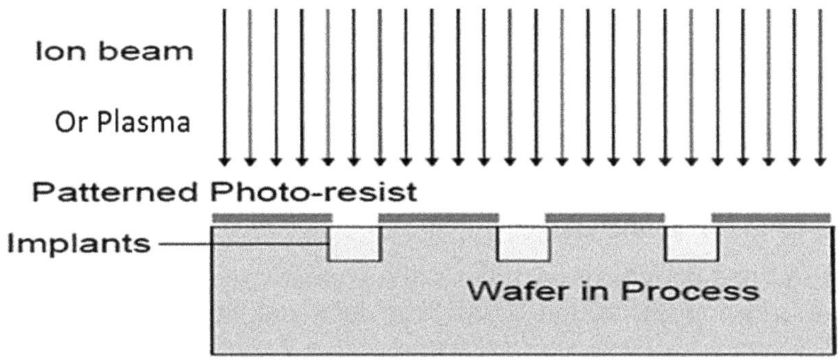

Figure 4.16. Ion etch.

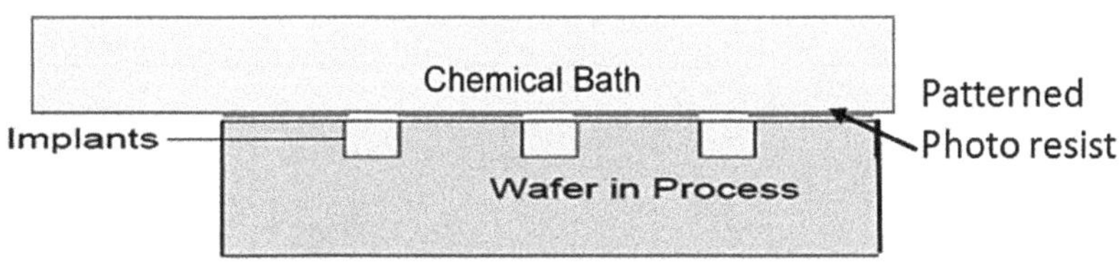

Figure 4.17. Wet etch.

or gallium arsenide, and include field effect transistors (FETs), bipolar junction transistors (BJTs), Gunn diodes, and IMPATT diodes. Microwave variations of BJTs include the hetero-junction bipolar transistor (HBT), and microwave variants of FETs include the MESFET, the HEMT (also known as HFET), and LDMOS transistor. Microwaves can be generated and processed using integrated circuits, MMICs. They are usually manufactured using gallium arsenide (GaAs) wafers, though silicon germanium (SiGe) and heavy-doped silicon are increasingly used. Vacuum tube based devices operate on the ballistic motion of electrons in a vacuum under the influence of controlling electric or magnetic fields, and include the magnetron, klystron, traveling wave tube (TWT), and gyrotron. These devices work in the density modulated mode, rather than the current modulated mode. This means that they work on the basis of clumps of electrons flying ballistically through them, rather than using a continuous stream.

4.4.7 MMIC examples and applications

Figure 4.18 presents a wideband mm wave power amplifier. The input power is divided by using a power divider. The RF signal is amplified by power amplifiers and combined by a power combiner to get the desired power at the device output.

Figure 4.19 presents a wideband mm wave up-converter. MMIC process cost is listed in table 4.7.

MMIC applications

- Ka band satellite communication.
- 60 GHz wireless communication.
- Automotive radars.
- Imaging in security.
- Gbit WLAN.

4.5 18–40 GHz front end

Development and design considerations of a compact wideband 18–40 GHz front end are described in this section. The RF modules and the system was designed by using ADS system software and momentum RF software. There is a good agreement between the computed and measured results.

Figure 4.18. Wideband power amplifier.

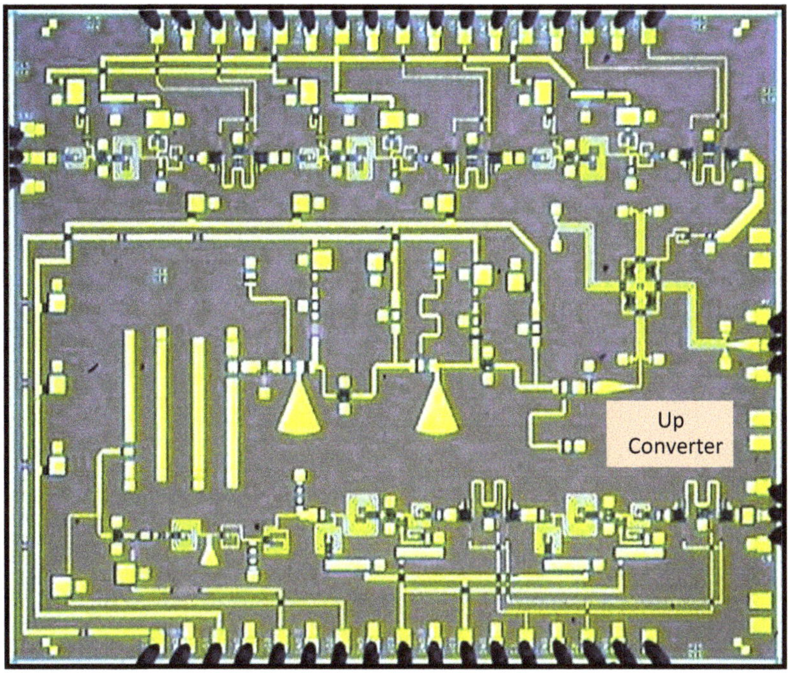

Figure 4.19. Ka band up-converter.

Table 4.7. MMIC cost

	Si CMOS	SiGe HBT	GaAs HEMT	InP HEMT
Chip cost ($/mm^2)	0.01	0.1–0.5	1–2	10
Mask cost (M$/mask set)	1.35	0.135	0.0135	0.0135

4.5.1 18–40 GHz front end requirements

The front end electrical specifications are listed in table 4.8. The front end design presented in this section meet the front end electrical specifications. Physical characteristics, interfaces and connectors are listed in table 4.9.

4.5.2 Front end design

A front end block diagram is shown in figure 4.20. The front end module consists of a limiter and a wideband 18–40 GHz. Filtronic, low noise amplifier LMA406. The LMA406 gain is around 12 dB with 4.5 dB noise figure and 14 dBm saturated output power. The LNA dimensions are 1.44 × 1.1 mm. We used a wideband PHEMT MMIC SPDT manufactured by Agilent, AMMC-2008. The SPDT insertion loss is lower than 2 dB. The isolation between the SPDT input port to the output ports is better than 25 dB. The SPDT 1dBc compression point is around 14 dBm. The SPDT dimensions are 1 × 0.7 × 0.1 mm. The front end electrical characteristics were evaluated by using

Table 4.8. Front end electrical specifications.

Parameter	Requirements	Performance
Frequency range	18–40 GHz	Comply
Gain	24/3 dB typical, switched by external control (−40 dB or lower for off state)	Comply
Gain flatness	±0.5 dB max for any 0.5 GHz BW in 18–40 GHz. ±2 dB max for any 4 GHz BW in 18–40 GHz. ±3 dB max for the whole range 18–40 GHz	Comply
Noise figure (high gain)	10 dB max for 40 °C baseplate temperature 11 dB over temperature	Comply
Input power range	−60 dBm to 10 dBm	Comply
Output power range	−39 dBm to 11 dBm not saturated, 13 dBm saturated	Comply
Linearity	Output 1 dB compression point at 12 dBm min. Third intercept point (Ip3) at 21 dBm. Single tone second harmonic power −25 dBc max for 10 dBm output	Comply
VSWR	2:1	Comply
Power input protection	No damage at +30 dBm CW and +47 dBm. Pulses (for average power higher than 30 dBm) input power at 0.1–40 GHz. Test for pulses: PW = 1 μs, PRF = 1 KHz	Comply
Power supply voltages	±5 V, ±15 V	Comply
Control logic	LVTTL standard '0' = 0–0.8 V; '1' = 2.0–3.3 V	Comply
Switching time	Less then 100 ns	Comply
Non-harmonic spurious (output)	−50 dBm max. (When it isn't correlative with the input signals)	Comply
Video leakage	Video leakage signals will be below the RF output level for the terminated input	Comply
Dimension	60 × 40 × 20 mm	Comply

Table 4.9. Physical characteristics—interface connectors.

Interface	Type
RF input	Wave guide WRD180 (Double Ridge)
RF output	K Connector
DC supply	D Type
Dimensions	60 × 40 × 20 mm
Control	D Type

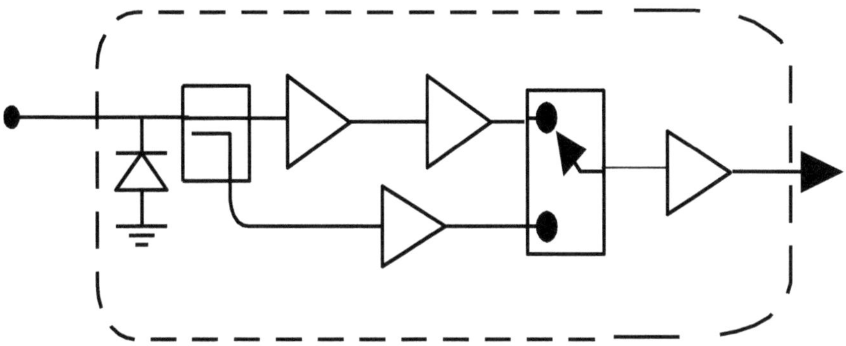

Figure 4.20. Front end block diagram.

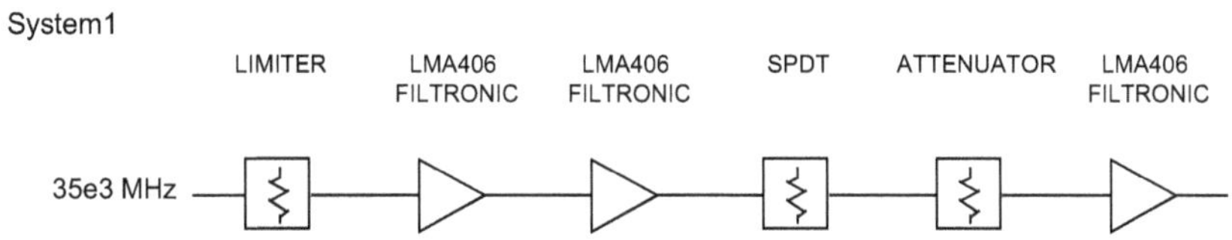

	LIMITER	LMA406 FILTRONIC	LMA406 FILTRONIC	SPDT	ATTENUATOR	LMA406 FILTRONIC	Total
NF (dB)	3.00	6.00	6.00	3.00	3.00	6.00	9.46
Gain (dB)	-3.00	10.00	10.00	-3.00	-3.00	10.00	21.00
OIP3 (dBm)	30.00	25.00	25.00	30.00	30.00	25.00	23.14

Input Pwr (dBm)	-60.00	System Temp (K)	290.00	IM Offset (MHz)	.025
OIP2 (dBm)	23.14	OIP3 (dBm)	23.14	Output P1dB (dBm)	11.37
IIP2 (dBm)	2.14	IIP3 (dBm)	2.14	Input P1dB (dBm)	-8.63
OIM2 (dBm)	-101.14	OIM3 (dBm)	-163.28	Compressed (dB)	0.00
ORR2 (dB)	62.14	ORR3 (dB)	124.28	Gain, Actual (dB)	21.00
IRR2 (dB)	31.07	IRR3 (dB)	41.43		
SFDR2 (dB)	61.34	SFDR3 (dB)	81.79	Gain, Linear (dB)	21.00
AGC Controlled Range:		Min Input (dBm)	N/A	Max Input (dBm)	N/A

Figure 4.21. Front end module design for a LNA NF = 6 dB.

ADS Agilent software and SYSCAL software. Figure 4.21 presents the front end module noise figure and gain for a LNA noise figure of 6 dB. The overall computed module noise figure is 9.46 dB. The module gain is 21 dB. Figure 4.22 presents the front end module noise figure and gain for a LNA noise figure of 5.5 dB. The overall computed module noise figure is 9.25 dB. The module gain is 21 dB.

The MMIC amplifiers and the SPDT are glued to the surface of the mechanical box. The MMIC chips are assembled on Covar carriers. During development it was found that the spacing between the front end carriers should be less than 0.03 mm in order to achieve the flatness requirements and a voltage standing wave ratio (VSWR) better than 2:1.

The front end voltage and current consumption are listed in table 4.10. The front end module has a high gain and low gain channels. The gain difference between the high gain and low gain channels is around 15–20 dB. The measured front end gain is presented in figure 4.23. The front end gain is around 20 ± 4 dB for the 18–40 GHz frequency range.

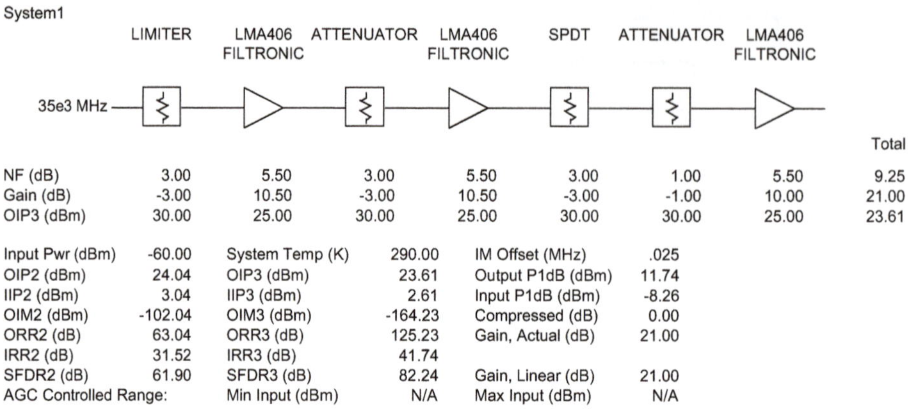

Figure 4.22. Front end module design for a LNA NF = 5.5 dB.

Table 4.10. Front end module voltage and current consumption.

Voltage (V)	3	5	−12	−5	Digital 5
Current (A)	0.25	0.15	0.1	0.1	0.1

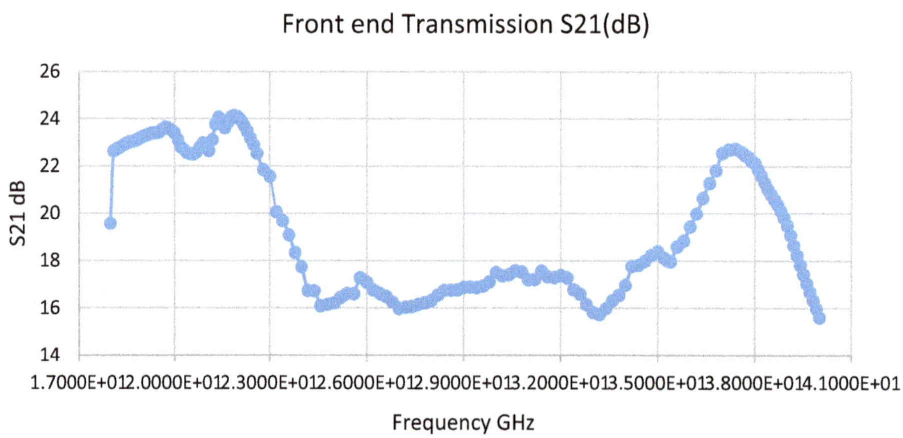

Figure 4.23. Measured front end gain.

4.5.3 High gain, front end module

To achieve a high gain, front end module, a medium power Hittite MMIC amplifier (HMC283) was added to the front end module presented in figure 4.24. The HMC283 gain is around 21 dB with 10 dB noise figure and 21 dBm saturated output power. The amplifier dimensions are 1.72 × 0.9 mm. The high gain, front end module block diagram is shown in figures 4.24 and 4.25. The front end module has

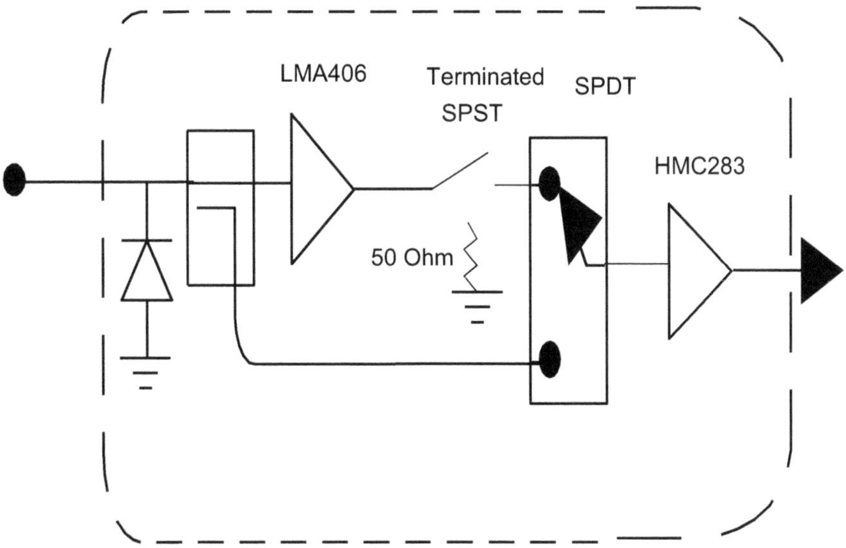

Figure 4.24. High gain, front end block diagram.

high gain and low gain channels. The gain difference between the high gain and low gain channels presented in figure 4.25 is around 15–20 dB. The gain difference between the high gain and low gain channels presented in figure 4.26 is around 10–15 dB. A detailed block diagram of a high gain module is shown in figure 4.27.

4.5.4 High gain, front end design

The high front end electrical characteristics were evaluated by using ADS Agilent software and SYSCAL software. Figure 4.27 presents the front end module noise figure and gain for a LNA noise figure of 9.5 dB. The overall computed module noise figure is 13.3 dB. The module gain is 32.48 dB. Figure 4.28 presents the front end module noise figure and gain for a LNA noise figure of 5 dB. The overall computed module noise figure is 10 dB. The module gain is 29.5 dB.

The measured results of front end modules are listed in table 4.11. HMC283 assembly is shown in figure 4.29. A photo of the front end is shown in figure 4.30. There is good agreement between the computed and measured results.

A photo of compact wideband 18–40 GHz RF modules is shown in figure 4.31.

4.6 MEMS technology

Micro-electro-mechanical systems (MEMS) is the integration of mechanical elements, sensors, actuators, and electronics on a common silicon substrate through micro-fabrication technology. These devices replace bulky actuators and sensors with a micron scale equivalent that can produce large quantities by a fabrication process used in integrated circuits in photolithography. They reduce cost, bulk, weight and power consumption while increasing performance, production volume and functionality by orders of magnitude.

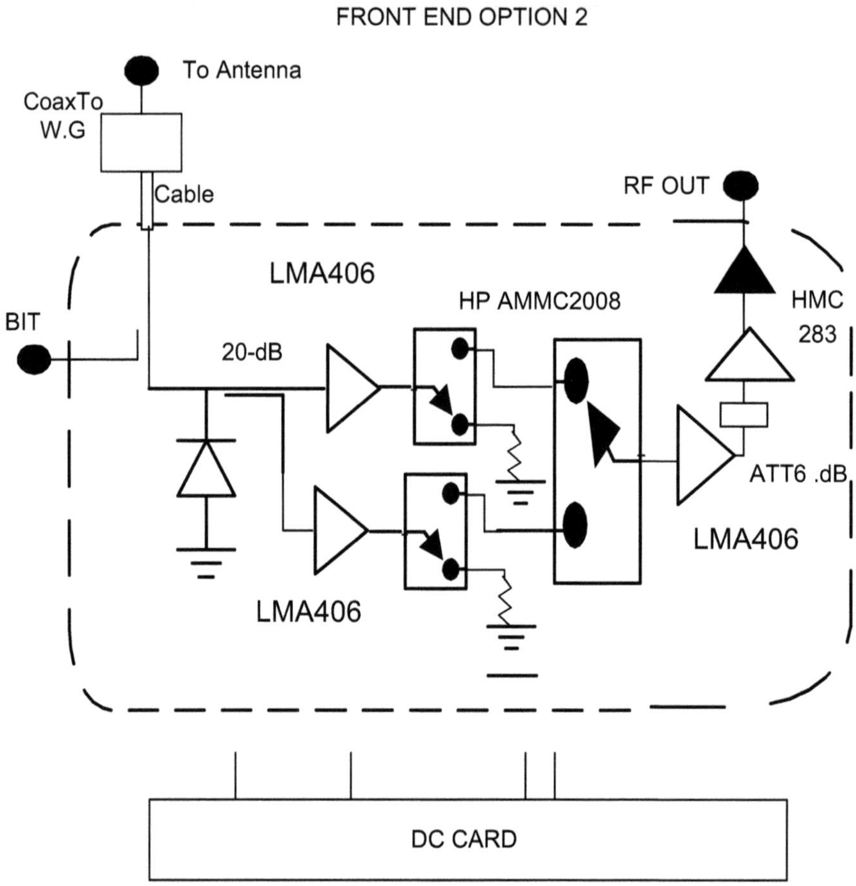

Figure 4.25. High gain, front end block diagram with an amplifier in the low gain channel.

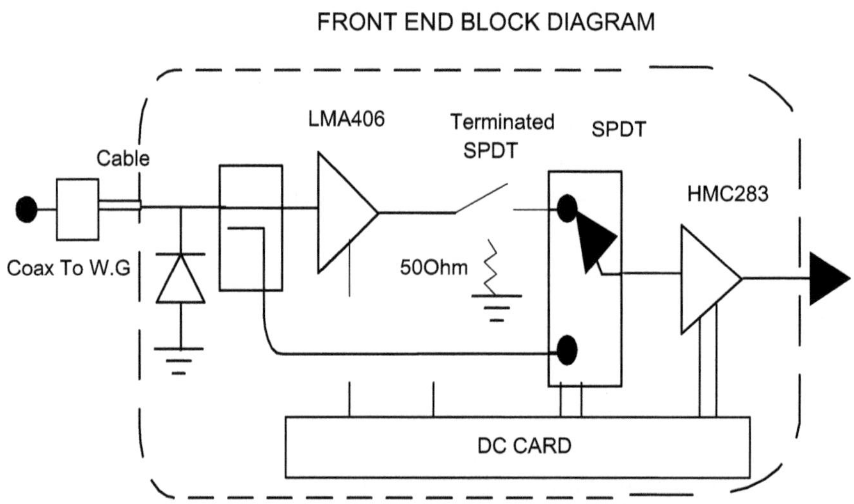

Figure 4.26. Detailed block diagram high gain, front end.

System1

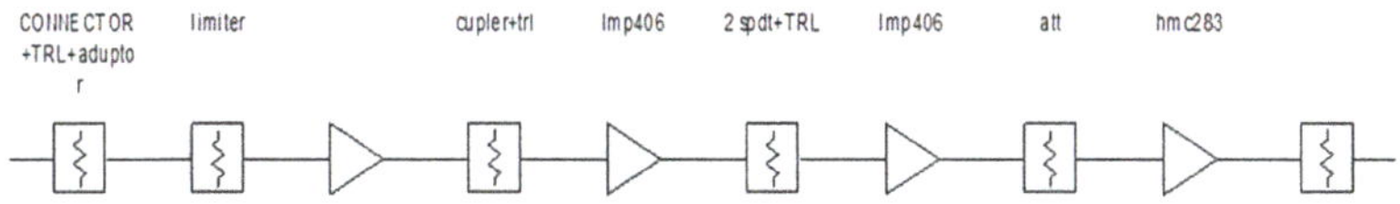

											Total
Gain (dB)	-3.00	0.50	9.50	-0.50	9.50	-6.00	9.50	-6.00	20.00	-0.70	32.48
OP1dB (dBm)	49.00	50.00	10.00	50.10	10.00	50.00	10.00	50.00	14.00	50.00	13.30
NF (dB)	3.00	0.50	5.00	0.50	5.00	6.00	5.00	6.00	13.00	0.70	8.94
Po (dBm)	-33.00	-32.50	-23.00	-23.50	-14.00	-20.16	-10.66	-16.82	3.18	2.48	
Pcomp (dB)	0.00	0.00	0.00	0.00	0.00	0.00	0.00	0.00	0.00	0.00	
NF+ (dB)	0.59		2.93		0.28	0.04	0.13	0.02	0.54		

Input Pwr (dBm) -30.00 System Temp (K) 298.00

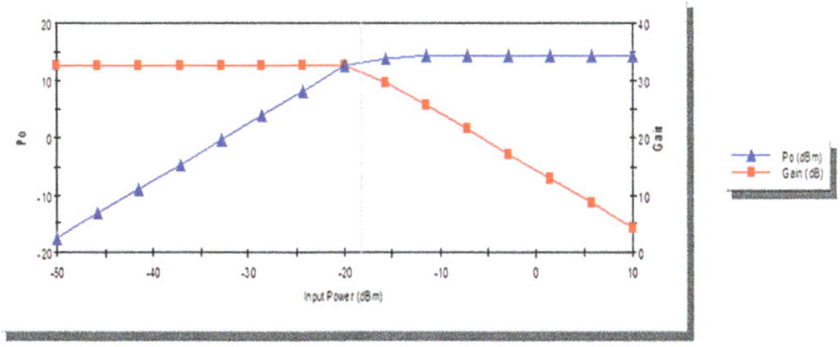

Figure 4.27. Front end module design for a LNA NF = 9.5 dB.

							Total
NF (dB)	3.50	5.00	5.00	5.00	6.00	9.00	10.02
Gain (dB)	-3.50	11.00	-5.00	11.00	-6.00	22.00	29.50
OIP3 (dBm)		20.00		20.00		20.00	19.87
NF+ (dB)		2.85		0.56		0.57	
Po (dBm)	-63.50	-52.50	-57.50	-46.50	-52.50	-30.50	

Input Pwr (dBm) -60.00 System Temp (K) 290.00

Figure 4.28. Front end module design for a LNA NF = 5 dB.

The electronics are fabricated using integrated circuit (IC) process sequences (e.g. CMOS, bipolar, or BICMOS processes), the micromechanical components are fabricated using compatible 'micromachining' processes that selectively etch away parts of the silicon wafer or add new structural layers to form the mechanical and electromechanical devices.

4.6.1 MEMS technology advantages

- Low insertion loss <0.1 dB.
- High isolation >50 dB.
- Low distortion.
- High linearity.
- Very high *Q*.

Table 4.11. Measured results of front end modules.

Parameter	DF6	DF4	DF3	DF2	DF1	OMNI02	OMNI01
High gain max	31	32	32.5	32.5	31	31.5	32
High gain min	26	26	27.5	27	26	28.5	28
High gain avg	29	29	29	29	29	30	30
Amp. Bal.	5	6	5	5	4	3	4
S_{11} (dB)	4.5	5	5	5	5	5	4
S_{22} (dB)	7.5	6	5	6	5	7	6
Isolation (dB)	9	9	10	10	6.5	21.5	22.5
Low gain max	19	18	17	17	17	16.5	18
Low gain min	13	10	7.5	12	12	10.5	11
Low gain avg	15	14	12	14	14	13.5	14.5
Amp. Bal.	6	8	9.5	5	5	6	7
P1 dB 30 GHz	11.6	11.93	11.7	11.4	10.9	14	15.96
P1 dB 40 GHz	13.96	14.5	15.58	15.28	14	14.48	16.8
NF 30 GHz	8.68	9.48	8.65	8.45	10.5	8.14	8.75
NF 40 GHz	9.28	10.1	8.64	9.17	10.24		8.75

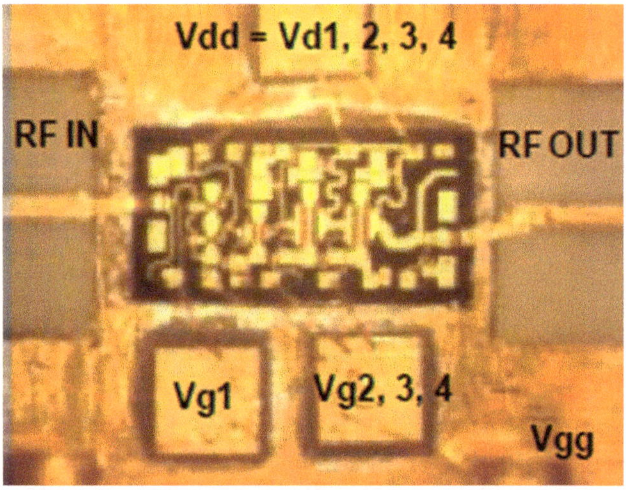

Figure 4.29. HMC283 assembly.

- Size reduction, system-on-a-chip.
- High power handling ~40 dBm.
- Low power consumption (~mW and no LNA).
- Low-cost, high volume fabrication.

4.6.2 MEMS technology process

Bulk micromachining fabricates mechanical structures in the substrate by using orientation dependent etching. A bulk micro-machined substrate is shown in figure 4.32.

Figure 4.30. 18–40 GHz front end module.

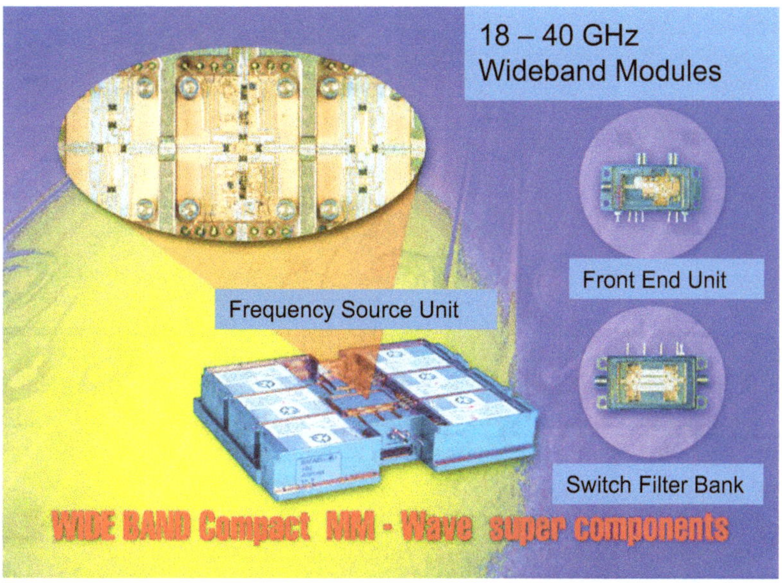

Figure 4.31. Photo of 18–40 GHz compact modules.

Surface micromachining fabricates mechanical structures above the substrate surface by using a sacrificial layer. A surface micro-machined substrate is presented in figure 4.33.

In the bulk micromachining process silicon is machined using various etching processes. Surface micromachining uses layers deposited on the surface of a

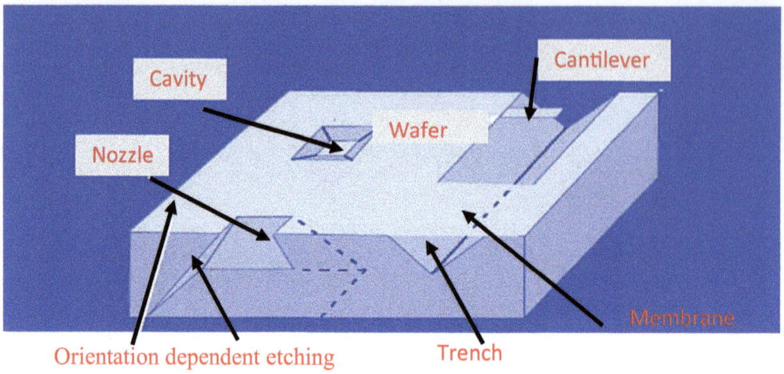

Figure 4.32. Bulk micromachining.

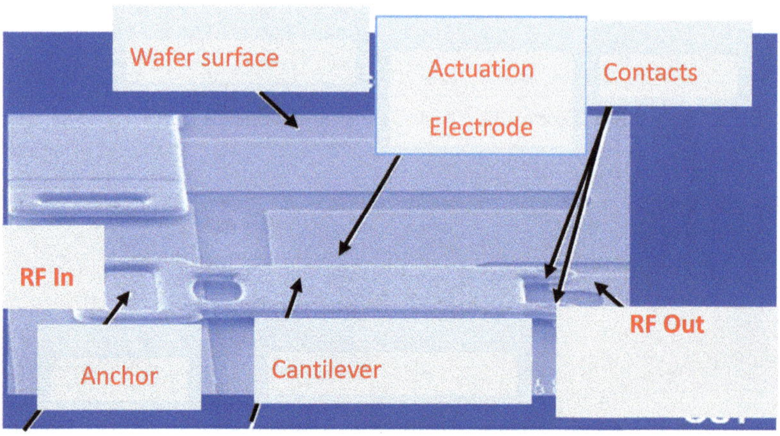

Figure 4.33. Surface micromachining.

substrate as the structural materials, rather than using the substrate itself. The surface micromachining technique is relatively independent of the substrate used, and therefore can be easily mixed with other fabrication techniques that modify the substrate first. One example is the fabrication of MEMS on a substrate with embedded control circuitry, in which MEMS technology is integrated with IC technology. This is being used to produce a wide variety of MEMS devices for many different applications. On the other hand, bulk micromachining is a subtractive fabrication technique, which converts the substrate, typically a single-crystal silicon, into the mechanical parts of the MEMS device. A MEMS device is first designed with a computer-aided design (CAD) tool. The design outcome is a layout and masks that are used to fabricate the MEMS device. In figure 4.34 a MEMS fabrication process is presented. A summary of MEMS fabrication technology is listed in table 4.12, and in figure 4.35 a block diagram of a MEMS bolometer coupled antenna array is presented.

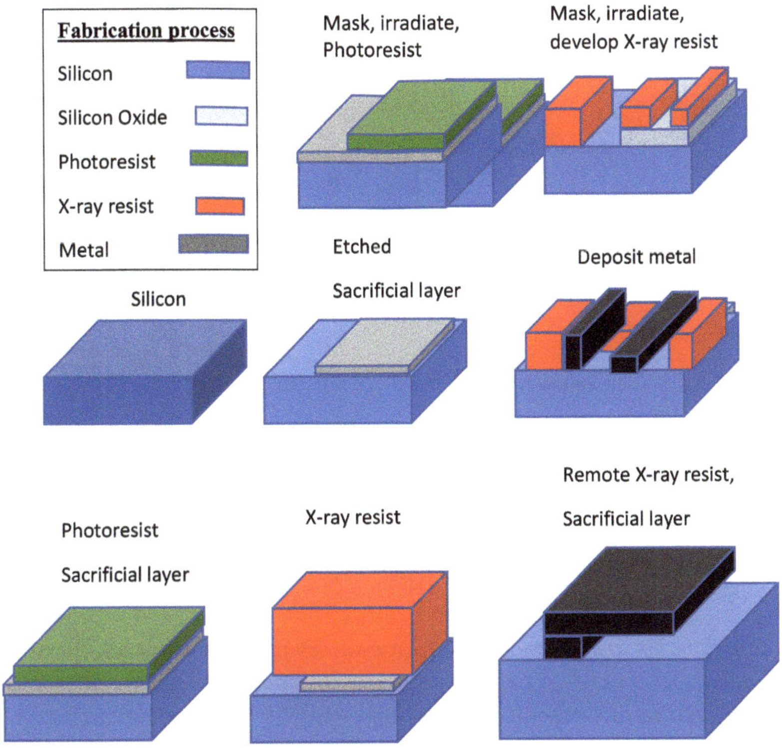

Figure 4.34. Fabrication process.

Table 4.12. Fabrication technology.

Fabrication technology	Process
Surface micromachining	Release and drying systems to realize free-standing microstructures
Bulk micromachining	Dry etching systems to produce deep 2D free-form geometries with vertical sidewalls in substrates. Anisotropic wet etching systems with protection for wafer front sides during etching. Bonding and aligning systems to join wafers and perform photolithography on the stacked substrates.

Packaging of the device tends to be more difficult, but structures with increased heights are easier to fabricate when compared to surface micromachining. This is because the substrates can be thicker resulting in relatively thick unsupported devices. Applications of RF MEMS technology:

- Tunable RF MEMS inductor.
- Low loss switching matrix.

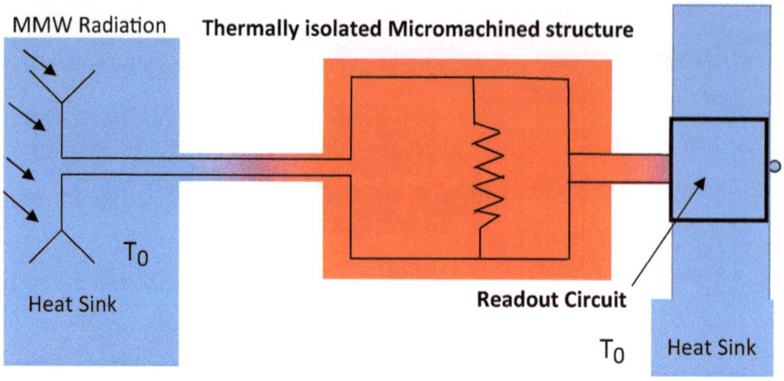

Figure 4.35. Bolometer coupled antenna array.

- Tunable filters.
- Bolometer coupled antenna array.
- Low-cost W band detection array.

4.6.3 MEMS components

MEMS components are categorized in one of several applications. Such as:

(1) **Sensors** are a class of MEMS that are designed to sense changes and interact with their environments. These classes of MEMS include chemical, motion, inertia, thermal, RF sensors and optical sensors. Micro sensors are useful because of their small physical size, which allows them to be less invasive.

(2) **Actuators** are a group of devices designed to provide power or stimulus to other components or MEMS devices. MEMS actuators are either electrostatically or thermally driven.

(3) **RF MEMS** are a class of devices used to switch or transmit high frequency, RF signals. Typical devices include: metal contact switches, shunt switches, tunable capacitors, antennas, etc.

(4) **Optical MEMS** are devices designed to direct, reflect, filter, and/or amplify light. These components include optical switches and reflectors.

(5) **Micro-fluidic MEMS** are devices designed to interact with fluid-based environments. Devices such as pumps and valves have been designed to move, eject, and mix small volumes of fluid.

(6) **Bio MEMS** are devices that, much like micro-fluidic MEMS, are designed to interact specifically with biological samples. Devices such as these are designed to interact with proteins, biological cells, medical reagents, etc, and can be used for drug delivery or other *in situ* medical analysis.

4.7 W band MEMS detection array

In this section, we present the development of a millimeter wave radiation detection array. The detection array may employ around 256–1024 patch antennas. These patches are coupled to a resistor. Optimization of the antenna structure, feed network dimensions and resistor structure allows us to maximize the power rate dissipated on the resistor. Design considerations of the detection antenna array are given in this section. Several imaging approaches are presented in the literature, [10–14]. The common approach is based on an array of radiators (antennas) that receives radiation from a specific direction by using a combination of electronic and mechanical scanning. Another approach is based on a steering array of radiation sensors at the focal plane of a lens of reflector. The sensor can be an antenna coupled to a resistor.

4.7.1 Detection array concept

Losses in the microstrip feed network are very high in the W band frequency range. In W band frequencies we may design a detection array. The array concept is based on an antenna coupled to a resistor. A direct antenna-coupling surface to a micro-machined micro bridge resistor is used for heating and sensing. The feed network determines the antenna's efficiency. The insertion loss of a gold microstrip line with a width of 1 μm and length of 188 μm is 4.4 dB at 95 GHz. The insertion loss of a gold microstrip line with a width of 10 μm and length of 188 μm is 3.6 dB at 95 GHz. The insertion loss of a gold microstrip line with width of 20 μm and length of 188 μm length is 3.2 dB at 95 GHz. To minimize losses the feed line dimensions were selected as 60 × 10 × 1 μm. An analog CMOS readout circuit may be employed as a sensing channel per pixel. Figure 4.36 presents a pixel block diagram.

4.7.2 The array principle of operation

The antenna receives effective mm wave radiation. The radiation power is transmitted to a thermally isolated resistor coupled to a Ti resistor. The electrical power raises the structure temperature with a short response time. The same resistor

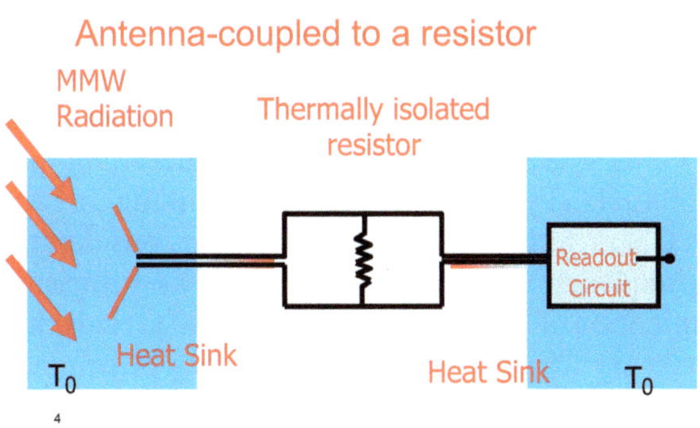

Figure 4.36. Antenna coupled to a resistor.

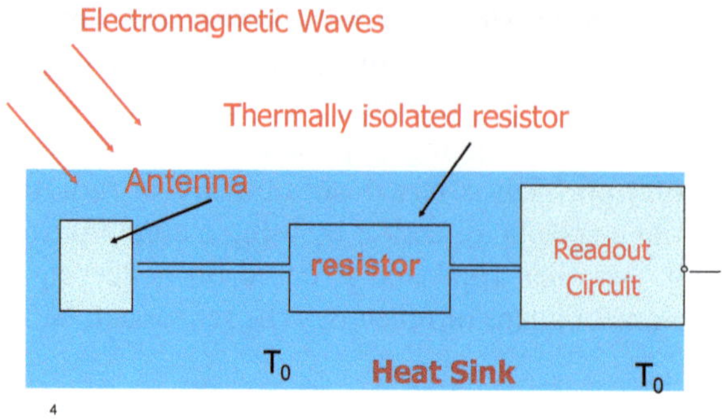

Figure 4.37. A single array pixel.

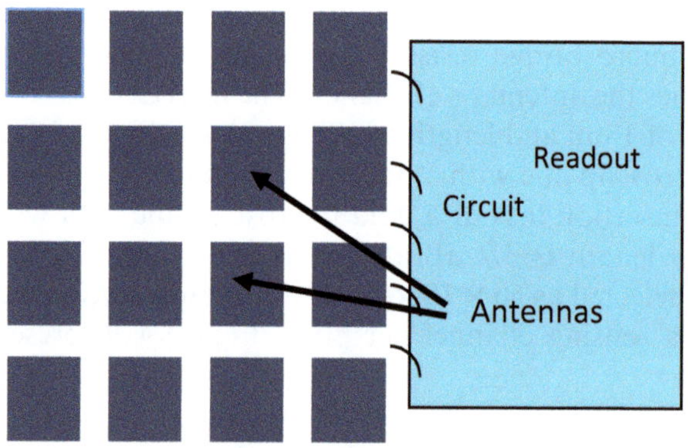

Figure 4.38. Array concept.

changes its temperature and therefore its electrical resistance. Figure 4.37 shows a single array pixel. The pixel consists of a patch antenna, a matching network, printed resistor and DC pads. The printed resistor consists of titanium lines, a titanium resistor coupled to an isolated resistor.

The operating frequency range of 92–100 GHz is the best choice. In the frequency range of 30–150 GHz there is a proven contrast between land, sky and high transmittance of clothes. Size and resolution considerations promote higher frequencies above 100 GHz. Typical penetration of clothing at 100 GHz is 1 dB and 5–10 dB at 1 THz. Characterization and measurement considerations promote lower frequencies. The frequency range of 100 GHz allows sufficient bandwidth when working with illumination. The frequency range of 100 GHz is the best compromise. Figure 4.38 presents the array concept. Several types of printed antennas may be used as the array element such as the bow-tie dipole, patch antenna and ring resonant slot.

4.7.3 W band antenna design

A bow-tie dipole and patch antenna have been considered as the array element. The computed results show that the directivity of the bow-tie dipole is around 5.3 dBi and the directivity of the patch antenna is around 4.8 dBi. However, the length of the bow-tie dipole is around 1.5 mm and the size of the patch antenna is around 700 × 700 μm. We used a quartz substrate with a thickness of 250 μm. The bandwidth of the bow-tie dipole is wider than that of the patch antenna. However, the patch antenna's bandwidth meets the detection array's electrical specifications. We chose the patch antenna as the array element since the patch size is significantly smaller than that of the bow-tie dipole. This feature allows us to design an array with a higher number of radiating elements. The resolution of a detection array with a higher number of radiating elements is improved. We also realized that the matching network between the antenna and the resistor is smaller for a patch antenna than for a bow-tie dipole. The matching network between the antenna and the resistor consists of microstrip open stubs. Figure 4.39 shows the 3D radiation pattern of the bow-tie dipole. Figure 4.40 presents the S_{11} parameter of the patch antenna. The electrical performance of the bow-tie dipole and the patch antenna was compared. The VSWR of the patch antenna is better than 2:1 for 10% bandwidth. Figure 4.41 presents the 3D radiation pattern of the patch antenna at 95 GHz.

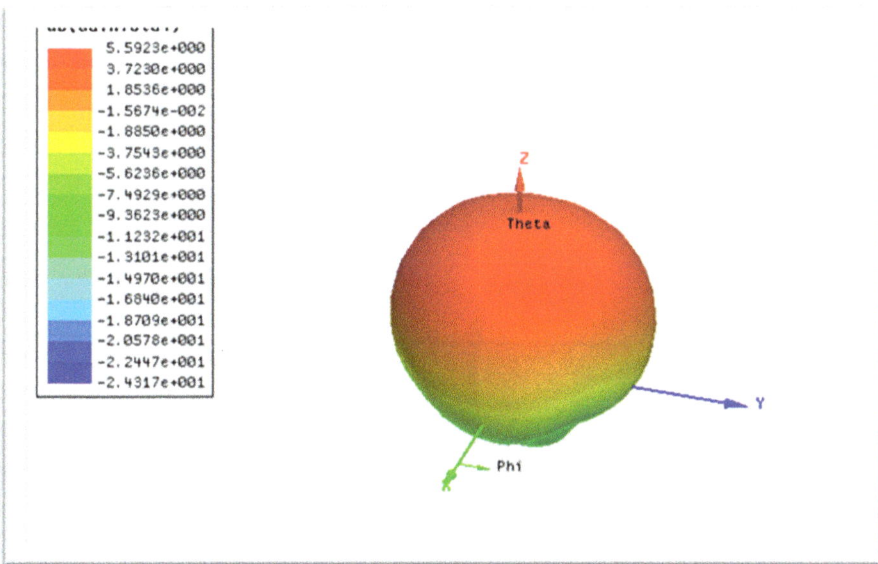

Figure 4.39. Dipole 3D radiation pattern.

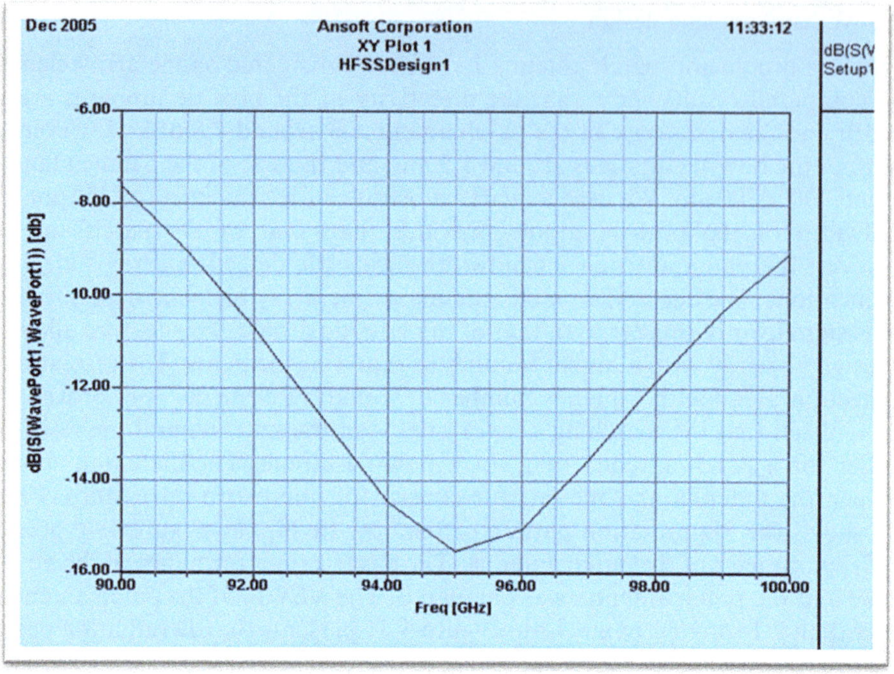

Figure 4.40. Patch S_{11} computed results.

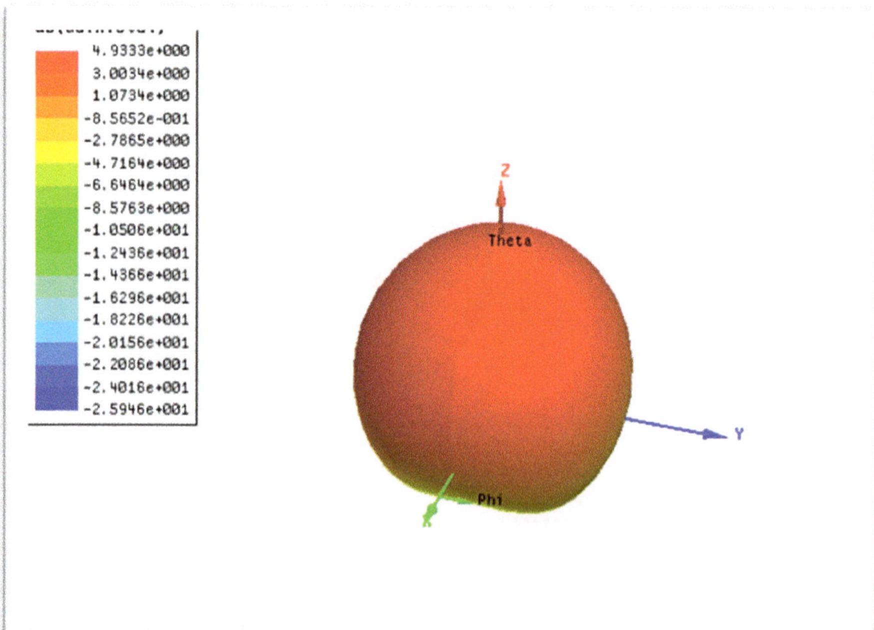

Figure 4.41. Patch 3D radiation pattern.

4.7.4 Resistor design

As described in [6], the resistor is thermally isolated from the patch antenna by using a sacrificial layer. Optimizations of the resistor structure maximize the power rate dissipated on the resistor. Ansoft HFSS software is employed to optimize the height of the sacrificial layer, the transmission line width and length. The dissipated power on a titanium resistor is higher than the dissipated power on a platinum resistor. The rate of the dissipated power on the titanium resistor is around 25%, but around 4% on the platinum resistor. The material properties are given in table 4.13.

The sacrificial layer's thickness may be 2–3 μm. Figure 4.42 shows the resistor configuration.

Figure 4.43 presents the MEMS bolometer layout.

The patch coupled to a bolometer is shown in figure 4.44.

4.7.5 Array fabrication and measurement

Nine masks are used to fabricate the detection array. The mask process and layer thickness is listed in table 4.14. Layer thickness has been determined as the best compromise between the technology limits and design considerations.

The dimensions of the detection array elements have been measured in several array pixels as part of a visual test of the array after fabrication and some of the

Table 4.13. Material properties.

Property	Units	siNi	Ti
Conductivity [K]	W m^{-1} K^{-1}	1.6	**7**
Capacity [C]	J^{-1} kg^{-1} K^{-1}	770	**520**
Density [ρ]	Gr cm^{-3}	2.85	**4.5**
Resistance	Ω^{-1}	$>1 \times 10^8$	**90**
Thickness	mm	0.1	0.1

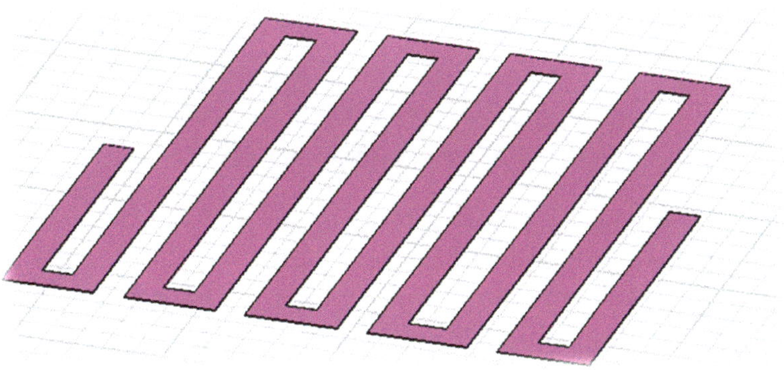

Figure 4.42. Resistor configuration.

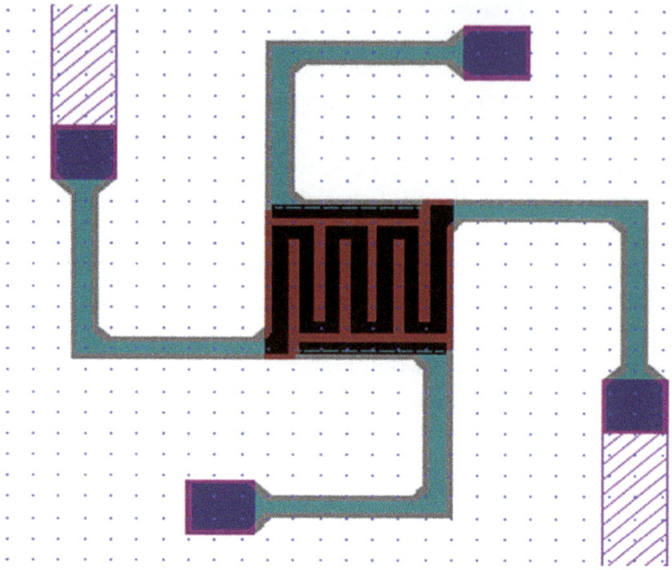

Figure 4.43. MEMS bolometer layout.

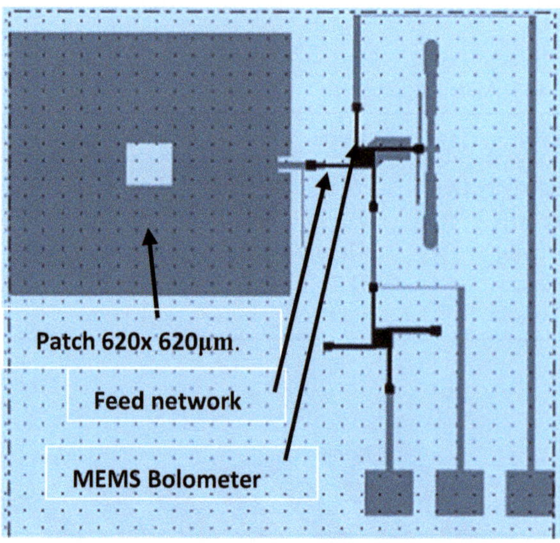

Figure 4.44. Patch antenna coupled to a bolometer.

measured results are listed in table 4.15. From the results listed in table 4.15 we may conclude that the fabrication process is very accurate. There is good agreement between the computed and measured results of the array's electrical and mechanical parameters.

Figure 4.45 shows how we measure the bolometer output voltage. V_{ref} is the bolometer output voltage when no radiated power is received by the detection array.

Table 4.14. Mask process.

Masks	Layer	Process	Layer thickness
1	L 1 Lift or Etch	Gold reflector Au	1 μm
2	L 2 Etch	Streets open SL	3 μm
3	L 3 Etch	SL contacts	
4	L 4 Etch	SiN + contacts	0.1 μm
5	L 5 Etch	Ti_1	0.1 μm
		SiN	0.15 μm
6	L 6 Etch	VOx	0.1 μm
7	L 7 Etch	Contacts for Ti_2	
		Ti_2	0.1 μm
	L 3 Lift	Metal cap	0.1–0.5 μm
8	L 8 Etch	Ti_2	
		SiN	0.1 μm
9	L 9 Etch	Membrane definition	

Table 4.15. Comparison of design and fabricated array dimensions.

Element	Design (μm)	Pixel 1 (μm)	Pixel 2 (μm)
Patch width	600	599.5	600.5
Patch length	600	600.3	600.5
Hole width	100	99.8	100
Hole length	100	100	99.8
Feed line	10	10	10
Feed line	10	9.8	10
Stub width	2	2	1.8
Tapered line	15	15.2	14.8
Stub width	2	1.8	2
Tapered line	25	25.3	25.2

The voltage difference between the bolometer voltage and V_{ref} is amplified by a low noise differential amplifier. The rate of the dissipated power on the titanium resistor is around 25%–30%. Figure 4.46 presents the operational concept of the detection array.

4.7.6 Mutual coupling effects between pixels

HFSS software has been used to compute the mutual coupling effects between pixels in the detection array as shown in figure 4.47. The computation results indicate that the power dissipated on the centered pixels in the array is higher by 1%–2% than the pixels located at the corners of the array.

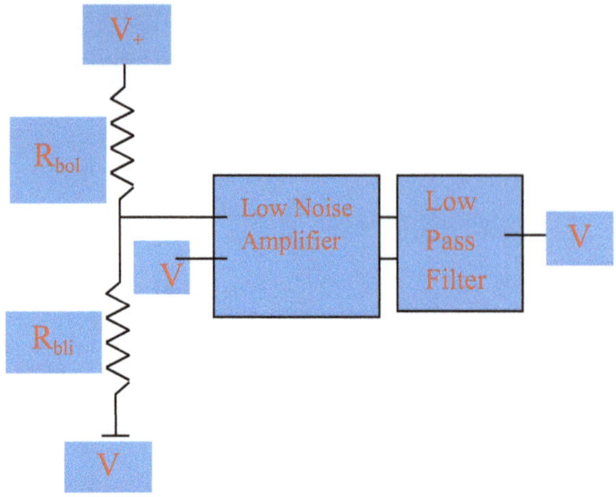

Figure 4.45. Measurements of the bolometer voltage.

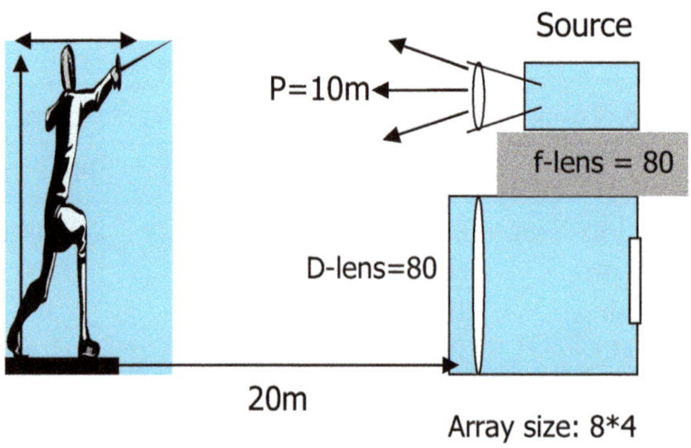

Figure 4.46. Detection array.

4.8 MEMS bow-tie dipole with a bolometer

A bow-tie dipole with a bolometer printed on a quartz substrate is shown in figure 4.48. The length of the bow-tie dipole is around 1.5 mm. The bolometer length is 0.6 mm.

Figure 4.49 presents the MEMS bow-tie dipole S11 computed results. Figure 4.50 presents the MEMS bow-tie dipole radiation pattern results.

4.9 LTCC and HTCC technology

Co-fired ceramic devices are monolithic, ceramic microelectronic devices where the entire ceramic support structure and any conductive, resistive, and dielectric materials are fired in a kiln at the same time. Typical devices include capacitors, inductors, resistors, transformers, and hybrid circuits. The technology is also used for a multi-layer packaging for the electronics industry, such as military electronics. Co-fired ceramic devices are made by processing a number of layers independently and assembling them into a device as a final step. Co-firing can be divided into low

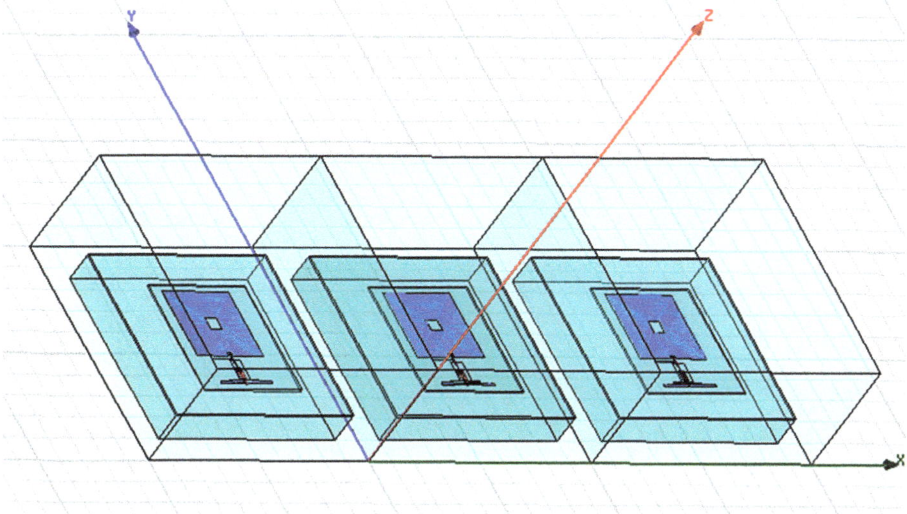

Figure 4.47. Computation of mutual coupling between pixels.

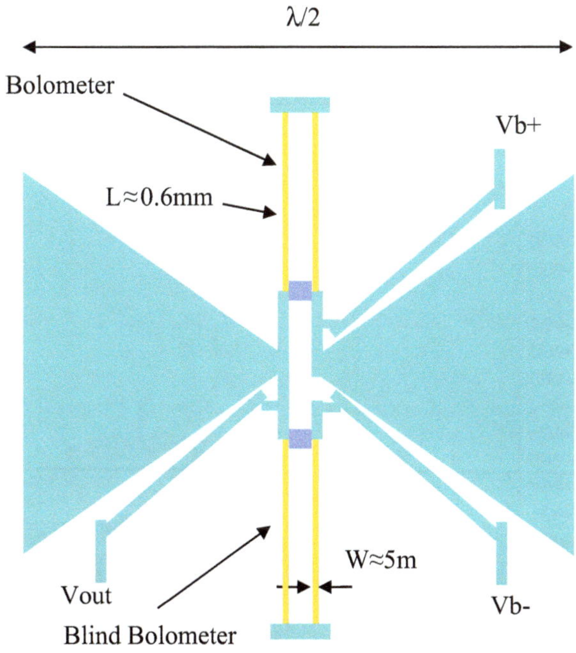

Figure 4.48. MEMS bow-tie dipole with bolometer.

temperature (LTCC) and high temperature (HTCC) applications: low temperature means that the sintering temperature is below 1000 °C (1830 °F), while high temperature is around 1600 °C (2910 °F). There are two types of raw ceramics to manufacture multi-layer ceramic (MLC) substrate:

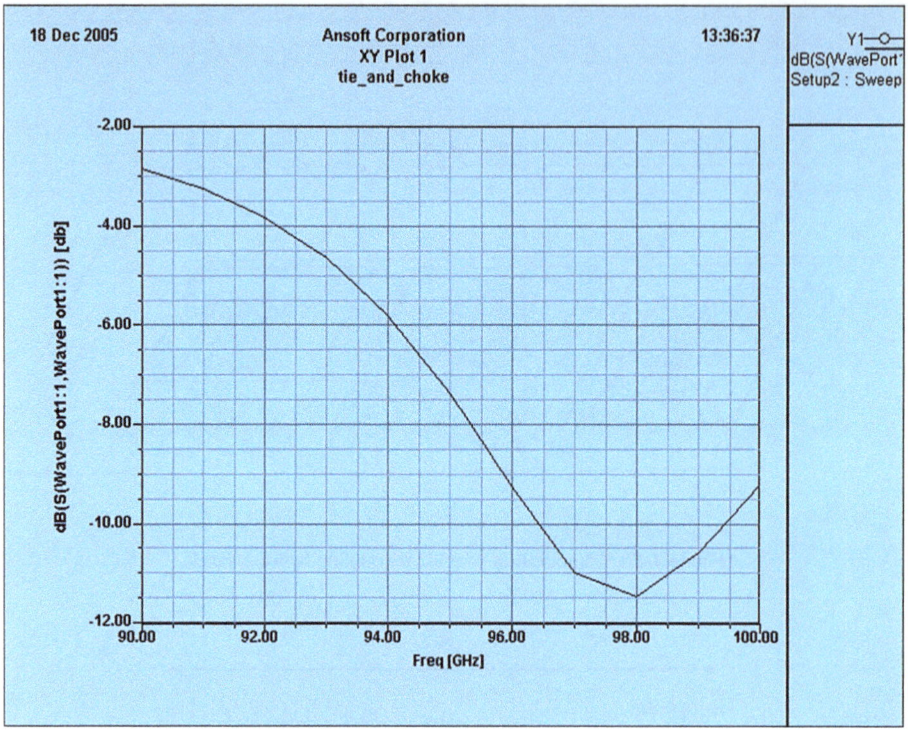

Figure 4.49. MEMS bow-tie dipole S_{11} computed results.

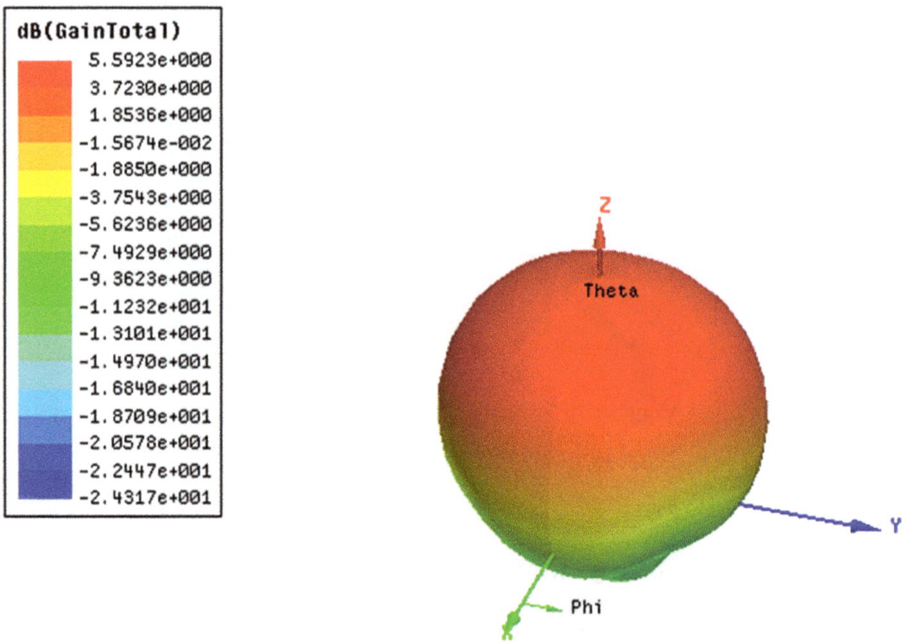

Figure 4.50. Bow-tie dipole 3D radiation pattern.

- Ceramics fired at high temperature ($T \geqslant 1500$ °C): High temperature co-fired ceramic (HTCC),
- Ceramics fired at low temperature ($T \leqslant 1000$ °C): Low temperature co-fired ceramic (LTCC).

The base material of a HTCC is usually Al_2O_3. HTCC substrates are a row of ceramic sheets. Because of the high firing temperature of Al_2O_3 the material of the embedded layers can only be high melting temperature metals: wolfram, molybdenum or manganese. The substrate is unsuitable to bury passive elements, although it is possible to produce thick-film networks and circuits on the surface of a HTCC.

The breakthrough for LTCC fabrication was when the firing temperature of a ceramic-glass substrate was reduced to 850 °C. The equipment for a conventional thick-film process could be used to fabricate LTCC devices. LTCC technology evolved from HTCC technology combined the advantageous features of thick-film technology. Because of the low firing temperature (850 °C) the same materials are used for producing buried and surface wiring and resistive layers as thick-film hybrid IC (i.e. Au, Ag, Cu wiring RuO_2 based resistive layers). It can be fired in an oxygen-rich environment unlike HTCC boards, where a reduced atmosphere is used. During co-firing the glass melts, the conductive and ceramic particles are sintered. On the surface of LTCC substrates hybrid integrated circuits can be realized, as shown in figure 4.51. Passive elements can be buried into the substrate, and we can place semiconductor chips in a cavity. The dielectric properties at 9 GHz of LTCC substrates are listed in table 4.16.

4.9.1 LTCC and HTCC technology process

- Low Temperature LTCC 875 °C.
- High Temperature HTCC 1400 °C–1600 °C.

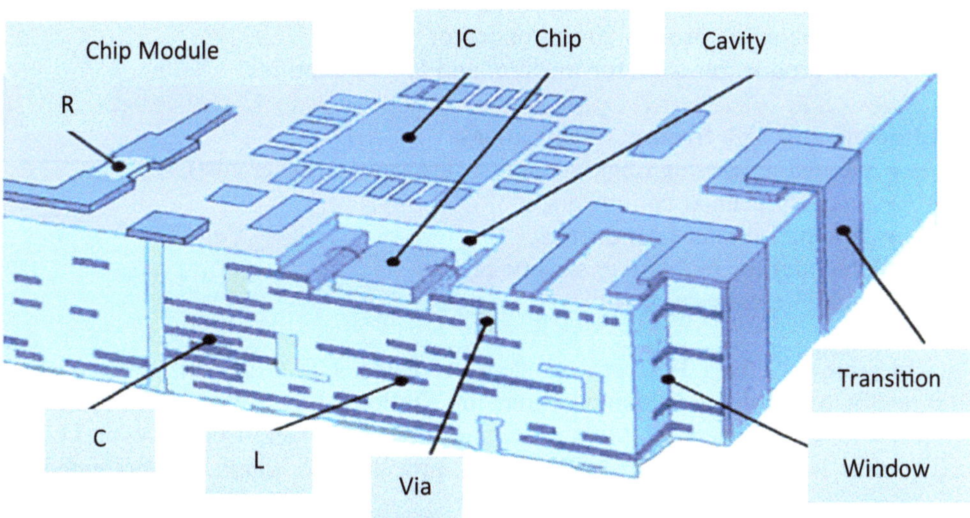

Figure 4.51. LTCC module.

Table 4.16. Dielectric properties at 9 GHz of LTCC substrates.

Material	ε_r	Tan $\delta \times 10{-}3$
99.5% AL	9.98	0.1
LTCC1	7.33	3.0
LTCC2	6.27	0.4
LTCC3	7.2	0.6
LTCC4	7.44	1.2
LTCC5	6.84	1.3
LTCC6	8.89	1.4

- Co-fired/Co-firing of (di)electric pastes.
  LTCC: precious metals (Au, Ag, Pd, Cu).
  HTCC: refractory metals (W, Mo, MoMn).
- Ceramic Mix of: alumina Al_2O_3.
  Glasses SiO_2–B_2O_3–CaO–MgO.
  Organic binders.
  HTTC: essentially Al_2O_3.

Advantages of LTCC

- low permittivity tolerance,
- good thermal conductivity,
- low TCE (adapted to silicon and GaAs),
- excellently suited for multi-layer modules,
- integration of cavities and passive elements such as R, L, and C-components,
- very robust against mechanical and thermal stress (hermetically sealed),
- composable with fluidic, chemical, thermal, and mechanical functionalities,
- low material costs for silver conductor paths,
- low production costs for medium and large quantities.

Advantages for high frequency applications:

- parallel processing (high yield, fast turnaround, lower cost),
- precisely defined parameters,
- high performance conductors,
- potential for multi-layer structures,
- high interconnect density.

In table 4.17 the LTCC process steps are listed. LTCC raw material comes as sheets or rolls. Material manufacturers are DuPont, ESL, Ferro and Heraeus. In table 4.18 several electrical, thermal and mechanical characteristics of several LTCC materials are listed. In figure 4.52 a LTCC process block diagram is presented.

Table 4.17. LTCC process list.

LTCC process
Tape casting
Sheet cutting
Laser punching
Printing
Cavity punching
Stacking
Bottom side printing
Pressing
Side hole formation
Side hole printing
Snap line formation
Pallet firing
Plating Ni–Au

Table 4.18. LTCC material characteristics.

Material	LTCC DP951	Al_2O_3 96%	BeO	AIN 98%
Electrical characteristics at 10 MHz				
Dielectric constant, ε_r	7.8	9.6	6.5	8.6
Dissipation factor, tan δ	0.00015	0.0003	0.0002	0.0005
Thermal characteristics				
Thermal expansion 10^{-6}/°C	5.8	7.1	7.5	4.6
Thermal conductivity W m^{-1} K^{-1} 25–300 °C	3	20.9	251	180
Mechanical characteristics				
Density	3.1	3.8	2.8	3.3
Flexural strength, MPa	320	274	241	340
Young's modulus, GPa	120	314	343	340

In table 4.19 LTCC line losses at 2 GHz are listed for several LTCC materials. For LTTC1 the material losses are 0.004 dB mm^{-1}.

4.9.2 Design of high pass LTCC filters

The trend in the wireless industry toward miniaturization, cost reduction and improved performance drives microwave designers to develop microwave components using LTCC technology. A significant reduction in the size and cost of microwave components may be achieved by using, low temperature co-fired ceramic LTCC, technology. In LTCC technology discrete surface-mounted components such as capacitors and inductors are replaced by integrated printed components.

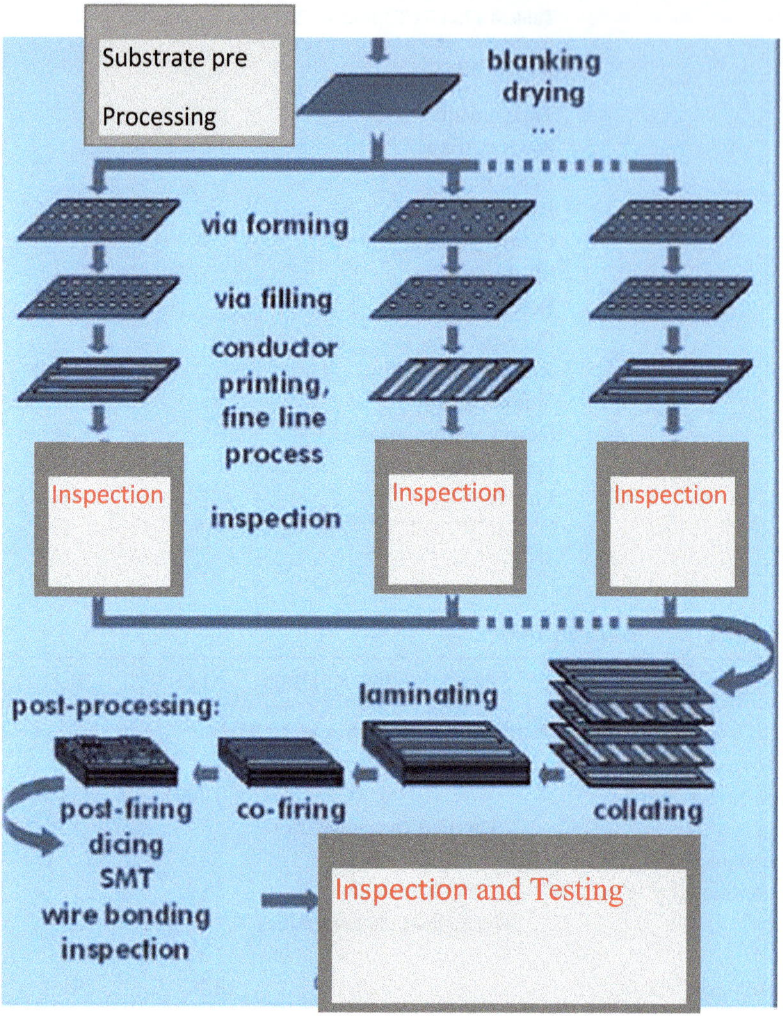

Figure 4.52. LTCC process.

Table 4.19. LTCC line loss.

Material	Dissipation factor, tan $\delta \times 10^{-3}$	Line loss dB mm^{-1}, at 2 GHz
LTTC1	3.8	0.004
LTTC2	2.0	0.0035
LTTC6-CT2000	1.7	0.0033
Alumina 99.5%	0.65	0.003

LTCC technology allows designers to use a multi-layer design if needed to reduce the size and cost of the circuit. However, multi-layer design results in more losses due to connections and parasitic coupling between different parts of the circuit. To improve the filter performance all of the filter parameters have been optimized. Package effects were taken into account in the design.

High pass filter specification

Frequency 1.5–2.5 GHz.	
Insertion loss 1.1 Fo	1 dB.
Rejection 0.9 Fo	3 dB.
Rejection 0.75 Fo	20 dB.
Rejection 0.5 Fo	40 dB.
VSWR	2:1
Case dimensions	700 × 300 × 25.5 mil inch.

The filters are realized by using lumped elements. The filters' inductor and capacitor parameters were optimized by using HP ADS software. The filter consists of five layers of 5.1 mil substrate with $\varepsilon_r = 7.8$. Package effects were taken into account in the design. Changes in the design were made to compensate for and minimize package effects. In figure 4.53 the filter layout is shown. S_{11} and S_{12} momentum simulation results can be seen in figure 4.54. In figure 4.55 the filter 2 layout is presented. S_{11} and S_{12} momentum simulation results are shown in figure 4.56. Simulation results of the tolerance check are shown in figure 4.57. The parameters that were tested in the tolerance check are: inductor and capacitor line width and length and spacing between capacitor fingers.

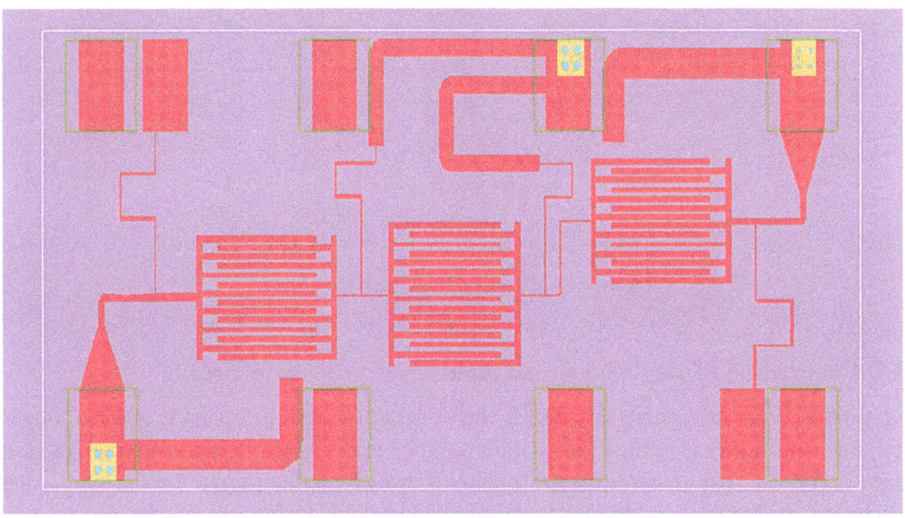

Figure 4.53. Layout of high pass filter No. 1.

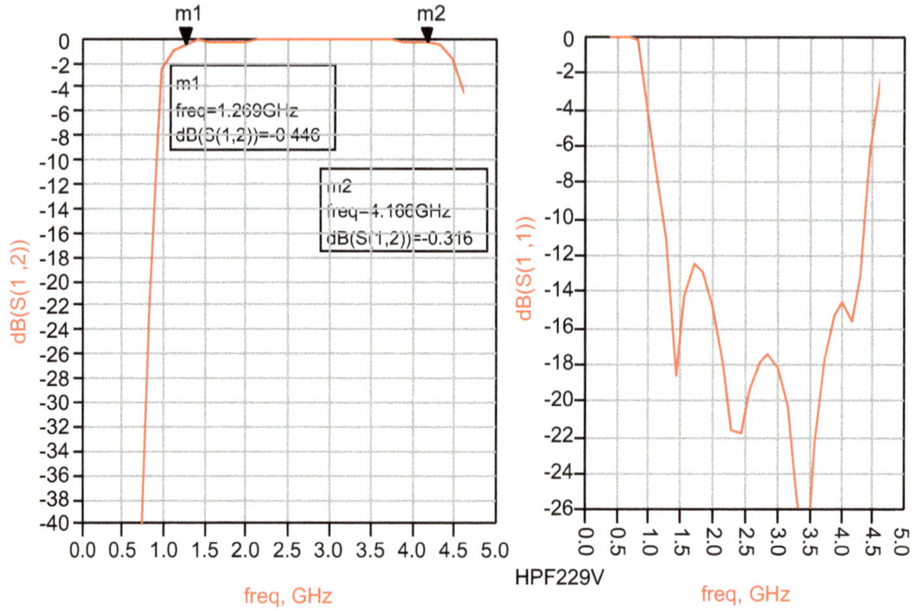

Figure 4.54. S_{12} and S_{11} results of high pass filter No. 1.

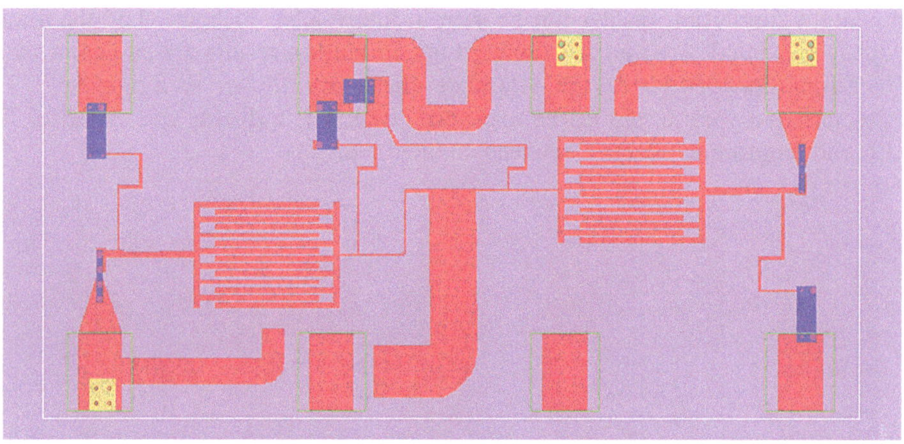

Figure 4.55. Layout of high pass filter No. 2.

4.9.3 Comparison of single-layer and multi-layer microstrip circuits

In a single-layer microstrip circuit all the conductors are in a single layer. Coupling between conductors is achieved through edge or end proximity (across narrow gaps). Single-layer microstrip circuits are cheap to produce. In figure 4.58 a single-layer microstrip edge coupled filter is shown.

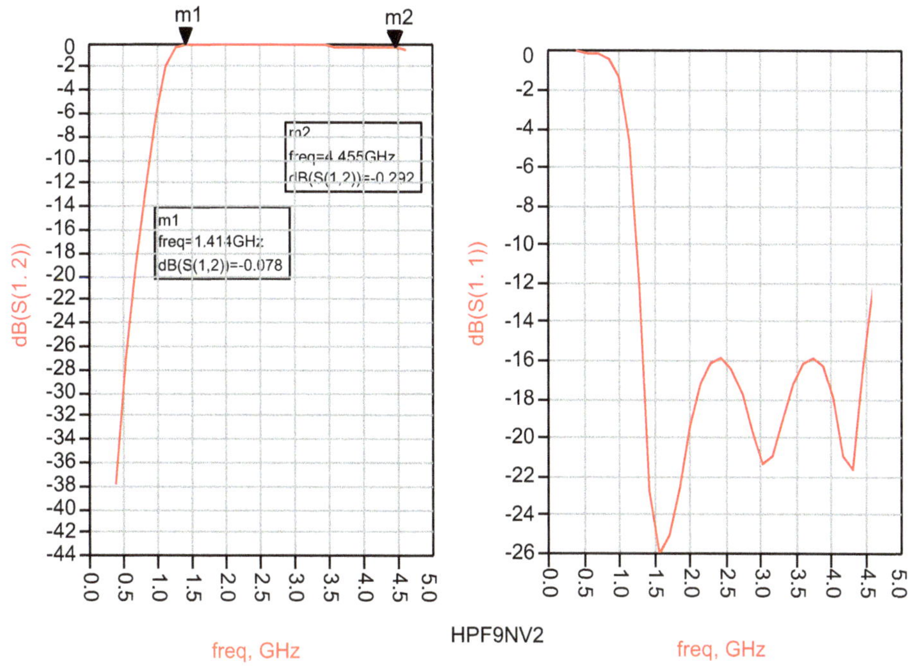

Figure 4.56. S_{12} and S_{11} results of high pass filter No. 2.

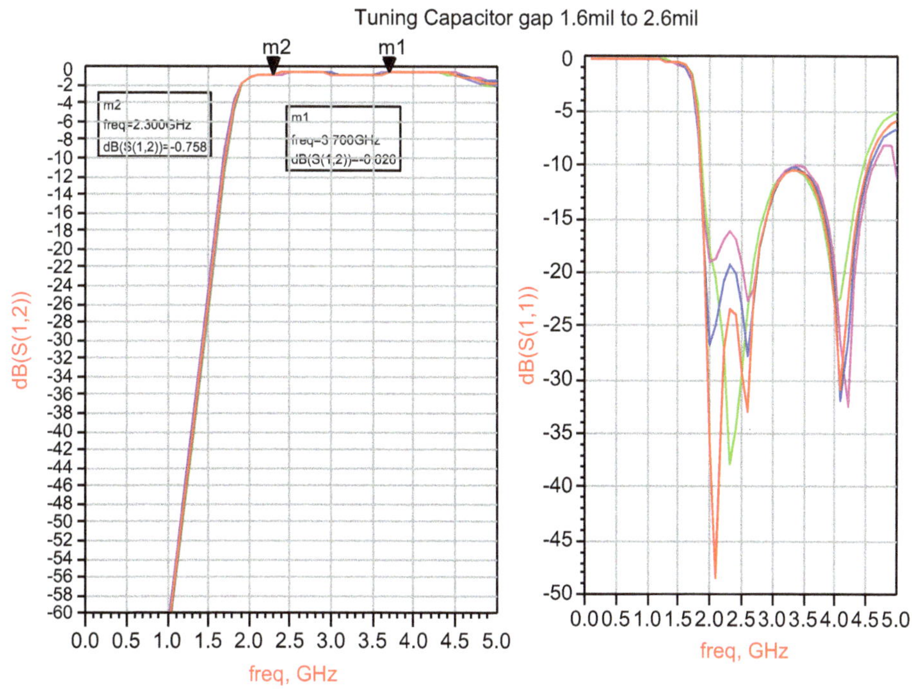

Figure 4.57. Tolerance simulation for spacing between capacitor fingers.

Figure 4.59 presents the layout of a single-layer microstrip directional coupler. Figure 4.60 presents the structure of a multi-layer microstrip coupler.

In multi-layer microwave circuits, conductors are separated by dielectric layers and stacked on different layers. This structure allows for (strong) broadside coupling. Registration between layers is not difficult to achieve as there are narrow

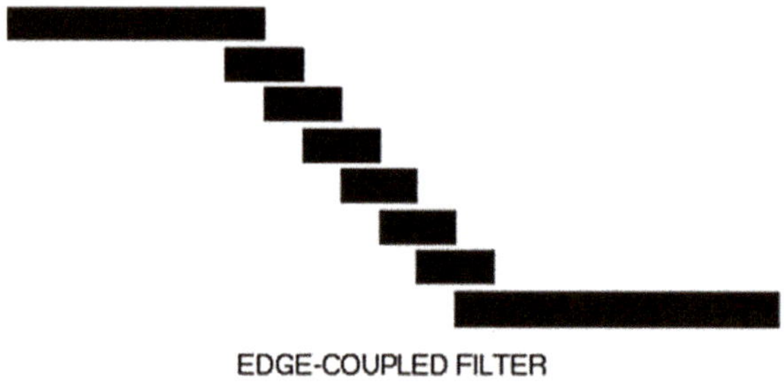

Figure 4.58. Edge coupled filter.

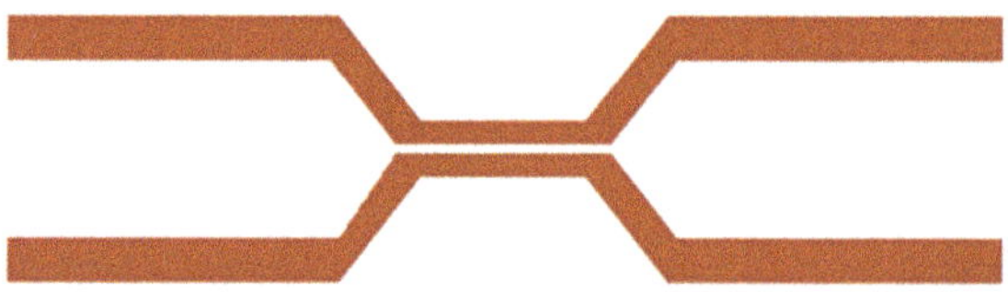

Figure 4.59. Single-layer microstrip directional coupler.

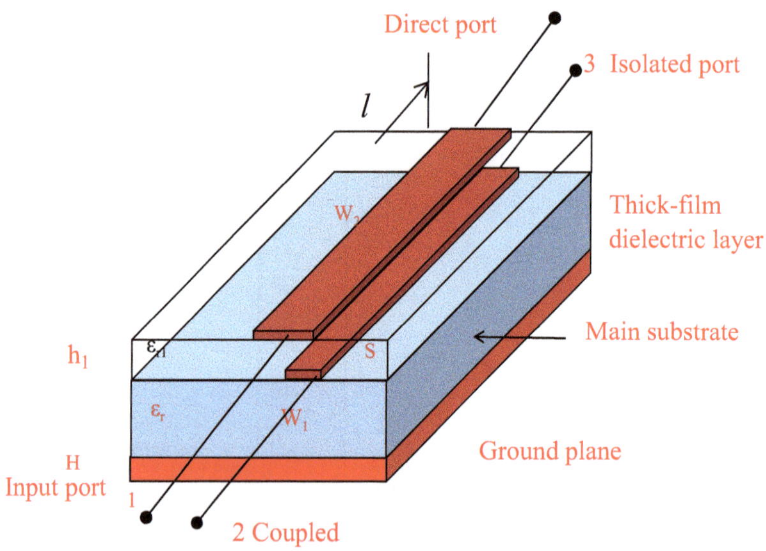

Figure 4.60. Multi-layer microstrip coupler.

gaps between strips in single-layer circuits. The multi-layer structure technique is well-suited to thick-film print technology and is also suitable for LTCC technology.

4.10 Conclusions

Dimensions and losses of microwave systems are minimized by using microwave integrated circuits, MMIC technology, MEMS and LTCC technology, and by using a multi-layer structure technique. This technique is well-suited to thick-film print technology and for LTCC technology. LTCC technology allows the integration of cavities and passive elements such as R, L, and C-components as part of the LTCCs. Sensors, actuators and RF switches may be manufactured by using MEMS technology. Losses of MEMS components are considerably lower than MIC and MMIC RF components. MMICs are circuits in which active and passive elements are formed on the same dielectric substrate. MMICs are dimensionally small (from around 1 mm^2 to 10 mm^2) and can be mass produced. Their components cannot be tuned. Accurate design is crucial in the design of MMIC circuits. The designer's goal for MMICs, MEMS and LTCCs is to comply with customer specifications in one design iteration.

References

[1] Rogers J and Plett C 2003 *Radio frequency Integrated Circuit Design* (Boston, MA: Artech House)

[2] Maluf N and Williams K 2004 *An Introduction to Microelectromechanical System Engineering* (Boston, MA: Artech House)

[3] Sabban A 2011 Microstrip antenna arrays *Microstrip Antennas* ed N Nasimuddin (Rijeka: InTech) 361–84

[4] Sabban A 2007 Applications of MM wave microstrip antenna arrays *ISSSE 2007 Conf. (Montreal, Canada)*

[5] Gauthier G P, Raskin G P, Rebiez G M and Kathei P B 1999 A 94GHz micro-machined aperture-coupled microstrip antenna *IEEE Trans. Antenna Propag.* **47** 1761–6

[6] Milkov M M 2000 Millimeter-wave imaging system based on antenna-coupled bolometer *MSc thesis* UCLA

[7] de Lange G *et al* 1999 A 3 × 3 mm-wave micro machined imaging array with sis mixers *Appl. Phys. Lett.* **75** 868–70

[8] Rahman A *et al* 1996 Micro-machined room temperature micro bolometers for MM-wave detection *Appl. Phys. Lett.* **68** 2020–2

[9] Mass S A 1997 *Nonlinear Microwave and RF Circuits* (Boston, MA: Artech House))

[10] Sabban A 2015 *Low Visibility Antennas for Communication Systems* (London: Taylor and Francis)

[11] Sabban A 1983 A new wideband stacked microstrip antenna *IEEE Antenna and Propagation Symp. (Houston, Texas, USA)*

[12] Sabban A 1981 Wideband microstrip antenna arrays *IEEE Antenna and Propagation Symp. MELCOM (TelAviv)*

[13] Sabban A 2013 New wideband printed antennas for medical applications *IEEE J. Trans. Antennas Propag.* **61** 84–91

[14] Sabban A 2016 *Wideband RF Technologies and Antenna in Microwave Frequencies* (New York: Wiley)
[15] Sabban A 2016 Small wearable meta materials antennas for medical systems *Appl. Comput. Electromagn. Soc. J.* **31** 434–43
[16] Sabban A 2016 Ultra-wideband RF modules for communication systems *PARIPEX Indian J. Res.* **5** 91–5
[17] Sabban A 2015 New compact wearable meta-material antennas *Global J. Res. Anal.* **IV** 268–71
[18] Sabban A 2015 New wideband meta materials printed antennas for medical applications *J. Adv. Med. Sci.* **3** 1–10
[19] Sabban A 2015 Wideband RF modules and antennas at microwave and MM wave frequencies for communication applications *J. Mod. Commun. Technol. Res.* **3** 89–97
[20] Sabban A 2014 W band MEMS detection arrays *J. Mod. Commun. Technol. Res.* **2** 9–13

IOP Publishing

Wearable Communication Systems and Antennas (Second Edition)
Design, efficiency, and miniaturization techniques
Albert Sabban

Chapter 5

RF components and module design for wearable communication systems

5.1 Introduction

MIC and MMIC technologies were presented in chapter 4. In the design of RF modules we may gain the advantages of each technology by using both in the development of RF transmitting and receiving modules, see [1–11]. For example, it is better to design filters using MIC technology. However, compact and low-cost modules in mass production are achieved by using MMIC design. Design considerations of passive and active devices will be presented in this chapter.

5.2 Passive elements

5.2.1 Resistors

Resistors, capacitors and inductors are basic passive elements that are used in the design of MIC and MMIC modules. The resistor schematic is shown in figure 5.1. The resistance may be computed by using equation (5.1). A resistor equivalent circuit is presented in figure 5.2.

$$R = R_{sh}\frac{S}{W}+2R_c \tag{5.1}$$

R_{sh}—the resistance of the metal film or doped semiconductor film.
S—separation between ohmic contacts.
W—the width of the conductor.
R_c—the resistance of the contacts.

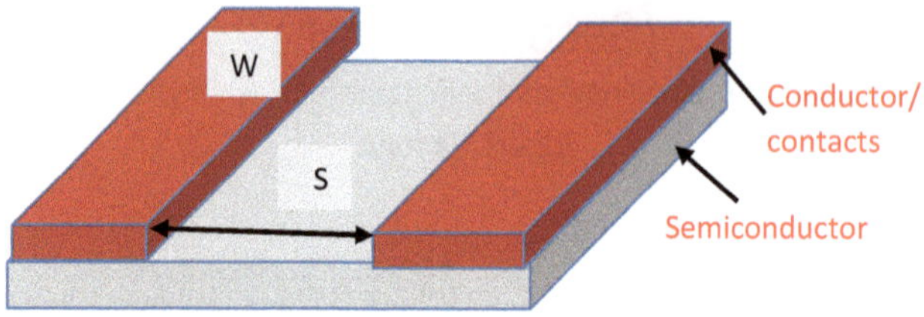

Figure 5.1. Schematic of a resistor.

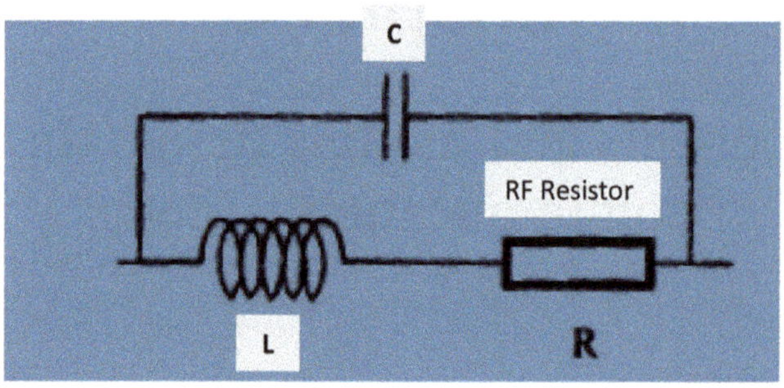

Figure 5.2. Resistor equivalent circuit.

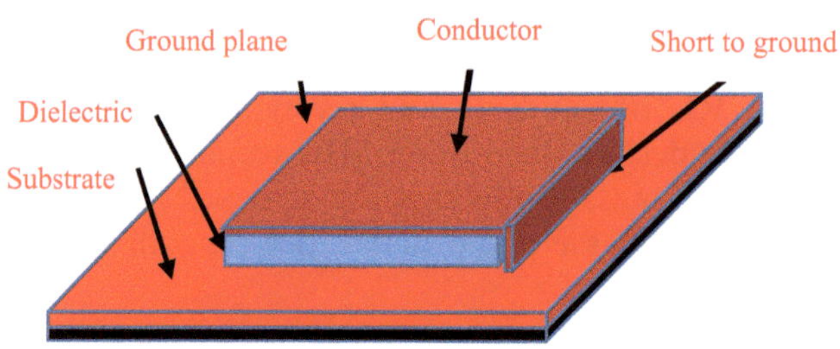

Figure 5.3. Capacitor.

5.2.2 Capacitor

A capacitor consists of two parallel plate conductors separated by a dielectric layer with dielectric constant ϵ_r, and thickness d. The conductor area is A. The capacitor schematic is shown in figure 5.3. The capacitance may be computed by using equation (5.2).

$$C = A\frac{\epsilon_0 \epsilon_r}{d} \tag{5.2}$$

where $\epsilon_0 = 8.85 \times 10^{-12}$ Fm^{-1}.

5.2.3 Inductor

A narrow short printed transmission line, as shown in figure 5.4, behaves as an inductor. The inductor equivalent circuit is presented in figure 5.5. The inductance L and the capacitance C can be calculated by equations (5.3) and (5.4).

Passive elements resistor, capacitor and inductor in MMIC design are shown in figure 5.6.

$$L = \frac{Z_0}{2\pi f}\sin\left(\frac{2\pi l}{\lambda}\right) \tag{5.3}$$

$$C = \frac{1}{2\pi f Z_0}\tan\left(\frac{\pi l}{\lambda}\right) \tag{5.4}$$

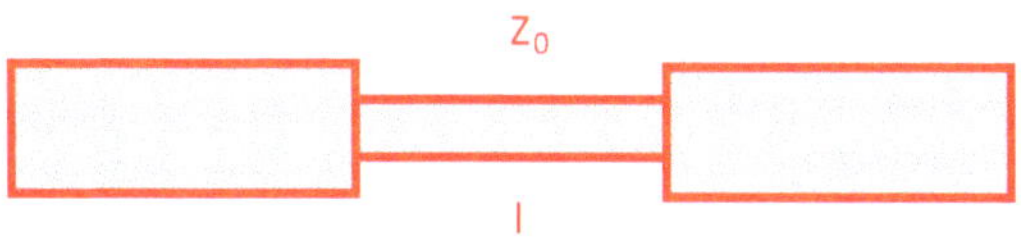

Figure 5.4. Short printed transmission line.

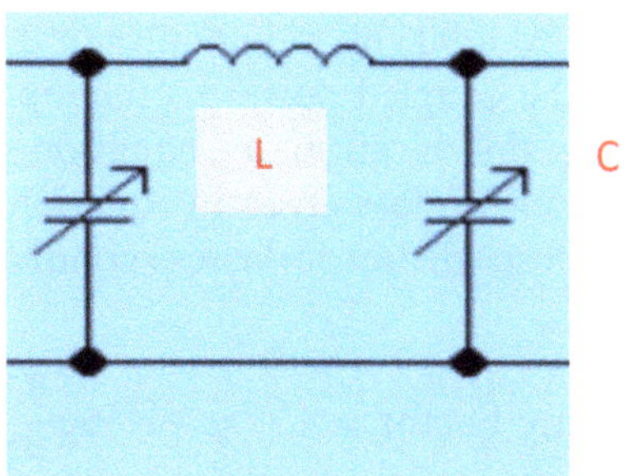

Figure 5.5. Inductor equivalent circuit.

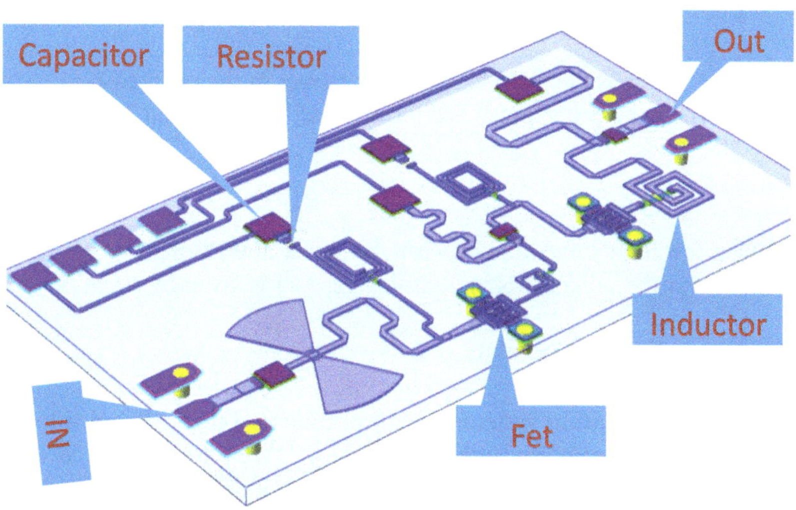

Figure 5.6. Passive and active elements in MMIC design.

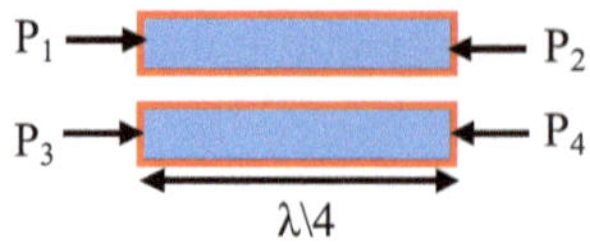

Figure 5.7. Coupled lines.

5.2.4 Couplers

Couplers are used to couple part of the power in the input port to a coupled port. Usually couplers consist of two coupled quarter-wave length transmission lines and have four ports, as shown in figure 5.7. P1 is the input port, P2 is the transmitted port, P3 is the coupled port, P4 is the isolated port. The coupled port may be used to obtain information about the signal such as frequency and power level without interrupting the main power flow in the device. If the power coupled to the coupled port is half of the input power, 3 dB below the input power level, the power on the main transmission line is also 3 dB below the input power and equals the coupled power. Such a coupler is called a 90° hybrid coupler or 3 dB coupler. The coupling factor is the ratio between the coupled power to the input power in dB, when the other ports are terminated; this is given by equation (5.5). Coupler losses are expressed via the coupler insertion loss given in equation (5.6). The actual coupler losses are due to coupling loss, dielectric losses, conductor losses, and matching losses.

$$\text{Coupling Factor} = CF = -10\log\frac{P3}{P1} \tag{5.5}$$

$$\text{Insertion Loss} = IL = 10\log\left(1 - \frac{P3}{P1}\right) \tag{5.6}$$

The isolation factor is the ratio between the power in the isolated port to the input power in dB, when the other ports are terminated, and is given by equation (5.7).

$$\text{Isolation} = -10\log\frac{P4}{P1} \tag{5.7}$$

The coupler directivity is the ratio between the power in the isolated port to the coupled power in dB, when the other ports are terminated, and is given by equation (5.8).

$$\text{Directivity} = D = -10\log\frac{P4}{P3} \tag{5.8}$$

In several communication systems the amplitude and phase balance between the system ports are important in signal processing. Amplitude balance defines the power difference in dB between two output ports of a 3 dB hybrid. In an ideal hybrid circuit, the difference should be 0 dB. However, in a practical device the amplitude balance is frequency dependent. The phase difference between two output ports of a hybrid coupler may be 0, 90, or 180° depending on the coupler type.

5.2.5 A wideband mm wave coupler

A wideband mm wave coupler is shown in figure 5.8 and is printed on alumina with 5.8 dielectric constant and 5.5 mil inch thickness. The coupler frequency range is 18 GHz to 40 GHz. The coupler was designed by Momentum ADS software. The coupling value is −13 dB and is shown in figure 5.9. The coupler insertion loss is 0.4 dB and is shown in figure 5.10. The coupler S_{11} parameter is better than −26 dB and is shown in figure 5.11.

5.3 Power dividers and combiners

Power dividers and combiners are used to distribute or combine microwave signals. Power dividers and combiners are passive reciprocal devices. There are several types

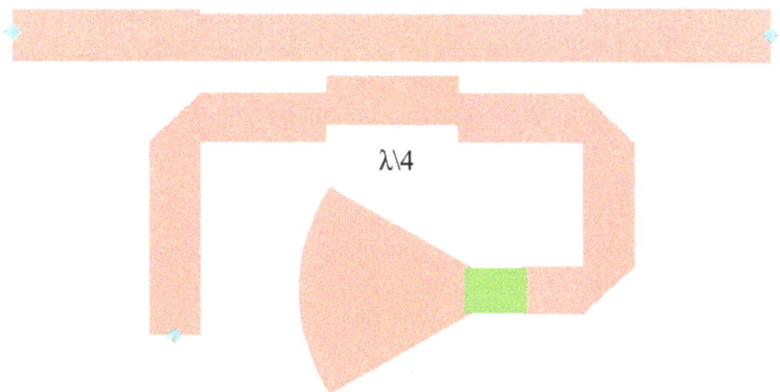

Figure 5.8. A wideband mm wave coupler.

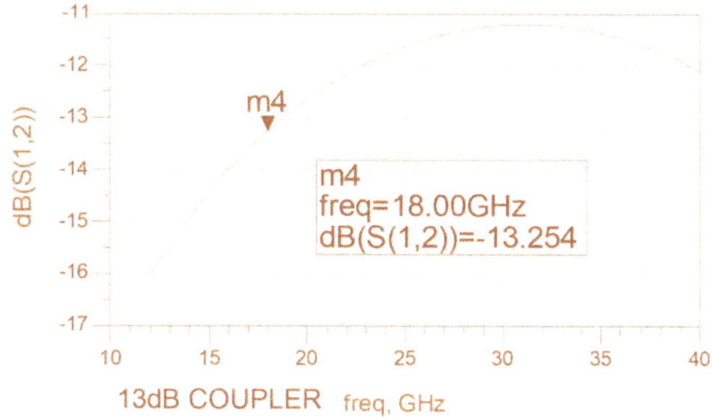

Figure 5.9. The coupler's coupling value.

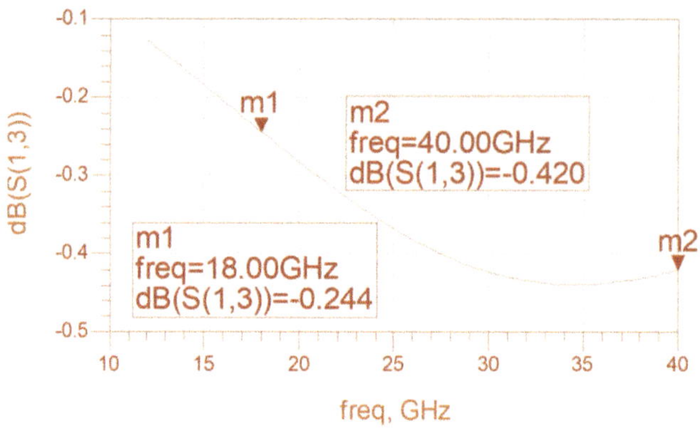

Figure 5.10. The coupler's insertion loss.

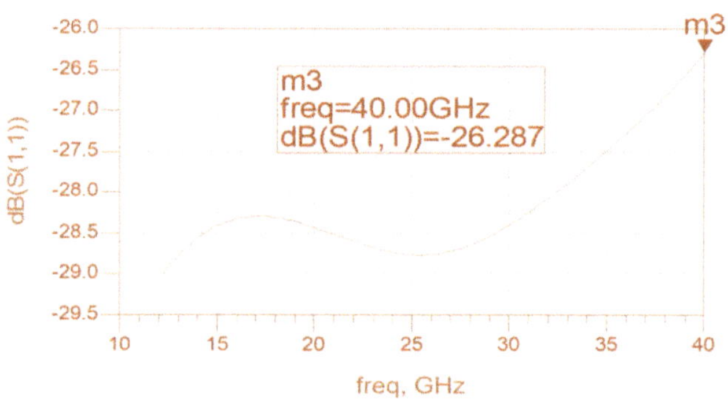

Figure 5.11. The coupler's S_{11} parameter.

of power dividers and combiners such as the Wilkinson power divider, rat-race power divider and Gysel power divider.

5.3.1 Wilkinson power divider

The Wilkinson power divider is shown in figure 5.12. The impedance of the input port is Z_0. At the junction point O the input power is half of the input power. The impedance at junction point O will be $2Z_0$. A quarter-wave transformer is used to match $2Z_0$ impedance to Z_0. The bandwidth of the Wilkinson power divider is around 20% for a voltage standing wave ratio (VSWR) better than 2:1. However, by using multiple section quarter-wave transformers we may achieve a bandwidth of 50% bandwidth for a VSWR better than 2:1. If a $2Z_0$ resistor is connected between the output ports the isolation between the output ports will be around 15 dB. If a resistor is not connected between the output ports the isolation between the output ports will be lower than 6 dB. The Wilkinson power divider may be used to get unequal power dividers.

For example, for a third of the input power at output port A, and two thirds of the input power at output port B, the impedance of the quarter-wave transformer connecting point O to output port A should be 86.6 Ω. The impedance of the the quarter-wave transformer connecting point O to output port B should be 61.23 Ω.

5.3.2 Rat-race coupler

A rat-race coupler is shown in figure 5.13. The rat-race circumference is 1.5 wavelengths. The distance from A to Δ port is 3λ/4. The distance from A to ∑ port is λ/4. For an equal-split rat-race coupler, the impedance of the entire ring is fixed at 1.41 × Z_0, or 70.7 Ω for Z_0 = 50 Ω. For an input signal V, the outputs at ports 2 and 4 are equal in magnitude, but 180° out-of-phase.

5.3.3 Gysel power divider

The Gysel power divider is shown in figure 5.14. The Gysel power divider is a wideband power divider, around 60% bandwidth for a VSWR better than 2:1. The Gysel power divider may transmit high power.

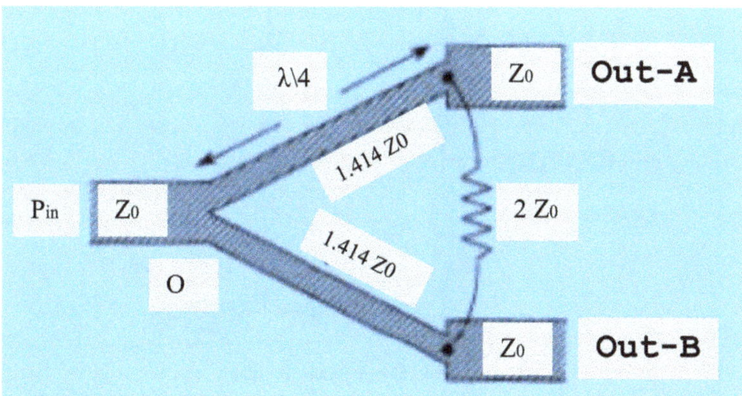

Figure 5.12. Wilkinson power divider.

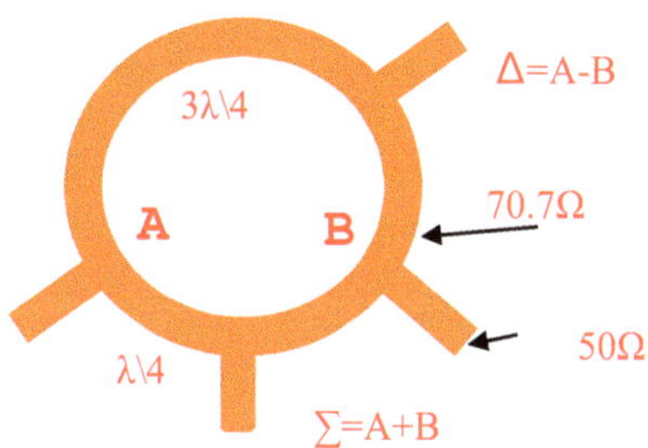

Figure 5.13. Rat-race coupler.

5.3.4 Unequal rat-race coupler

The rat-race coupler may be used to get unequal power dividers. The unequal rat-race power divider is shown in figure 5.15. The power ratio $\frac{P_A}{P_B}$ may be achieved by realizing the impedances Z_{0A} and Z_{0B} as given in equations (5.9) and (5.10).

$$Z_{0A} = Z_0\left[\frac{1+\frac{P_A}{P_B}}{\frac{P_A}{P_B}}\right]^{0.5} \tag{5.9}$$

$$Z_{0B} = Z_0\left[1+\frac{P_A}{P_B}\right]^{0.5} \tag{5.10}$$

5.3.5 Wideband three-way unequal power divider

In figure 5.16 a 6 GHz to 18 GHz three-way power divider is presented. The power divider is printed on RT-Duroid with a dielectric constant of 2.2 and 10 mil inch

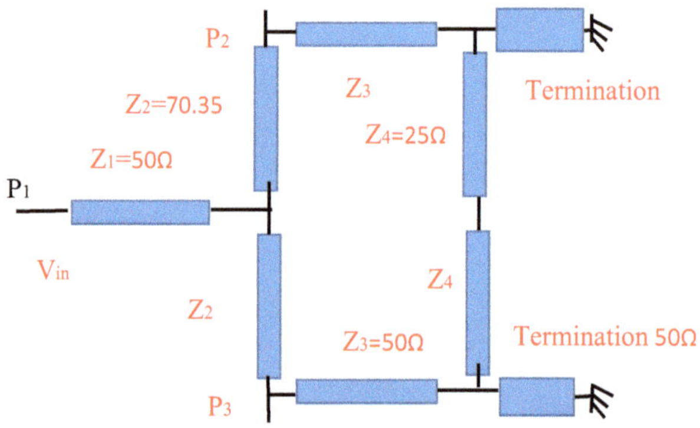

Figure 5.14. Gysel power divider.

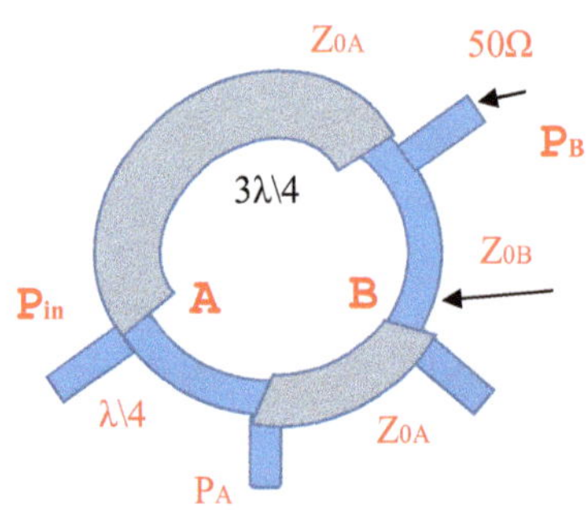

Figure 5.15. Unequal rat-race power divider.

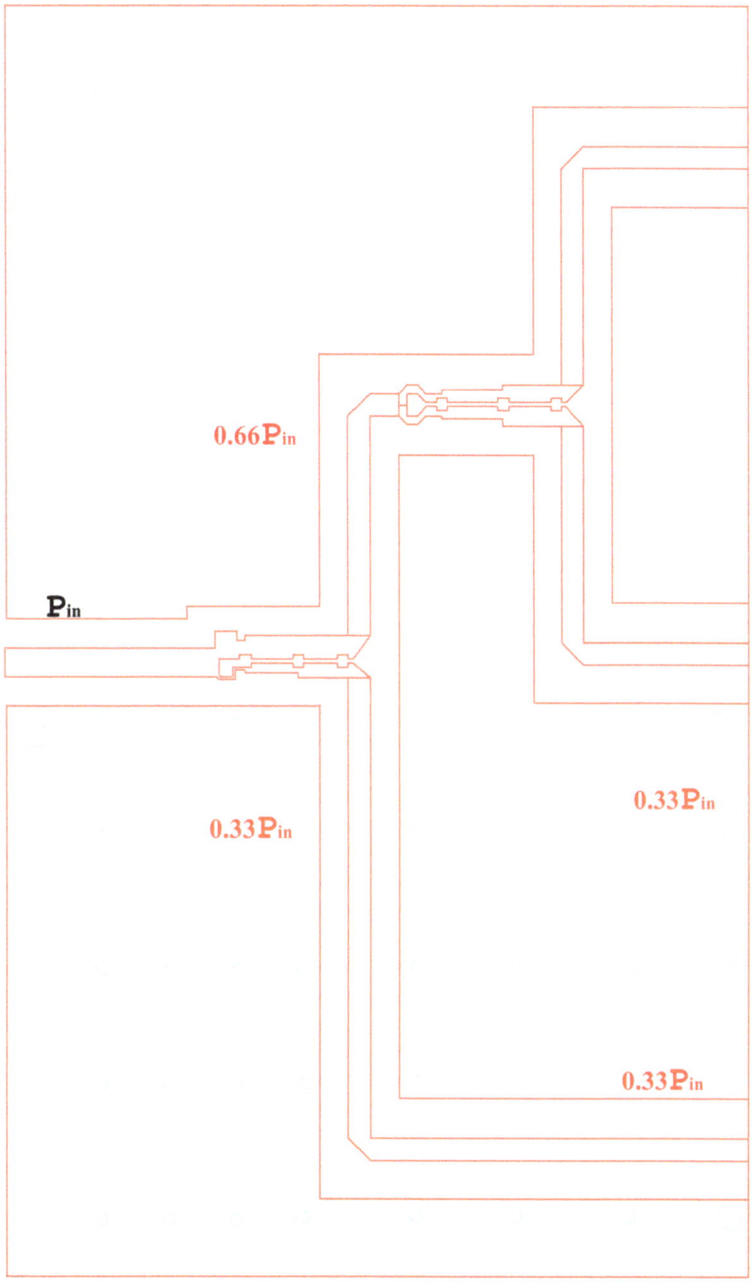

Figure 5.16. Wideband three-way unequal power divider.

thickness. The dimensions of the power divider are around 40 × 24 × 0.25 mm. The power divider consists of three quarter-wave transformer sections. The input power is divided into two thirds and a third of the input port power. The two thirds of the input port power are divided by an equal power divider to a third of the input port power. The power divider's insertion loss is around 1 dB. The power divider's S_{11} parameter is better than −10 dB in the frequency range of 6 GHz to 18 GHz. The power divider was designed and optimized by Momentum ADS software.

5.3.6 Wideband six-way unequal power divider

In figure 5.17 a 6 GHz to 18 GHz six-way power divider is presented. The power divider is printed on RT-Duroid with a dielectric constant of 2.2 and 10 mil inch thickness. The dimensions of the power divider are around 68.79 × 23.4 × 0.25 mm. The power divider consists of three quarter-wave transformer sections. The input power is divided into two thirds and a third of the input port power. The two thirds of the input port power are divided by an equal power divider to a third of the input port power. The power divider's insertion loss is around 1.5 dB. The power divider's

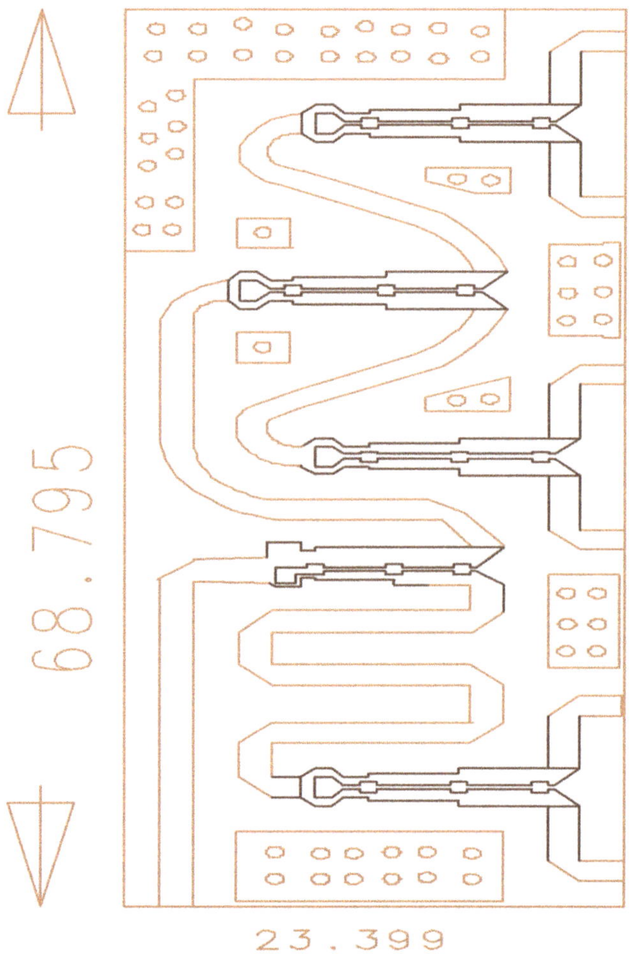

Figure 5.17. Wideband six-way unequal power divider.

S_{11} parameter is better than −10 dB in the frequency range of 6 GHz to 18 GHz. The power divider was designed and optimized by Momentum ADS software.

5.3.7 Wideband five-way unequal power divider

In figure 5.18 a 6 GHz to 18 GHz five-way power divider is presented. The power divider is printed on RT-Duroid with a dielectric constant of 2.2 and 10 mil inch thickness. The dimensions of the power divider are around 25.5 × 22.1 × 0.25 mm. The input power is divided into two fifths and three fifths of the input port power. The two fifths of the input port power are divided by an equal power divider to a fifth of the input port power. The three fifths of the input port power are divided by an unequal power divider to a fifth of the input port power and to two fifths of the power. The two fifths of the input port power are divided by an equal power divider to a fifth of the input port power. The power at each output port is a fifth of the input power. The power divider's insertion loss is around 1.5 dB. The power divider's S_{11} parameter is better than −10 dB in the frequency range of 6 GHz to 18 GHz.

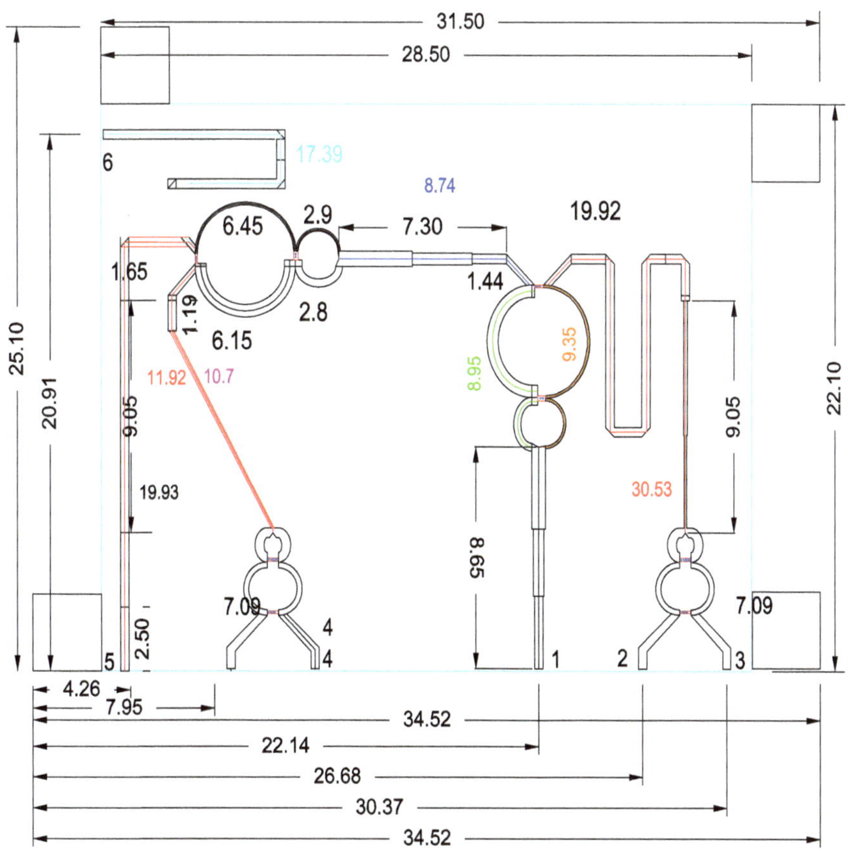

Figure 5.18. Wideband five-way unequal power divider.

5.4 RF amplifiers

Amplifiers are specified by their gain, frequency, efficiency, linearity, power and size. Power amplifiers are designed to transmit high power values. However, low noise amplifiers (LNAs) are designed to get low noise figure values. Gain amplifiers are designed to get higher gain values. A typical transceiver block diagram is shown in figure 5.19. The transceiver consists of a power amplifier in the transmitting channel, a LNA in the receiving channel and IF amplifiers.

There are several definitions to amplifier gain. Power gain is the ratio of power dissipated in the load, Z_L, to the power delivered to the input of the amplifier. Available gain is the ratio of the amplifier output power to the available power from the generator source, it depends on Z_G but it is not dependent on Z_L. Exchangeable gain is the ratio of the output exchangeable power to the input exchangeable power as written in equation (5.11). Insertion gain is the ratio of the output power that would be dissipated in the load if the amplifier was not connected. There is a problem in applying this definition to mixers of parametric up-converters where the input and output frequencies are different.

$$P = \frac{|V|^2}{4\Re\{Z_G\}} \quad \Re\{Z_G\} \neq 0 \tag{5.11}$$

Transducer gain is the ratio of the power delivered to the load to the available power from the source. A two port equivalent network is presented in figure 5.20. This definition depends on both Z_G and Z_L. It gives positive gain for negative resistance amplifiers as well. Since the characteristics of real amplifiers' impedance change when either the load or generator impedance is changed, thus the transducer power gain definition is found to be the most useful and is given in equation (5.12).

Load impedance is $\rightarrow V_2 = -I_2 Z_L$

Input impedance is $\rightarrow Z_{in} = \frac{V_1}{I_1} = z_{11} - \frac{z_{12}z_{21}}{z_{22} + Z_L}$

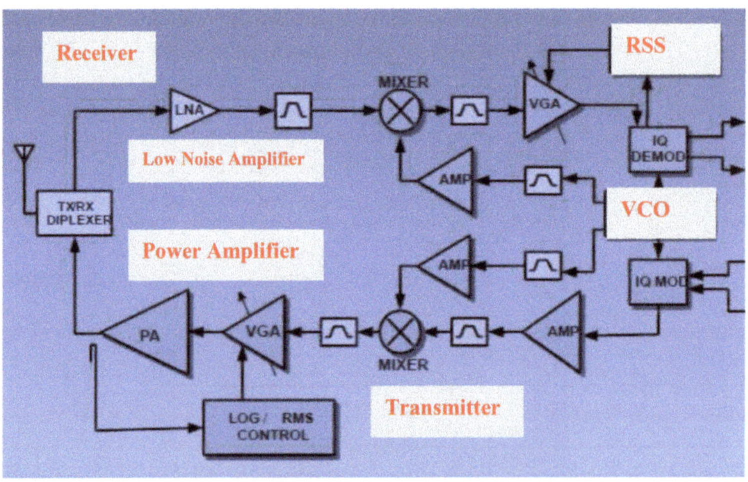

Figure 5.19. Typical transceiver block diagram.

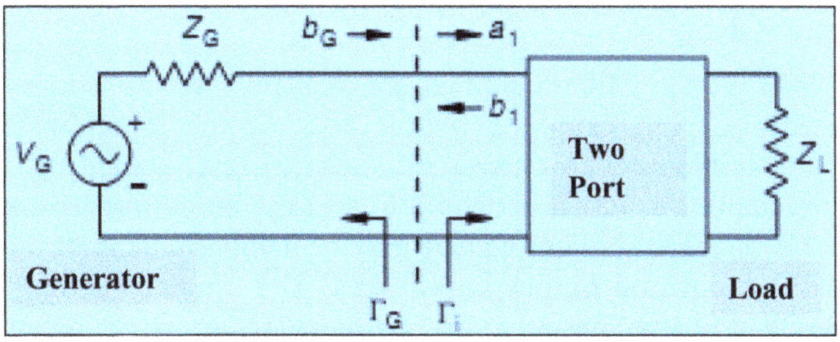

Figure 5.20. Two port equivalent network.

Power delivered to the load is $\rightarrow P_2 = \frac{1}{2}\,|I_2|^2\Re\{Z_L\}$

Power available from the source is $\rightarrow P_{1a} = \frac{|V_G|^2}{8\Re\{Z_G\}}$

$$\underbrace{G_T}_{\substack{\text{Transducer}\\ \text{Power}\\ \text{Gain}}} = \frac{P_2}{P_{1a}} = \frac{4\Re\{Z_L\}\Re\{Z_G\}|\,z_{21}\,|^2}{|(Z_G + z_{11})(Z_L + z_{22}) - z_{21}z_{21}\,|^2} \tag{5.12}$$

The available power, P_a, is written in equation (5.13).

$$P_a = \frac{\frac{1}{2}\,|\,b_G\,|^2}{1 - |\,\Gamma_G\,|^2} \tag{5.13}$$

The power delivered to the load is written in equation (5.14).

$$P_L = \frac{1}{2}\,|\,b_2\,|\left(1 - |\,\Gamma_L\,|^2\right) \tag{5.14}$$

The transducer gain is written in equation (5.15). In equation (5.16) the ratio b_2/b_G is written. The transducer gain G_T is written in equation (5.17).

$$G_T = \frac{|\,b_2\,|^2}{|\,b_G\,|^2}\left(1 - |\,\Gamma_L\,|^2\right)\left(1 - |\,\Gamma_G\,|^2\right) \tag{5.15}$$

$$b_2 = \frac{\frac{b_G}{1 - S_{11}\Gamma_G}}{1 - \frac{S_{21}S_{12}\Gamma_G\Gamma_L}{(1 - S_{11}\Gamma_G)(1 - S_{22}\Gamma_L)}}\frac{S_{21}}{(1 - S_{22}\Gamma_L)} \tag{5.16}$$

$$G_T = \frac{|S_{21}|^2(1 - |\Gamma_G|^2)(1 - |\Gamma_L|^2)}{|(1 - S_{11}\Gamma_G)(1 - S_{22}\Gamma_L) - \Gamma_G\Gamma_L S_{21}S_{12}|^2} \tag{5.17}$$

5.4.1 Amplifier stability

A stable amplifier is an amplifier that does not oscillate. Oscillations may occur when there is a feedback path from the output back to the input or under certain environmental conditions. The criteria for unconditional stability is required for any passive terminating loads. It is helpful to find the borderline between the stable and the unstable regions. The Smith chart may be helpful in defining stability circles for the amplifier input and output impedances. The center and the radius of the stability circles for the amplifier input and output ports are given in equations (5.18) and (5.19).

$$C_L = \frac{\Delta S_{11} - S_{22}}{|\Delta|^2 - |S_{22}|^2}\text{(Load Center) } C_G = \frac{\Delta S_{22} - S_{11}}{|\Delta|^2 - |S_{22}|^2}\text{(Generator Center)} \quad (5.18)$$

$$r_L = \left|\frac{S_{21}S_{12}}{|\Delta|^2 - |S_{22}|^2}\right|\text{(Stability radius) } r_G = \left|\frac{S_{21}S_{12}}{|\Delta|^2 - |S_{11}|^2}\right| \quad (5.19)$$

where $|\Delta| = |S_{11}S_{22} - S_{12}S_{21}|$.

These two stability circles represent the borderline between stability and instability and they are drawn on a Smith chart. Unconditional stability requires that both $|\Gamma_i| < 1$ and $|\Gamma_o| < 1$ for any passive load impedance and generator impedance connected to the ports.

The Rollett criteria determines if a given transistor is unconditionally stable for any passive load impedance and generator impedance connected to the ports.

There are two conditions written in equations (5.20) and (5.21).

$$k = \frac{1 - |S_{11}|^2 - |S_{22}|^2 + |\Delta|^2}{2|S_{12}S_{21}|} \geqslant 1 \quad (5.20)$$

$$|\Delta| \leqslant 1$$
where
$$\{|\Delta| = |S_{11}S_{22} - S_{12}S_{21}| < |S_{11}S_{22}| + |S_{12}S_{21}|\} \quad (5.21)$$

5.4.2 Stabilizing a transistor amplifier

There are a variety of approaches to stabilize an amplifier. We can stabilize the amplifier by choosing the amplifier terminating impedances inside the stable regions at all frequencies. We can also load the amplifier with an additional shunt or series resistor on either the generator or load ports. However, it is better to load the output port to optimize the amplifier NF. Another approach is to introduce an external feedback path to minimize the internal feedback of the transistor.

5.4.3 Class A amplifiers

Class A amplifiers amplify the incoming signal in a linear fashion. A linear class A amplifier will introduce low harmonic frequency components and low (IMD) intermodulation distortion. The amplifier output voltage is given by equation (5.22).

$$\frac{V}{2}\cos(\omega_c + \omega_m)t + \frac{V}{2}\cos(\omega_c - \omega_m)t \quad (5.22)$$

where ω_c is the high-frequency carrier frequency. Where ω_m is the low-frequency modulation. Frequency intermodulation distortion would result in frequencies at $\omega_c \pm n\omega_m$. Harmonic distortion would cause frequency generation at $k\omega_c + n\omega_m$. Reduction of IMD depends on efficient power combining methods and careful design of the amplifier. An example of a common emitter class A amplifier is presented in figure 5.21. The amplifier currents and output voltage are given in equations (5.23) and (5.24).

$$i_c = i_{dc} - i_O \sin \omega t \quad (5.23)$$

where $i_{dc} = i_Q = i_{outmax}$
i_Q—quiescent current.

$$V_{ce} = V_{cc} + V_0 \sin \omega t \quad (5.24)$$

where $V_{outmax} = i_{outmax}{}^{*}R_L = V_{cc}$.

The maximum efficiency of the class A amplifier shown in figure 5.21 is around 25%. By replacing the collector resistor by an inductor the efficiency may be improved up to 50%.

5.4.4 Class B amplifier

The class B amplifier is biased so that the transistor is on only during half of the incoming cycle. The other half of the cycle is amplified by another transistor so that at the output the full wave is reconstituted. While each transistor is clearly operating in a nonlinear mode, the total input wave is directly replicated at the output. The class B amplifier is therefore classed as a linear amplifier. In this case the bias current, IC = 0.

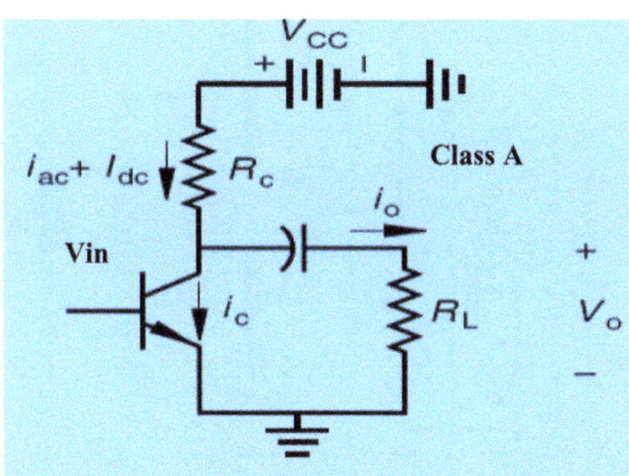

Figure 5.21. Common emitter class A amplifier.

Since only one of the transistors is cut off when the total voltage is less than 0, only the positive half of the wave is amplified. The conduction angle is 180°.

Figure 5.22 shows a complementary type of class B amplifier. In this case, transistor Q1 is biased so that it is in active mode when the input voltage, $V_{IN} > 0.7$, and cut off when the input signal $V_{IN} < 0.7$. The other half of the signal is amplified by transistor Q2 when $V_{IN} < 0.7$. When no input signal is present, no power is drawn from the bias supply through the collectors of Q1 or Q2, so the class B operation is attractive when low standby power consumption is an important consideration. There is a small region of the input signal for which neither Q1 nor Q2 is on. The amplifier output will suffer some distortion. One of the primary problems in using this type of class B amplifier is the requirement for obtaining two equivalent complementary transistors. The problem fundamentally arises because of the greater mobility of electrons by over a 3:1 factor over that of holes in silicon. The symmetry of the gain in this circuit depends on the two output transistors having the same short circuit base to collector current gain, $\beta = ic/ib$. When it is not possible to obtain a high β pnp transistor, it is sometimes possible to use a composite transistor connection. A high power npn transistor, Q1, is connected to a low-power low β pnp transistor, Q2.

The maximum efficiency of a class B amplifier is found by calculating the ratio of the output power delivered to the load to the required dc power from the bias voltage supply. By calculating efficiency in this way, power losses caused by nonzero base currents and crossover distortion compensation circuits are neglected. Furthermore, the power efficiency rather than the power added efficiency is calculated so as to form a basis for comparison for alternative circuits. It is sufficient to do the calculation during part of the cycle when Q1 is on and Q2 is off. The load resistance is R_L.

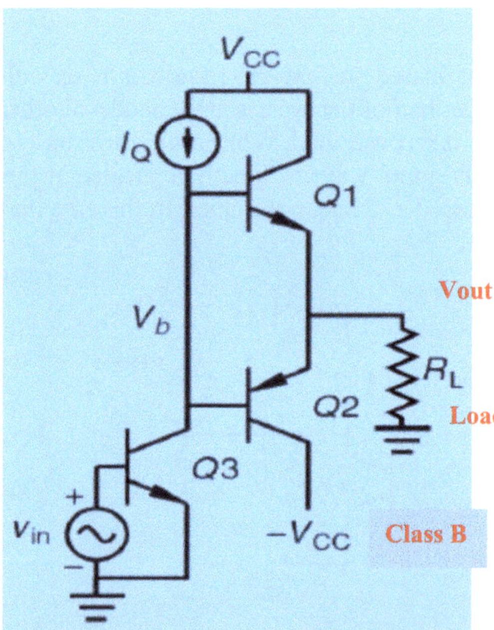

Figure 5.22. Class B amplifier.

The maximum output power occurs when the output voltage is $V_{cc} - V_{ce(sat)}$ and is given in equation (5.25).

$$P_{L(max)} = 0.5\frac{(V_{cc} - V_{ce(sat)})^2}{R_L} \tag{5.25}$$

$$\eta_{max} = \frac{\pi}{4}\frac{V_{cc} - V_{ce(sat)}}{V_{cc}} \approx 78.5\% \tag{5.26}$$

The efficiency for the class B amplifier is around 78.5%, given in equation (5.26), and should be compared with the maximum efficiency of a class A amplifier, where η_{max} is 25% when the bias to the collector is supplied through a resistor and η_{max} is 50% when the bias to the collector is supplied through an RF choke.

5.4.5 Class C amplifier

The class C amplifier is useful for providing a high power continuous wave (CW) or frequency modulation (FM) output. When it is used in amplitude modulation schemes, the output variation is done by varying the bias supply. There are several characteristics that distinguish the class C amplifier from the class A or B amplifiers. The class C amplifier is biased so that the transistor conduction angle is <180°. Consequently, the class C amplifier is clearly nonlinear in that it does not directly replicate the input signal like the class A and B amplifiers. The class A amplifier requires one transistor, the class B amplifier requires two transistors, and the class C amplifier uses one transistor. Topologically it looks similar to class A except for the dc bias levels. It was noted that in the class B amplifier, an output filter is used optionally to help clean up the output signal. In the class C amplifier, such a tuned output is necessary in order to recover the sine wave. Finally, a class C operation is capable of higher efficiency than class A and class B. Class C amplifiers, for the appropriate signal types, are very attractive as power amplifiers.

5.4.6 Class D amplifier

In a class D operation the transistors act as near ideal switches that are on half of the time and off half of the time. The input ideally is excited by a square wave. If the transistor switching time is near zero, then the maximum drain current flows while the drain–source voltage, $V_{ds} = 0$. As a result, 100% efficiency is theoretically possible. In practice, the switching speed of a bipolar transistor is not sufficiently fast for square waves above a few MHz, and the switching speed for field effect transistors is not adequate for frequencies above a few tens of megahertz.

5.4.7 Class F amplifier

A class F amplifier is characterized by a load network that has resonances at one or more harmonic frequencies as well as at the carrier frequency. The class F amplifier was one of the early methods used to increase amplifier efficiency and has recently attracted some renewed interest. When the transistor is excited by a sinusoidal source, it is on for approximately half the period of time and off for half of the period of time.

5.4.8 Feedforward amplifiers

The feedback approach is an attempt to correct an error after it has occurred. A 180° phase difference in the forward and reverse paths in a feedback system can cause unwanted oscillations. In contrast, the feedforward design is based on a cancellation of amplifier errors in the same time frame in which they occur. Signals are handled by wideband analog circuits, so multiple carriers in a signal can be controlled simultaneously. Feedforward amplifiers are inherently stable, but this comes at the price of a somewhat more complicated circuit. Consequently, feedforward circuitry is sensitive to changes in ambient temperature, input power level, and supply voltage variation. Nevertheless, feedforward offers many advantages that have brought it increased interest. The major source of distortion, such as harmonics, intermodulation distortion, and noise, in a transmitter is the power amplifier. This distortion can be greatly reduced using a feedforward design.

5.5 Linearity of RF amplifiers, active devices

Any continuous analytical function may described as given in equation (5.27).

$$V_O = \sum_{i=1}^{\infty} K_i V_{IN}^i \tag{5.27}$$

In the linear zone the function V_O may be approximated by the first three elements of the Taylor series as written in equation (5.28).

$$V_O \approx K_1 V_{IN} + K_2 V_{IN}^2 + K_3 V_{IN}^3 \tag{5.28}$$

For a sinusoidal signal the second order element in equation (5.28) influences the DC and the second harmonic at the output. In this analysis we assume that the magnitude of the input signal does not change the component parameters. K_1 is the first order parameter and usually represents gain or attenuation. K_2 is the second order parameter (figure 5.23).

K_3 is the third order parameter. For a narrow band frequency range V_O may be written as in equation (5.29).

$$V_{O\text{-NarrowBand}} \approx K_1 V_{IN} + K_3 V_{IN}^3 \tag{5.29}$$

In active RF components when the input power is increased we get a deviation from the linearity of the output signal. If the input power is increased by X (dB) the output power is increased by Y (dB). Where $Y < X$. The third order component causes a deviation from linearity and compresses the output signal. Compression is obtained if $K_1 K_3 < 0$.

For a narrow band frequency range V_O may be written as in equation (5.30).

$$| V_O | \approx | K_1 | \, V_{IN} - | K_3 | \, V_{IN}^3 \tag{5.30}$$

After normalization of equation (5.30) by K_1, V_O may be written as in equation (5.31).

$$\left| \hat{V}_O \right| \approx V_{IN} - K \, V_{IN}^3 \qquad \text{where } K = \frac{K_3}{K_1} \tag{5.31}$$

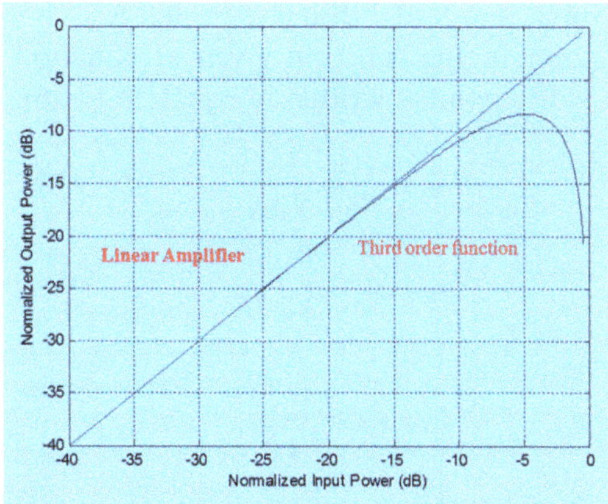

Figure 5.23. Third order function versus linear function.

The output power of a linear device with gain of $G_{[dB]}$ is given in equation (5.32).

$$\begin{aligned} P_{O-Lin[dBm]} &= P_{IN[dBm]} + G_{[dB]} \\ G_{[dB]} &= 20\log(K_1) \end{aligned} \tag{5.32}$$

However, in an active device the output power of the device with gain of $G_{[dB]}$ is given in equation (5.33).

$$\begin{aligned} P_{out[dBm]} &= P_{O-Lin[dBm]} + 20\log(1 - K\, P_{in[W]}) \\ &\text{where } K = \left| \frac{K_3}{K_1} \right| \end{aligned} \tag{5.33}$$

The compression may be written as in equation (5.34).

$$Comp_{[dB]} = 20\log(1 - L\, p_{in[W]}) \tag{5.34}$$

Example

The amplifier's small signal gain is 12 dB. The output 1 dBc power is 16 dBm. What is the input power at the 1 dBc point?

Solution

The amplifier gain at the compression point is 11 dB.

$$P_{in} = P_{out} - G = +5 \text{ dBm}$$

$$\begin{aligned} At \quad & 1 \text{ dB } Comp\ 0.89125 \approx 1 - K\, p_{in-Comp[W]} \\ & Comp_{[dB]} \approx -20\log(1 - 0.109d) \\ & \text{where } d = 10^{\Delta/10} \\ & \text{and } \Delta = P_{in[dBm]} - P_{in-Comp[dBm]} \end{aligned}$$

where $P_{in\text{-}Comp}$ is the input power at 1 dBc power, and $\Delta_{[dB]}$ is called the input back-off. The maximum point of the function given at equation (5.30) is written in equation (5.35). The 1 dBc point is written in equation (5.36).

$$V_{IN-MAX} = \sqrt{\frac{1}{3K}}$$
$$V_{O-MAX} = \frac{2}{3}\sqrt{\frac{1}{3K}} \tag{5.35}$$

$$V_{IN-CP} \approx \frac{0.33}{\sqrt{K}}$$
$$V_{O-CP} \approx \frac{0.29}{\sqrt{K}} \tag{5.36}$$

From equations (5.35) and (5.36) we may conclude that for the maximum saturated output power the amplifier is 3.52 dB compressed, relative to the linear line. The maximum saturated output power the amplifier is 2.34 dB from 1 dBc output power. We have to increase the input power by 4.87 dB from the 1 dBc input power to reach to the maximum saturated output power as shown in figure 5.24.

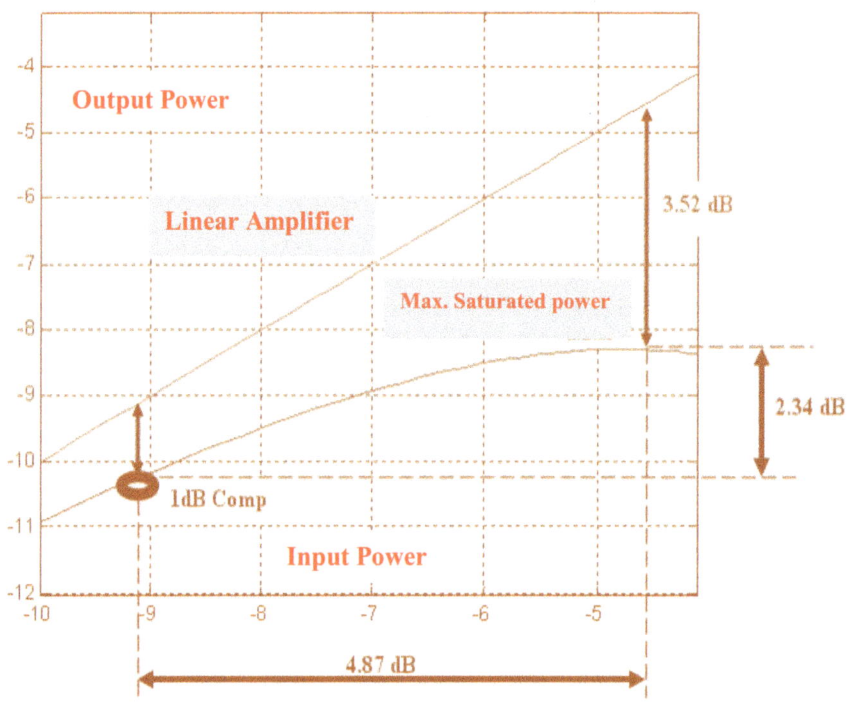

Figure 5.24. Saturation region in amplifiers.

5.5.1 Third order model for a single tone

A third order model for a single tone was presented in equation (5.28). The input signal is presented by equation (5.37). The output signal consists of a V_{dc} component and the fundamental signal V_{fund} and two harmonics h_2 and h_3 as written in equation (5.38).

$$V_{IN} = a\cos(\omega_0 t) \tag{5.37}$$

$$V_O = V_{dc} + V_{fund} + h_2 + h_3 \tag{5.38}$$

where the output signals are written in equations (5.38)–(5.40).

$$V_{dc} = \frac{1}{2}K_2 a^2 \tag{5.39}$$

$$V_{fund} = \left(K_1 + \frac{3}{4}K_3 a^2\right)V_{IN} \tag{5.40}$$

$$h_2 = \frac{1}{2}K_2 a^2 \cos(2\omega_0 t) \tag{5.41}$$

$$h_3 = \frac{1}{4}K_3 a^3 \cos(3\omega_0 t) \tag{5.42}$$

Figure 5.25 presents the output spectrum for a single tone.
The voltage gain is given in equation (5.43). The gain $G_{[dB]}$ is written in equation (5.44).

$$\sqrt{g} = \left|\frac{V_{fund}}{V_{IN}}\right| = |K_1|\left(1 - \frac{3}{4}K\,a^2\right) \quad \text{Where } K = \left|\frac{K_3}{K_1}\right| \tag{5.43}$$

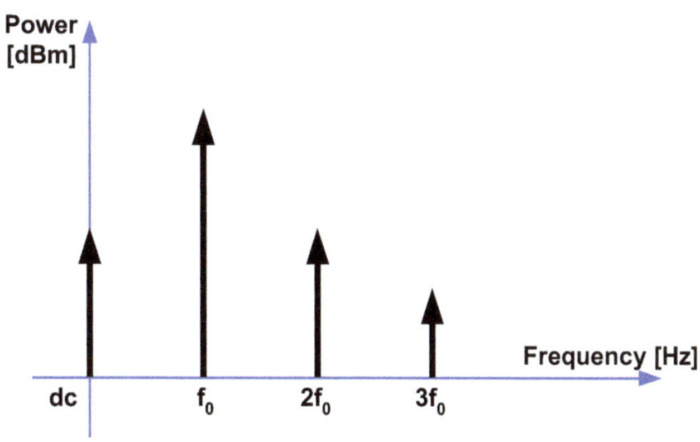

Figure 5.25. Output spectrum for a single tone.

$$G_{[dB]} = 10 \log \mid g \mid \tag{5.44}$$

The small signal gain (SSG) is written in equation (5.45). The gain G_{dB} is written in equation (5.46).

$$SSG_{[dB]} = 20 \log \mid K_1 \mid \tag{5.45}$$

$$\begin{aligned} G_{dB} &= SSG - Comp \\ Comp &= 20 \log\left(1 - \frac{3}{4}K\, a^2\right) \end{aligned} \tag{5.46}$$

where$P_{in[W]} = \dfrac{a^2}{2}$.

The gain $G_{[dB]}$ may be written as in equation (5.47).

$$\begin{aligned} G_{[dB]} &= SSG_{[dB]} - Comp_{[dB]} \\ Comp_{[dB]} &= 20 \log\left(1 - \frac{3}{2}K\, p_{in[W]}\right) \end{aligned} \tag{5.47}$$

5.5.2 Third order model for two tones

For two tones for the input signal is written as $V_{IN} = A + B$. For two tones the output signal is given by equation (5.48).

$$\begin{aligned} V_O &= V_{out-a} + V_{out-b} + V_{im2} + V_{im3} \\ V_{out-a} &= K_1 A + K_2 A^2 + K_3 A^3 \\ V_{out-b} &= K_1 B + K_2 B^2 + K_3 B^3 \\ V_{im2} &= 2K_2\, A\, B \\ V_{im3} &= 3K_3\, A^2 B + 3K_3\, A\, B^2 \end{aligned} \tag{5.48}$$

where IM_2 is the 2nd order intermodulation and IM_3 is the 3nd order intermodulation.

If the magnitude level of the two input tones is equal, the ratio between the voltage level of the second order intermodulation and the voltage of the second harmonic is two. The power level of the second order intermodulation is greater by 6 dB from the power level of the second harmonic. If the magnitude level of the two input tones is equal, the ratio between the voltage level of the third order intermodulation and the voltage of the third harmonic is three. The power level of the third order intermodulation is greater by 5.54 dB from the power level of the third harmonic as given in equation (5.49).

$$\begin{aligned} IM_{2[dBm]} &= h_{2[dBm]} + 6\ \text{dB} \\ IM_{3[dBm]} &= h_{3[dBm]} + 9.54\ \text{dB} \end{aligned} \tag{5.49}$$

For the input signals given in equation (5.50) the output frequency spectrum is shown in figure 5.26.

$$
\begin{aligned}
V_{IN} &= A + B \\
A &= E\cos(\omega_1 t) \\
B &= E\cos(\omega_2 t)
\end{aligned}
\tag{5.50}
$$

For a decrease of $\Delta_{[dB]}$ in the signals A and B the level of the second intermodulation signal will decrease by $2\Delta_{[dB]}$ as given in equation (5.51). The intermodulation signal IM_2 is written in equation (5.52). The second order intermodulation calculation is shown in figure 5.27.

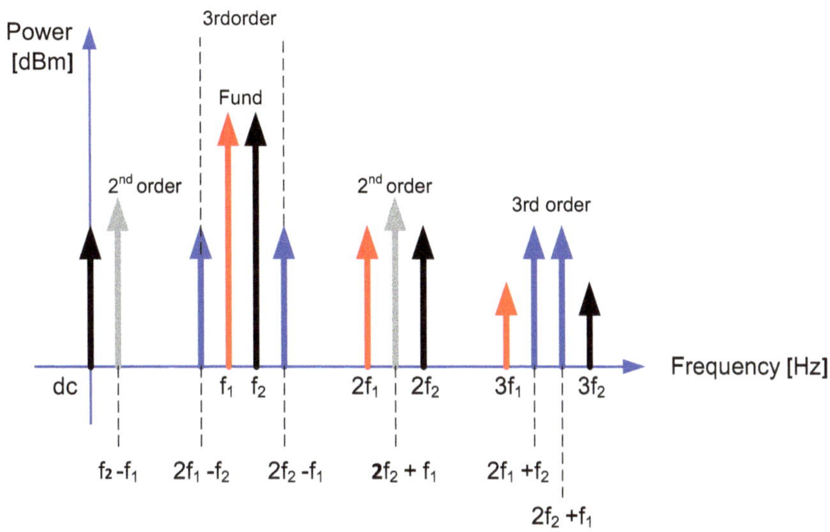

Figure 5.26. Two tones output spectrum.

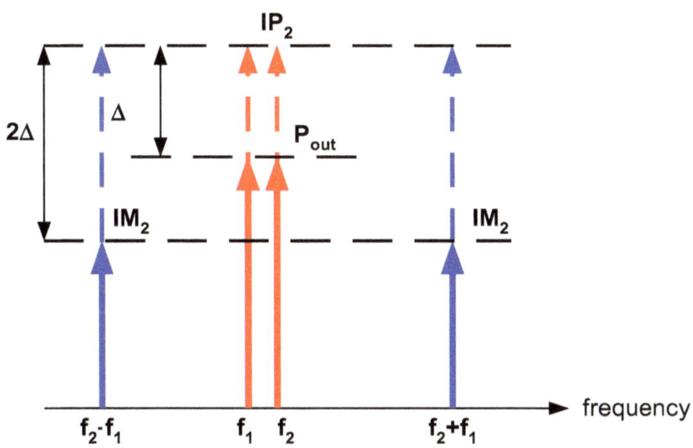

Figure 5.27. Second order intermodulation.

$$IP_{2[dBm]} - P_{out[dBm]} = \Delta_{[dB]}$$
$$IP_{2[dBm]} - IM_{2[dBm]} = 2\Delta_{[dB]} \tag{5.51}$$

$$IM_{2[dBm]} = 2P_{out[dBm]} - IP_{2[dBm]} \tag{5.52}$$

5.5.3 Third order intercept point

For a decrease of $\Delta_{[dB]}$ in the signals A and B, the level of the third intermodulation signals will decrease by $3\Delta_{[dB]}$ as given in equation (5.53). The intermodulation signal IM_3 is written in equation (5.54). The third order intermodulation calculation is shown in figure 5.28.

$$IP_{3[dBm]} - P_{out[dBm]} = \Delta_{[dB]}$$
$$IP_{3[dBm]} - IM_{3[dBm]} = 2\Delta_{[dB]} \tag{5.53}$$

$$IM_{3[dBm]} = 3P_{out[dBm]} - 2IP_{3[dBm]} \tag{5.54}$$

5.5.4 IP_2 and IP_3 measurements

The setup for the IP_2 and IP_3 measurements is shown in figure 5.29. Second order intermodulation results are shown in figure. IP_2 may be computed by equation (5.55).

$$IP_{2[dBm]} = P_{out[dBm]} + \Delta_{[dB]} \tag{5.55}$$

Second order intermodulation results are shown in figure 5.30.

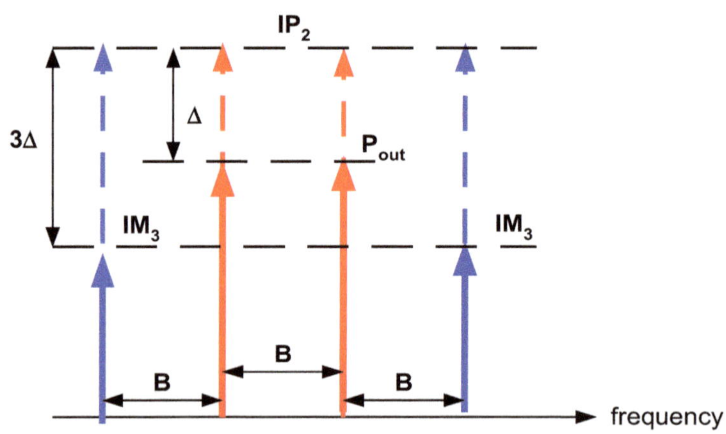

Figure 5.28. Third order intermodulation.

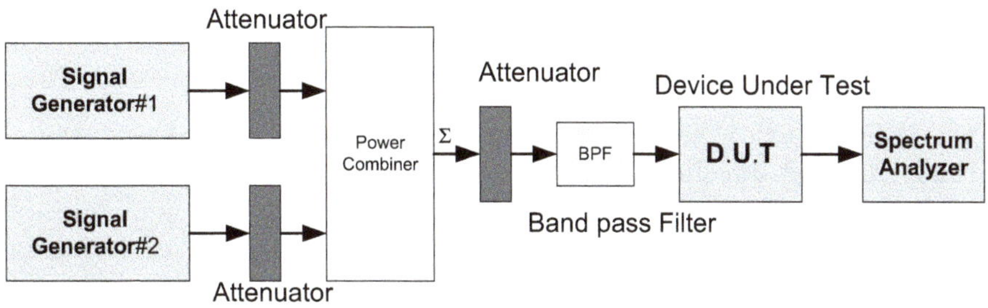

Figure 5.29. Setup for IP_2 and IP_3 measurements.

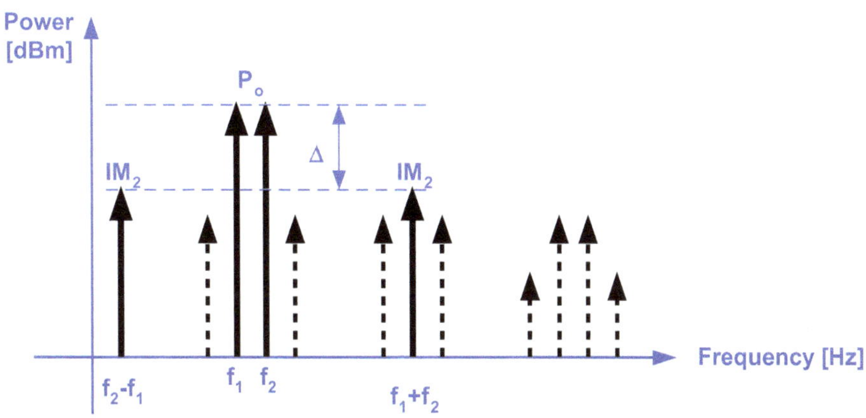

Figure 5.30. Second order intermodulation results.

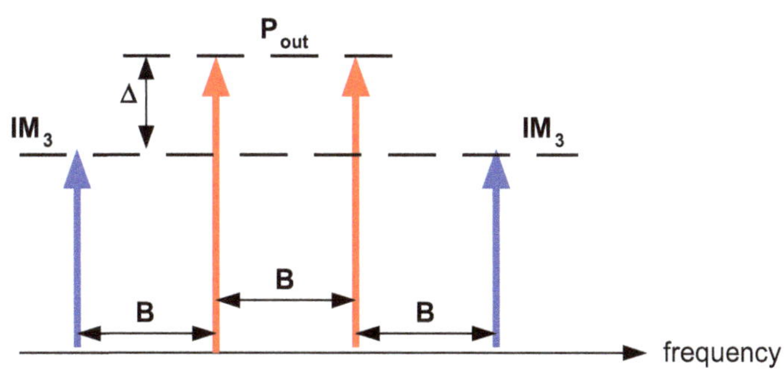

Figure 5.31. Third order intermodulation results.

Third order intermodulation results are shown in figure 5.31. IP_3 may be computed by equation (5.56).

$$IP_{3[dBm]} = P_{out[dBm]} + \frac{\Delta_{[dB]}}{2} \tag{5.56}$$

5.6 Wideband phased array direction finding system

5.6.1 Introduction

The wideband direction finding system consists of an active phased array and a wideband receiving direction finding system. A block diagram of the system is shown in figure 5.32. The system consists of eleven receiving and transmitting modules for vertical scanning. The received signals in the six receiving antennas are amplified by LNA amplifiers and down-converted to 70 MHz IF frequency. Input

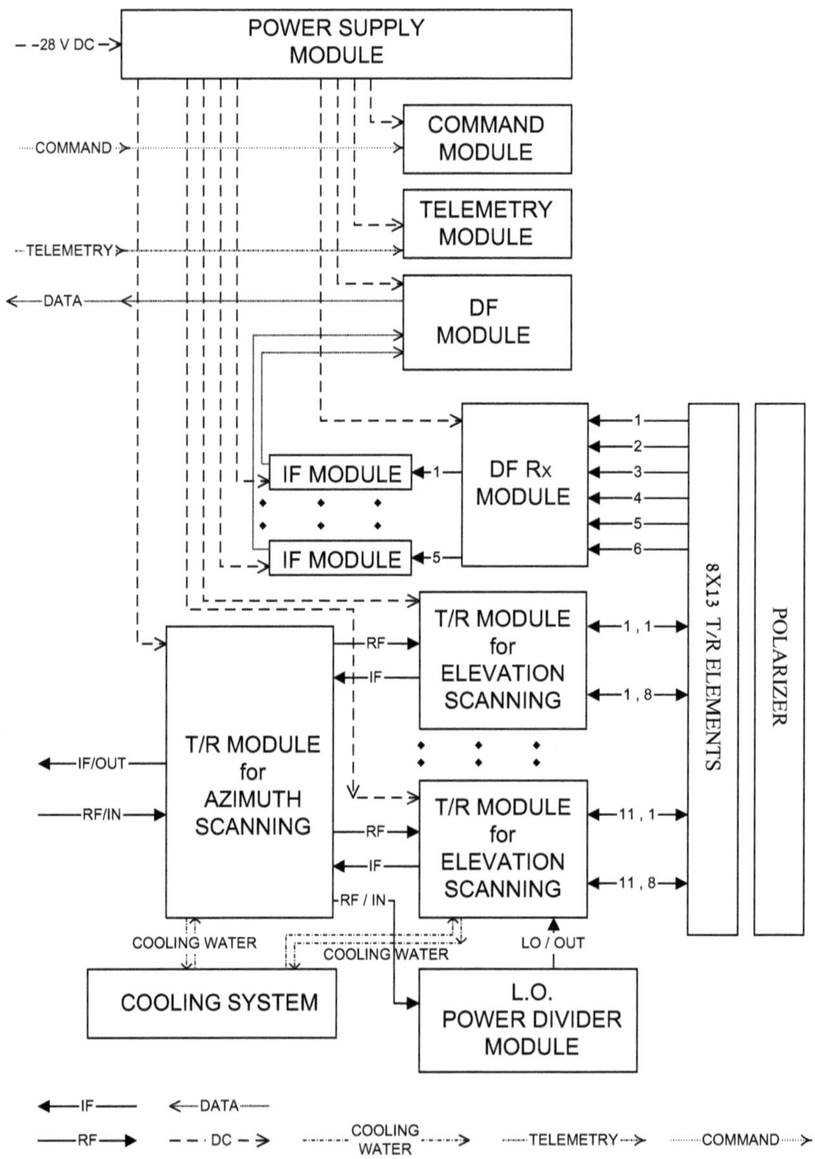

Figure 5.32. Wideband phased array direction finding system block diagram.

port number six provides the local oscillator (LO) signal to the module. The LO signal is amplified by a LNA amplifier and is routed to the single side band (SB) unit. A 70 MHz voltage controlled oscillator supplies a RF signal to the SSB unit. The LO signal is amplified to 10 dBm. A wideband six-way power divider supplies the LO signals to the mixers. The antennas are wideband notch antennas, 6 GHz to 18 GHz, with a VSWR better than 3:1.

5.6.2 Wideband receiving direction finding system

The system consists of eleven receiving identical modules for vertical scanning. The receiving channel module consists of LNA amplifiers and down-converter mixers to 70 MHz IF frequency. The received signals in the six receiving ports are amplified by LNA amplifiers and down-converted to 70 MHz IF frequency. Input port number six provides the LO signal to the module. The LO signal is amplified by a LNA amplifier and is routed to the SSB unit. A 70 MHz voltage controlled oscillator supplies a RF signal to the SSB unit. The LO signal is amplified to 10 dBm. A wideband six-way power divider supplies the LO signals to the mixers. The module dimensions are 279 × 130 × 15.5 mm. The module hot spots are cooled by a cold water cooling system that is located on the back side of the module.

Module specifications

The wideband receiving direction finding system specifications are listed in table 5.1.

Table 5.1. Wideband receiving direction finding system specifications.

Parameter	Specifications
RF frequency	6–18 GHz
IF frequency	70 MHz
Gain	4 dB
Gain flatness	±2 dB
Phase flatness	±6°
Noise figure	7.5 dB
Output power	0 dBm
Maximum input RF power	−7 dBm
Maximum input LO power	−7 dBm
60 MHz input RF power	10 dBm
IF ports	5
VSWR	3:1
Impedance	50 Ω
Dimensions	279 × 130 × 15.5 mm
Temperature	71 °C–54 °C
Weight	700 Gr
DC power	20 W

5.6.3 Receiving channel design

The wideband receiving direction finding system block diagram is shown in figure 5.33.

The RF components are located on the upper side of the module.

The received signals in the six receiving ports are amplified by LNA amplifiers and down-converted to 70 MHz IF frequency. The LNA amplifiers are packed MMICS with 1 dB noise figure. The mixers are compact double balance mixers with 7 dB insertion loss. Input port number six provides the LO signal to the module. The LO signal is amplified by a LNA amplifier and is routed to the SSB unit. A 70 MHz voltage controlled oscillator supplies a RF signal to the SSB unit. The LO signal is amplified to 10 dBm. A wideband six-way power divider supplies the LO signals to the mixers. The IF and DC components are located on the back side of the module.

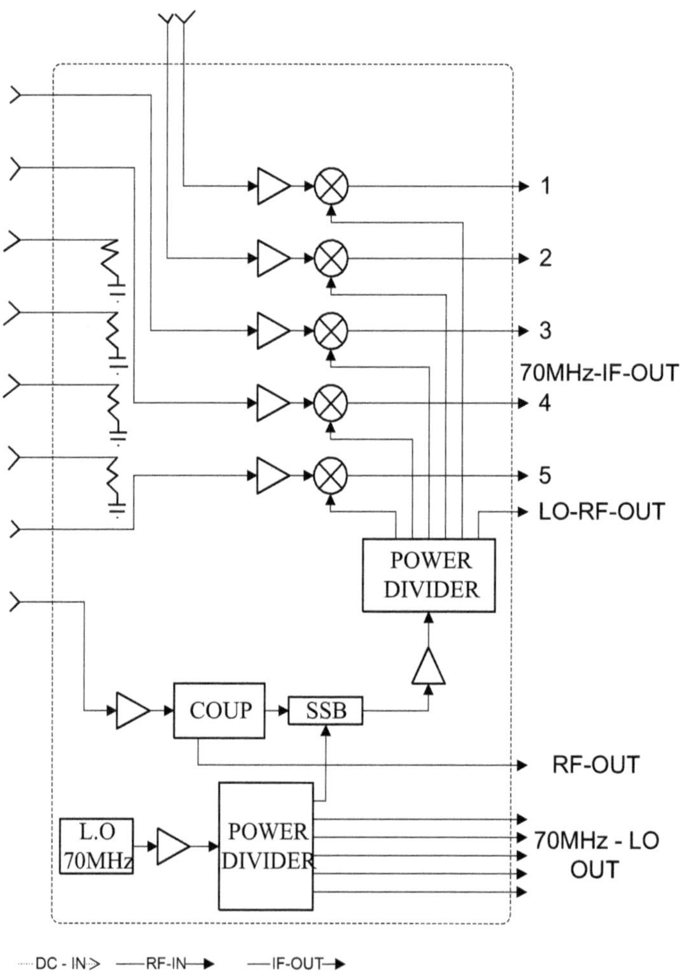

Figure 5.33. Wideband receiving direction finding system block diagram.

Figure 5.34 presents the receiving direction finding system's analysis results. The analysis results prove that the module specifications may be satisfied.

Cooling system

The cooling system is located on the back side of the amplifier. The cooling system consists of heat sink pipe lines. The cooling liquid flows via the heat sink pipe lines and return to the system through a cooling return connector.

DC unit

A 24-pins DC connector supplies the DC voltage to the receiving system. The connector consists of 17 DC pins and 7 RF connectors. A list of the DC connector pins and DC power consumption is given in table 5.2.

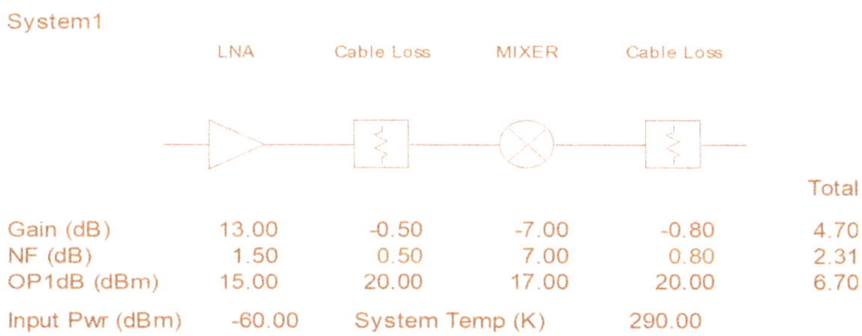

Figure 5.34. Wideband receiving direction finding system analysis design.

Table 5.2. DC connector.

Pin	Function	Voltage	Current
1	VCO	15 V	0.9 A
2	LNA	6 V	0.1 A
3	LNA	6 V	0.1 A
4	LNA	6 V	0.1 A
5	IF AMP	15 V	0.1 A
6	Limiter	−15 V	
7		NC	
8		NC	
9		NC	
10		NC	
11	Limiter	15 V	
12		NC	
13		GND	
14		NC	
15		NC	
16		NC	
17		NC	

5.6.4 Measured results of the receiving channel

Measured results versus requirements of the receiving module are listed in table 5.3.

A photo of the wideband receiving direction finding system is shown in figure 5.35.

A photo of the wideband receiving module with the module cover is shown in figure 5.36. Absorbing material is glued on the cover to improve isolation between the module components. A photo of the wideband receiving direction finding system's interface and connectors is shown in figure 5.37.

The RF module dimensions are 279 × 130 × 15.5 mm. The compact dimensions were achieved by using a compact packed LNA and mixers. However, a much more compact module may be designed by combining LNAs, mixers and a power divider on the same MMIC module. The electrical performance may degrade but a MMIC module can be designed to meet the electrical specifications. The cost of a MMIC module in mass production will be cheaper than the presented module in figure 5.35.

5.7 Conclusions

The design of MIC and MMIC components and modules was presented in this chapter.

In the design of RF modules we may gain the advantages of each technology by using both in the development of RF transmitting and receiving modules. For

Table 5.3. Measured results versus requirements of the receiving module.

Parameter	Specifications	Measured results
RF frequency	6–18 GHz	6–18 GHz
IF frequency	70 MHz	70 MHz
Gain	4 dB	4 dB
Gain flatness	±2 dB	±2 dB
Phase flatness	±6°	±6°
Noise figure	7.5 dB	7.5 dB
Output power	0 dBm	0 dBm
Maximum input RF power	−7 dBm	−7 dBm
Maximum input LO power	−7 dBm	−7 dBm
60 MHz input RF power	10 dBm	10 dBm
1 dBc Compression point	−10 dBm	−9 dBm
IP_3	4 dBm	4 dBm
Spurious level	−40 dB	−45 dB
VSWR	3:1	3:1
Impedance	50 Ω	50 Ω
Dimensions	279 × 130 × 15.5 mm	279 × 130 × 15.5 mm
Temperature	71 °C–54 °C	71 °C–54 °C
Weight	700 Gr	700 Gr
DC power	20 W	20 W

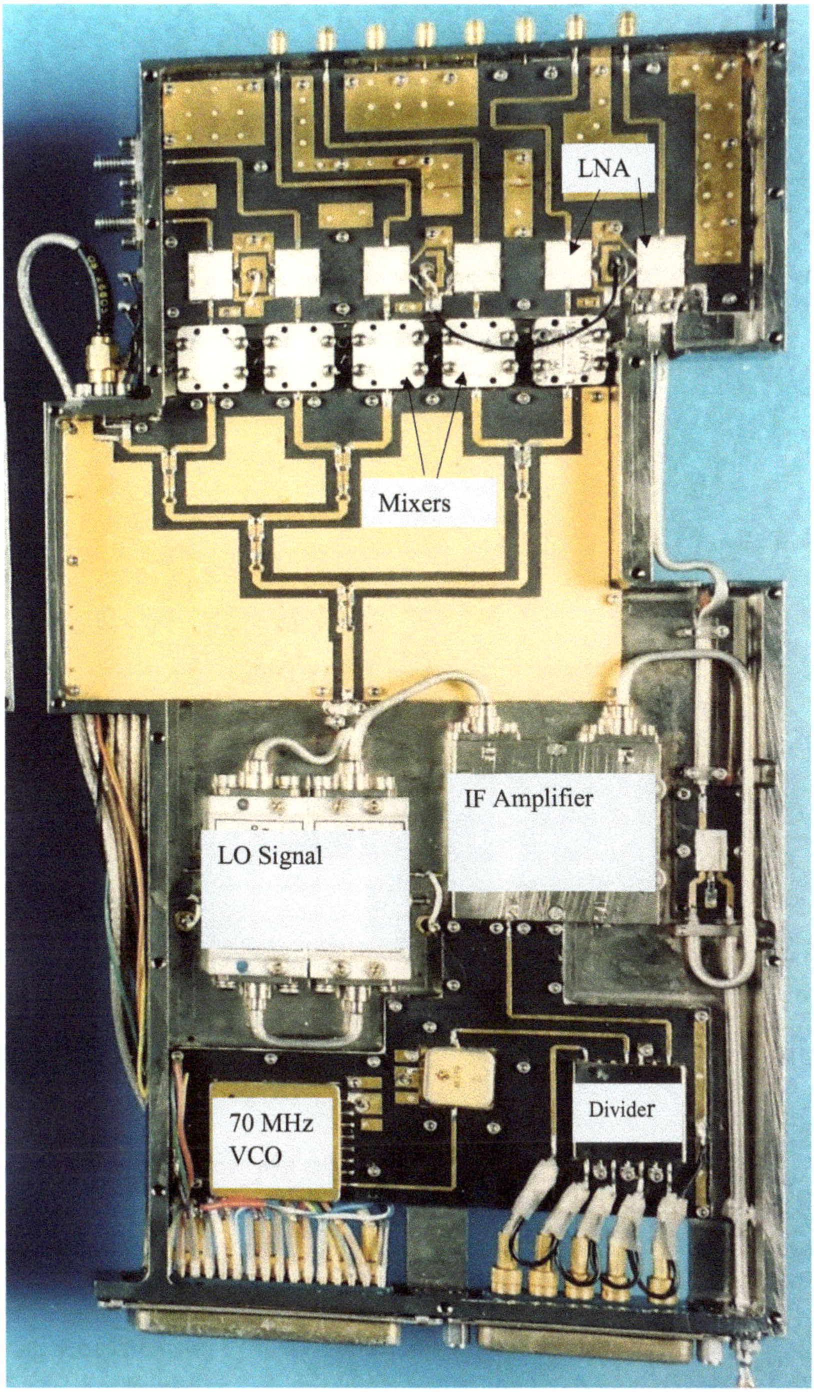

Figure 5.35. Photo of the wideband receiving direction finding system.

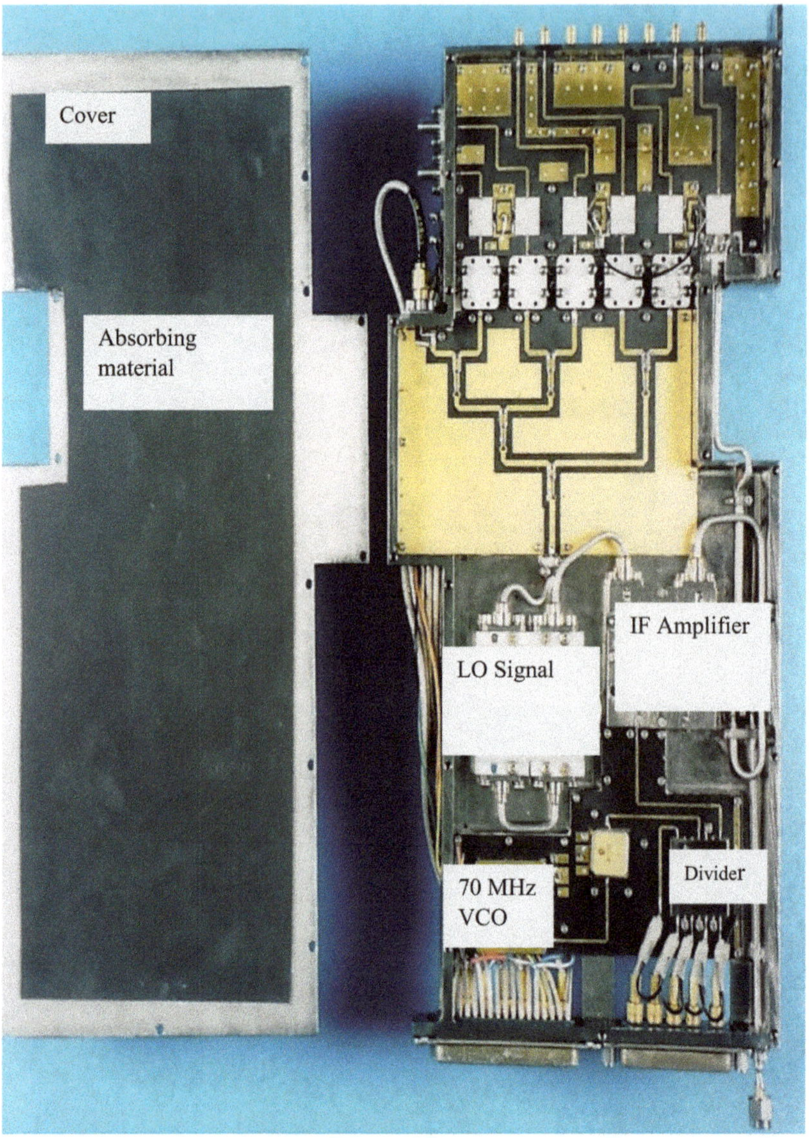

Figure 5.36. Photo of the wideband receiving module with the module cover ports.

example, it is better to design filters by using MIC technology. However, compact and low-cost modules in mass production is achieved by using MMIC design. The design considerations of passive and active devices were presented in this chapter.

The cost of MMIC modules in mass production will be much cheaper than MIC modules. However, there are technological limitations in designing LNAs and high power amplifiers.

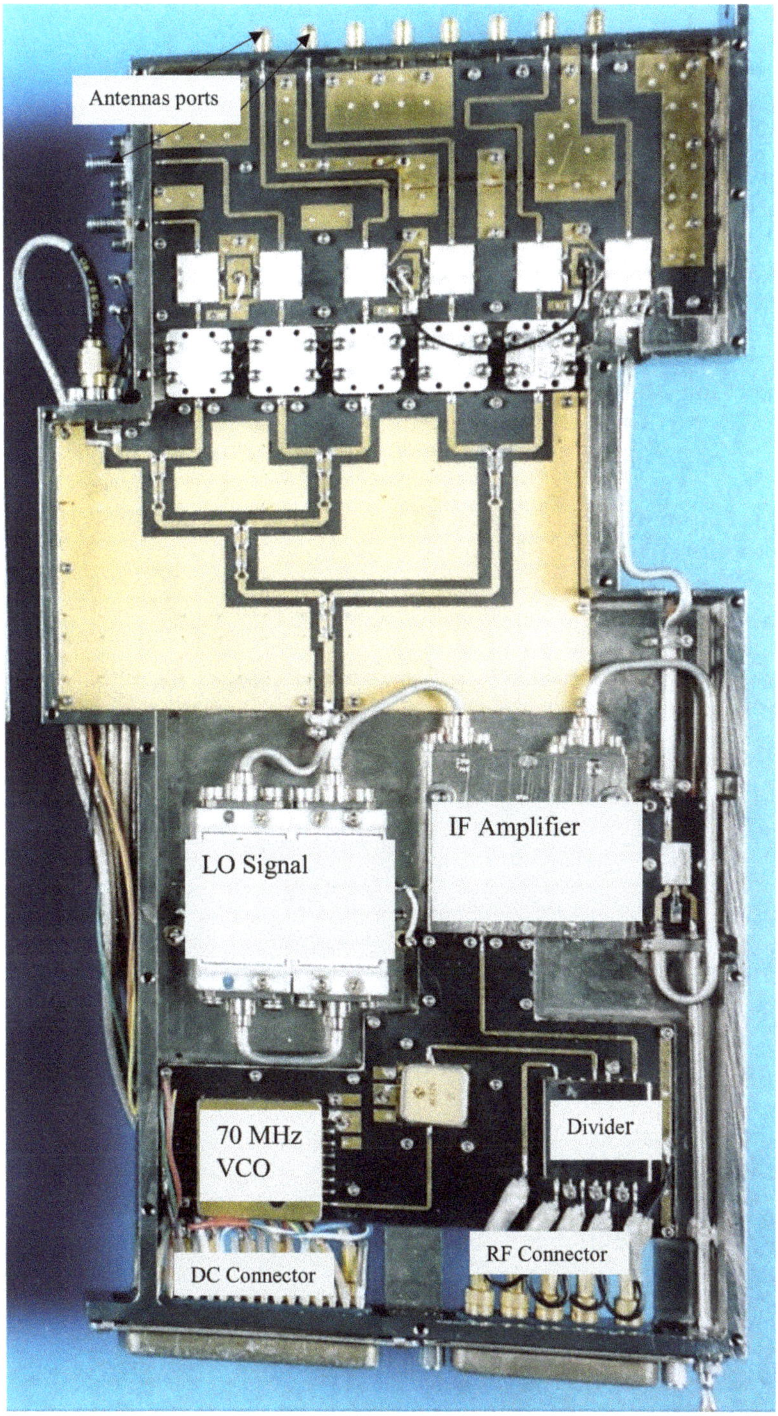

Figure 5.37. Photo of the wideband receiving direction finding system's interface.

References

[1] Rogers J and Plett C 2003 *Radio frequency Integrated Circuit Design* (Boston, MA: Artech House)

[2] Maluf N and Williams K 2004 *An Introduction to Microelectromechanical System Engineering* (Boston, MA: Artech House)

[3] Sabban A 2011 Microstrip antenna arrays *Microstrip Antennas* ed N Nasimuddin (Rijeka: InTech) pp 361–84

[4] Sabban A 2007 Applications of MM wave microstrip antenna arrays *ISSSE 2007 Conf. (Montreal, Canada)*

[5] Gauthier G P, Raskin G P, Rebiez G M and Kathei P B 1999 A 94GHz micro-machined aperture-coupled microstrip antenna *IEEE Trans. Antenna Propag.* **47** 1761–6

[6] Milkov M M 2000 Millimeter-wave imaging system based on antenna-coupled bolometer *MSc Thesis* UCLA

[7] de Lange G *et al* 1999 A 3×3 mm-wave micro machined imaging array with sis mixers *Appl. Phys. Lett.* **75** 868–70

[8] Rahman A *et al* 1996 Micro-machined room temperature micro bolometers for MM-wave detection *Appl. Phys. Lett.* **68** 2020–2

[9] Mass S A 1997 *Nonlinear Microwave and RF Circuits* (Boston, MA: Artech House)

[10] Sabban A 2015 *Low Visibility Antennas for Communication Systems* (New York: Taylor and Francis)

[11] Sabban A 2016 *Wideband RF Technologies and Antenna in Microwave Frequencies* (New York: Wiley)

IOP Publishing

Wearable Communication Systems and Antennas (Second Edition)
Design, efficiency, and miniaturization techniques
Albert Sabban

Chapter 6

System engineering of body-area networks, BAN communication systems

In the last few years, personal communication and the biomedical industry have continuously grown. Due to huge progress in the development of communication systems in the last decade, development of low-cost wearable communication systems is not a risky venture. However, development of compact efficient wearable antennas is one of the major challenges in the development of wearable communication and medical systems. Low profile compact antennas and transceivers are crucial in the development of wearable human communication and biomedical systems. The development of wearable antennas and compact transceivers for communication and biomedical systems will be described in this chapter. Design considerations, computational results and measured results of wearable compact transceivers will also be presented. The main goal of wireless BANs (WBANs) is to provide constant medical data to physicians. Systems engineering methodology should be employed in the development of wearable communication systems to optimize the design and production process. This methodology will be presented in this chapter.

6.1 Introduction

Wearable systems have several applications in personal communication devices and medical devices as presented in [1–8]. The biomedical industry has continuously grown in the last decade. Many medical devices and systems have been developed to monitor patient health as presented in several books and papers [1–43]. Wearable technology provides a powerful new tool to medical and surgical rehabilitation services. Wireless body-area networks (WBANs) can record electrocardiograms, measure body temperature, blood pressure, heart rate, arterial blood pressure, electro-dermal activity and other healthcare parameters. The recorded and collected

doi:10.1088/978-0-7503-5224-6ch6

data may be stored and analyzed by employing cloud storage and cloud computing services.

6.2 Cloud storage and computing services for wearable body-area networks

Cloud storage is a service package in which data is stored, managed, backed up remotely and made available to users over a network and internet services. This service is useful for wearable communication systems. Medical data can be stored in cloud data centers. Cloud storage is based on a virtualized infrastructure with accessible interfaces. Cloud-based data is stored in servers located in data centers managed by a cloud provider. A file and its associated metadata are stored in the server by using an object storage protocol. The server assigns an identification number (ID) to each stored file. When a file needs to be retrieved, the user presents the ID to the system and the content is assembled with all its metadata, authentication and security. The most common use of cloud services is cloud backup, disaster recovery and archiving infrequently accessed data. Cloud storage providers are responsible for keeping the data available and accessible, and the physical environment protected and running. Customers buy or lease storage capacity from providers to store and archive data files. Cloud storage services may be accessed via cloud computers and web services that use an application programming interface (API) such as cloud desktop storage and cloud storage gateways.

Advantages of cloud storage

- Cloud storage can provide the benefits of greater accessibility, reliability, rapid deployment, strong protection for data backup, archival and disaster recovery purposes.
- Cloud storage is used as a natural disaster proof backup. Usually, there are at least two backup servers located in different places around the world.
- Cloud storage can be used for copying virtual machine images from the cloud to a desired location or to import a virtual machine image from any designated location to the cloud image library. Cloud storage can also be used to move virtual machine images between user accounts or between data centers.
- Cloud storage provides users with immediate access to a broad range of resources hosted in the infrastructure of another organization via a web service interface.
- By using cloud storage companies can cut computing expenses such us storage maintenance tasks, purchasing additional storage capacity and cut their energy consumption.
- Storage availability and data protection are provided by cloud storage services. So, depending on the application, the additional technology effort and cost to ensure availability and protection of data storage can be eliminated.

Disadvantages of cloud storage

- Increasing the risk of unauthorized physical access to the data.
- In cloud-based architecture, data is replicated and moved frequently so the risk of unauthorized data recovery increases dramatically.
- Decrease in the security level of the stored data.
- It increases the number of networks over which the data travels. Instead of just a local area network (LAN) or storage area network. Data stored on a cloud requires a wide area network to connect them both.
- A cloud storage company has many customers and thousands of servers. Therefore, there is a larger team of technical staff with physical and electronic access to almost all the data at the entire facility. Encryption keys that are kept by the service user, as opposed to the service provider, limit access to the data by service provider employees. Many keys must be distributed to users via secure channels for decryption. The keys must be securely stored and managed by the users in their devices. Storing these keys requires expensive secure storage.
- Cloud storage companies are not permanent and the services and products they provide can change.
- Cloud storage companies can be purchased by other foreign larger companies, can go bankrupt and suffer from an irrecoverable disaster.
- Cloud storage is a rich resource for both hackers and national security agencies. The cloud stores data from many different users. Hackers see it as a very valuable target.
- Cloud storage sites have faced lawsuits from the owners of the intellectual property uploaded and shared in the site. Piracy and copyright problems may be enabled by sites that permit file sharing.

There are three main cloud-based storage architecture models: public, private and hybrid.

Public cloud storage services provide a multi-customer storage environment that is most suited for data storage. Data is stored in global data centers with storage data spread across multiple regions or continents.

Private cloud storage provides local storage services to a dedicated environment protected behind an organization's firewall. Private clouds are appropriate for users who need customization and more control over their data.

Hybrid cloud storage is a mix of private cloud and third-party public cloud services with synchronization between the platforms. The model offers businesses flexibility and more data deployment options. An organization might, for example, store actively used and structured data in a local cloud, and unstructured and archival data in a public cloud. In recent years, a greater number of customers have adopted the hybrid cloud model. Despite its benefits, a hybrid cloud presents technical, business and management challenges. For example, private workers must access and interact with public cloud storage providers, so compatibility and solid network connectivity are very important factors.

Cloud computing

Cloud computing is a type of internet computing service that provides shared computer processing resources and data to computers and other devices on demand. Cloud computing enables on-demand access to a shared pool of configurable computing resources such as computer networks, servers, storage, applications and services. Cloud computing relies on sharing of computing resources. Cloud computing services can be rapidly provisioned and released with minimal management effort. Cloud computing and storage solutions provide users and enterprises with various capabilities to store and process their data in privately owned data centers. Cloud computing allows companies to avoid high infrastructure costs such as servers and expensive software. It also allows organizations to focus on their core business instead of spending time and money on computer networks. Cloud computing allows companies to get their applications up and running faster, with improved manageability and less maintenance costs. Information technology teams can rapidly adjust resources to meet unpredictable business demands. Cloud computing applies high-performance computing power to perform tens of trillions of computations per second.

6.3 Wireless body area networks (WBANs) systems and applications

Wearable devices for medical applications are presented in figure 6.1. The main goal of WBANs is to provide continuously biofeedback data. A WBAN, with various wearable devices for medical applications is presented in figure 6.2. A WBAN health monitoring system is presented in figure 6.3. The recorded and monitoring data may be stored and analyzed by employing cloud storage and cloud computing services.

WBANs can measure and record body temperature, electrocardiograms, blood pressure, heartbeat rate, arterial blood pressure, electro-dermal activity and other healthcare parameters in an efficient way. For example, accelerometers can be used to sense heartbeat rate, movement or even muscular activity. Body area networks employ the applications and communication devices using wearable and implantable wireless networks. A sensor network that senses health parameters is called a body sensor network (BSN). A WBAN is a special purpose wireless-sensor network that incorporates different networks and wireless devices to enable remote monitoring in various locations.

In medical centers where many patients are constantly being monitored, a WBAN system may be particularly needed. Wireless monitoring of physiological signals of many patients is needed to deploy a complete WSN, Wearable Sensor Network, in hospitals. Human health monitoring is emerging as a crucial application of the embedded sensor networks. A WBAN can monitor vital signs, providing real-time feedback to allow many patient diagnostics procedures using continuous monitoring

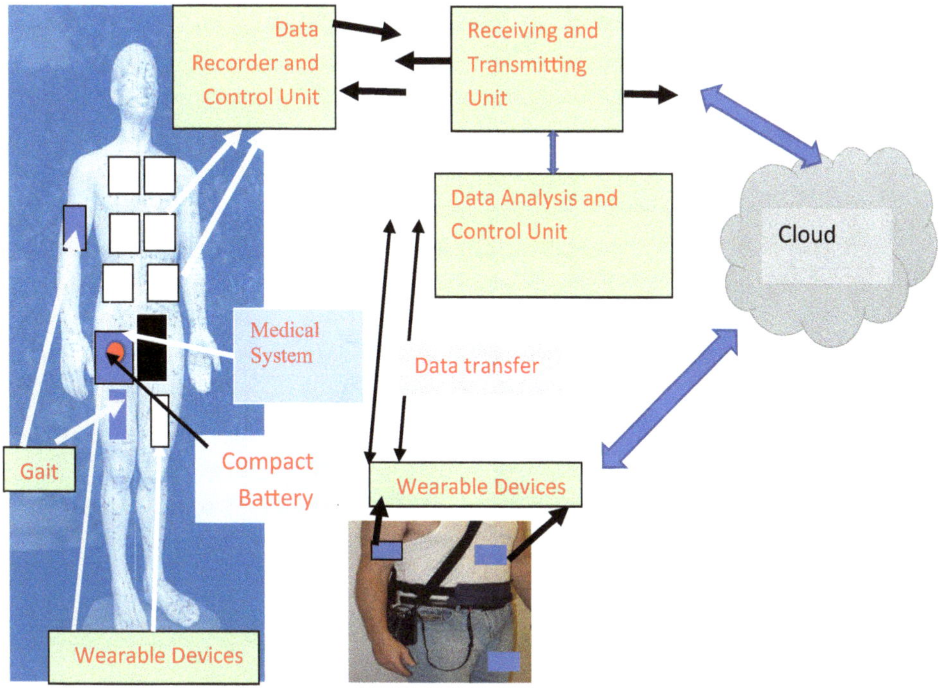

Figure 6.1. Wearable devices for various medical applications.

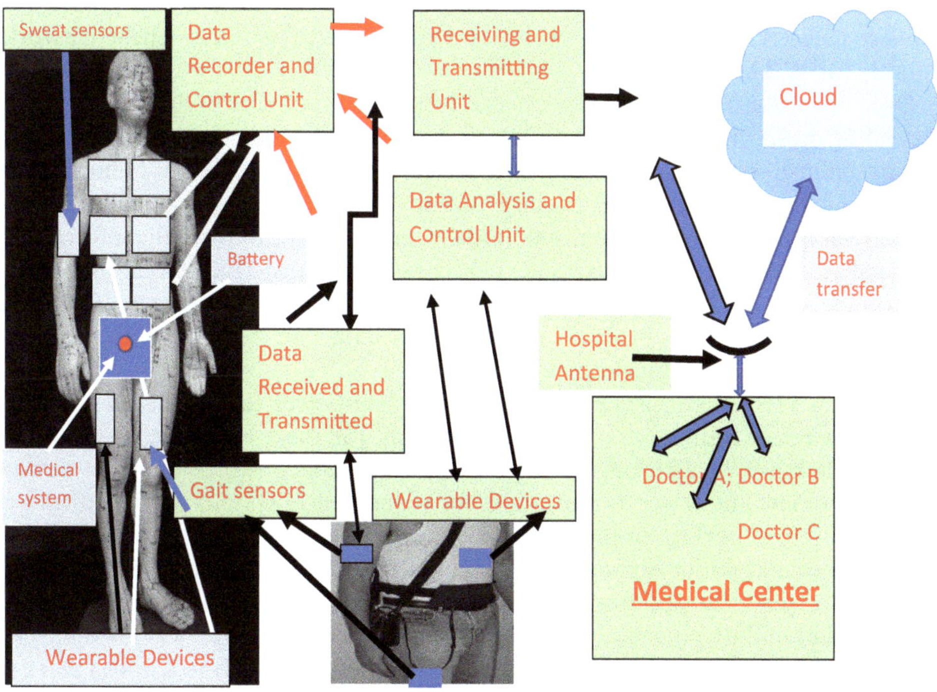

Figure 6.2. Wearable body-area network for various medical applications.

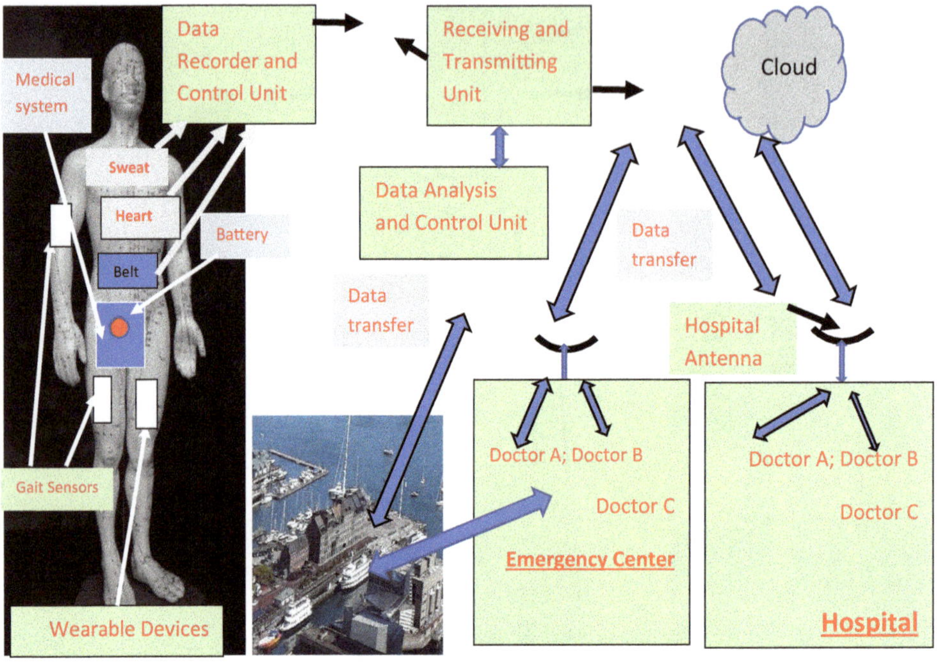

Figure 6.3. Wearable body-area network health monitoring system.

of chronic conditions, or progress of recovery from an illness. The progress in wireless networking technology promises a new class of wireless sensor networks suitable for WBAN systems.

Data acquisition in WBAN devices can be point-to-point or multipoint-to-point, depending on specific applications. Detection of an athlete's health condition would require point-to-point data sharing across various on-body sensors. Human body monitoring of vital signs will require one to route data from several wearable sensors, multipoint-to-point, to a sink node, which in turn can relay the information wirelessly to an out-of-body computer. Data may be transferred in real-time mode or non-real-time. Human body monitoring applications require real-time data transfer. Monitoring an athlete's physiological data can be collected offline for processing and analysis purposes by physicians.

A typical WBAN consists of a few compact low-power sensing devices, control unit and wireless transceivers. The power supply for these components should be compact, lightweight and long lasting as well. WBANs consist of compact devices with low volume and fewer opportunities for redundancy. To improve the efficiency of WBAN it is important to minimize the number of nodes in the network. Adding more devices and path redundancy for solving node failure and network problems cannot be a practical option in WBAN systems. WBANs receive and transmit a large amount of data constantly. Data processing must be hierarchical and efficient. WBANs in a medical area consist of wearable and implantable sensor nodes that can sense biological information from the human body and transmit it over a short distance wirelessly to a control device worn on the patient body. The sensor electronics must be miniaturized, low-power and detect medical signals such as pulse rate, electroencephalography, electrocardiograms, pressure, and temperature. The gathered data from the control devices are then transmitted to remote destinations in a wireless body-area network for diagnostic and therapeutic purposes by including other wireless networks for long-range transmission. A wireless control unit is used to collect information from sensors through wires and transmits it to a remote station for monitoring. The recorded and collected data may be stored and analyzed by employing cloud storage and cloud computing services.

6.4 Wearable wireless body area network (WWBAN) systems and applications

A wireless wearable body area network (WWBAN) health monitoring system is presented in figure 6.4. Wearable health monitoring devices and systems allow the person to follow closely changes in important health parameters and provide feedback for maintaining optimal patient health. Wireless communication systems offer a wide range of applications to medical centers, patients, physicians and sport centers by continuous measuring and monitoring of medical information, early detection of abnormal conditions, supervised rehabilitation, and potential discovery of knowledge through data analysis of the collected information. Long-term

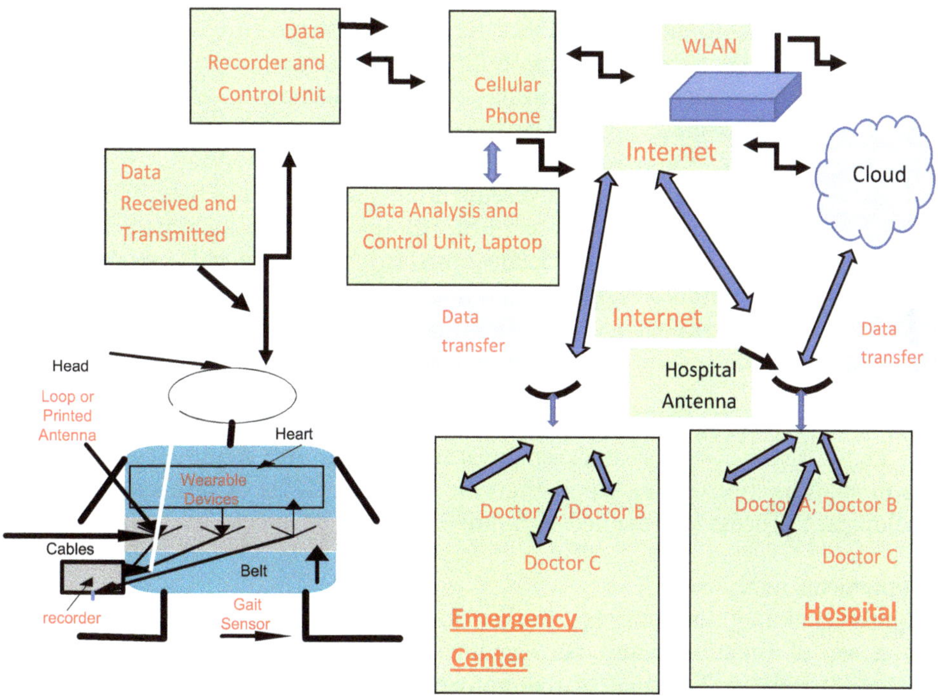

Figure 6.4. Wireless wearable body-area network (WWBAN) health monitoring system.

monitoring can confirm adherence to treatment guidelines or help monitor the effects of drug therapy. Health monitors can be used to monitor physical rehabilitation of patients during stroke rehabilitation, or brain trauma rehabilitation and after hip or knee surgeries. If the WWBAN is part of the telemedicine system, the medical system can alert medical personnel when life-threatening events occur. Patients and physicians may benefit from continuous long-term monitoring as part of a diagnostic procedure. Patients and physicians may achieve optimal maintenance of a chronic condition and can monitor the recovery period after the acute event or surgical procedure. The collected medical data may be a good indicator of cardiac recovery of patients after heart surgery. Many people are using WBAN devices such as wearable heart rate monitors, respiration rate monitors and pedometers for medical reasons or as part of a fitness regime. A WBAN may be attached to cotton shirts to measure respiratory activity, electrocardiograms, electromyograms and body posture. The recorded patient data may be stored and analyzed by employing cloud storage and cloud computing services.

6.5 Systems engineering methodology for wearable medical systems

Systems engineering design is the process of developing a system, component, or process to meet the desired customer requirements, see [44–48]. It is an iterative decision-making process, in which basic science, mathematics, physics, biology, biomedical engineering and other engineering sciences are applied to convert resources optimally to meet the specifications of the system. Among the fundamental steps of the design process are the establishment of requirements and criteria, synthesis, analysis, construction, testing and evaluation of the system's characteristics. The systems engineering process transforms needs and requirements into a set of system products and process descriptions, and generates information for investors and customers. The process is applied sequentially, one level at a time, top-down, to develop a system by integrated teams. For example, for wearable medical systems the customers are patients, physicians, medical staff, medical centers' system developers and distributors. The requirements of patients and physicians should be translated by employing functional analysis tools to a system's technical specifications.

Steps in a systems engineering process

The major steps in the systems engineering process are: requirements analysis, system analysis control, functional analysis, design synthesis. A systems engineering development process is presented in figure 6.5

Requirements analysis

Requirements analysis examines, evaluates, and translates customer requirements into a set of functional and performance requirements that are the basis for functional analysis. The goal of requirements analysis is to determine the system specifications that assists the system engineers in developing a system that satisfies the overall customer requirements. Requirements analysis is used to establish system

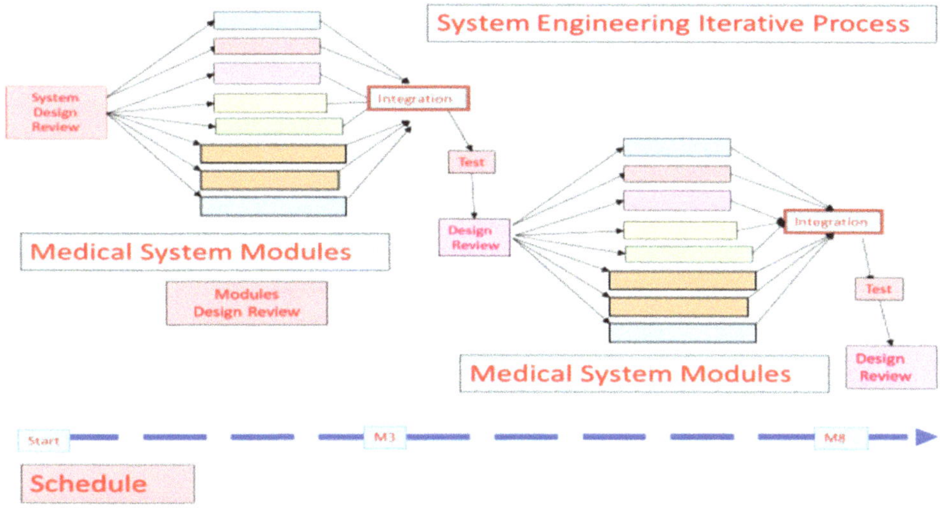

Figure 6.5. Systems engineering development process.

functions, system specifications and design constraints. The output of this process generates functional requirements, system specifications and system architecture. This process is an iterative decision-making process that examines original system requirements. This process results in some modification of the original system requirements and reflects the customer's needs that can be impractical or excessively costly. A quality function deployment (QFD) method is employed to optimize this process. The output of the requirements analysis is a set of design requirements and functional definitions that are the starting milestones in the system's development process. This iterative process determines how firm the requirements are for items that affect significantly the system cost, schedule, performance and risk. At the final stage of this process, detailed system characteristics are compared against the customer requirements to verify that the major customer needs are being met. A systems engineering iterative development process for a wearable system, first cycle, is presented in figure 6.6. A systems engineering development process for a wearable system, second cycle, is presented in figure 6.7.

System analysis control

System analysis and control manages and controls the system development process. This activity determines the work required to develop the system and the development schedule. This process estimates the cost of the system development and manufacturing. It helps the project manager to coordinate all activities and ensures that all the project activities meet the same set of requirements, agreements and design constraints. This process results in a final system engineering plan of all the system activities. It determines when the result of one activity requires the action of another activity. The system engineering plan can be updated as needed during the development process. As the system engineering process progresses, trade-off studies and system cost-effectiveness analyses are performed to support the evaluation and

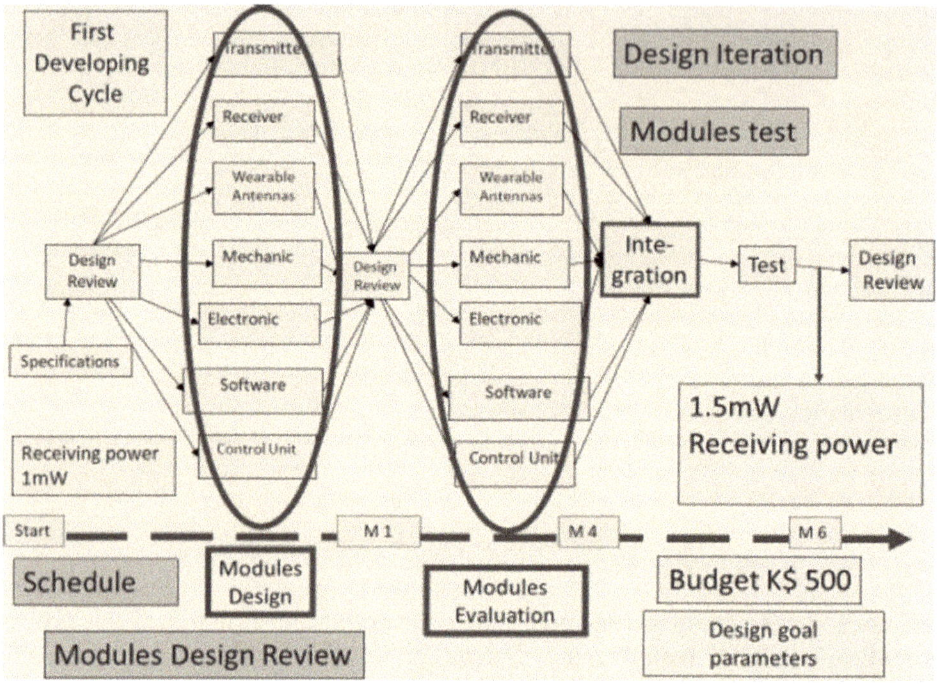

Figure 6.6. Systems engineering iterative development process for a wearable system, first cycle.

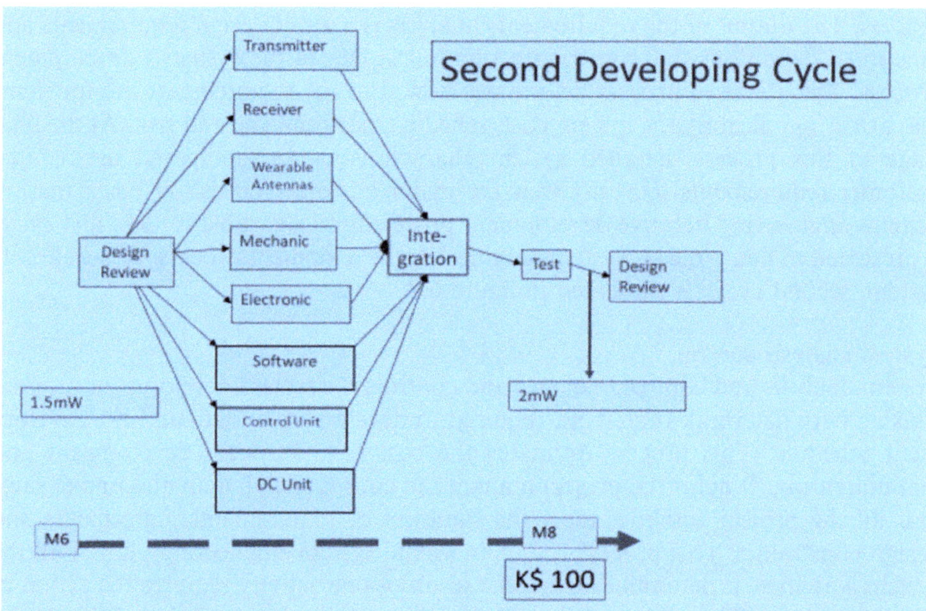

Figure 6.7. Systems engineering development process for a wearable system, second cycle.

selection processes of all the development activities. Risk identification studies are performed as part of a risk management plan. System analysis and control performs configuration management and identifies critical milestones to be used in the project progress control. In this process, interface management and data management are performed. The system engineer specifies the performance parameters to be tracked in the progress control plan. The system control plan consists of design reviews, progress reports, risk management, budget control and schedule management. A project design goals list is given in table 6.1. A system development plan is given in table 6.2.

Table 6.1. Project design goals list for a wearable medical system.

#	Parameter	Value [units]	Testing conditions
1	Transmitted power	27 dBm	25 °C
2	Received power	−10 dBm	25 °C
3	DC voltage	5 Vt	25 °C
4	Medical tests	$10 000	Test on 20 patients
5	Food and drug administration approval process	$10 000	24 months
6	Fabrication cost	$200	Fabrication of 1000 units

Table 6.2. System development plan.

#	Task	Time (H)	Materials costs $	Schedule
1	Requirements analysis and customer survey	100		1.1.2018
2	System design review	50	5000	1.2.2018
3	System analysis control	100	1000	1.3.2018
4	Functional analysis	100	1000	1.4.2018
5	System design	300		1.6.2018
6	Critical design review	100		15.6.2018
7	Material purchase	50	10000	1.8.2018
8	Modules fabrication	100	6000	15.9.2018
9	Modules test	200	1000	1.10.2018
10	Modules improvements	100	2000	1.11.2018
11	Modules design review	50		15.11.2018
12	Modules integration	100	2000	1.12.2018
13	System tests	200		5.12.2018
14	Final design review	50	2000	15.12.2018
15	Production of 100 units	1000	10 000	1.4.2019
	Total	2600	40 000	

Functional analysis

Functional analysis is a process of translating system requirements and customer needs into detailed functional and performance design specifications. The result of the process is a defined architectural model that identifies system functions and their interactions. A functional analysis process is presented in figure 6.8. It defines how the functions will operate together to perform the system functions. More than one architecture can satisfy the customer needs. However, each architecture has its own set of associated requirements, different cost values, schedule, performance, and risk implications. A system engineering modular architecture for a wearable BAN system is presented in figure 6.9. The functional architecture is used to define functional and evaluation development tests. The initial step in the process of functional analysis is to identify the basic functions required to perform the various system missions. As this is accomplished, the system specifications are defined, and functional architectures are developed for each module and for the system. These activities are continually validated and optimized to get the best system architecture.

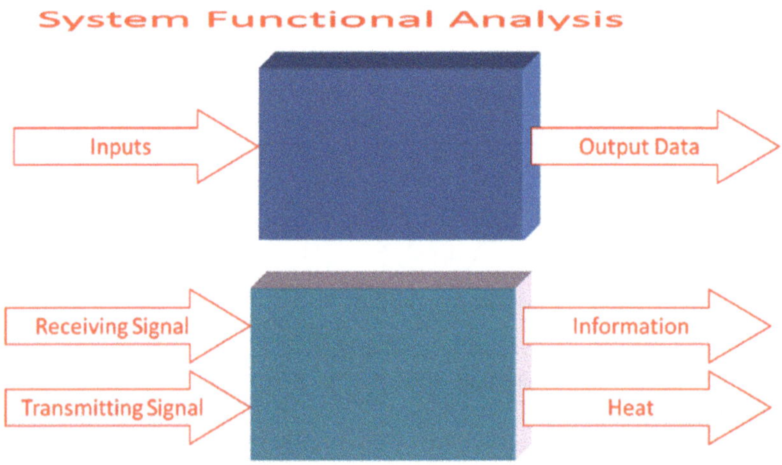

Figure 6.8. System functional analysis.

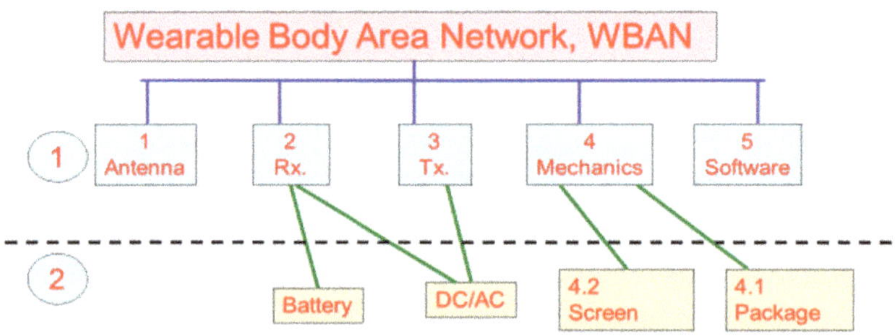

Figure 6.9. System engineering modular architecture for a wearable BAN system.

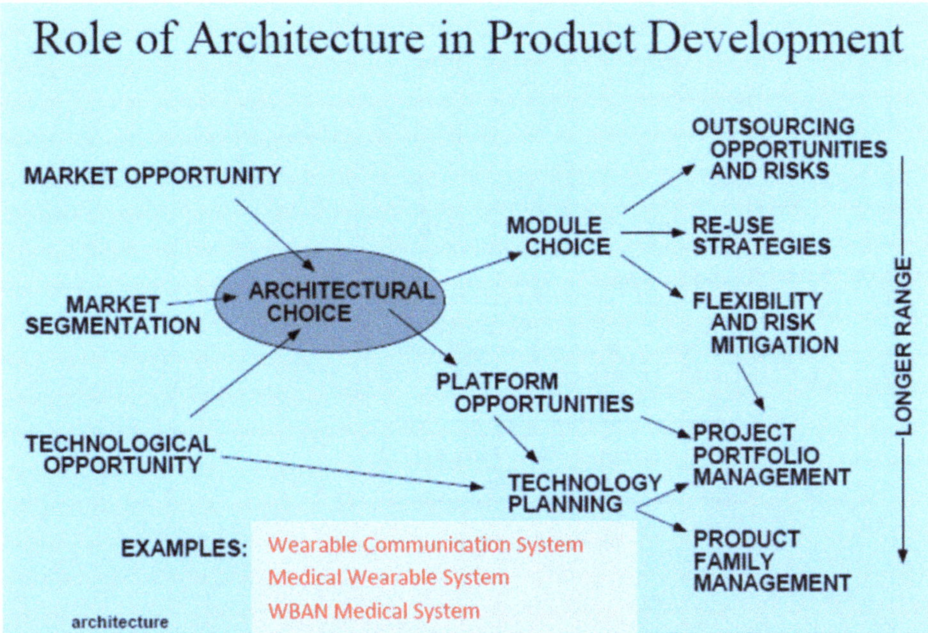

Figure 6.10. Role of architecture in system and product development.

The functional system architecture and the functional system requirements are the input to the synthesis and development process.

Design synthesis

The synthesis process translates functional architectures and requirements into physical architectures and to one or more physical sets of hardware, software and personnel solutions. The design results are compared to the original requirements, developed during the requirements analysis process, to verify that the developed system meets all of the system specifications. The output of this activity is a checklist that verifies that the system specifications are met and a database that documents the design process. The role of system architecture in system and product development is presented in figure 6.10. System architecture affects the technology that we will use in the development process. System architecture affects almost every stage of the system life cycle such as design, marketing, technology and system production.

6.6 System engineering tools for the development of wearable medical systems

The continuous growth in system complexity in the last 50 years has created the demand to develop system engineering tools that help system engineers to optimize the development process of products and systems. A system engineer transfers customer needs and requirements into technical specifications for new product and service development. Mitsubishi and Toyota engineers developed a QFD method

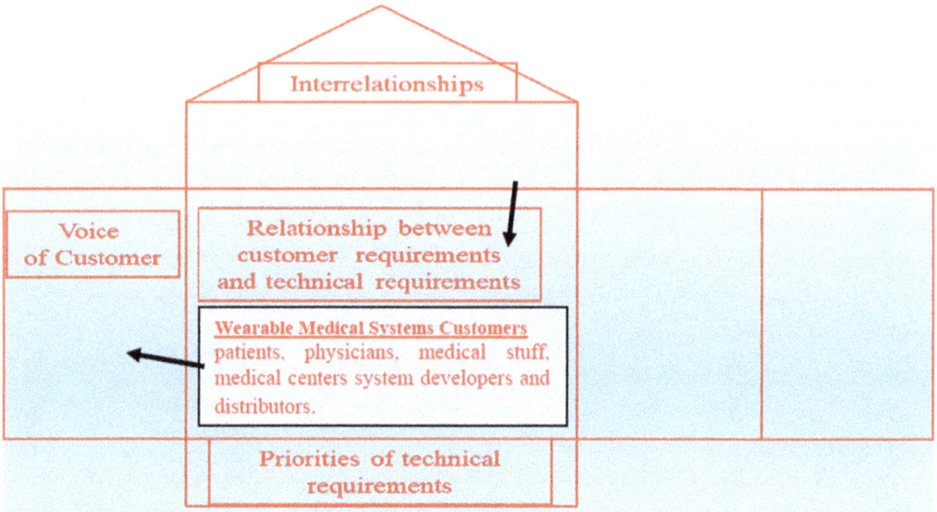

Figure 6.11. The house of quality.

that transfers customer needs and requirements into technical specifications for new product and service development.

6.6.1 Quality function deployment (QFD) method

QFD is a method that transfers customer needs and requirements into technical specifications for new product and service development. The first task in this method is to define who is a customer and how we understand what the customer needs are. A 'house of quality' image is presented in figure 6.11.

Customer definition

- Users who are concerned with functionality.
- Management who are concerned with financial and strategic issues.
- Distribution and purchasing agents who are concerned with purchase transaction and availability issues.
- Internal workers who are concerned with how the product will affect the quality of their work life.

Understanding customer needs

- Customer interviews.
- Focus groups.
- Customer visits.
- Customer telephone survey.

QFD step 1, understanding customer needs for a wearable medical system are presented in figure 6.12.

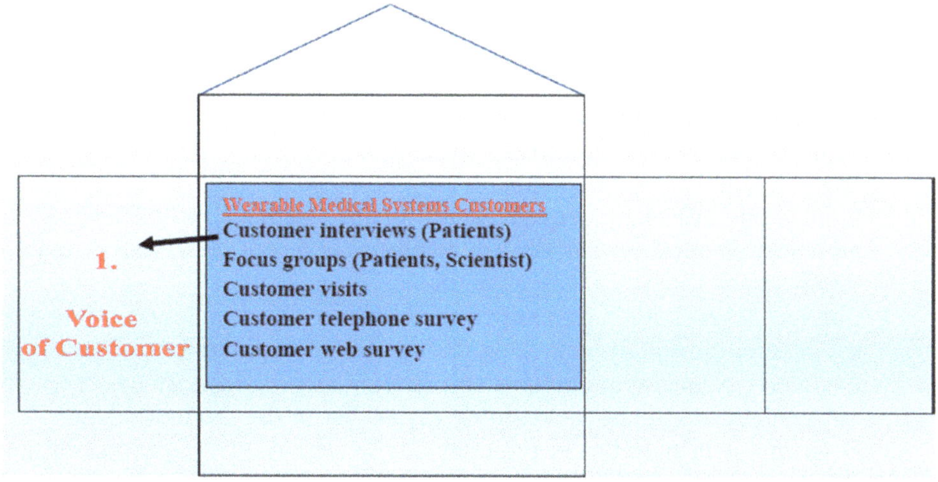

Figure 6.12. The 'house of quality', step 1, voice of customer.

Applying customer data

- Importance to the customer.
- Our current product.
- Competitors' products.
- Future products.
- Improvement factor.
- Overall importance.
- Percent importance.

QFD definition and goals

- QFD is a quality system method that implements elements of systems thinking and psychology (understanding customer needs and how customers or end users become interested and satisfied from the system performance).
- QFD is a quality deployment method of good knowledge that helps the system engineer to understand the needs of the customer and to decide what features to include in the design.
- QFD maximizes positive qualities that add value to the system. The QFD process seeks out spoken and unspoken customer requirements, translates them into technical requirements, prioritizes them and directs us to optimize those features that will bring the greatest competitive advantage.
- QFD evaluates technical requirements compared to competitive products and services.
- The final goal of many QFD projects is to set the target values for the design measures.

Result of the QFD process

- To develop a comprehensive optimal product specification.
- Answers the question:

'What actions do we need to take to achieve the targets that we have set in order to satisfy our customers?'

Six steps to build the house of quality

1. Identify customer requirements and needs.
2. Identify technical requirements.
3. Relate the customer requirements to the technical requirements, as presented in figure 6.13. Illustrates customer needs observed in market surveys.
4. Conduct an evaluation of competing products or services, concentrating on key relationships and minimizing the number of requirements.
5. Evaluate technical requirements and develop targets. Identify where technical requirements support or impede each other in the product design.
6. Determine which technical requirements to deploy in the production and delivery process. QFD steps 1 to 6, for wearable medical systems requirements, are presented in figure 6.14.

QFD process—four phase approach as presented in figure 6.15.

- Translate customer needs into product characteristics.
- Translate product characteristics into modules specifications.
- Modules specifications into system specifications.
- Product production.

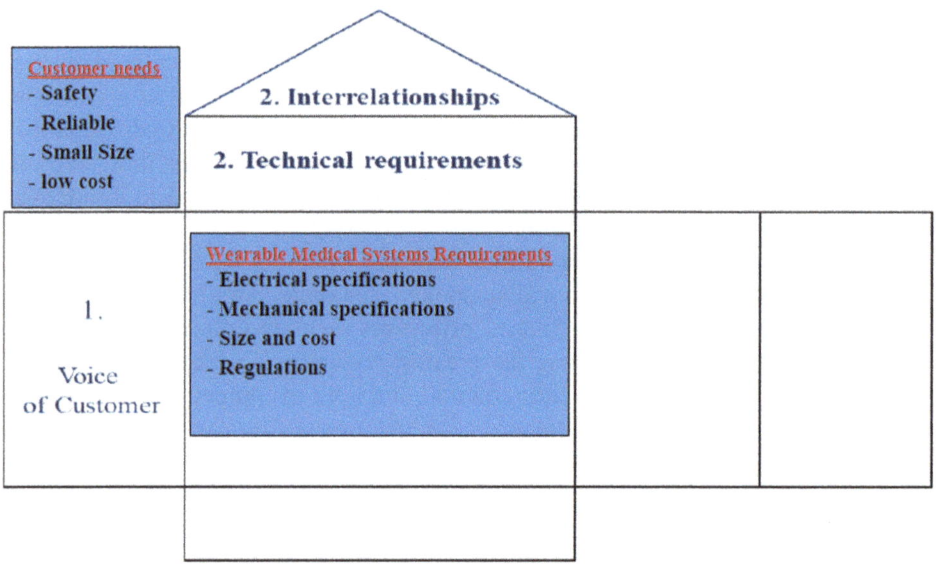

Figure 6.13. QFD step 2, wearable medical systems requirements.

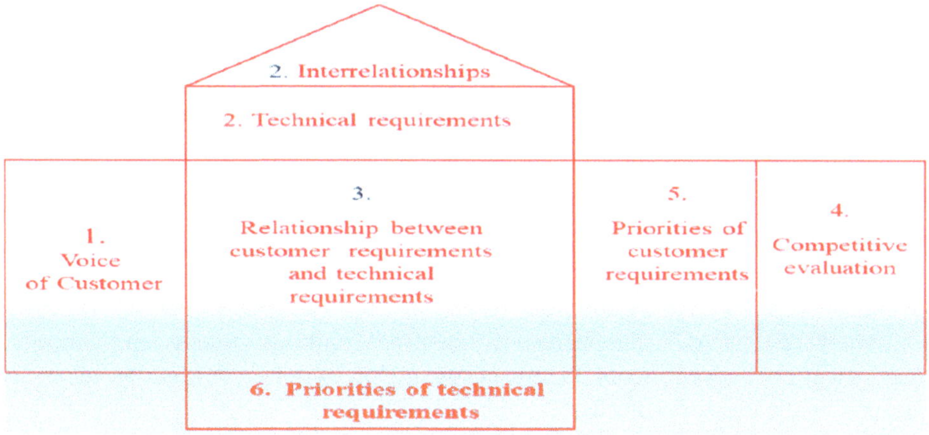

Figure 6.14. QFD step 1 to step 6, wearable medical systems requirements.

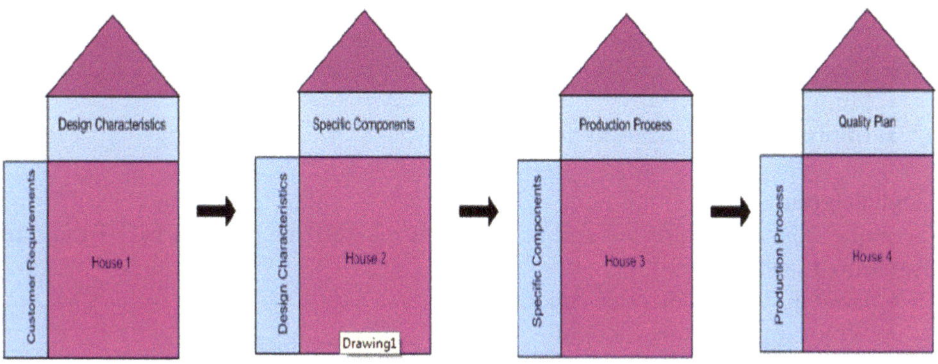

Figure 6.15. Different modes of QFD applications, the four phase approach.

QFD advantages

- Improved system quality.
  Build and deliver a quality product or service by focusing on customer satisfaction.
- Increased customer satisfaction. Customer voice is heard.
- Reduced time to market of system development and production.
- Reduction in design changes.
- Decreased design and manufacturing costs. Optimization of system development and production process.

6.6.2 Functional analysis system techniques (FAST)

The functional analysis system technique (FAST) was presented in 1963 by Charles W Bytheway. The function analysis system technique helps in thinking about the problem objectively and in identifying the scope of the system by showing the logical relationships between functions. Drawing a FAST diagram presents the system functions and problems. A FAST diagram enables the participants to identify all of the required functions. The FAST diagram can be used to verify if, and illustrate how, a proposed solution achieves the needs of the system, and to identify unnecessary, duplicated or missing functions. The function analysis system technique diagram is usually prepared in a project staff meeting and led by an experienced project manager. Input for the diagram is received from the meeting participants.

Advantages of the FAST Technique

Evaluation of a FAST diagram is a creative thinking process that supports communication between team members and helps to generate creative ideas and solutions to the customer needs.

The development of a FAST diagram helps project development teams to:

- Develop a common understanding of the project.
- Define, simplify and clarify the system problems.
- Identify missing functions of the system.
- Organize and understand the relationships between functions.
- Identify the basic function of the system, process or product.
- Stimulate creativity.
- Improve communication and consensus.

FAST diagram

Three questions are addressed and answered in a FAST diagram. A FAST diagram is presented in figure 6.16.

- How do we achieve this function?
- Why do we do this function?
- When we do this function, what other functions must we do?

Constructing a FAST diagram

- Prepare a list of all functions. Use a verb and noun to define a function.
- Expand the functions in the 'How' and 'Why' directions.
- Build along the 'How' path by asking 'how is the function achieved?' Place the answer to the right in terms of an active verb and measurable noun.
- Involve the whole team in the diagraming exercise.
- Test the logic in the direction of the 'Why' path (right to left) by asking 'why is this function undertaken?'
- When the logic does not work, identify any missing or redundant functions.
- To identify functions that happen at the same time, ask 'when this function is done, what else is done or caused by the function?'
- The higher order functions (functions towards the left on the FAST diagram) describe what is being accomplished and lower order functions (functions towards the right on the FAST diagram) describe how they are being accomplished.

- ‘When’ does not refer to time as measured by a clock, but functions that occur together with or because of each other.

6.6.3 Pugh method to evaluate the best system concept

The Pugh method is a creative design idea or concept evaluation technique that uses criteria derived from the ‘voice of the customer’ in an advantage–disadvantage matrix. Each concept is evaluated against a datum using a three-way evaluation scheme. The goal of the Pugh process is to evaluate the best concept and not just to choose the best concept. The Pugh matrix was invented by Stuart Pugh in 1981. A Pugh method evaluation process flow chart is presented in figure 6.17. A Pugh

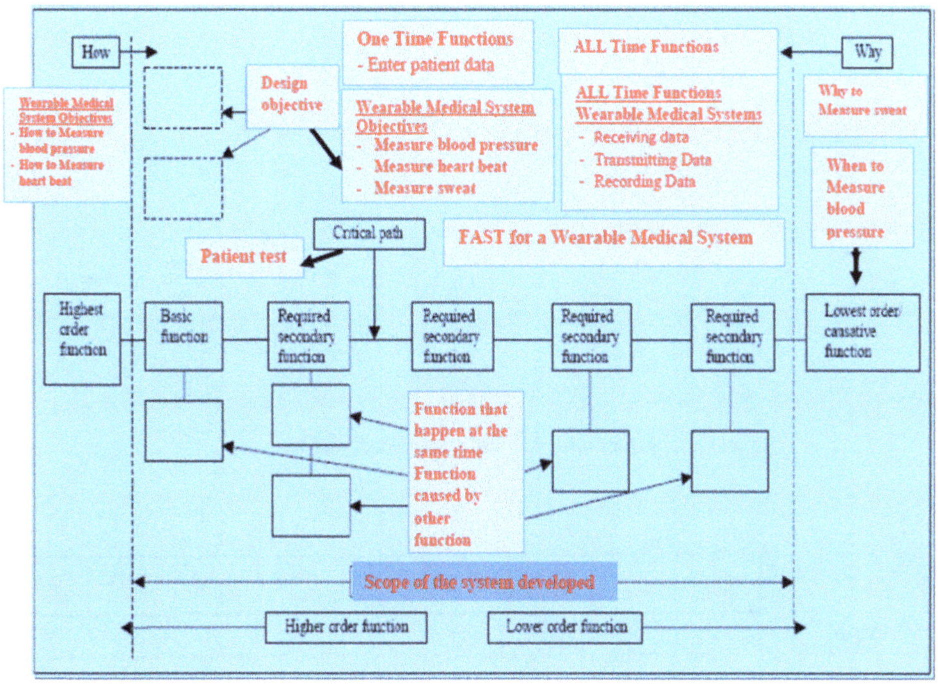

Figure 6.16. FAST diagram.

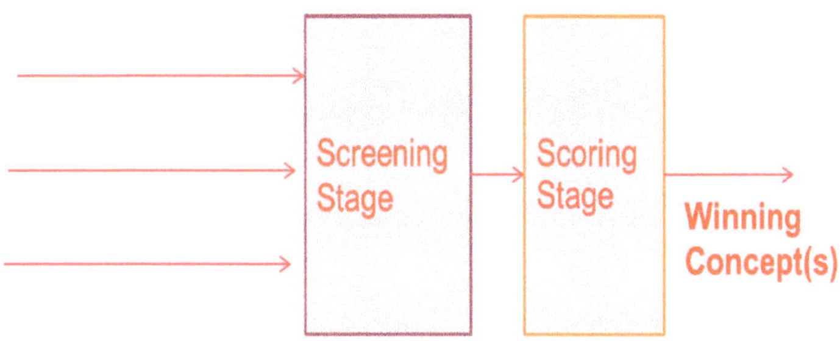

Figure 6.17. Pugh method evaluation process flow chart.

process matrix is shown in figure 6.18, and a Pugh method evaluation process can be seen in figure 6.19.

Six steps in the Pugh process

- concept generation based on customer requirements.
- concept screening. Use a candidate whose performance is well known as a baseline.
- concept scoring.

Evaluation Criteria		Score	Concepts 1	2	3	4
Schedule	Medical Systems Design		-	+	+	+
	Production		-	+	+	+
Development cost	Wearable Antenna		-	S	S	+
	Package		-	-	+	+
	Recorder		+	-	S	+
	Sensor		+	S	-	S
Size	Wearable Antenna		+	-	-	S
	Recorder		+	-	-	-
Cost	Part cost		-	-	-	-
	Production cost		S	S	-	S
	Service cost		+	-	S	-
	Technology cost		-	+	+	+
Reliability			-	-	S	+
		Σ -	7	7	5	3
		Σ S	1	3	4	3
		Σ +	5	3	4	7
		Position	3	4	2	1

Figure 6.18. Medical system Pugh matrix.

Criteria \ Concept	1	2	3	4	5	6	7	8	9
A	+	−	+	−	+	−	D	−	+
B	+	S	+	S	−	−		+	−
C	−	+	−	−	S	S	A	+	S
D	−	+	+	−	S	+		S	−
E	+	−	+	−	S	+	T	S	+
F	−	−	S	+	+	−		+	−
Sum +	3	2	4	1	2	2	U	3	2
Sum -	3	3	1	4	1	3		1	3
Total Sum	0	1	1	1	3	1	M		1

Figure 6.19. Pugh matrix evaluation.

Evaluation scale

Means substantially better, +.
Means clearly worse (or flawed), —.
Means almost the same, *S*.
- concept testing. Choose the 'winner' and the 'loser'.
- Hybrid solution. Combine the best concept from each alternative.
- Choose the best hybrid solution that makes sense.

What to do when it is difficult to evaluate the best solution

- Refine criteria and requirements.
- Gain more information.
- Add 'weight' to the criteria.
- Make a sensitivity test.
- Use a different decision method.

Failure of the Pugh process

- Incorrect selection criteria.
- Incomplete selection criteria.
- Inadequate selection criteria.
- Granularity of pairwise scale.
- Insufficient experience in teams.

Success of the Pugh process

- Small team size between four to eight team members.
- Choose an experienced independent team member
- Add 'experts' with 'newbies' in the team.
- Plan the Pugh process schedule to around four hours.
- Define clearly what the team is trying to do.
- Have the 'customer requirements' available during the Pugh process.
- Make sure the team understands the system functions and requirements.
- Document the Pugh process and all the debates.

6.7 ICDM—integrated, customer-driven, conceptual design method

ICDM is a system, ten step integrated, customer-driven, conceptual design method. Figure 6.20 presents the ten steps in a customer-driven conceptual design. ICDM flow charts and tools are presented in figure 6.20. ICDM ten steps and tools are listed in table 6.3.

The ICDM combines and employs system tools such as QFD, FAST and Pugh. The optimal final system architecture and concept is optimized and evaluated by using brainstorming techniques. NGT, a **nominal group technique,** is a brainstorming technique. This technique is a structured variation of small group discussion methods. The process prevents the domination of a discussion by a single person, and encourages the more passive group members to participate, resulting in a set of prioritized solutions or recommendations.

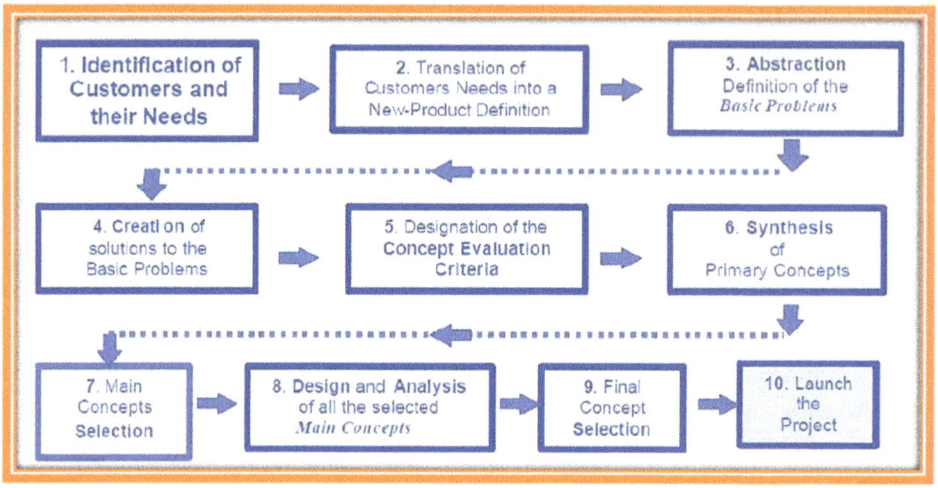

Figure 6.20. ICDM—integrated, customer-driven, conceptual design method in ten steps.

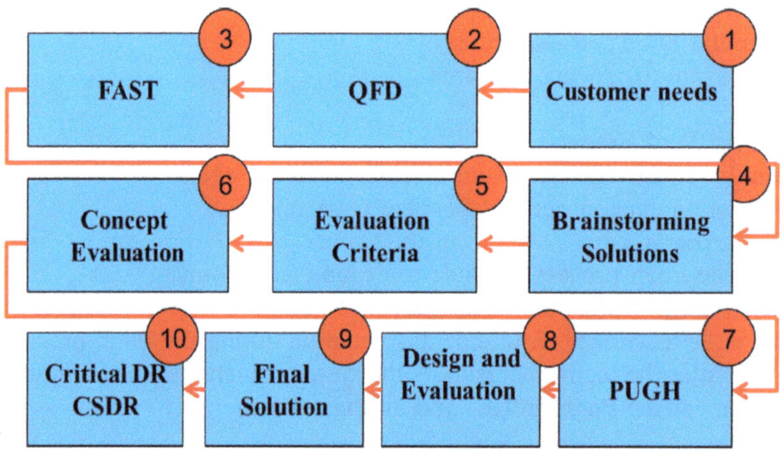

Figure 6.21. ICDM flow charts and tools.

<u>**NGT procedure steps**</u>

- The meeting participants are first presented by the session moderator. The session moderator presents the topic area to be evaluated and discussed.
- The meeting participants are directed to reflect individually on the topic.
- The group moderator asks a participant to state one of the responses he has arrived at.
- The next stage involves consolidation and review of the ideas.
- They are then requested to establish the relative importance that should be accorded to each of the response ideas.
- The final stage is the compilation of the results and anonymous voting.

TRIZ—theory of solving problems inventively

TRIZ is a theory of solving problems inventively. It is a systematic methodology for reducing creativity and innovation to a set of principles and algorithms. TRIZ systematizes successful methods of solving technological problems, and reveals regularities in the evolution of technological systems. It is more structured and based on logic and data, not intuition or brainstorming, and uses a creative solution to overcome a system conflict or contradiction.

6.8 434 MHz receiving channel for communication and medical systems

A medical system may be implanted or inserted into the human body as a swallowed capsule. The medical device will transmit medical data to a recorder. The medical data may be analyzed by medical staff online or stored as medical data about the patient. The receiving channel is part of the recorder and consists of receiving wearable antennas, a RF head and a signal processing unit. A block diagram of the receiver is shown in figure 6.22. The receiving channel's main specifications are listed in table 6.4.

Table 6.3. ICDM ten steps and tools.

#	Step	Tools	Remarks
1	Identify customer requirements and needs	Survey	
2	Identify technical requirements	QFD	
3	Functional analysis	FAST	
4	Creation of solutions to the basic needs	Brainstorming techniques	NGT, TRIZ
5	Criteria evaluation	PUGH	
6	Concept synthesis and evaluation	PUGH	
7	Main concept selection	PUGH	
8	Design and analysis of all system modules	Design tools	
9	Final concept selection	Design reviews	
10	System critical design reviews. Launch project at the of the CDR	Design reviews	Launch project

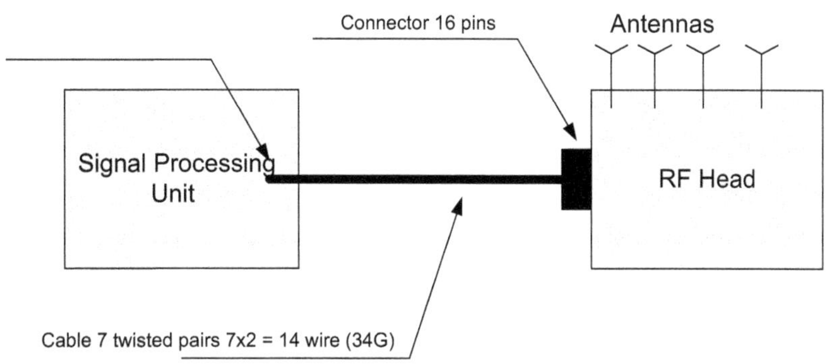

Figure 6.22. Recorder block diagram.

Table 6.4. Receiving channel's main specifications.

Requirement	Specification
Frequency range uplink	430–440 MHz
Return loss (dB)	−9
Group delay	Maximum 50 ns for 12 MHz BW
Input power	−30 dBm to −60 dBm
SNR	>20 dB
Current consumption (mA)	50
Dimensions (cm)	12 × 12 × 5
Frequency range downlink	13.56 MHz
Received power downlink	−8 dBm ± 1 dB
Current consumption (mA)	100–110

A block diagram of the receiving channel is shown in figure 6.23. The receiving channel consists of an uplink channel at 434 MHz and a downlink at 13.56 MHz.

The uplink channel consists of a switching matrix, low noise amplifier and filter. The switching matrix losses are around 2 dB. The LNA noise figure is around 1 dB with 21 dB gain. The downlink channel consists of a transmitting antenna, antenna matching network and a differential amplifier. The downlink channel transmits commands to a medical system. A receiving channel's gain and noise figure budget is shown in figure 6.24. The receiving channel's noise figure is around 3.5 dB. A receiving channel with lower noise figure values is shown in figure 6.25. The LNA amplifier is connected to the receiving antenna.

A receiving channel's gain and noise figure budget is shown in figure 6.26. The receiving channel noise figure is around 3.5 dB.

Four folded dipole or loop antennas may be assembled in a belt and attached to a patient's stomach as shown in figures 6.27(a) and (b). The cable from each antenna is connected to a recorder. The received signal is routed to a switching matrix. The signal with the highest level is selected during the medical test. The antennas receive a signal that is transmitted from various positions in the human body. Wearable antennas may be also attached on a patient's back in order to improve the level of the received signal from different locations in the human body.

6.9 Conclusions

Wearable technology provides a powerful new tool for medical and surgical rehabilitation services. The wearable body-area network (WBAN) is emerging as an important option for medical centers and patients. Wearable technology provides a convenient platform that may quantify the long-term context and physiological response of individuals. Wearable technology will support the

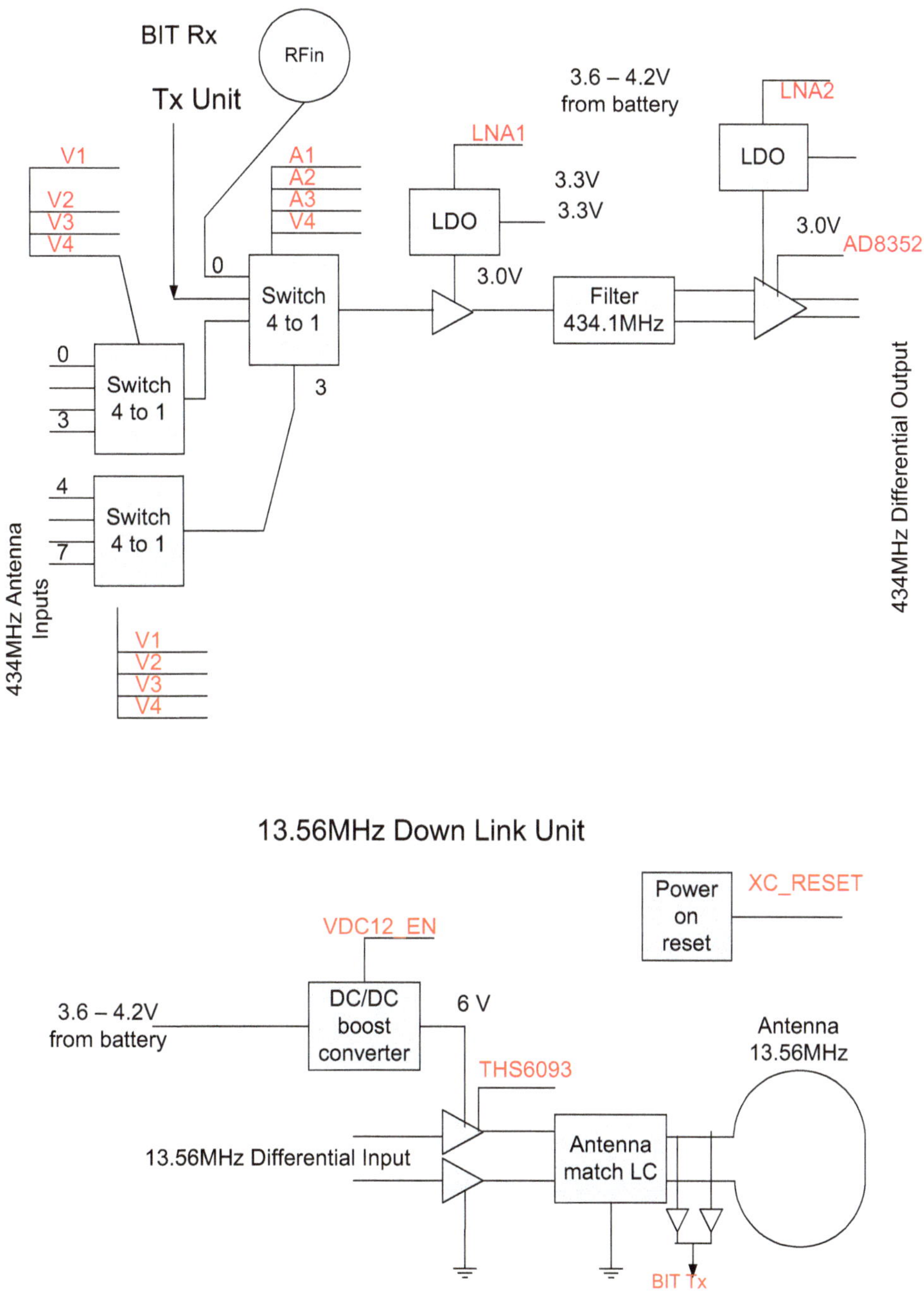

Figure 6.23. Receiving channel block diagram.

System1

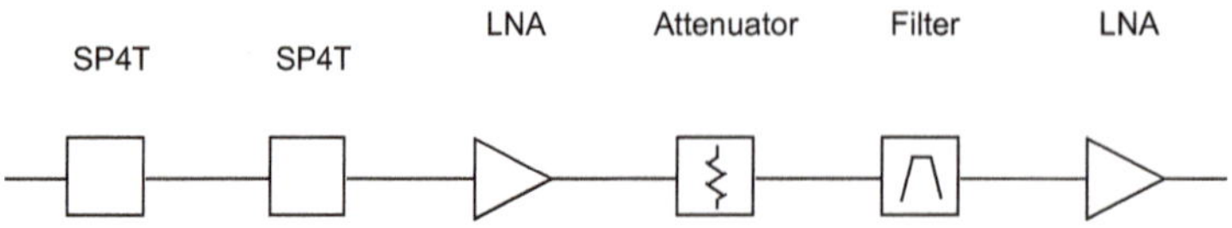

	SP4T	SP4T	LNA	Attenuator	Filter	LNA	Total
Gain (dB)	-1.10	-1.10	21.00	-0.50	-0.80	21.00	38.50
NF (dB)	1.10	1.10	1.10	0.50	0.80	1.50	3.32
OP1dB (dBm)	20.00	20.00	17.00	20.00		17.00	17.00

Input Pwr (dBm) -60.00 System Temp (K) 290.00

Figure 6.24. Receiving channel's gain and noise figure budget.

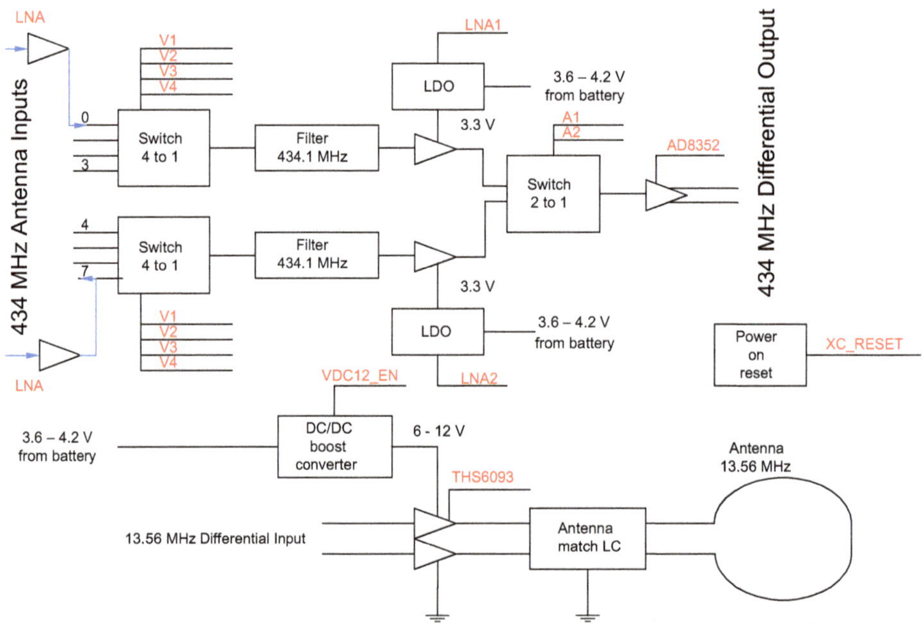

Figure 6.25. Receiving channel block diagram with LNA connected to the antennas.

System1

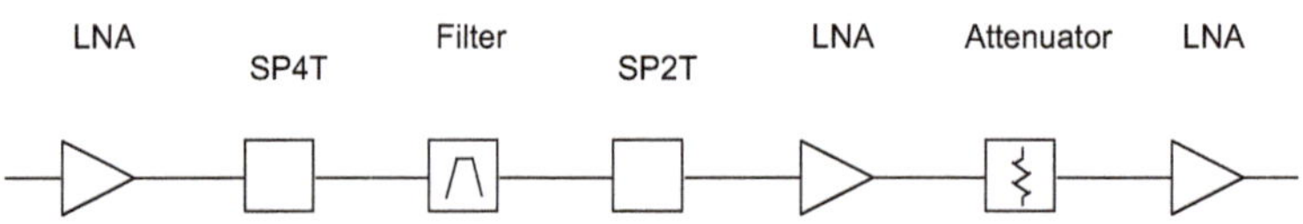

	LNA	SP4T	Filter	SP2T	LNA	Attenuator	LNA	Total
Gain (dB)	21.00	-1.20	-0.80	-1.00	17.00	-1.00	15.50	49.50
NF (dB)	1.10	1.20	0.80	2.00	1.10	1.00	2.00	1.16
OP1dB (dBm)	17.00	20.00		20.00	17.00	20.00	17.00	17.00

Input Pwr (dBm) -60.00 System Temp (K) 290.00

Figure 6.26. Receiving channel with a LNA connected to the antennas, gain and noise figure budget.

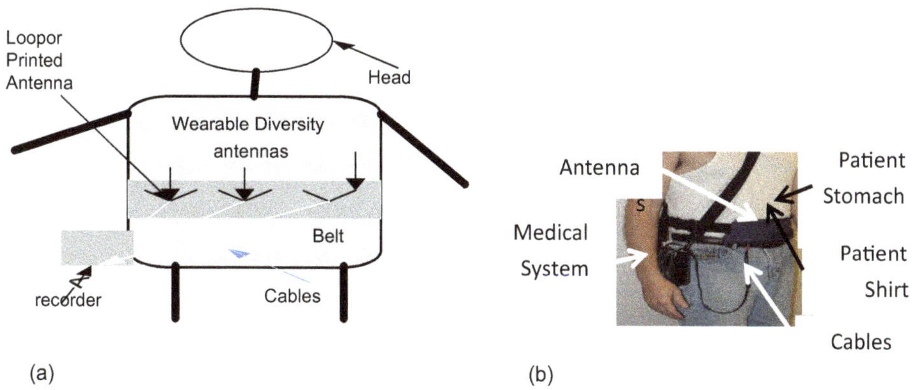

Figure 6.27. (a) Wearable medical system. (b) Medical system on a patient.

development of individualized treatment systems with real-time feedback to help promote the health of patients. Wearable medical systems and sensors can measure body temperature, heart rate, blood pressure, sweat rate, perform gait analysis and other physiological parameters of the person wearing the medical device. Gait analysis is a useful tool, both in clinical practice and biomechanical research. Gait analysis using wearable sensors provides quantitative and repeatable results over extended time periods with low cost and good portability, showing better prospects and making great progress in recent years. At present, commercialized wearable sensors have been adopted in various applications of gait analysis. System engineering tools must be used in the development of wearable medical systems. Requirements analysis examines, evaluates, and translates customer requirements into a set of functional and performance requirements that are the basis for functional analysis. The goal of requirements analysis is to determine system specifications, which assists system engineers to develop a system that satisfies the overall customer requirements. Requirements analysis is used to establish system functions, system specifications and design constraints. The requirements of patients and physicians should be translated by employing functional analysis tools to system technical specifications.

References

[1] Mukhopadhyay S C (ed) 2015 *Wearable Electronics Sensors* (Cham: Springer)

[2] Sabban A 2016 *Wideband RF Technologies and Antenna in Microwave Frequencies* (New York: Wiley)

[3] Sabban A 2015 *Low Visibility Antennas for Communication Systems* (New York: Taylor and Francis)

[4] Sabban A 2016 Small wearable meta materials antennas for medical systems *Appl. Comput. Electromagn. Soc. J.* **31** 434–43

[5] Sabban A 2011 Microstrip antenna arrays *Microstrip Antennas* ed N Nasimuddin (Rijeka: InTech) 361–84

[6] Sabban A 2013 New wideband printed antennas for medical applications *IEEE J. Trans. Antennas Propag.* **61** 84–91

[7] Sabban A 2012 *Dual polarized dipole wearable antenna* US Patent number: 8203497

[8] Bonfiglio A and De Rossi D (ed) 2011 *Wearable Monitoring Systems* (New York: Springer)

[9] Gao T, Greenspan D, Welsh M, Juang R R and Alm A 2005 Vital signs monitoring and patient tracking over a wireless network *Proc. of IEEE-EMBS 27th Annual Int. Conf. of the Engineering in Medicine and Biology (Shanghai, China, 1–5 September)* pp 102–5

[10] Otto C A, Jovanov E and Milenkovic E A 2006 WBAN-based system for health monitoring at home *Proc. of IEEE/EMBS Int. Summer School, Medical Devices and Biosensors (Boston, MA, USA, 4–6 September)* pp 20–3

[11] Zhang G H, Poon C C Y, Li Y and Zhang Y T 2009 A biometric method to secure telemedicine systems *Proc. of the 31st Annual Int. Conf. of the IEEE Engineering in Medicine and Biology Society (Minneapolis, MN, USA)* pp 701–4

[12] Bao S, Zhang Y and Shen L 2005 Physiological signal based entity authentication for body area sensor networks and mobile healthcare systems *Proc. of the 27th Annual Int. Conf. of the IEEE EMBS (Shanghai, China, 1–4 September)* pp 2455–8

[13] Ikonen V and Kaasinen E 2008 Ethical assessment in the design of ambient assisted living *Proc. of Assisted Living Systems—Models, Architectures and Engineering Approaches (Schloss Dagstuhl, Germany)* pp 14–7

[14] Srinivasan V, Stankovic J and Whitehouse K 2008 Protecting your daily in home activity information from a wireless snooping attack *Proc. of the 10th Int. Conf. on Ubiquitous Computing (Seoul, Korea, 21–24 September)* pp 202–11

[15] Casas R, Blasco M R, Robinet A, Delgado A R, Yarza A R, Mcginn J, Picking R and Grout V 2008 User modelling in ambient intelligence for elderly and disabled people *Proc. of the 11th Int. Conf. on Computers Helping People with Special Needs (Linz, Austria)* pp 114–22

[16] Jasemian Y 2008 Elderly comfort and compliance to modern telemedicine system at home *Proc. of the Second Int. Conf. on Pervasive Computing Technologies for Healthcare, (Tampere, Finland, 30 January–1 February)* pp 60–3

[17] Atallah L, Lo B, Yang G Z and Siegemund F 2008 Wirelessly accessible sensor populations (WASP) for elderly care monitoring *Proc. of the Second Int. Conf. on Pervasive Computing Technologies for Healthcare (Tampere, Finland, 30 January–1 February)* pp 2–7

[18] Hori T, Nishida Y, Suehiro T and Hirai S 2000 SELF-Network: Design and implementation of network for distributed embedded sensors *Proc. of IEEE/RSJ Int. Conf. on Intelligent Robots and Systems (Takamatsu, Japan, 30 October–5 November)* pp 1373–8

[19] Mori Y, Yamauchi M and Kaneko K 2000 Design and implementation of the Vital Sign Box for home healthcare *Proc. of IEEE EMBS Int. Conf. on Information Technology Applications in Biomedicine (Arlington, VA, USA)* pp 104–9

[20] Lauterbach C, Strasser M, Jung S and Weber W 2002 Smart clothes self-powered by body heat *Proc. of Avantex Symp. (Frankfurt, Germany)* pp 5259–63

[21] Marinkovic S and Popovici E 2009 Network coding for efficient error recovery in wireless sensor networks for medical applications *Proc. of Int. Conf. on Emerging Network Intelligence (Sliema, Malta, 11–16 October)* pp 15–20

[22] Schoellhammer T, Osterweil E, Greenstein B, Wimbrow M and Estrin D 2004 Lightweight temporal compression of microclimate datasets *Proc. of the 29th Annual IEEE Int. Conf. on Local Computer Networks (Tampa, FL, USA, 16–18 November)* pp 516–24

[23] Barth A T, Hanson M A, Powell Jr H C and Lach J 2009 Tempo 3.1: A body area sensor network platform for continuous movement assessment *Proc. of the 6th Int. Workshop on Wearable and Implantable Body Sensor Networks (Berkeley, CA, USA)* pp 71–6

[24] Gietzelt M, Wolf K H, Marschollek M and Haux R 2008 Automatic self-calibration of body worn triaxial-accelerometers for application in healthcare *Proc. of the Second Int. Conf. on Pervasive Computing Technologies for Healthcare (Tampere, Finland)* pp 177–80

[25] Gao T, Greenspan D, Welsh M, Juang R R and Alm A 2005 Vital signs monitoring and patient tracking over a wireless network *Proc. of the 27th Annual Int. Conf. of the IEEE EMBS (Shanghai, China, 1–4 September)* pp 102–5 Purwar A, Jeong D U and Chung W Y 2007 Activity monitoring from realtime triaxial accelerometer data using sensor network *Proc. of Int. Conf. on Control, Automation and Systems (Hong Kong, 21–23 March)* pp 2402–6

[26] Baker C *et al* 2007 Wireless sensor networks for home health care *Proc. of the 21st Int. Conf. on Advanced Information Networking and Applications Workshops (Niagara Falls, Canada, 21–23 May)* pp 832–7

[27] Schwiebert L, Gupta S K S and Weinmann J 2001 Research challenges in wireless networks of biomedical sensors *Proc. of the 7th Annual Int. Conf. on Mobile Computing and Networking (Rome, Italy, 16–21 July)* pp 151–65

[28] Aziz O, Lo B, King R, Darzi A and Yang G Z 2006 Pervasive body sensor network: An approach to monitoring the postoperative surgical patient *Proc. of Int. Workshop on Wearable and implantable Body Sensor Networks (BSN 2006),(Cambridge, MA, USA)* pp 13–8

[29] Kahn J M, Katz R H and Pister K S J 1999 Next century challenges: Mobile networking for smart dust *Proc. of the ACM MobiCom'99 (Washington, DC, USA, August)* pp 271–8

[30] Noury N, Herve T, Rialle V, Virone G, Mercier E, Morey G, Moro A and Porcheron T 2000 Monitoring behavior in home using a smart fall sensor *Proc. of IEEE-EMBS Special Topic Conf. on Micro-technologies in Medicine and Biology (Lyon, France, 12–14 October)* pp 607–10

[31] Kwon D Y and Gross M 2005 Combining body sensors and visual sensors for motion training *Proc. of the 2005 ACM SIGCHI Int. Conf. on Advances in Computer Entertainment Technology (Valencia, Spain, 15–17 June)* pp 94–101

[32] Boulgouris N K, Hatzinakos D and Plataniotis K N 2005 Gait recognition: A challenging signal processing technology for biometric identification *IEEE Signal Process. Mag.* **22** 78–90

[33] Kimmeskamp S and Hennig E M 2001 Heel to toe motion characteristics in Parkinson patients during free walking *Clin. Biomech.* **16** 806–12

[34] Turcot K, Aissaoui R, Boivin K, Pelletier M, Hagemeister N and de Guise J A 2008 New accelerometric method to discriminate between asymptomatic subjects and patients with medial knee osteoarthritis during 3-D gait *IEEE Trans. Biomed. Eng.* **55** 1415–22

[35] Furnée H 1997 Real-time motion capture systems *Three-Dimensional Analysis of Human Locomotion* ed P Allard, A Cappozzo, A Lundberg and C L Vaughan (Chichester: Wiley) pp 85–108

[36] Bamberg S J M, Benbasat A Y, Scarborough D M, Krebs D E and Paradiso J A 2008 Gait analysis using a shoe-integrated wireless sensor system *IEEE Trans. Inf. Technol. Biomed.* **12** 413–23

[37] Choi J H, Cho J, Park J H, Eun J M and Kim M S 2008 An efficient gait phase detection device based on magnetic sensor array *Proc. of the 4th Kuala Lumpur Int. Conf. on Biomedical Engineering (Kuala Lumpur, Malaysia, 25–28 June)* **vol 21** pp 778–81

[38] Hidler J 2004 Robotic-assessment of walking in individuals with gait disorders *Proc. of the 26th Annual Int. Conf. of the IEEE Engineering in Medicine and Biology Society (San Francisco, CA, USA, 1–5 September)* **vol 7** pp 4829–31

[39] Wahab Y and Bakar N A 2011 Gait analysis measurement for sport application based on ultrasonic system *Proc. of the 2011 IEEE 15th Int. Symp. on Consumer Electronics (Singapore, 14–17 June)* pp 20–4

[40] De Silva B, Jan A N, Motani M and Chua K C 2008 A real-time feedback utility with body sensor networks *Proc. of the 5th Int. Workshop on Wearable and Implantable Body Sensor Networks (BSN 08) (Hong Kong, 1–3 June)* pp 49–53

[41] Salarian A, Russmann H, Vingerhoets F J G, Dehollain C, Blanc Y, Burkhard P R and Aminian K 2004 Gait assessment in Parkinson's disease: Toward an ambulatory system for long-term monitoring *IEEE Trans. Biomed. Eng.* **51** 1434–43

[42] Atallah L, Jones G G, Ali R, Leong J J H, Lo B and Yang G Z 2011 Observing recovery from knee-replacement surgery by using wearable sensors *Proc. of the 2011 Int. Conf. on Body Sensor Networks (Dallas, TX, USA, 23–25 May)* pp 29–34

[43] ElSayed M, Alsebai A, Salaheldin A, El Gayar N and ElHelw M 2010 Ambient and wearable sensing for gait classification in pervasive healthcare environments *Proc. of the 12th IEEE Int. Conf. on e-Health Networking Applications and Services (Healthcom) (Lyon, France, 1–3 July)* pp 240–5

[44] Cross N 2000 *Engineering Design Methods—Strategies for Product Design* 3rd edn (New York: Wiley)

[45] Buede D M 1999 *The Engineering Design of Systems Models and Methods* (New York: Wiley)

[46] Delbecq A L and VandeVen A H 1971 A group process model for problem identification and program planning *J. Appl. Behav. Sci.* **VII** 466–91

[47] Rechtin E 2000 *The Art of Systems Architecting* 2nd edn ed M W Maier (Boca Raton, FL: CRC Press)

[48] Blanchard B S and Fabrycky W J 2005 *Systems Engineering and Analysis* 4th edn (Englewood Cliffs, NJ: Prentice-Hall)

Chapter 7

Wearable antennas for wireless communication systems

Wireless communication and the medical industry have continuously grown in the last few years. Low profile compact antennas are crucial in the development of wearable human biomedical systems. Printed antennas are the perfect antenna solution for wireless and medical communication systems. Printed antennas possess attractive features such as being low profile, lightweight, with small volume and low production costs. These features are crucial for wireless compact wearable communication systems. In addition, the benefit of a compact low-cost feed network is attained by integrating the feed structure with the radiating elements on the same substrate. Printed antennas are used in communication systems that employ MIC and MMIC technologies.

7.1 Introduction to antennas

Antennas are a major component in wearable communication and medical systems [1–16]. Mobile antenna systems are presented in [10]. Physically, an antenna is an arrangement of one or more conductors. However, there several dielectric antennas. In transmitting mode, an alternating current is created in the elements by applying a voltage at the antenna terminals, causing the elements to radiate an electromagnetic field. In receiving mode, an electromagnetic field from another source induces an alternating current in the elements and a corresponding voltage at the antenna's terminals. Some receiving antennas (such as parabolic and horn) incorporate shaped reflective surfaces to receive the radio waves striking them and direct or focus them onto the actual conductive elements. Antennas are used to radiate efficiently electromagnetic energy in desired directions. Antennas match radio frequency systems, sources of electromagnetic energy, to space. All antennas may be used to receive or radiate electromagnetic waves at radio frequencies. Antennas convert

doi:10.1088/978-0-7503-5224-6ch7

electromagnetic radiation into electric current, or vice versa. Antennas are a necessary part of all communication systems and RF devices. Antennas are used in systems such as radio, television broadcasting, point-to-point radio communication, wireless LAN, cell phones, radar, medical systems, wearable systems, wireless wearable systems and spacecraft communication. Antennas are employed in air or outer space but can also be operated under water. Antennas are implanted on and inside the human body. Antennas can also be operated through soil and rock at low frequencies for short distances.

7.2 Antenna definitions

Radiation pattern—A radiation pattern is the antenna's radiated field as a function of the direction in space. It is a way of plotting the radiated power from an antenna. This power is measured at various angles at a constant distance from the antenna.
Radiator—This is the basic element of an antenna. An antenna can be made up of multiple radiators.
Bore sight—The direction in space to which the antenna radiates maximum electromagnetic energy.
Range—Antenna range is the radial range from an antenna to an object in space.
Azimuth (AZ)—The angle from left to right from a reference point, from 0° to 360°.
Elevation (EL)—EL angle is the angle from a horizontal (x, y) plane, from −90° (down) to +90° (up).
Main beam—Main beam is the region around the direction of maximum radiation, usually the region that is within 3 dB of the peak of the main lobe.
Beamwidth—Beamwidth is the angular range of the antenna pattern in which at least half of the maximum power is emitted. This angular range, of the major lobe, is defined as the points at which the field strength falls around 3 dB with regard to the maximum field strength.
Side lobes—Side lobes are smaller beams that are away from the main beam. Side lobes present radiation in undesired directions. The side lobe level is a parameter used to characterize the antenna radiation pattern. It is the maximum value of the side lobes away from the main beam and is usually expressed in decibels.
Radiated power—Total radiated power when the antenna is excited by a current or voltage of known intensity.
Isotropic radiator—Theoretical lossless radiator that radiates, or receives, equal electromagnetic energy in free space to all directions.
Wearable antenna—Antenna worn on the human body.
Directivity—The ratio between the amounts of energy propagating in a certain direction compared to the average energy radiated to all directions over a sphere.

$$D = \frac{P(\theta, \phi)_{\text{maximal}}}{P(\theta, \phi)_{\text{average}}} = 4\pi \frac{P(\theta, \phi)_{\text{maximal}}}{P_{rad}} \tag{7.1}$$

$$P(\theta, \phi)_{\text{average}} = \frac{1}{4\pi} \iint P(\theta, \phi) \sin\theta \, d\theta \, d\phi = \frac{P_{rad}}{4\pi} \tag{7.2}$$

$$D \sim \frac{4\pi}{\theta E \times \theta H} \tag{7.3}$$

θE – Beamwidth in radian in ELplane

θH – Beamwidth in radian in AZ plane

Antenna effective area (A_{eff})—The antenna area that contributes to the antenna's directivity.

$$D = \frac{4\pi A_{eff}}{\lambda^2} \tag{7.4}$$

Antenna gain (G)—The ratio between the amounts of energy propagating in a certain direction compared to the energy that would be propagating in the same direction if the antenna were not a directional, isotropic radiator, is known as its gain.

Radiation efficiency (α)—Radiation efficiency is the ratio of power radiated to the total input power. The efficiency of an antenna takes into account losses, and is equal to the total radiated power divided by the radiated power of an ideal lossless antenna.

$$G = \alpha D \tag{7.5}$$

Types of wearable antennas

Wire-type and printed antennas

Dipoles,
Monopoles,
Slot antennas,
Bi-conical antennas,
Loop antennas,
Helical antennas,
Printed antennas.

Aperture-type antennas

Horn and open waveguide,
Reflector antennas,
Arrays and phased arrays,
Microstrip and printed antenna arrays,
Slot antenna arrays.

A comparison of the directivity and gain values for several antennas is given in table 7.1.

$$\text{For small antennas}\left(l < \frac{\lambda}{2}\right) G \cong \frac{41\,000}{\theta E^{\circ} \times \theta H^{\circ}} \tag{7.6a}$$

Table 7.1. Antenna directivity versus antenna gain.

Antenna type	Directivity (dBi)	Gain (dBi)
Isotropic radiator	0	0
Dipole $\lambda/2$	2	2
Dipole above ground plane	6–4	6–4
Microstrip antenna	7–8	6–7
Yagi antenna	6–18	5–16
Helix antenna	7–20	6–18
Horn antenna	10–30	9–29
Reflector antenna	15–60	14–58

$$\text{For medium size antennas } (\lambda < l < 8\lambda)\ G \cong \frac{31\,000}{\theta E^{\circ} \times \theta H^{\circ}} \tag{7.6b}$$

$$\text{For big antennas } (8\lambda < l)\ G \cong \frac{27\,000}{\theta E^{\circ} \times \theta H^{\circ}} \tag{7.6c}$$

Antenna impedance—Antenna impedance is the ratio of voltage at any given point along the antenna to the current at that point. Antenna impedance depends upon the height of the antenna above the ground and the influence of the surrounding objects. The impedance of a quarter-wave monopole near a perfect ground is approximately 36 ohms. The impedance of a half-wave dipole is approximately 75 ohms.
Antenna array—An array of antenna elements is a set of antennas used for transmitting or receiving electromagnetic waves.
Active antenna—Antenna consists of a radiating element and active elements such as amplifiers.
Phased arrays—Phased array antennas are electrically steerable. The physical antenna can be stationary. Phased arrays, smart antennas, incorporate active components for beam steering.

Steerable antennas

- Arrays with switchable elements and partially mechanically and electronically steerable arrays.
- Hybrid antenna systems—to fully electronically steerable arrays. Such systems can be equipped with phase and amplitude shifters for each element, or the design can be based on digital beam forming (DBF).
- This technique, in which the steering is performed directly on a digital level, allows the most flexible and powerful control of the antenna beam.

7.3 Dipole antenna

A dipole antenna is a small wire antenna. It consists of two straight conductors excited by a voltage fed via a transmission line as shown in figure 7.1. Each side of the transmission line is connected to one of the conductors. The most common

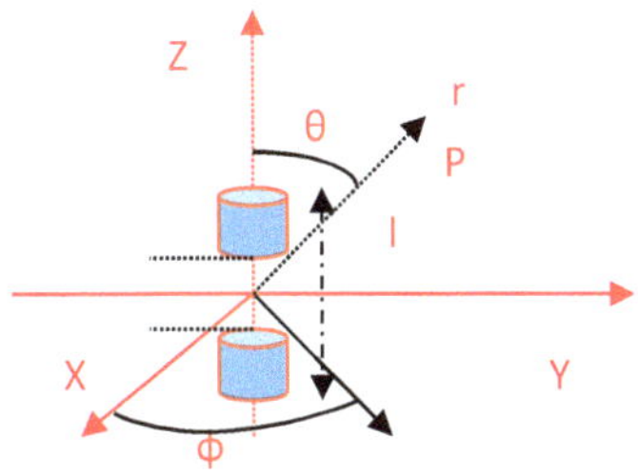

Figure 7.1. Dipole antenna.

dipole is the half-wave dipole, in which each of the two conductors is approximately a quarter-wavelength long, so the length of the antenna is a half-wavelength.

We can calculate the fields radiated from the dipole by using the potential function. The electric potential function is φ_l. The electric potential is A. The potential function is given in equation (7.7).

$$\begin{aligned} \varphi_l &= \frac{1}{4\pi\varepsilon_0}\int_c \frac{\rho_l e^{j(\omega t-\beta R)}}{R}dl \\ A_l &= \frac{\mu_0}{4\pi}\int_c \frac{ie^{j(\omega t-\beta R)}}{R}dl \end{aligned} \tag{7.7}$$

Radiation from a small dipole

The length of a small dipole is small compared to the wavelength and is called an elementary dipole. We may assume that the current along the elementary dipole is uniform. We can solve the wave equation in spherical coordinates by using the potential function given in equation (7.7). The electromagnetic fields in a point $P(r, \theta, \varphi)$ are given in equation (7.8). The electromagnetic fields in equation (7.8) vary as $\frac{1}{r}, \frac{1}{r^2}, \frac{1}{r^3}$. For $r \ll 1$, the dominant component of the field varies as $\frac{1}{(r)^3}$ and is given in equation (7.9). These fields are the dipole near-fields. In this case, the waves are standing waves and the energy oscillates in the antenna's near zone and are not radiated to the open space. The real part of the Poynting vector is equal to zero. For $r \gg 1$, the dominant component of the field varies as $1/r$ as given in equation (7.10). These fields are the dipole far-fields.

$$\begin{aligned} E_r &= \eta_0\frac{lI_0\cos\theta}{2\pi r^2}\left(1-\frac{j}{\beta r}\right)e^{j(\omega t-\beta r)} \\ E_\theta &= j\eta_0\frac{\beta lI_0\sin\theta}{4\pi r}\left(1-\frac{j}{\beta r}-\frac{1}{(\beta r)^2}\right)e^{j(\omega t-\beta r)} \\ H_\varphi &= j\frac{\beta lI_0\sin\theta}{4\pi r}\left(1-\frac{j}{\beta r}\right)e^{j(\omega t-\beta r)} \\ H_r &= 0 \qquad H_\theta = 0 \qquad E_\phi = 0 \\ I &= I_0\cos\omega t \end{aligned} \tag{7.8}$$

$$E_r = -j\eta_0 \frac{lI_0 \cos\theta}{2\pi\beta r^3} e^{j(\omega t - \beta r)}$$
$$E_\theta = -j\eta_0 \frac{lI_0 \sin\theta}{4\pi\beta r^3} e^{j(\omega t - \beta r)} \tag{7.9}$$
$$H_\varphi = \frac{lI_0 \sin\theta}{4\pi r^2} e^{j(\omega t - \beta r)}$$

$$E_r = 0$$
$$E_\theta = j\eta_0 \frac{l\beta I_0 \sin\theta}{4\pi r} e^{j(\omega t - \beta r)} \tag{7.10}$$
$$H_\varphi = j\frac{l\beta I_0 \sin\theta}{4\pi r} e^{j(\omega t - \beta r)}$$

$$\frac{E_\theta}{H_\varphi} = \eta_0 = \sqrt{\frac{\mu_0}{\varepsilon_0}} \tag{7.11}$$

In the far-fields the electromagnetic fields vary as $\frac{1}{r}$ and sin θ. The wave impedance in free space is given in equation (7.11).

Dipole radiation pattern

The antenna radiation pattern represents the radiated fields in space at a point $P(r, \theta, \varphi)$ as function of θ, φ. The antenna radiation pattern is three-dimensional. When φ is constant and θ varies we get the E plane radiation pattern. When φ varies and θ is constant, usually $\theta = \pi/2$, we get the H plane radiation pattern.

Dipole *E* plane radiation pattern

The dipole E plane radiation pattern is given in equation (7.12) and presented in figure 7.2.

$$|E_\theta| = \eta_0 \frac{l\beta I_0\, |\sin\theta|}{4\pi r} \tag{7.12}$$

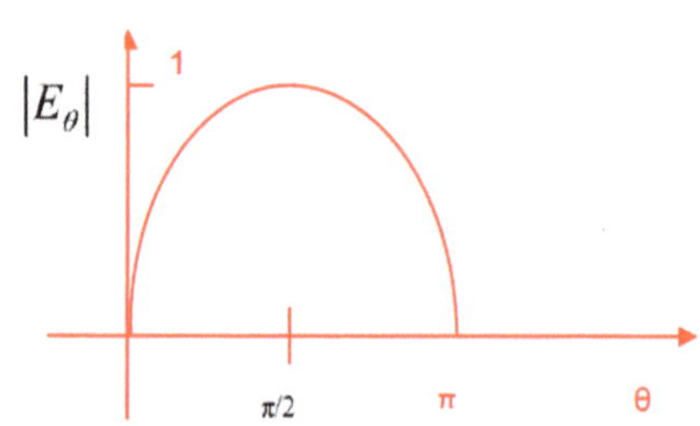

Figure 7.2. Dipole E plane radiation pattern.

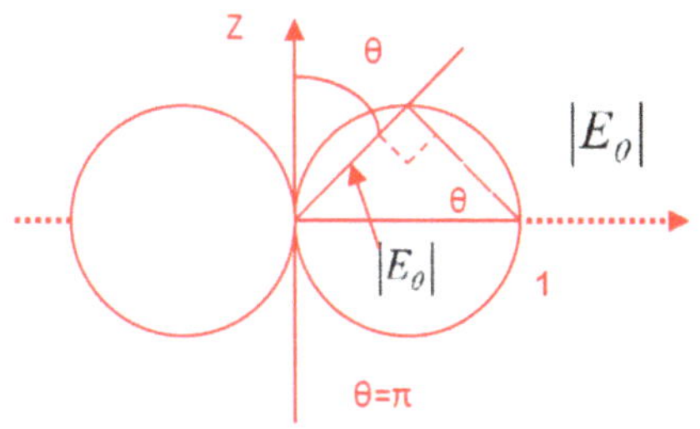

Figure 7.3. Dipole E plane radiation pattern in a spherical coordinate system.

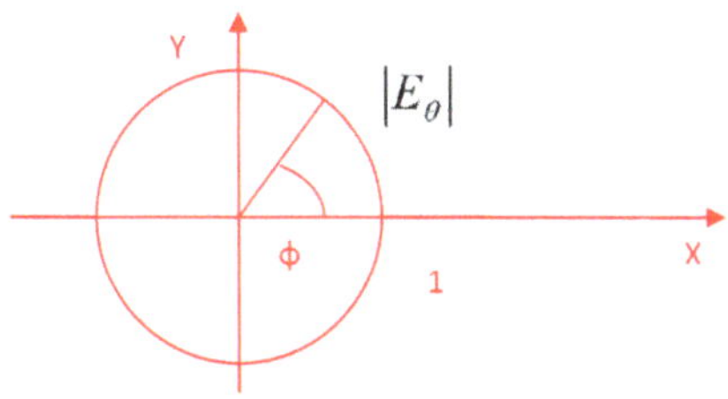

Figure 7.4. Dipole H plane radiation pattern for $\theta = \pi/2$.

At a given point $P(r, \theta, \varphi)$, the dipole E plane radiation pattern is given in equation (7.13).

$$\begin{aligned} |E_\theta| &= \eta_0 \frac{l\beta I_0\, |\sin\theta|}{4\pi r} = A|\sin\theta| \\ \text{Choose} \quad & A = 1 \\ |E_\theta| &= |\sin\theta| \end{aligned} \tag{7.13}$$

The dipole E plane radiation pattern in a spherical coordinate system is shown in figure 7.3.

Dipole *H* plane radiation pattern

For $\theta = \pi/2$, the dipole H plane radiation pattern is given in equation (7.14) and presented in figure 7.4.

$$|E_\theta| = \eta_0 \frac{l\beta I_0}{4\pi r} \tag{7.14}$$

At a given point $P(r, \theta, \varphi)$, the dipole H plane radiation pattern is given in equation (7.15).

$$\begin{aligned} |E_\theta| &= \eta_0 \frac{l\beta I_0\, |\sin\theta|}{4\pi r} = A \\ \text{Choose} \quad & A = 1 \\ |E_\theta| &= 1 \end{aligned} \tag{7.15}$$

The dipole H plane radiation pattern in the xy plane is a circle with $r = 1$.

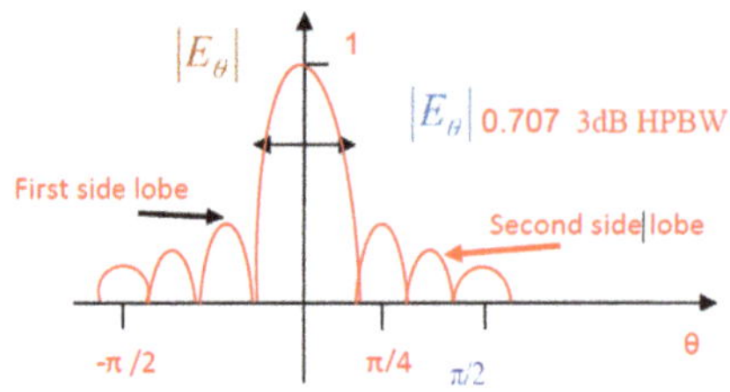

Figure 7.5. Antenna's typical radiation pattern.

The radiation pattern of a vertical dipole is omnidirectional. It radiates equal power in all azimuthal directions perpendicular to the axis of the antenna. A dipole *H* plane radiation pattern in a spherical coordinate system is shown in figure 7.4.

Antenna radiation pattern

A typical antenna radiation pattern is shown in figure 7.5. The antenna's main beam is measured between the points that the maximum relative field intensity *E* decays to 0.707*E*. Half of the radiated power is concentrated in the antenna's main beam. The antenna's main beam is 3 dB beamwidth. Radiation to undesired directions is concentrated in the antenna's side lobes.

For a dipole the power intensity varies as ($\sin^2\theta$). At $\theta = 45°$ and $\theta = 135°$ the radiated power equals half the power radiated toward $\theta = 90°$. The dipole's beamwidth is $\theta = (135 - 45) = 90°$.

Dipole directivity

Directivity is defined as the ratio between the amounts of energy propagating in a certain direction compared to the average energy radiated to all directions over a sphere as written in equations (7.16) and (7.17).

$$D = \frac{P(\theta, \phi)_{\text{maximal}}}{P(\theta, \phi)_{\text{average}}} = 4\pi\frac{P(\theta, \phi)_{\text{maximal}}}{P_{rad}} \tag{7.16}$$

$$P(\theta, \phi)_{\text{average}} = \frac{1}{4\pi}\iint P(\theta, \phi)\sin\theta\, d\theta\, d\phi = \frac{P_{rad}}{4\pi} \tag{7.17}$$

The radiated power from a dipole is calculated by computing the Poynting vector *P* as given in equation (7.18).

$$\begin{aligned} P &= 0.5(E \times H^*) = \frac{15\pi I_0^2 l^2 \sin^2\theta}{r^2\lambda^2} \\ W_T &= \int_s P \cdot ds = \frac{15\pi I_0^2 l^2}{\lambda^2}\int_0^{\pi} \sin^3\theta\, d\theta \int_0^{2\pi} d\varphi = \frac{40\pi^2 I_0^2 l^2}{\lambda^2} \end{aligned} \tag{7.18}$$

The overall radiated energy is W_T. W_T is computed by integration of the power flow over an imaginary sphere surrounding the dipole. The power flow of an isotropic radiator is equal to W_T divided by the surrounding area of the sphere, $4\pi r^2$, as given in equation (7.19). The dipole directivity at $\theta = 90°$ is 1.5 or 1.76 dB as shown in equation (7.20).

$$\begin{aligned} \oint_s ds &= r^2 \int_0^{\pi} \sin\theta \, d\theta \int_0^{2\pi} d\phi = 4\pi r^2 \\ P_{iso} &= \frac{W_T}{4\pi r^2} = \frac{10\pi I_0^2 l^2}{r^2 \lambda^2} \\ D &= \frac{P}{P_{iso}} = 1.5 \sin^2\theta \end{aligned} \tag{7.19}$$

$$G_{dB} = 10 \log_{10} G = 10 \log_{10} 1.5 = 1.76 \text{ dB} \tag{7.20}$$

For small antennas or for antennas without losses, $D = G$, losses are negligible. For a given θ and φ for small antennas the approximate directivity is given by equation (7.21).

$$\begin{aligned} D &= \frac{41\,253}{\theta_{3dB}\phi_{3dB}} \\ G &= \xi D \qquad \xi = \text{Efficiency} \end{aligned} \tag{7.21}$$

Antenna losses degrade the antenna efficiency. Antenna losses consist of conductor loss, dielectric loss, radiation loss and mismatch losses. For resonant small antennas $\xi = 1$. For reflector and horn antennas the efficiency varies between $\xi = 0.5$ to $\xi = 0.7$. The beamwidth of a small dipole, 0.1λ long, is around 90°. The 0.1λ dipole impedance is around 2 Ω. The beamwidth of a dipole, 0.5λ long, is around 80°. The impedance of a 0.5λ dipole is around 73 Ω.

Antenna impedance

Antenna impedance determines the efficiency of transmitting and receiving energy in antennas. The dipole impedance is given in equation (7.22).

$$\begin{aligned} R_{rad} &= \frac{2W_T}{I_0^2} \\ \text{For a dipole: } R_{rad} &= \frac{80\pi^2 l^2}{\lambda^2} \end{aligned} \tag{7.22}$$

Impedance of a folded dipole

A folded dipole is a half-wave dipole with an additional wire connecting its two ends. If the additional wire has the same diameter and cross section as the dipole, two nearly identical radiating currents are generated. The resulting far-field emission pattern is nearly identical to the one for the single-wire dipole described above, but at resonance its feed point impedance R_{rad-f} is four times the radiation resistance of a dipole. This is because for a fixed amount of power, the total radiating current I_0 is equal to twice the current in each wire and thus equal to twice the current at the feed point. Equating the average radiated power to the average power delivered at the

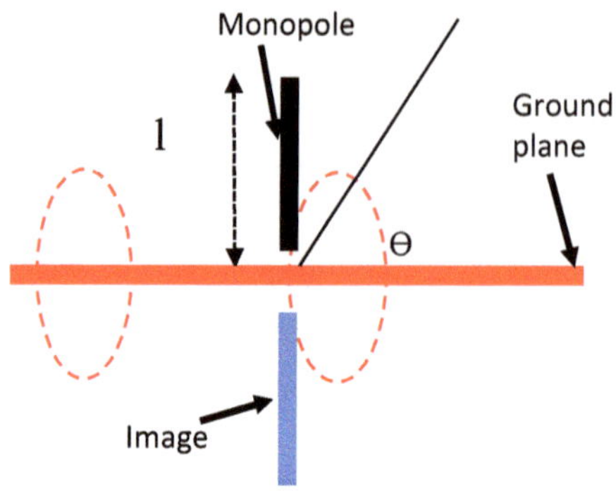

Figure 7.6. Monopole antenna.

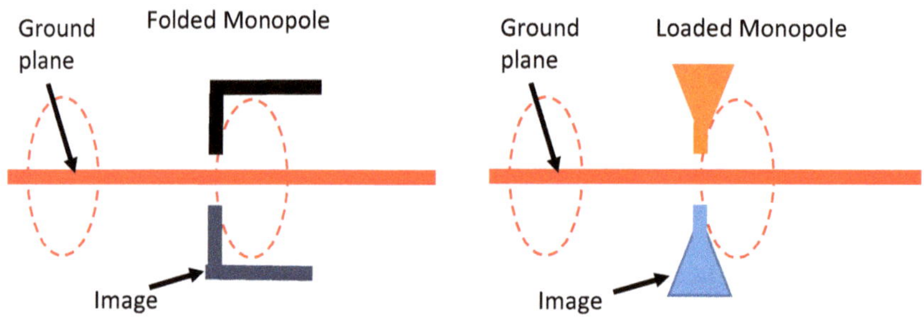

Figure 7.7. (a) Inverted monopole antenna. (b) Loaded monopole antenna.

feed point, we find that $R_{rad-f} = 4\ R_{rad} = 300\ \Omega$. The folded dipole has a wider bandwidth than a single dipole.

7.4 Monopole antenna for wearable communication systems

A monopole antenna is usually a one-quarter-wavelength long conductor mounted above a ground plane or the Earth. Based on the image theory, the monopole image is located behind the ground plane. The monopole antenna and the monopole image form a dipole antenna (figure 7.6).

A monopole antenna is half a dipole that radiates electromagnetic fields above the ground plane. The impedance of a 0.5λ monopole antenna is around 37 Ω. The beamwidth of a monopole, 0.25λ long, is around 40°. The directivity of a monopole, 0.25λ long, is around 3 dBi to 5 dBi. Usually in wireless communication systems very short monopole antennas are employed. The impedance of a 0.05λ monopole antenna is around 1 Ω with capacitive reactance. The beamwidth of a monopole, 0.05λ long, is around 45°. The directivity of a monopole, 0.05λ long, is around 3 dBi.

An inverted monopole antenna is shown in figure 7.7(a), and a loaded monopole antenna is shown in figure 7.7(b). Monopole antennas may be printed on a dielectric substrate as part of wearable communication devices.

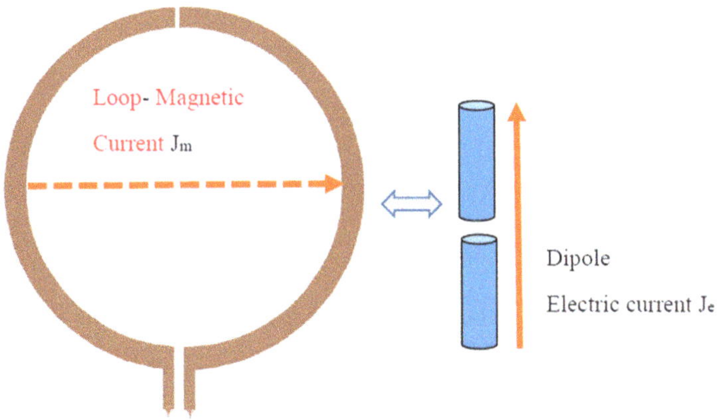

Figure 7.8. Duality relationship between dipole and loop antennas.

7.5 Loop antennas for wireless communication systems

Loop antennas are compact, low profile and low-cost antennas. Loop antennas are employed in wearable wireless communication systems.

7.5.1 Duality relationship between dipole and loop antennas

The loop antenna is referred to as the dual of the dipole antenna, see figure 7.8. A small dipole has a magnetic current flowing (as opposed to an electric current as in a regular dipole), the fields would resemble that of a small loop. The short dipole has a capacitive impedance (the imaginary part of the impedance is negative). The impedance of a small loop is inductive (positive imaginary part).

Duality means combing two different things that are closely linked. In antennas, duality theory means that it is possible to write the fields of one antenna from the field expressions of the other antenna by interchanging linked parameters. Electric current may be interchanged by an equivalent magnetic current. The variation of electromagnetic waves as a function of time may be written as $e^{j\omega t}$. The derivative as a function of time is $j\omega e^{j\omega t}$. Maxwell's equations for system 1 may be written as in equation (7.23):

$$\begin{aligned} \nabla X E_1 &= -j\omega\mu_1 H_1 \\ \nabla X H_1 &= (\sigma + j\omega\varepsilon_1)E = J_e + j\omega\varepsilon_1 E_1 \end{aligned} \tag{7.23}$$

System 2 is a dual system to system 1. Maxwell's equations for system 2 may be written as in equation (7.24):

$$\begin{aligned} \nabla X H_2 &= j\omega\varepsilon_2 E_2 \\ \nabla X E_2 &= -J_m - j\omega\mu_2 H_2 \end{aligned} \tag{7.24}$$

System 1 electric current source	System 2 magnetic current source
J_e	J_m
E_1	H_2
μ_1	ε_2
H_1	$-E_2$
ε_1	μ_2

By using the duality principle the far electromagnetic fields of the loop antenna are given in equation (7.25).

$$\begin{aligned} &E_r = H_\varphi = E_\theta = 0 \\ &H_\theta = j\eta_0 \frac{l\beta I_m \sin\theta}{4\pi r} e^{j(\omega t - \beta r)} \\ &E_\varphi = -j\frac{l\beta I_m \sin\theta}{4\pi r} e^{j(\omega t - \beta r)} \end{aligned} \tag{7.25}$$

The directivity of a loop antenna with a circumference 0.5λ long is around 1 dBi. The directivity of a loop antenna with a circumference 1λ long is around 4 dBi. An H plane 3D radiation pattern of a loop antenna with a circumference of 0.45λ in free space is shown in figure 7.9.

7.5.2 Medical applications of printed loop antennas

Loop antennas have low radiation resistance and high reactance. It is difficult to match the antenna to a transmitter. Loop antennas are most often used as receiving

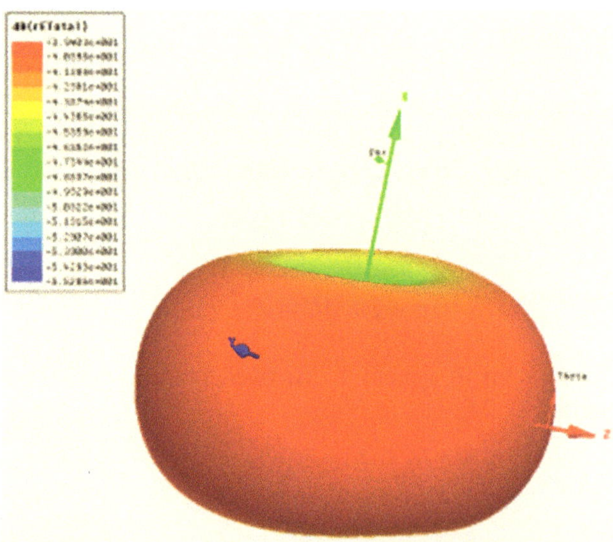

Figure 7.9. H plane radiation pattern of loop antenna in free space.

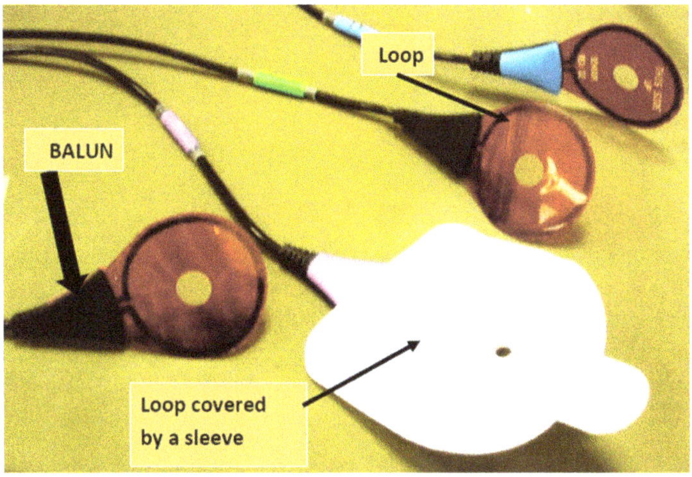

Figure 7.10. Photo of loop antennas.

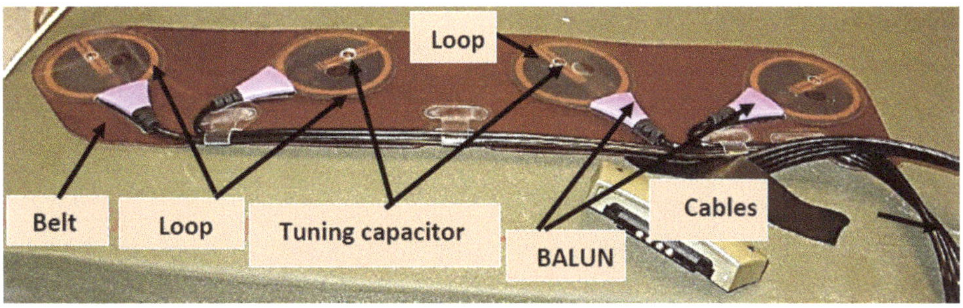

Figure 7.11. Photo of a loop antenna array inside a belt.

antennas, where impedance mismatch loss can be accepted. Small loop antennas are used in medical devices as field strength probes, in pagers and wireless measurements.

A Balun transformer is connected to the loop feed-lines. A Balun transformer is a device that converts signals between a balanced transmission line and an unbalanced transmission line. The transformer may be used to match the loop antenna to the communication system. A photo of loop antennas is shown in figure 7.10. The loop antenna may be inserted into a sleeve as shown in figure 7.10. The sleeve's electrical properties were chosen to match the loop antenna to the human body. The sleeve also provides protection from the environment to the loop antenna. A photo of a four loop antenna array inside a belt is shown in figure 7.11. The loop antenna may be tuned by adding a capacitor or a varactor as shown in figure 7.11.

7.6 Wearable printed antennas

Printed antennas possess attractive features such as being low profile, flexible, lightweight, with small volume and low production costs. Printed antennas may be

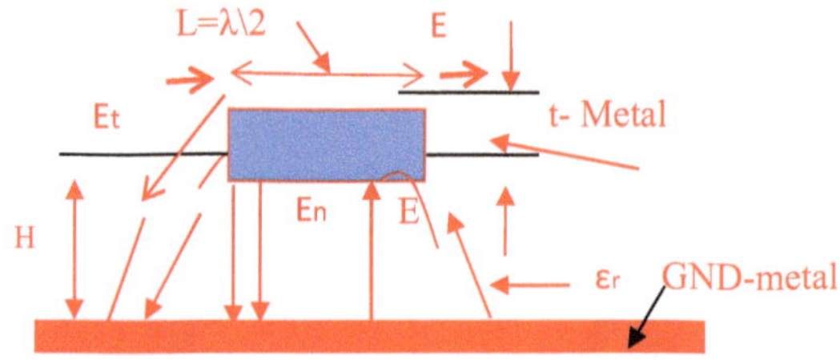

Figure 7.12. Cross section of a microstrip antenna.

used as wearable antennas. Printed antennas have been widely discussed in books and papers in the last decade [1–19]. The most popular type of printed antennas are microstrip antennas. However, PIFA, slot and dipole printed are widely used in communication systems. Printed antennas may be employed in communication links, seekers and medical systems.

Applications of wearable antennas

- Medical.
- Wireless communication.
- WLAN.
- HIPER LAN.
- GPS.
- Military applications.

7.6.1 Wearable microstrip antennas

Microstrip antennas are printed on a dielectric substrate with low dielectric losses. A cross section of the microstrip antenna is shown in figure 7.12. Microstrip antennas are thin patches etched on a dielectric substrate ε_r. The substrate thickness, H, is less than 0.1λ. Microstrip antennas are widely discussed in [1–7]. A printed antenna may be attached to the human body or inserted inside a wearable belt.

Advantages of microstrip antennas:

- Low-cost to fabricate.
- Conformal structures are possible.
- Easy to form a large uniform array with half-wavelength spacing.
- Lightweight and low volume.

These features are crucial for wearable communication systems.

Disadvantages of microstrip antennas:

- Limited bandwidth (usually 1%–5%), but much more is possible with increased complexity.
- Low power handling.

The electric field along the radiating edges is shown in figure 7.13. The magnetic field is perpendicular to the E field according to Maxwell's equations. At the edge of the strip ($X/L = 0$ and $X/L = 1$) the H field drops to zero, because there is no conductor

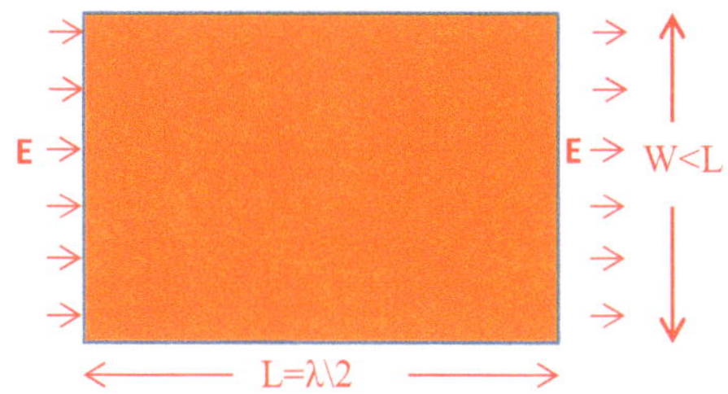

Figure 7.13. Rectangular microstrip antenna.

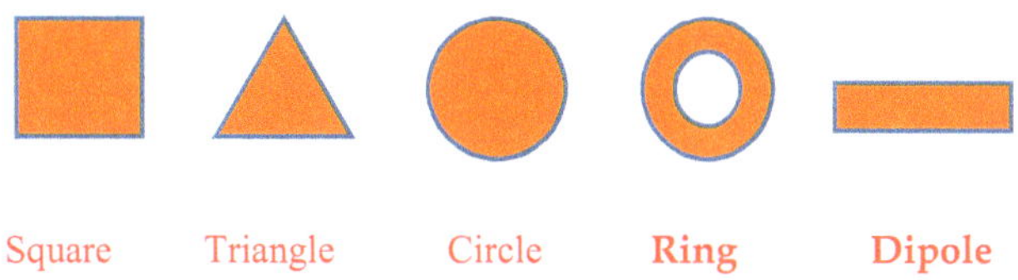

Figure 7.14. Microstrip antenna shapes.

to carry the RF current, it is maximum in the center. The E field intensity is at maximum magnitude (and opposite polarity) at the edges ($X/L = 0$ and $X/L = 1$) and zero at the center. The ratio of E to H field is proportional to the impedance that we see when we feed the patch. Microstrip antennas may be fed by a microstrip line or by a coaxial line or probe feed. By adjusting the location of the feed point between the center and the edge, we can get any impedance, including 50 Ω. A microstrip antenna's shape may be square, rectangular, triangular, circular or any arbitrary shape as shown in figure 7.14.

The dielectric constant that controls the resonance of the antenna is the effective dielectric constant of the microstrip line. The antenna's dimension W is given by equation (7.26).

$$W = \frac{c}{2f\sqrt{\epsilon_{eff}}} \tag{7.26}$$

The antenna's bandwidth is given in equation (7.27).

$$BW = \frac{H}{\sqrt{\epsilon_{eff}}} \tag{7.27}$$

The gain of the microstrip antenna is between 0 dBi to 7 dBi. The microstrip antenna's gain is a function of the antenna's dimensions and configuration. We may increase printed antenna gain by using a microstrip antenna array configuration. In a microstrip antenna array the benefit of a compact low-cost feed network is attained by integrating the RF feed network with the radiating elements on the same substrate. Microstrip antenna feed networks are presented in figure 7.15. Figure 7.15(a) presents a parallel feed network, and figure 7.15(b) presents a parallel series feed network.

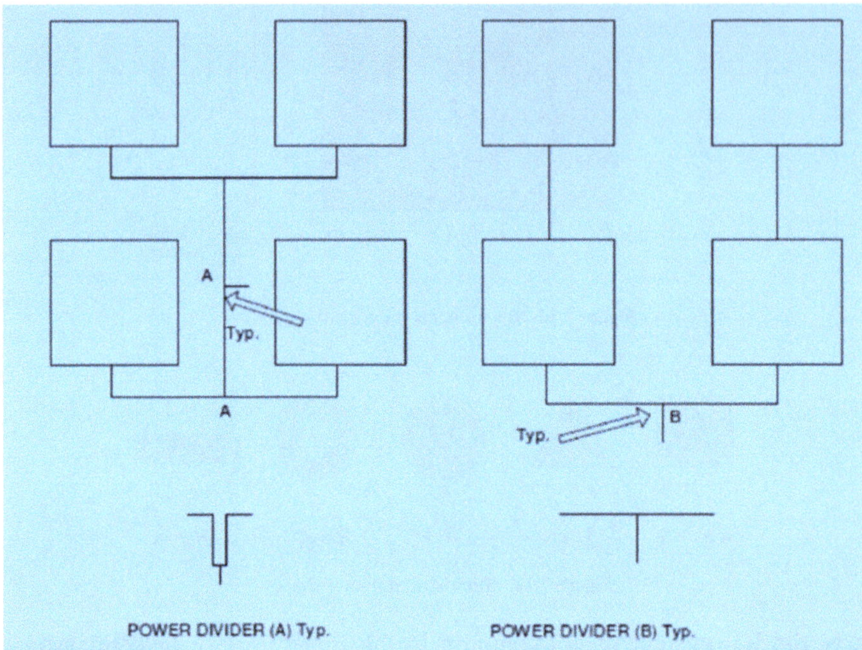

Figure 7.15. Configuration of microstrip antenna array. (a) Parallel feed network. (b) Parallel series feed network.

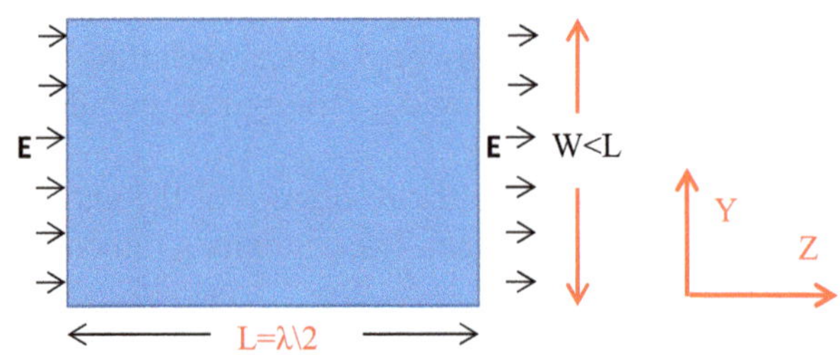

Figure 7.16. Transmission line model of patch microstrip antennas.

7.6.2 Transmission line model of microstrip antennas

In the transmission line model (TLM) of patch microstrip antennas, the antenna is represented as two slots connected by a transmission line. A TLM is presented in figure 7.16. A TLM is not an accurate model; however, it gives a good physical understanding of patch microstrip antennas. The electric field along and underneath the patch depend on the z coordinate, see equation (7.28). In the design of wearable patch antennas the human body's electrical parameters should be considered in the design.

$$E_x \sim \cos\left(\frac{\pi z}{L_{eff}}\right) \tag{7.28}$$

At $z = 0$ and $z = L_{eff}$ the electric field is maximum. At $z = \frac{L_{eff}}{2}$ the electric field equal zero. For $\frac{H}{\lambda_0} < 0.1$ the electric field distribution along the x-axis is assumed to be uniform. The slot admittance may be written as given in equations (7.29) and (7.30).

$$G = \frac{W}{120\lambda_0}\left[1 - \frac{1}{24}\left(\frac{2\pi H}{\lambda_0}\right)^2\right] for\ \frac{H}{\lambda_0} < 0.1 \tag{7.29}$$

$$B = \frac{W}{120\lambda_0}\left[1 - 0.636\ln\left(\frac{2\pi H}{\lambda_0}\right)^2\right] for\ \frac{H}{\lambda_0} < 0.1 \tag{7.30}$$

B represents the capacitive nature of the slot. $G = 1/R$. R represents the radiation losses. When the antenna is resonant the susceptances of both slots cancel out at the feed point for any position of the feed point along the patch. However, the patch admittance depends on the feed point position along the z-axis as given in equation (7.31). At the feed point the slot admittance is transformed by the equivalent length of the transmission line. The width W of the microstrip antenna controls the input impedance. Larger widths can increase the bandwidth. For a square patch antenna fed by a microstrip line, the input impedance is around 300 ohms. By increasing the width, the impedance can be reduced.

$$\begin{aligned} Y(l_1) &= Z_0\frac{1 + j\frac{Z_L}{Z_0}\tan\beta l_1}{\frac{Z_L}{Z_0} + j\tan\beta l_1} = Y_1 \\ Y_{in} &= Y_1 + Y_2 \end{aligned} \tag{7.31}$$

7.6.3 Higher-order transmission modes in microstrip antennas

In order to prevent higher-order transmission modes we should limit the thickness of the microstrip substrate to 10% of a wavelength. The cutoff frequency of the higher-order mode is given in equation (7.32).

$$f_c = \frac{c}{4H\sqrt{\varepsilon - 1}} \tag{7.32}$$

7.6.4 Effective dielectric constant

Part of the field in the microstrip antenna's structure exists in air and the other part of the field exists in the dielectric substrate. The effective dielectric constant is somewhat less than the substrate's dielectric constant. The effective dielectric constant of the microstrip line may be calculated by equations (7.33a) and (7.33b) as a function of W/H:

$$\text{For}\left(\frac{W}{H}\right) < 1 \tag{7.33a}$$

$$\varepsilon_e = \frac{\varepsilon_r+1}{2} + \frac{\varepsilon_r-1}{2}\left[\left(1 + 12\left(\frac{H}{W}\right)\right)^{-0.5} + 0.04\left(1-\left(\frac{W}{H}\right)\right)^{2}\right]$$

$$\text{For}\left(\frac{W}{H}\right) \geqslant 1 \tag{7.33b}$$

$$\varepsilon_e = \frac{\varepsilon_r+1}{2} + \frac{\varepsilon_r-1}{2}\left[\left(1 + 12\left(\frac{H}{W}\right)\right)^{-0.5}\right]$$

This calculation ignores strip thickness and frequency dispersion, but their effects are negligible.

7.6.5 Losses in microstrip antennas

Losses in a microstrip line are due to conductor loss, radiation loss and dielectric loss.

Conductor loss

Conductor loss may be calculated by using equation (7.34).

$$\begin{aligned} \alpha_c &= 8.686 \log(R_S/(2WZ_0)) \qquad dB/\text{length} \\ R_S &= \sqrt{\pi f \mu \rho} \qquad \text{skin resistance} \end{aligned} \tag{7.34}$$

Conductor losses may also be calculated by defining an equivalent loss tangent δc, given by $\delta c = \delta s/h$, where $\delta s = \sqrt{2/\omega\mu\sigma}$. Where σ is the strip conductivity, h is the substrate height and μ is the free space permeability.

Dielectric loss

Dielectric loss may be calculated by using equation (7.35).

$$\begin{aligned} \alpha_d &= 27.3\frac{\varepsilon_r}{\sqrt{\varepsilon_{eff}}}\frac{\varepsilon_{eff} - 1}{\varepsilon_r - 1}\frac{tg\delta}{\lambda_0} \text{ dB cm}^{-1} \\ tg\delta &= \text{dielectric loss coefficent} \end{aligned} \tag{7.35}$$

7.6.6 Patch radiation pattern

The patch width, W, controls the antenna's radiation pattern. The coordinate system is shown in figure 7.17. The normalized radiation pattern is approximately given by:

$$\begin{aligned} E_\theta &= \frac{\sin\left(\frac{k_0 W}{2}\sin\theta\sin\varphi\right)}{\frac{k_0 W}{2}\sin\theta\sin\varphi}\cos\left(\frac{k_0 L}{2}\sin\theta\cos\varphi\right)\cos\varphi \\ k_0 &= 2\pi/\lambda \end{aligned} \tag{7.36}$$

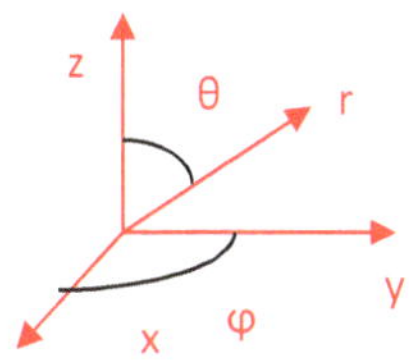

Figure 7.17. Coordinate system.

$$E_\varphi = \frac{\sin\left(\frac{k_0 W}{2}\sin\theta\sin\varphi\right)}{\frac{k_0 W}{2}\sin\theta\sin\varphi}\cos\left(\frac{k_0 L}{2}\sin\theta\cos\varphi\right)\cos\theta\sin\varphi \tag{7.37}$$

$$k_0 = 2\pi/\lambda$$

The magnitude of the fields, given by:

$$f(\theta, \varphi) = \sqrt{E_\theta^2 + E_\varphi^2} \tag{7.38}$$

7.7 Two-layer wearable stacked microstrip antennas

Two-layer microstrip antennas were presented first in [1–7]. The major disadvantage of single layer microstrip antennas is the narrow bandwidth. By designing a double layer microstrip antenna we can get a wider bandwidth. Two-layer microstrip antennas may be the best choice for wideband wearable systems.

In the first layer the antenna's feed network and a resonator are printed. In the second layer the radiating element is printed. The electromagnetic field is coupled from the resonator to the radiating element. The resonator and the radiating element shape may be square, rectangular, triangular, circular or any arbitrary shape. The distance between the layers is optimized to get the maximum bandwidth with the best antenna efficiency. The spacing between the layers may be air or a foam with low dielectric losses.

A circular polarization double layer antenna was designed at 2.2 GHz. The resonator and the feed network were printed on a substrate with a relative dielectric constant of 2.5 with thickness of 1.6 mm. The resonator is a square microstrip resonator with dimensions $W = L = 45$ mm. The radiating element was printed on a substrate with a relative dielectric constant of 2.2 with thickness 1.6 mm. The radiating element is a square patch with dimensions $W = L = 48$ mm. The patch was designed as a circular polarized antenna by connecting a 3 dB 90° branch coupler to the antenna feed-lines, as shown in figure 7.18. The antenna's bandwidth is 10% for a VSWR better than 2:1. The measured antenna's beamwidth is around 72°, and the gain is 7.5 dBi. This antenna may be applied in wideband wireless wearable systems. The measured results of the stacked microstrip wearable antennas are listed in table 7.2. The antennas listed in table 7.2 may be used in wearable communication systems. The results in table 7.2 indicate that the bandwidth of two-layer microstrip antennas may be around 9%–15% for a VSWR better than 2:1. In figure 7.19 a stacked microstrip antenna is shown. The antenna's feed network is printed on FR4

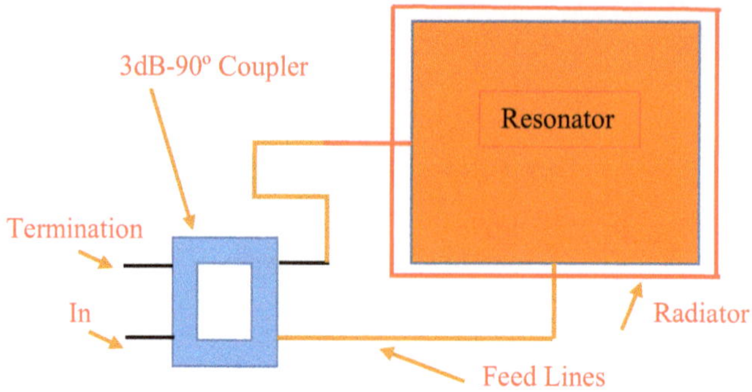

Figure 7.18. Circular polarized microstrip stacked patch antenna.

Table 7.2. Measured results of stacked microstrip antennas.

Antenna	F (GHz)	Bandwidth %	Beamwidth °	Gain dBi	Side lobe dB	Polarization
Square	2.2	10	72	7.5	−22	Circular
Circular	2.2	15	72	7.9	−22	Linear
Annular disc	2.2	11.5	78	6.6	−14	Linear
Rectangular	2.0	9	72	7.4	−25	Linear
Circular	2.4	9	72	7	−22	Linear
Circular	2.4	10	72	7.5	−22	Circular

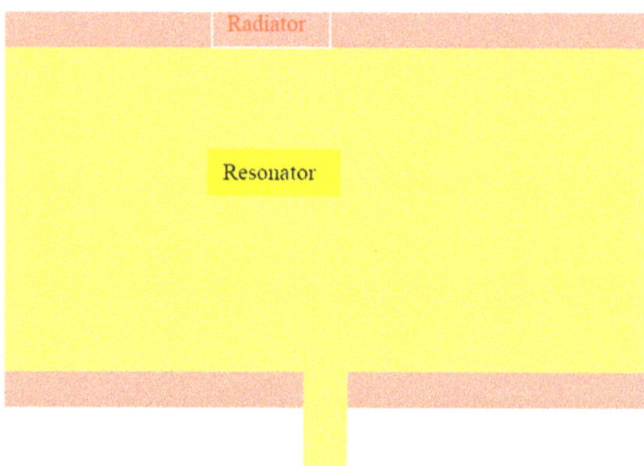

Figure 7.19. A microstrip stacked patch antenna.

dielectric substrate with a dielectric constant 4 and 1.6 mm thick. The radiator is printed on RT-Duroid 5880 dielectric substrate with a dielectric constant 2.2 and 1.6 mm thick. The antenna's electrical parameters were calculated and optimized by using ADS software. The dimensions of the microstrip stacked patch antenna shown in figure 7.19 are 33 × 20 × 3.2 mm. The computed S_{11} parameters are presented in

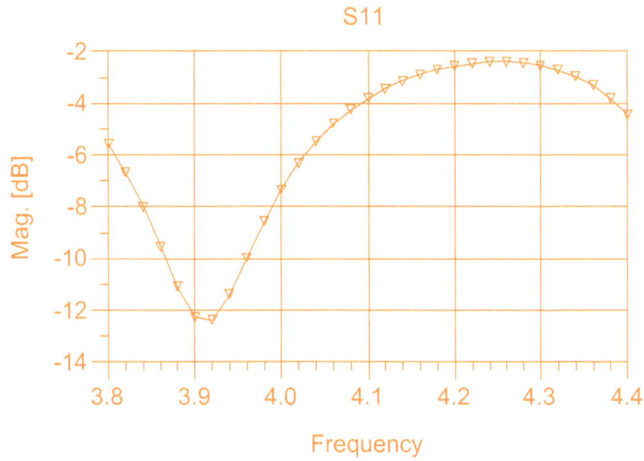

Figure 7.20. Computed S_{11} of the microstrip stacked patch.

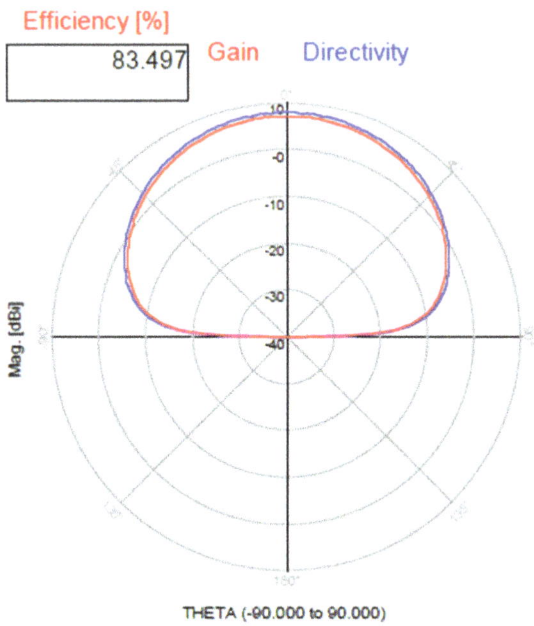

Figure 7.21. Radiation pattern of the microstrip stacked patch.

figure 7.20. The radiation pattern of the microstrip stacked patch is shown in figure 7.21. The antenna's bandwidth is around 5% for a VSWR better than 2.5:1.

The antenna's bandwidth is improved to 10% for a VSWR better than 2.0:1 by adding 8 mm air spacing between the layers. The antenna's beamwidth is around 72°, and the gain is around 7 dBi.

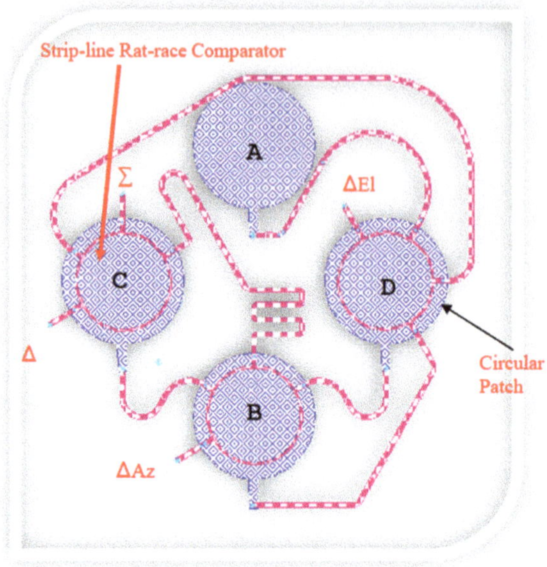

Figure 7.22. A microstrip stacked mono-pulse antenna.

7.8 Stacked mono-pulse Ku band patch antenna

A mono-pulse double layer antenna was designed at 15 GHz. The mono-pulse double layer antenna consists of four circular patch antennas as shown in figure 7.22. The resonator and the feed network was printed on a substrate with a relative dielectric constant of 2.5 with thickness 0.8 mm. The resonator is a circular microstrip resonator with diameter $a = 4.2$ mm. The radiating element was printed on a substrate with a relative dielectric constant of 2.2 with thickness 0.8 mm. The radiating element is a circular microstrip patch with diameter $a = 4.5$ mm. The four circular patch antennas are connected to three 3 dB 180° rat-race couplers via the antenna feed-lines, as shown in figure 7.22. The comparator consists of three strip-line 3 dB 180° rat-race couplers printed on a substrate with a relative dielectric constant of 2.2 with thickness of 0.8 mm. The comparator has four output ports: a sum port $\sum$, difference port $\mathbf{\Delta}$, elevation difference port $\mathbf{\Delta}_{El}$ and azimuth difference port $\mathbf{\Delta}_{Az}$ as shown in figure 7.16. The antenna's bandwidth is 10% for a VSWR better than 2:1. The antenna's beamwidth is around 36°, the measured gain is around 10 dBi, and the comparator losses are around 0.7 dB.

Rat-race coupler

A rat-race coupler is shown in figure 7.23. The rat-race circumference is 1.5 wavelengths. The distance from A to Δ port is $3\lambda/4$. The distance from A to $\sum$ port is $\lambda/4$.

For an equal-split rat-race coupler, the impedance of the entire ring is fixed at $1.41 \times Z_0$, or 70.7 Ω for $Z_0 = 50$ Ω. For an input signal V, the outputs at ports 2 and 4 are equal in magnitude, but 180 degrees out of phase.

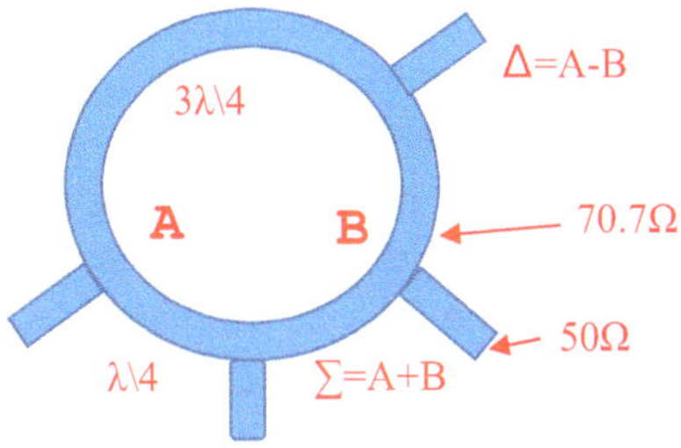

Figure 7.23. Rat-race coupler.

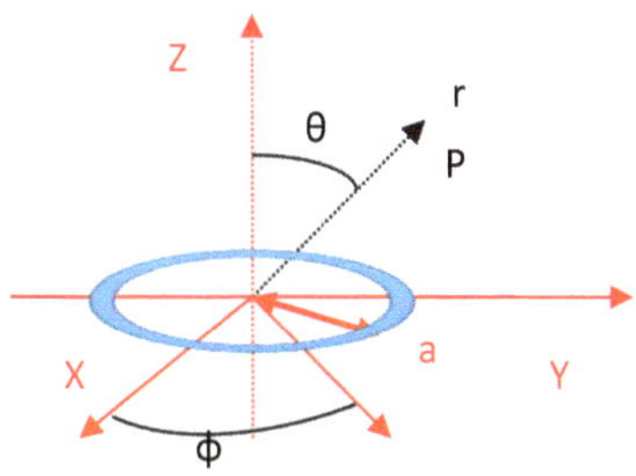

Figure 7.24. Small loop antenna.

7.9 Wearable loop antennas

Loop antennas are used as receiving antennas in wearable communication and medical systems. Loop antennas may be printed on a dielectric substrate or manufactured as a wired antenna. In this section we present several loop antennas.

7.9.1 Small wearable loop antenna

A small loop antenna is shown in figure 7.24. The shape of the loop antenna may be circular or rectangular. These antennas have low radiation resistance and high reactance. It is difficult to match the antenna to a transmitter. Loop antennas are most often used as receiving antennas, where impedance mismatch loss can be accepted. Small loop antennas are used as field strength probes, in pagers and in wireless measurements. The loop lies in the x–y plane. The radius a, of the small loop antenna is smaller than a wavelength ($a \ll \lambda$). A loop antenna's electric field is given equation (7.39), and the magnetic field is given in equation (7.40).

$$E_{\phi} = \eta \frac{(ak)^2 I_0 \sin\theta}{4r} e^{j(\omega t - \beta r)} \tag{7.39}$$

$$H_{\theta} = -\frac{(ak)^2 I_0 \sin\theta}{4r} e^{j(\omega t - \beta r)} \tag{7.40}$$

The variation of the radiation pattern with direction is $\sin\theta$, the same as the dipole antenna. The fields of a small loop have the E and H fields switched relative to that of a short dipole. The E field is horizontally polarized in the x–y plane.

The small loop is often referred to as the dual of the dipole antenna, because if a small dipole had a magnetic current flowing (as opposed to an electric current as in a regular dipole), the fields would resemble that of a small loop. The short dipole has a capacitive impedance (the imaginary part of the impedance is negative). The impedance of a small loop is inductive (positive imaginary part). The radiation resistance (and ohmic loss resistance) can be increased by adding more turns to the loop. If there are N turns of a small loop antenna, each with a surface area S, the radiation resistance for small loops can be approximated as given in equation (7.41).

$$R_{rad} = \frac{31\,329 N^2 S^2}{\lambda^4} \tag{7.41}$$

For a small loop, the reactive component of the impedance can be determined by finding the inductance of the loop. For a circular loop with radius a, and wire radius r, the reactive component of the impedance is given by equation (7.42).

$$X = 2\pi a f\ \mu \left(\ln\left(\frac{8a}{r}\right) - 1.75 \right) \tag{7.42}$$

Loop antennas behave better in the vicinity of the human body than dipole antennas. The reason is that the electric near-fields in dipole antennas are very strong. For $r \ll 1$, the dominant component of the field varies as $1/r^3$. These fields are the dipole near-fields. In this case, the waves are standing waves and the energy oscillates in the antenna's near zone and are not radiated to the open space. The real part of the Poynting vector is equal to zero. Near the body the electric fields decay rapidly. However, the magnetic fields are not affected near the body. The magnetic fields are strong in the near-field of the loop antenna. These magnetic fields give rise to the loop antenna radiation. The loop antenna radiation near the human body is stronger than the dipole radiation near the human body. Several loop antennas are used as 'wearable antennas'.

7.9.2 Wearable printed loop antenna

The diameter of a printed loop antenna is around half a wavelength. A loop antenna is dual to a half-wavelength dipole. Several loop antennas were designed for medical systems at a frequency range between 400 MHz and 500 MHz. In figure 7.25(a) a printed loop antenna is shown. A photo of a printed loop antenna with a Balun transformer is presented in figure 7.25(b). The antenna was printed on FR4 with 0.5 mm thickness. The loop diameter is 45 mm. The loop antenna's VSWR is around 4:1. The printed loop antenna's radiation pattern at 435 MHz is shown figure 7.26. The gain is around 1.8 dBi. The antenna with a tuning capacitor is shown in figure 7.27. The loop antenna's VSWR without the tuning capacitor was 4:1. This loop antenna may be tuned by adding a capacitor or varactor as shown in figure 7.28. Matching stubs are employed to tune the antenna to the resonant

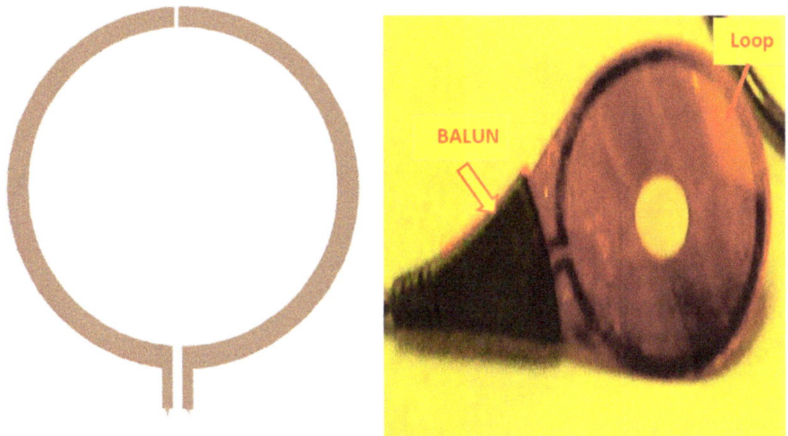

Figure 7.25. (a) Printed loop antenna. (b) Photo of a printed loop antenna.

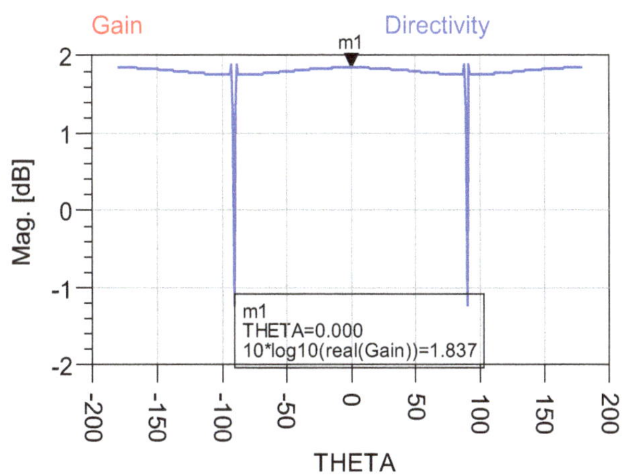

Figure 7.26. Printed loop antenna's radiation pattern at 435 MHz.

frequency. Tuning the antenna allows us to work in a wider bandwidth as shown in figure 7.28. Loop antennas are used as receiving antennas in medical systems. The loop antenna's radiation pattern on the human body is shown figure 7.29.

The computed 3D radiation pattern and the coordinate used in this chapter are shown in figure 7.30.

7.9.3 Wired loop antenna

A wire seven-turn loop antenna is shown in figure 7.31. The antenna's length is $l = 4.5$ mm. The loop's diameter is 3.5 mm. Equation (7.43) is an approximation to calculate the inductance value for the air coil loop antenna with N turns, diameter r and length l.

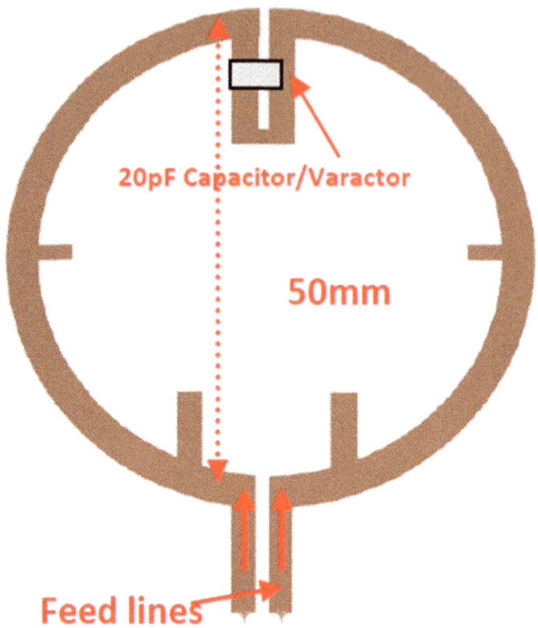

Figure 7.27. Tunable loop antenna without a ground plane.

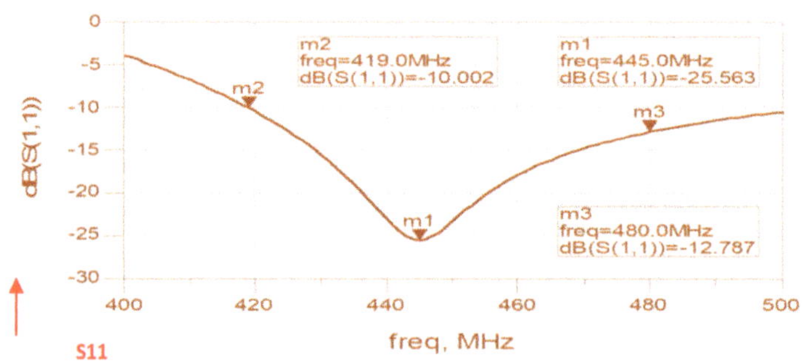

Figure 7.28. Computed S_{11} of a loop antenna, without a ground plane, with a tuning capacitor.

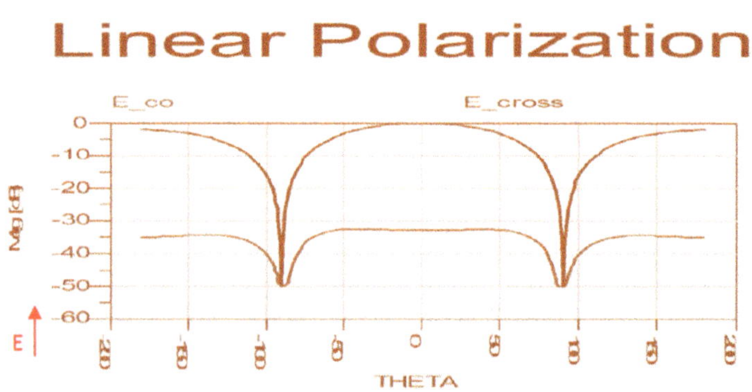

Figure 7.29. Radiation pattern of a loop antenna on the human body.

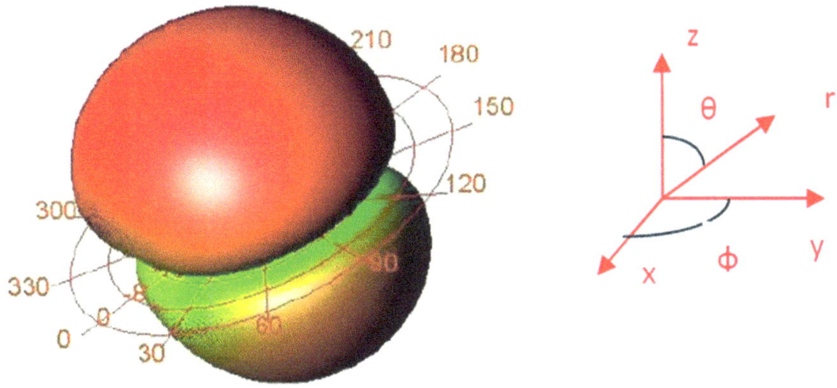

Figure 7.30. Loop antenna's 3D radiation pattern.

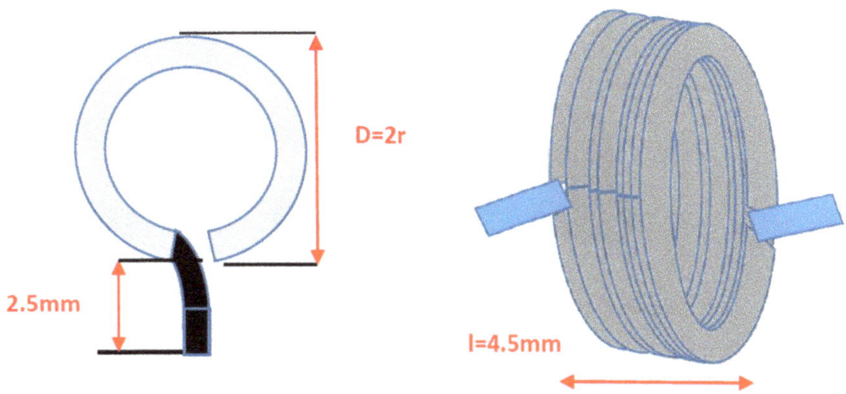

Figure 7.31. Wire seven-turn loop antenna.

$$L(nH) = \frac{3.94 r_{mm} N^2}{0.9\frac{\ell}{r} + 1} = \frac{0.1 r_{mil} N^2}{0.9\frac{\ell}{r} + 1} \tag{7.43}$$

An approximation to calculate the inductance value for the air coil loop antenna with N turns, diameter r and length l is given in equation (7.44).

$$L(nH) \approx \frac{r_{mm}^2 N^2}{2l + r} \tag{7.44}$$

For length $l = 4.5$ mm, wire diameter 0.6 mm, loop diameter 3.5 mm and $N = 7$ the inductance is around $L = 52nH$. The quality factor of this seven-turn wire loop is around 100.

Figure 7.32. Wire loop antenna on a PCB board.

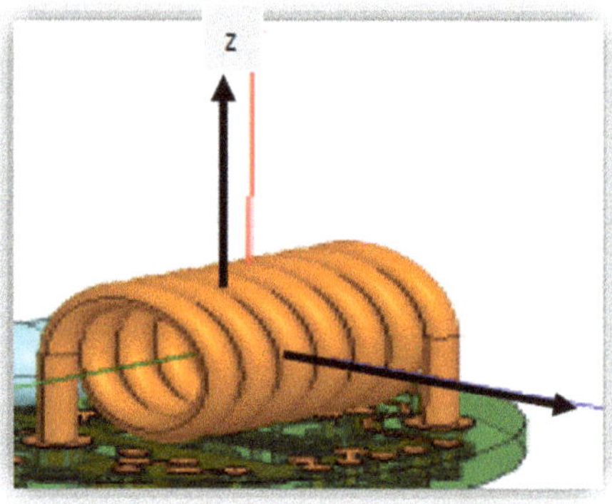

Figure 7.33. Seven-turns loop antenna on a PCB board.

For length $l = 1.7$ mm, wire diameter 0.6 mm, loop diameter 6.5 mm and $N = 2$ the inductance is around $L = 45nH$.

For length $l = 2$ mm, wire diameter 0.6 mm, loop diameter 7.62 mm and $N = 2$ the inductance is around $L = 42.1nH$.

For length $l = 0.5$ mm, wire diameter 0.6 mm, loop diameter 7 mm and $N = 2$ the inductance is around $L = 87nH$.

For length $l = 2.5$ mm, wire diameter 0.6, loop diameter 5 mm and $N = 2$ the inductance is around $L = 20.7nH$.

The wire loop antenna has very low radiation efficiency. The amount of power radiated is a small fraction of the input power. The antenna efficiency is around 0.01%, −41 dB. The ratio between the antenna's dimension and wavelength is around 1:100. Small antennas are characterized by radiation resistance R_r. The radiated power is given by I^2R_r, where I is the current through the coil. Remember that the current through the coil is Q times the current through the antenna. For example, for a current of 2 mA and Q of 20, if $R_r = 10^{-3}$ ohm, the radiated power is around −32 dBm.

Figure 7.32 presents a wire loop antenna with 2.5 turns on a PCB board. Figure 7.33 presents a wire loop antenna with 7 turns on a PCB board.

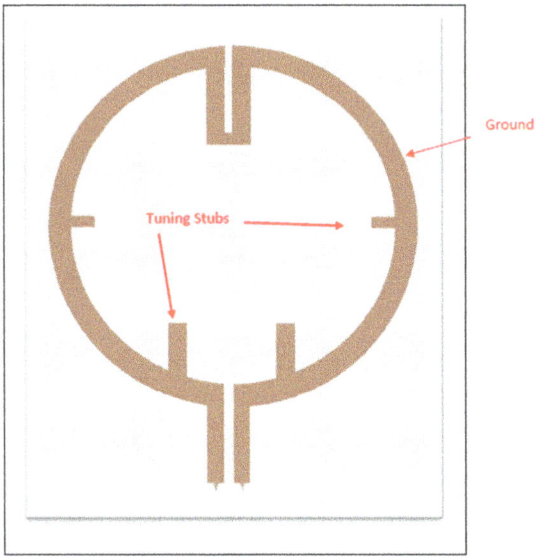

Figure 7.34. Printed loop antenna with a ground plane.

7.9.4 Wearable loop antennas with a ground plane

A new loop antenna with a ground plane has been designed on Kapton substrates with a relative dielectric constant of 3.5 and thickness of 0.25 mm. The antenna is presented in figure 7.34. Matching stubs are employed to tune the antenna to the resonant frequency. The loop antenna with a ground plane is 45 mm in diameter. The antenna was designed by employing ADS software. The antenna's center frequency is 427 MHz. The antenna's bandwidth for a VSWR better than 2:1 is around 12% as shown in figure 7.35. The printed loop antenna's radiation pattern at 435 MHz is shown figure 7.36. The loop antenna's gain is around 1.8 dBi. The loop antenna with a ground plane beamwidth is around 100°.

A printed loop antenna with a ground plane and shorter tuning stubs is shown in figure 7.37. The diameter of the loop antenna with ground, and a shorter tuning stubs plane, is 45 mm. The antenna's center frequency is 438 MHz. The antenna's bandwidth for a VSWR better than 2:1 is around 12% as shown in figure 7.38. The printed loop antenna with a shorter tuning stubs radiation pattern at 435 MHz is shown in figure 7.39. The loop antenna's gain is around 1.8 dBi. The loop antenna (with a ground plane) has a beamwidth around 100°. The printed loop antenna with a shorter tuning stubs 3D radiation pattern at 430 MHz is shown figure 7.40.

Table 7.3 compares the electrical performance of a loop antenna with a ground plane with a loop antenna without a ground plane.

7.9.5 Radiation pattern of a loop antenna near a metal sheet

An *E* and *H* plane 3D radiation pattern of a wire loop antenna in free space is shown in figure 7.41.

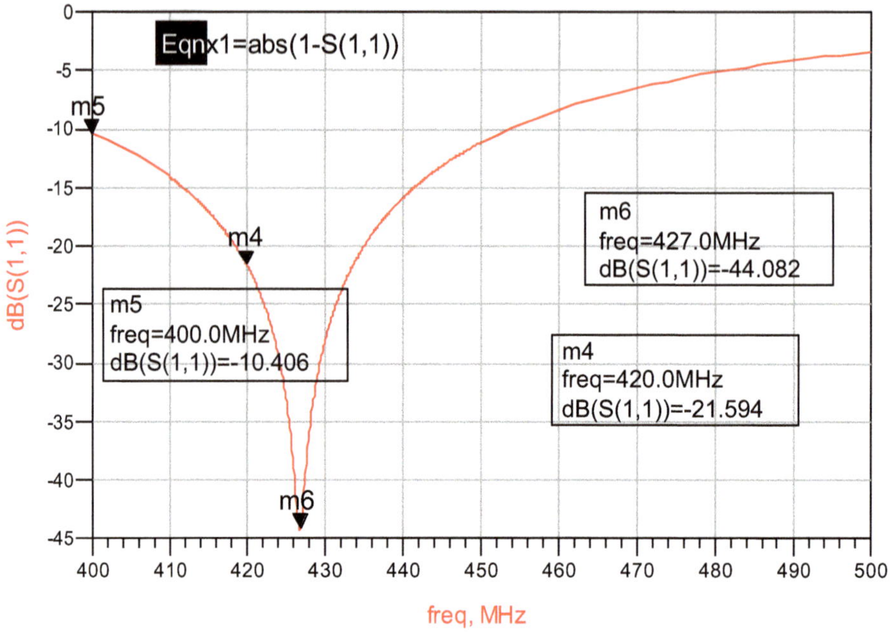

Figure 7.35. Computed S_{11} of a loop antenna with a ground plane.

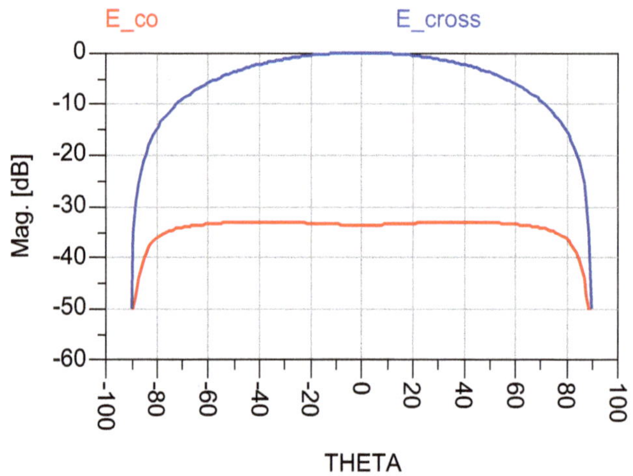

Figure 7.36. Radiation pattern of a loop antenna with a ground plane.

An *E* and *H* plane 3D radiation pattern of a wire loop antenna near a metal sheet as shown in figure 7.42 was computed by employing HFSS software.

Figure 7.43 presents the *E* and *H* plane radiation pattern of a loop antenna for a distance of 30 cm from a metal sheet (figure 7.44). We can see that a metal sheet in the vicinity of the antenna splits the main beam and creates holes of around 20 dB in the radiation pattern.

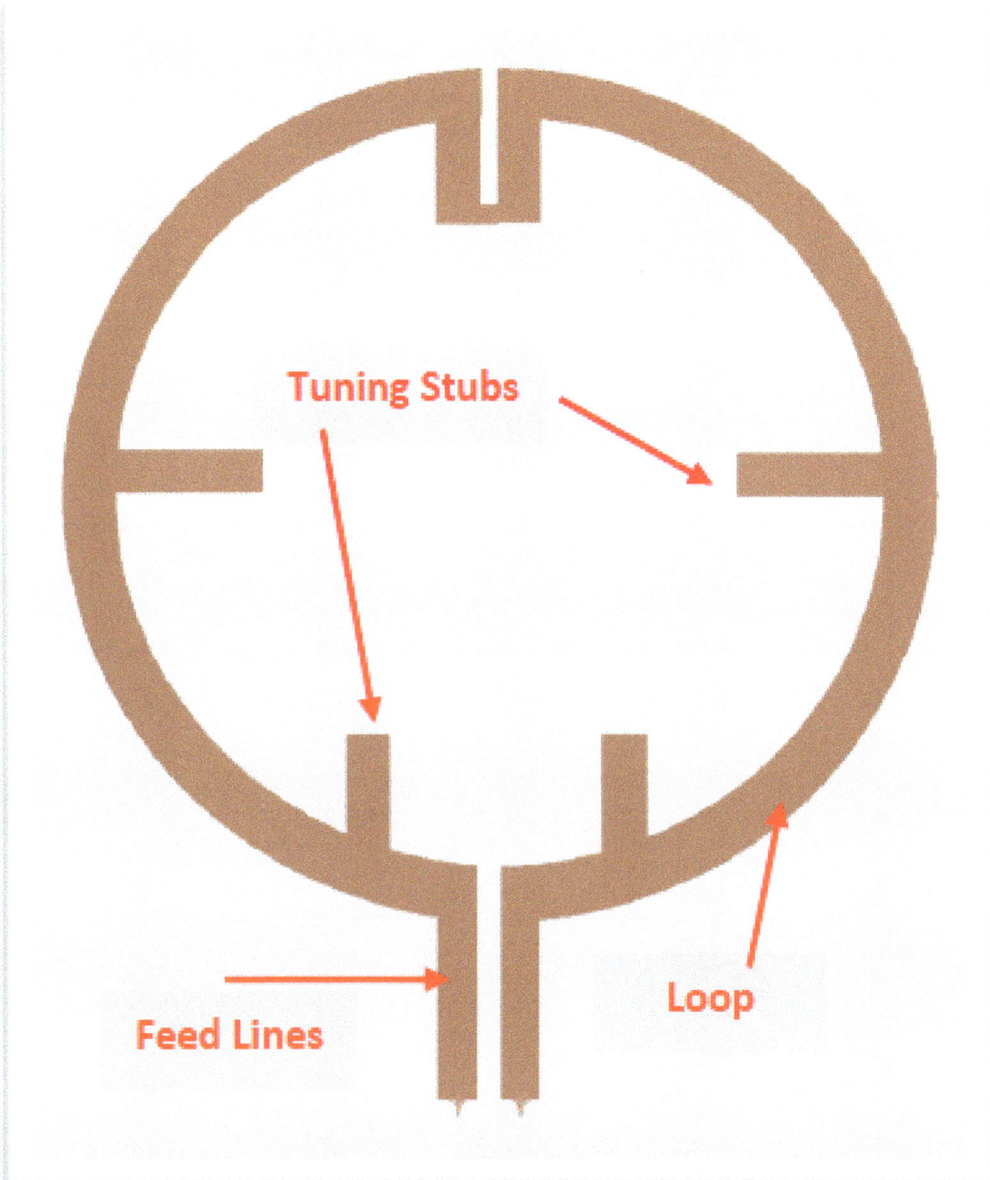

Figure 7.37. Printed loop antenna with a ground plane and short tuning stubs.

Figure 7.45 presents the *E* and *H* plane radiation pattern of a loop antenna located 10 cm from a metal sheet as presented in figure 7.46. We can see that a metal sheet in the vicinity of the antenna splits the main beam and creates holes up to 15 dB in the radiation pattern.

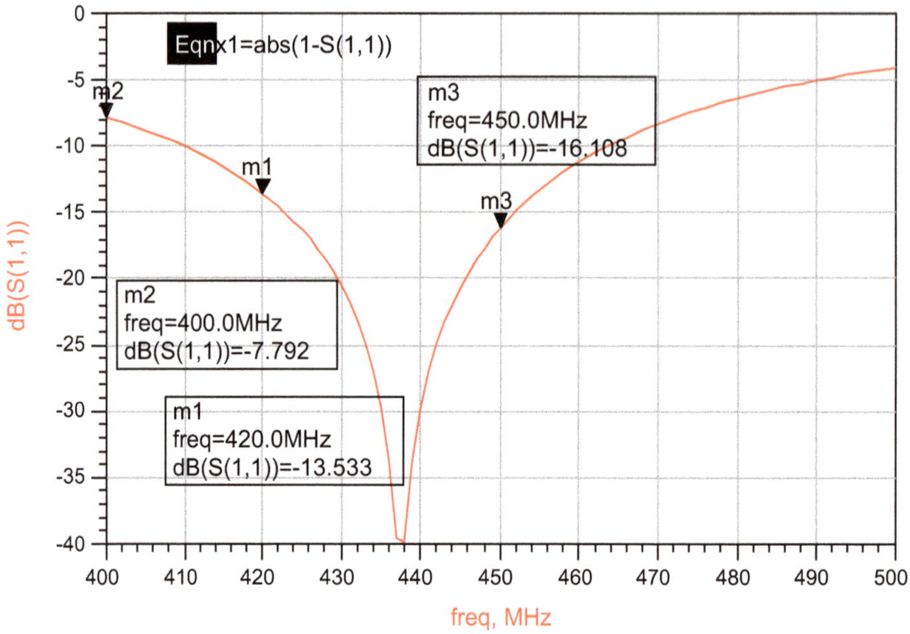

Figure 7.38. Computed S_{11} of a loop antenna with a ground plane and short tuning stubs.

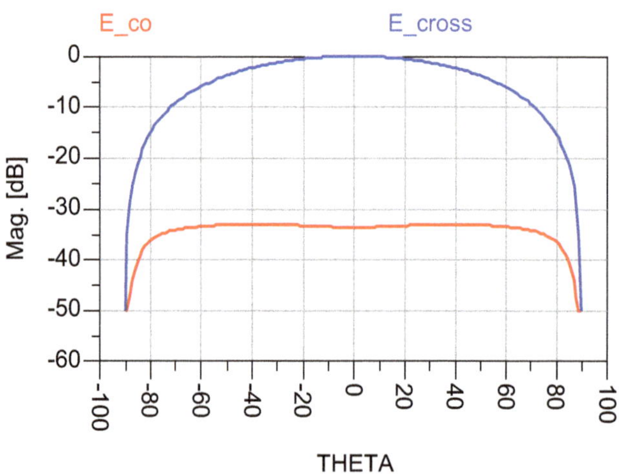

Figure 7.39. Radiation pattern of a loop antenna with a ground plane and short tuning stubs.

7.10 Planar wearable inverted-F antenna (PIFA)

The planar inverted-F antenna possesses attractive features such as having a low profile, small size, and low fabrication costs [16–18]. A PIFA antenna's bandwidth is higher than the bandwidth of the conventional patch antenna, because

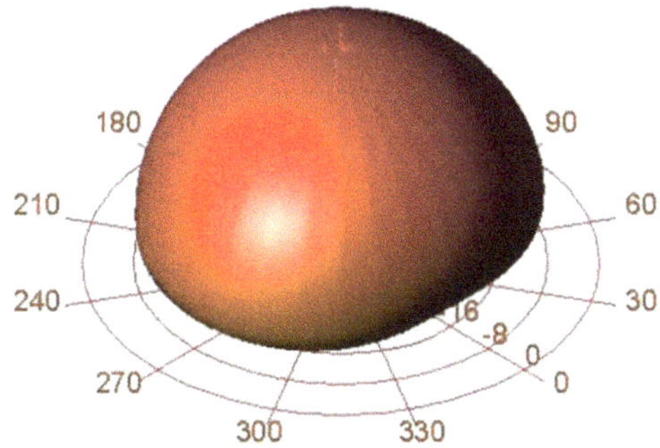

Figure 7.40. 3D Radiation pattern of a loop antenna with a ground plane.

Table 7.3. Electrical performance of several loop antenna configurations.

Antenna with no tuning capacitor	Beamwidth 3 dB	Gain dBi	VSWR
Loop no GND	100°	0	4:1
Loop with tuning capacitor (no GND)	100°	0	2:1
Wearable loop with GND	100°	0–2	2:1
Loop with GND in free space	100° to 110°	−3	5:1

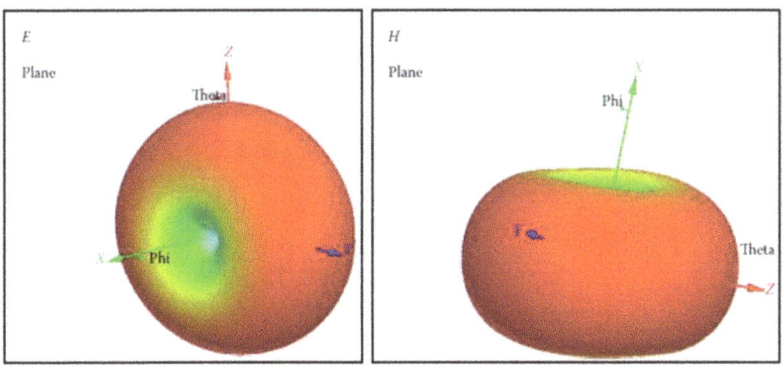

Figure 7.41. *E* and *H* plane radiation pattern of a loop antenna in free space.

the PIFA antenna thickness is higher than the thickness of patch antennas. The conventional PIFA antenna is a grounded quarter-wavelength patch antenna. The antenna consists of ground plane, a top plate radiating element, feed wire and a shorting plate or via holes from the radiating element to the ground plane as shown in figure 7.46.

The patch is shorted at the end. The fringing fields, which are responsible for radiation are shorted on the far end, so only the fields nearest the transmission line radiate. Consequently, the gain is reduced, but the patch antenna maintains the same basic properties as a half-wavelength patch. However, the antenna's length is

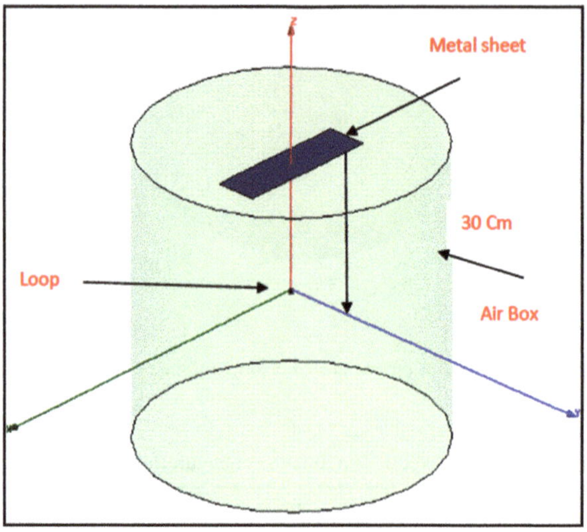

Figure 7.42. Loop antenna located near a metal sheet.

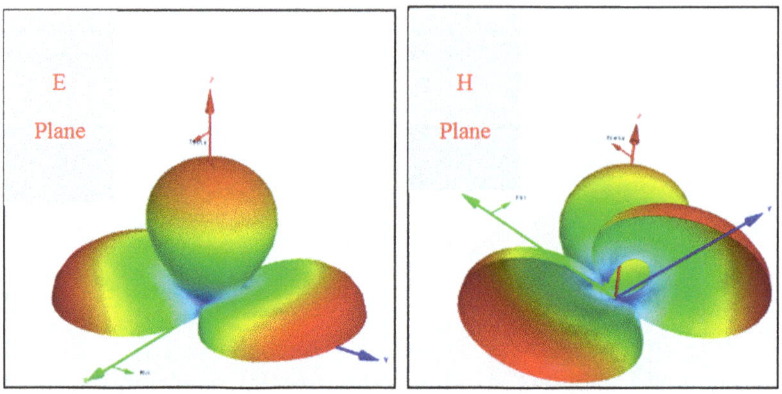

Figure 7.43. *E* and *H* plane radiation pattern of a loop antenna for a distance of 30 cm from a metal sheet.

reduced by 50%. The feed location may be placed between the open and shorted end. The feed location controls the antenna's input impedance.

7.10.1 Grounded quarter-wavelength patch antenna

A grounded quarter-wavelength patch antenna was designed on FR4 substrate with a relative dielectric constant of 4.5 and 1.6 mm thickness at 3.85 GHz. The antenna is shown in figure 7.47. The antenna was designed by using Momentum software. The dimensions are 34 × 17 × 1.6 mm. S_{11} results of the antenna are shown in figure 7.48. The antenna's bandwidth is around 6% for a VSWR better than 3:1 without a matching network. The radiation pattern is shown in figure 7.49. The antenna's beamwidth is around 76°. The grounded quarter-wavelength patch antenna's gain is

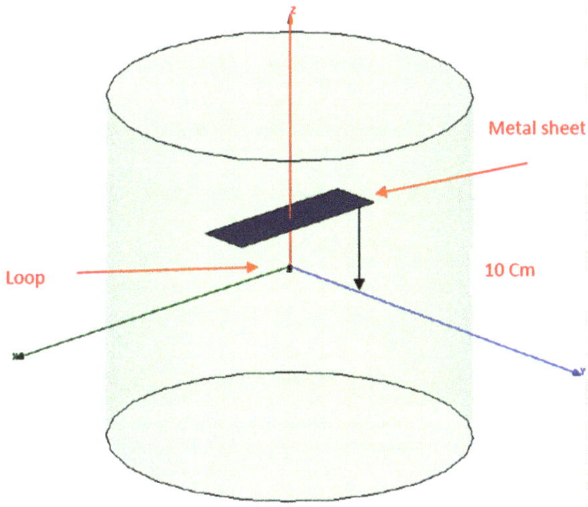

Figure 7.44. Loop antenna located 10 cm from a metal sheet.

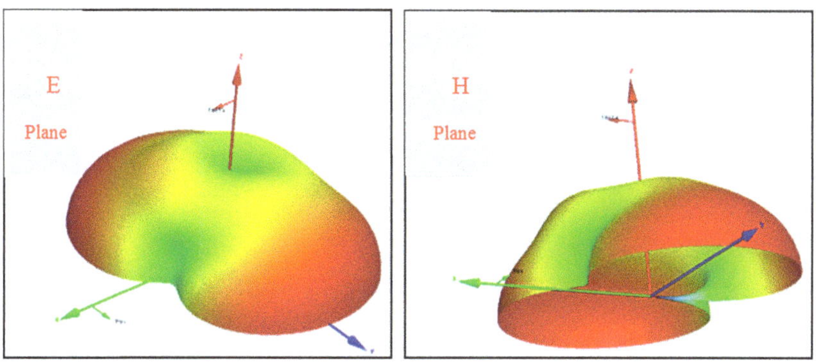

Figure 7.45. *E* and *H* plane radiation pattern of a loop antenna for a distance of 10 cm from a metal sheet.

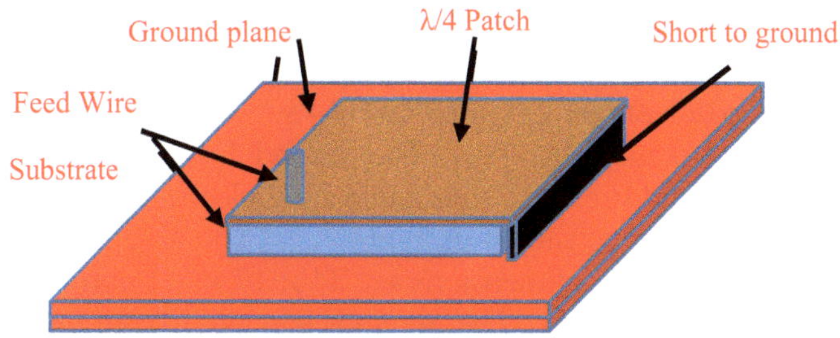

Figure 7.46. Conventional PIFA antenna.

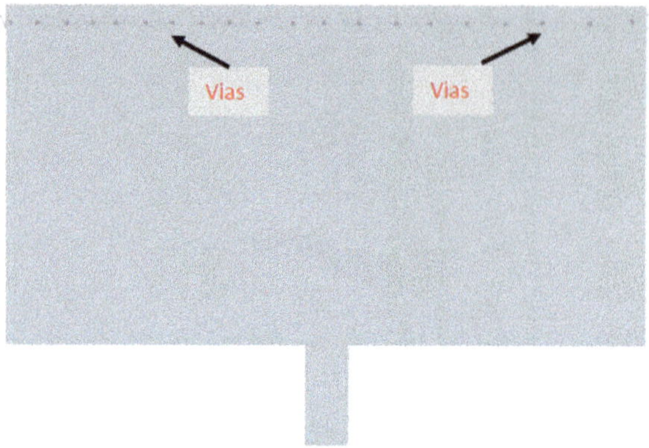

Figure 7.47. Grounded quarter-wavelength patch antenna.

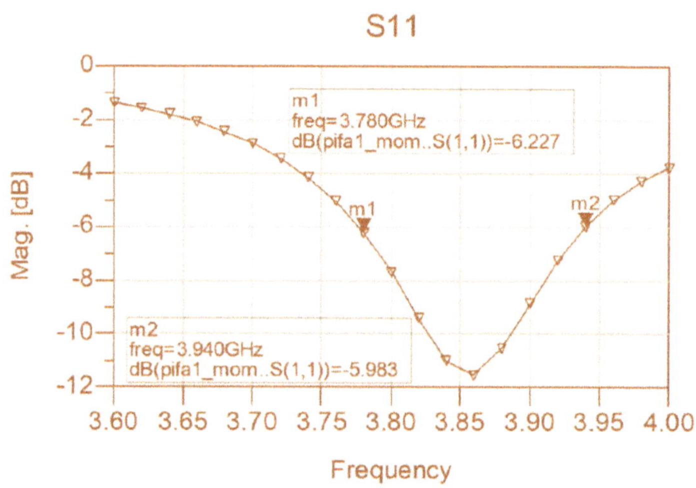

Figure 7.48. Grounded quarter-wavelength patch antenna S_{11} results.

around 6.7 dBi, and the antenna's efficiency is around 92%. The compact PIFA antenna may be attached to the human body or inserted inside a wearable belt.

7.10.2 A wearable double layer PIFA antenna

A new double layer PIFA antenna was designed. The first layer is a grounded quarter-wavelength patch antenna printed on FR4 substrate with a relative dielectric constant of 4.5 and 1.6 mm thickness. The second layer is a rectangular patch antenna printed on Duroid substrate with a relative dielectric constant of 2.2 and 1.6 mm thickness. The antenna is shown in figure 7.50. The antenna was designed by employing ADS software. The dimensions are 34 × 17 × 3.2 mm. S_{11} results of the antenna are shown in figure 7.51. The antenna is a dual band antenna. The first resonant frequency is 3.48 GHz. The second resonant frequency is 4.02 GHz. The

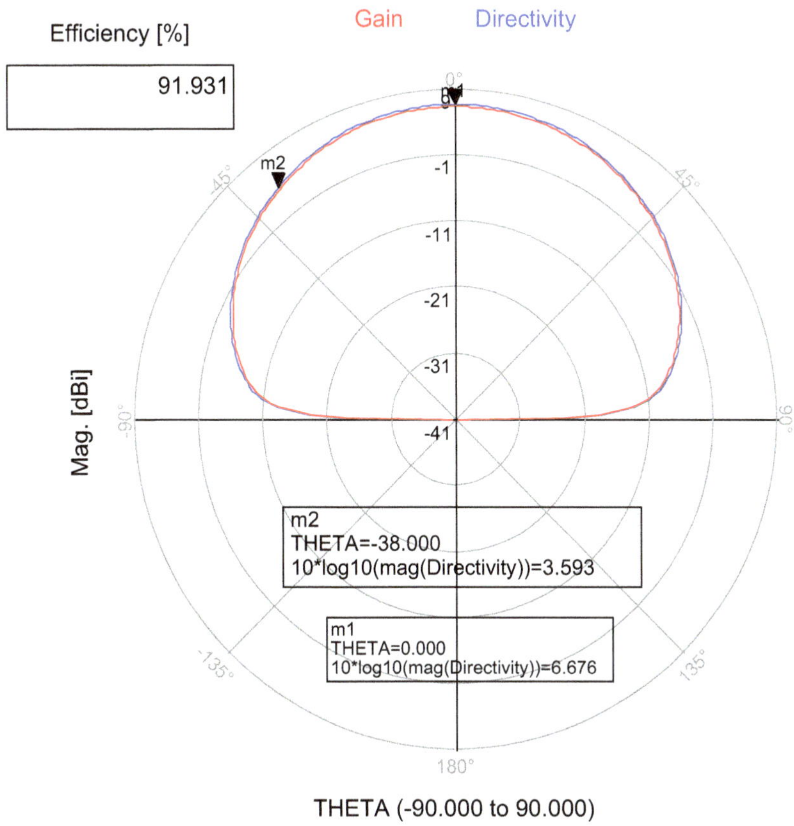

Figure 7.49. Grounded quarter-wavelength patch antenna's radiation pattern.

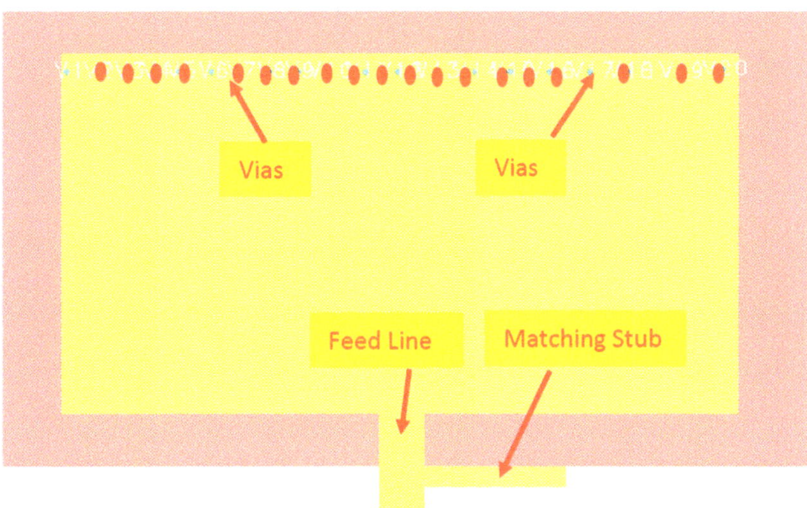

Figure 7.50. Double layer PIFA antenna.

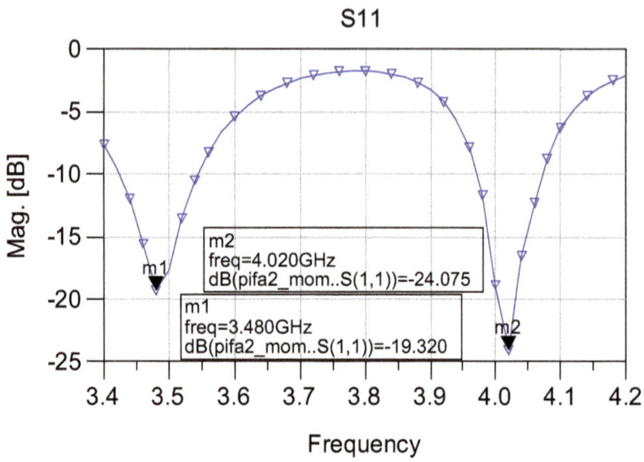

Figure 7.51. Double layer PIFA antenna S_{11} results.

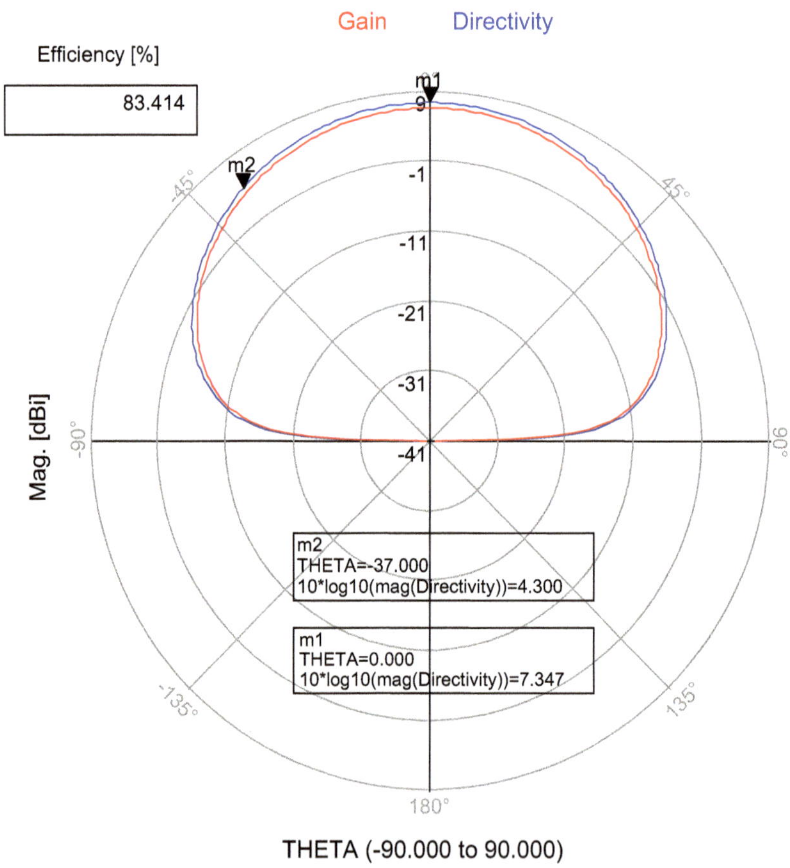

Figure 7.52. Double layer PIFA antenna's radiation pattern.

radiation pattern is shown in figure 7.52. The antenna's beamwidth is around 74°, the gain is around 7.4 dBi, and the efficiency is around 83.4%.

7.11 Conclusions

This chapter presents several wideband wearable antennas with high efficiency for medical and sport applications. The design considerations, and computed and measured results of several wearable printed antennas are presented in this chapter. Antenna dimensions may vary from 26 × 6 × 0.16 cm to 5 × 5 × 0.05 cm according to the medical system's specification. The antenna's bandwidth is around 10% for a VSWR better than 2:1. The beamwidth is around 100°, and the gain varies from 0 to 4 dBi. Wideband tunable microstrip antennas with high efficiency for medical applications have been presented in this chapter. The dimensions may vary from 26 × 6 × 0.16 cm to 5 × 5 × 0.05 cm according to the medical system's specification. The bandwidth is around 10% for a VSWR better than 2:1, and the beamwidth is around 100°. The gain varies from 0 to 2 dBi.

A varactor is employed to compensate for variations in an antenna's resonant frequency at different locations on the human body.

References

[1] Sabban A 2016 *Wideband RF Technologies and Antenna in Microwave Frequencies* (New York: Wiley)

[2] James J R, Hall P S and Wood C 1981 *Microstrip Antenna Theory and Design* (London: Peter Peregrinus)

[3] Sabban A and Gupta K C 1991 Characterization of radiation loss from microstrip discontinuities using a multiport network modeling approach *IEEE Trans. Microwave Theory Tech.* **39** 705–12

[4] Sabban A 1983 A new wideband stacked microstrip antenna *IEEE Antennas and Propagation Symp (Houston, Texas, USA)*

[5] Sabban A 2012 *Dual polarized dipole wearable antenna US Patent number:* 8203497

[6] Sabban A 1981 Wideband microstrip antenna arrays *IEEE Antenna and Propagation Symp. MELCOM (Tel Aviv)*

[7] Sabban A 2011 Microstrip antenna arrays *Microstrip Antennas* ed N Nasimuddin (Rijeka: InTech) pp 361–84

[8] Chirwa L C, Hammond P A, Roy S and Cumming D R S 2003 Electromagnetic radiation from ingested sources in the human intestine between 150 MHz and 1.2 GHz *IEEE Trans. Biomed. Eng.* **50** 484–92

[9] Werber D, Schwentner A and Biebl E M 2006 Investigation of RF transmission properties of human tissues *Adv. Radio Sci.* **4** 357–60

[10] Fujimoto K and James J R (ed) 1994 *Mobile Antenna Systems Handbook* (Boston, MA: Artech House)

[11] Gupta B, Sankaralingam S and Dhar S 2010 Development of wearable and implantable antennas in the last decade *Microwave Mediterranean Symp. (MMS) (Guzelyurt, Turkey)* pp 251–67

[12] Thalmann T, Popovic Z, Notaros B M and Mosig J R 2009 Investigation and design of a multi-band wearable antenna *3rd European Conf. on Antennas and Propagation, EuCAP (Berlin, Germany)* pp 462–5

[13] Salonen P, Rahmat-Samii Y and Kivikoski M 2004 *Wearable antennas in the vicinity of human body IEEE Antennas and Propagation Society Symp.* **1** *(Monterey, CA)* pp 467–70

[14] Kellomaki T, Heikkinen J and Kivikoski M 2006 Wearable antennas for FM reception *First European Conference on Antennas and Propagation, EuCAP* (*The Hague, Netherlands*) pp 1–6

[15] Sabban A 2009 Wideband printed antennas for medical applications *APMC 2009 Conf. (Singapore)*

[16] Lee Y 2003 *Antenna Circuit Design for RFID Applications* (AZ, USA: Microchip Technology Inc.) Microchip AN 710c

[17] Sabban A 2015 *Low Visibility Antennas for Communication Systems* (New York: Taylor and Francis)

[18] Sabban A 1991 *Multiport network model for evaluating radiation loss and coupling among discontinuities in microstrip circuits PhD Thesis* University of Colorado at Boulder

[19] Keysight software, http://www.keysight.com/en/pc-1297113/advanced-design-system-ads?c.c.=IL&lc=eng

Chapter 8

Wideband wearable antennas for communication and medical applications

Wearable compact antennas are a major part of every wearable communication and biomedical system. Compact efficient wideband antennas significantly affect the electrical performance of wearable communication and biomedical systems. The antennas presented in this chapter may be linearly or dually polarized. Design considerations, computational and measured results on the human body of several compact wideband microstrip antennas with high efficiency are presented. For example, the dimensions of a compact dually polarized antenna are 5 × 5 × 0.05 cm. The antenna's beamwidth is around 100°, and the gain is around 0 dBi to 4 dBi. The proposed antennas may be used in wearable communication and Medicare RF systems. Varactors may be used to tune the antennas' resonant frequency. The S_{11} results for different belt thicknesses, shirt thicknesses and air spacing between the antennas and the human body are presented in this chapter.

8.1 Introduction

Microstrip antennas are widely employed in communication systems and seekers. Microstrip antennas possess attractive features that are crucial to communication and medical systems. They have features such as being low profile, flexible, lightweight, with low production costs. In addition, the benefit of a compact, low-cost feed network is attained by integrating the RF front end with the radiating elements on the same substrate. Printed antennas have been widely discussed in books and papers in the last decade [1–9]. RF transmission properties of human tissues have been investigated in several articles [10, 11]. However, the effect of the human body on the electrical performance of wearable antennas at 434 MHz is not presented [12, 13]. Several wearable antennas have been studied in the last decade [1–20]. A review of wearable and body mounted antennas designed and developed

doi:10.1088/978-0-7503-5224-6ch8

for various applications at different frequency bands over the last ten years can be found in [13]. In [14] meander wearable antennas near the human body are presented in the frequency range between 800 MHz and 2700 MHz. In [15] a textile antenna performance near the human body is presented at 2.4 GHz. In [16] the effect of the human body on wearable 100 MHz portable radio antennas is studied. In [16] the authors concluded that wearable antennas need to be shorter by 15%–25% from the antenna length in free space. Measurement of the antenna gain in [16] shows that a wide dipole (116 × 10 cm) has −13 dBi gain. The antennas presented in [12–16] were developed mostly for cellular applications. Requirements and the frequency range for medical applications are different from those for cellular applications. Several wearable sensors for medical applications are presented in [17–58].

In this chapter, a new class of wideband compact wearable antennas for communication and medical RF systems are presented. Numerical results in free space and in the presence of the human body are discussed.

8.2 Printed wearable dual polarized dipole antennas

In several communication and medical systems, the polarization of the received signal is not known. The polarization of the received signal may be vertical, horizontal or circularly polarized. In these systems it is crucial to use dual polarized receiving antennas. Two wearable antennas are presented in this section. The first is a dual polarized printed dipole. The second antenna is a dual polarized folded printed microstrip dipole. A compact microstrip loaded dipole antenna has been designed to provide horizontal polarization. The antenna's dimensions have been optimized to operate on the human body by employing Keysight Advanced Design System (ADS) software [21]. The antenna consists of two layers. The first layer consists of RO3035 0.8 mm dielectric substrate. The second layer consists of RT-Duroid 5880 0.8 mm dielectric substrate. The substrate thickness determines the antenna's bandwidth. However, thinner antennas are flexible. Thicker antennas have been designed with a wider bandwidth. The printed slot antenna provides a vertical polarization. In several medical systems the required polarization may be vertical or horizontal. The proposed antenna is dually polarized. The printed dipole and the slot antenna provide dual orthogonal polarizations. The dimensions and current distribution of the dual polarized wearable antenna are presented in figure 8.1. The antenna's dimensions are 26 × 6 × 0.16 cm. The antenna may be used as a wearable antenna on a human body. The antenna may be attached to a patient's shirt, stomach, or back area. The antenna has been analyzed by using Keysight Momentum software [21]. There is good agreement between the measured and computed results. The antenna's bandwidth is around 10% for a VSWR better than 2:1. The antenna's beamwidth is around 100°, and the gain is around 2 dBi. The computed S_{11} and S_{22} parameters are presented in figure 8.2. Figure 8.3 presents the antenna's measured S_{11} parameters. The computed radiation patterns are shown in figure 8.4. The co-polar radiation pattern belongs to the *yz* plane, and the cross-polar radiation pattern belongs to the *xz* plane. The antenna's cross-polarized field strength may be adjusted by varying the slot feed location. The dimensions and current distribution of the folded dually polarized antenna are presented in figure 8.5. The dimensions are 6 × 5 × 0.16 cm. Figure 8.6 presents the antenna's computed S_{11} and S_{22} parameters. The computed radiation patterns of the folded dipole are shown in figure 8.7. The antenna's radiation characteristics on a human body have been measured by using a phantom. The phantom's electrical characteristics represent a human body's electrical characteristics.

The phantom has a cylindrical shape with a 40 cm diameter and length of 1.5 m. The phantom contains a mix of 55% water, 44% sugar, and 1% salt. The antenna under test was placed on the phantom during the measurements of the antenna's radiation characteristics. S_{11} and S_{12} parameters were measured directly on a human body by using a network analyzer. The measured results were compared to a known reference antenna.

8.3 Printed wearable loop antenna

Wearable loop antennas with a ground plane are presented in this section. A wearable loop antenna with a ground plane has been designed on Kapton substrates with a thickness of 0.25 mm and 0.4 mm. An antenna without a ground plane is

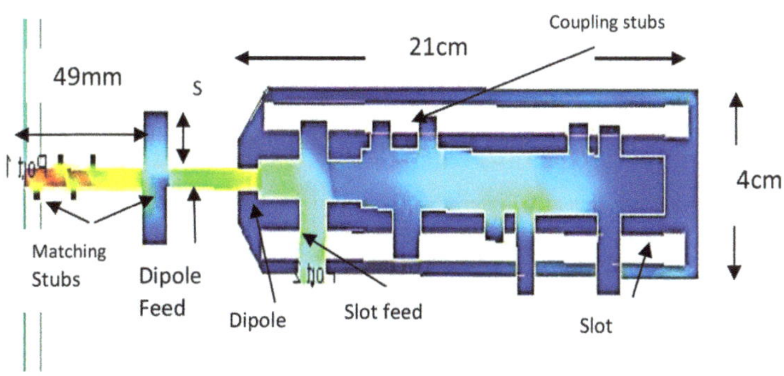

Figure 8.1. Current distribution of a wearable antenna.

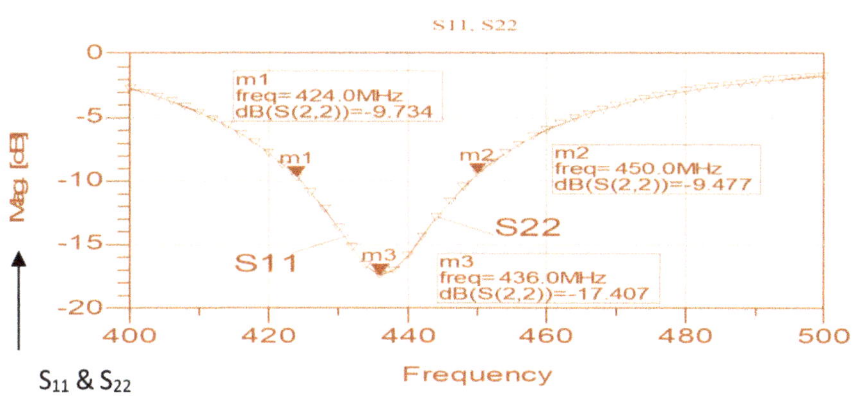

Figure 8.2. Computed S_{11} and S_{22} results on a human body.

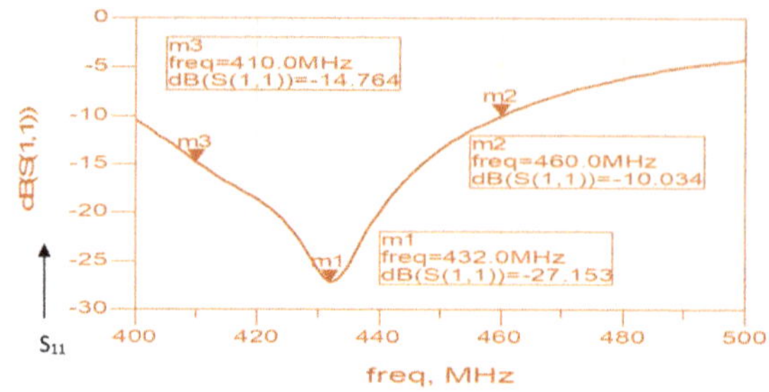

Figure 8.3. Measured S_{11} of the dual polarized antenna on a human body.

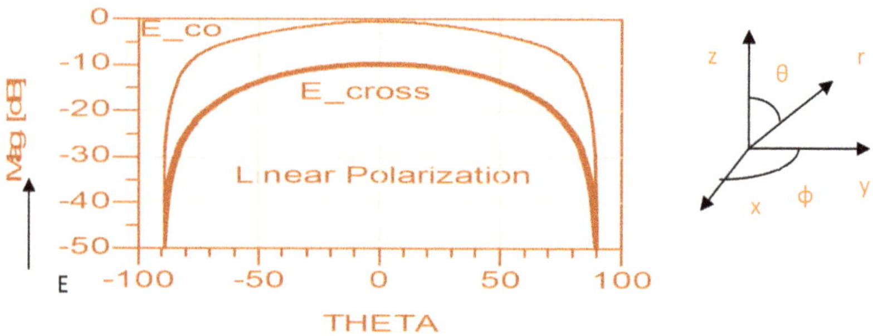

Figure 8.4. Wearable antenna's radiation patterns (antenna shown in figure 8.1).

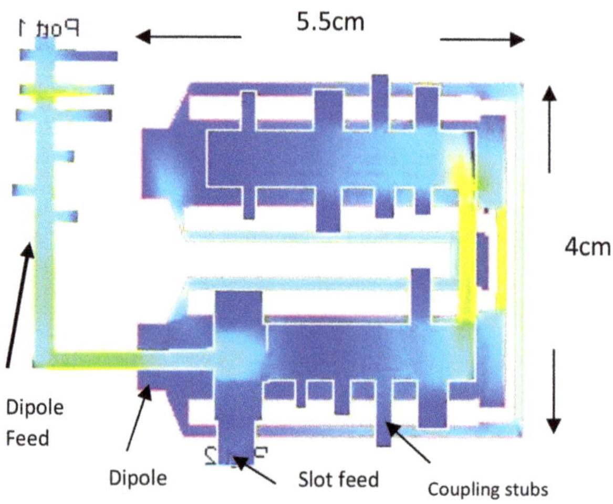

Figure 8.5. Current distribution of a folded dipole antenna, 6 × 5 × 0.16 cm.

shown in figure 8.8(a). A photo of wearable antennas is presented in figure 8.8(b). The loop antenna's VSWR without a tuning capacitor was 4:1. This loop antenna may be tuned by adding a capacitor or varactor as shown in figure 8.9. Tuning the antenna allows us to work in a wider bandwidth. Figure 8.10 presents the computed

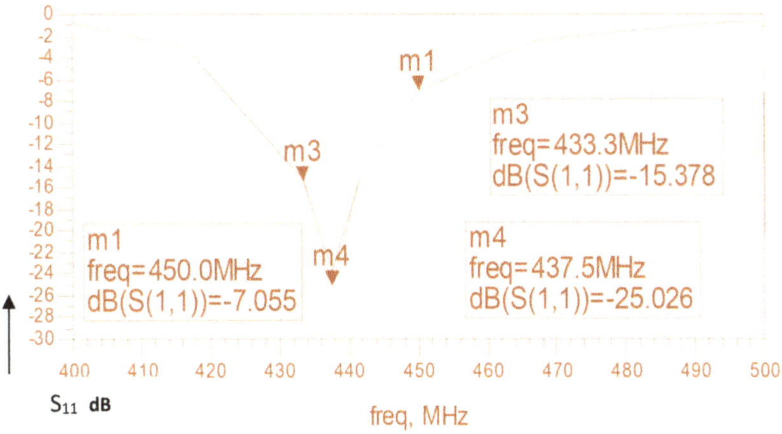

Figure 8.6. Folded antenna's computed S_{11} and S_{22} results on the human body.

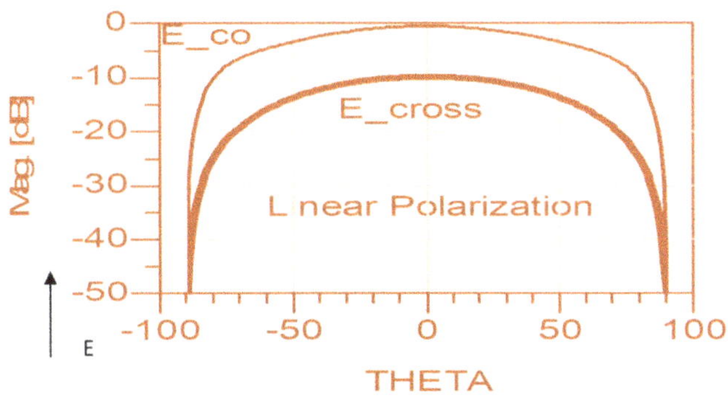

Figure 8.7. Folded antenna's radiation patterns.

S_{11} of a loop antenna with a ground plane on a human body. There is good agreement between the measured and computed results for several loop antennas' electrical parameters on a human body.

The results presented in table 8.1 indicate that the loop antenna with a ground plane is matched to a human body's environment, without the tuning capacitor, better than the loop antenna without a ground plane. A comparison of the electrical performance of several wearable printed antennas is given in table 8.1.

The computed S_{11} parameters of a loop antenna with ground, in free space, are shown in figure 8.11. A loop antenna with a ground radiation pattern on a human body is presented in figure 8.12. The computed 3D radiation pattern and the coordinate used in this chapter are shown in figure 8.13. The computed S_{11} of the loop antenna with a tuning capacitor is given in figure 8.14. Figure 8.15 presents the radiation pattern of a loop antenna without ground on a human body. Figure 8.16 presents a loop antenna with a ground plane printed on a 0.4 mm thick substrate.

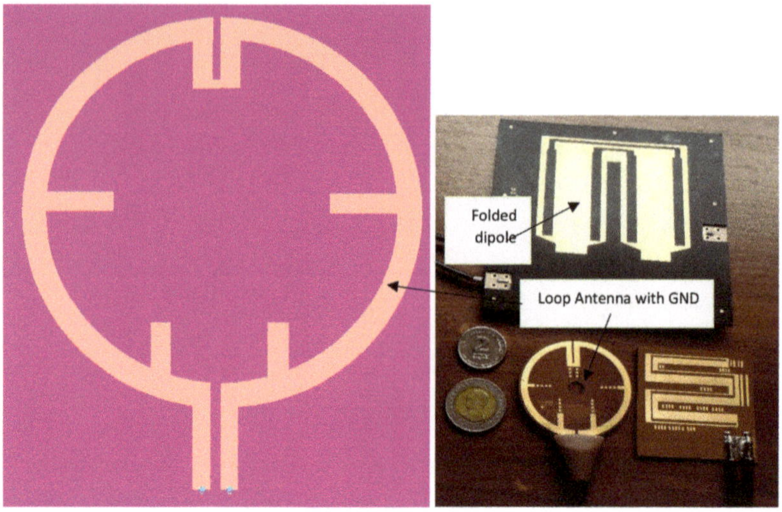

(a) (b)

Figure 8.8. (a) Printed loop antenna with a ground plane. (b) Photo of wearable antennas.

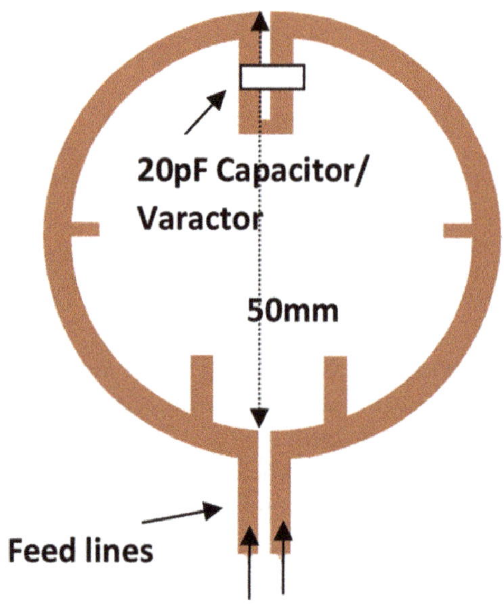

Figure 8.9. Tunable loop antenna without ground.

Figure 8.17 presents computed S_{11} of the loop antenna on a human body. Loop antennas printed on thicker substrate have a wider bandwidth as presented in figure 8.17. Figure 8.18 presents the radiation pattern of a loop antenna, printed on a 0.4 mm thick substrate.

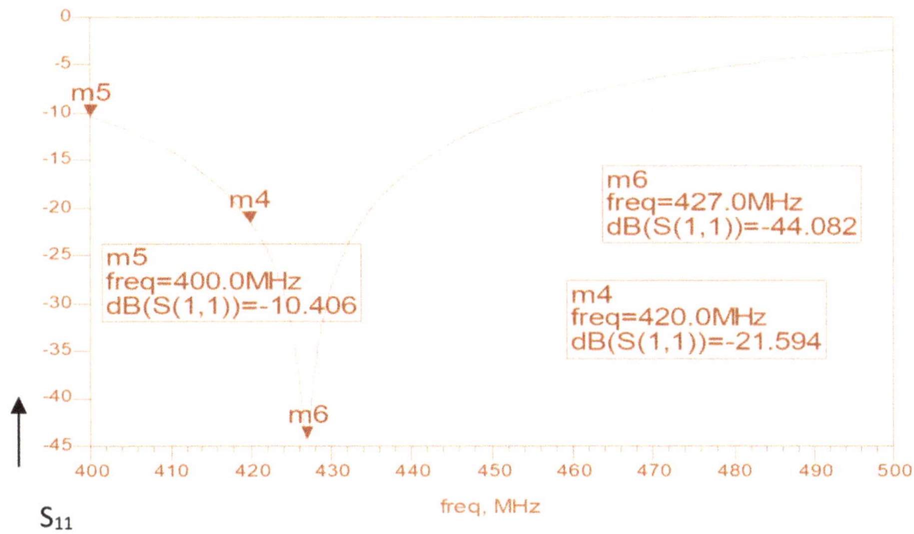

Figure 8.10. Computed S_{11} of a new loop antenna, presented in figure 8.8(a).

Table 8.1. Comparison of the electrical performance of several printed antennas.

Parameter	Beamwidth 3 dB	Gain dBi	VSWR	Dimensions cm
Printed loop antenna	100°	0	4:1	6 × 5 × 0.05
Printed microstrip dipole	100°	2	2:1	26 × 6 × 0.16
Wearable loop with ground	100°	0–2	2:1	6 × 5 × 0.05
Folded printed microstrip dipole	100°	2–3	2:1	6 × 5 × 0.16

8.4 Compact dual polarized wearable antennas

A compact microstrip loaded dipole antenna has been designed. The antenna consists of two layers. The first layer consists of FR4 0.25 mm dielectric substrate. The second layer consists of Kapton 0.25 mm dielectric substrate. The substrate thickness determines the antenna's bandwidth. However, with a thinner substrate we may achieve better flexibility. The proposed antenna is dual polarized. The printed dipole and the slot antenna provide dual orthogonal polarizations. The dual polarized antenna and antennas for medical applications are shown in figures 8.19 and 8.20. The dimensions are 5 × 5 × 0.05 cm as presented in figure 8.21. The antenna may be attached to a patient's shirt on their stomach or back areas. The antenna has been analyzed by using electromagnetic full wave software.

There is good agreement between the measured and computed results. The antenna's bandwidth is around 10% for a VSWR better than 2:1. The antenna's beamwidth is around 100°, and the gain is around 0 dBi. The computed S_{11}

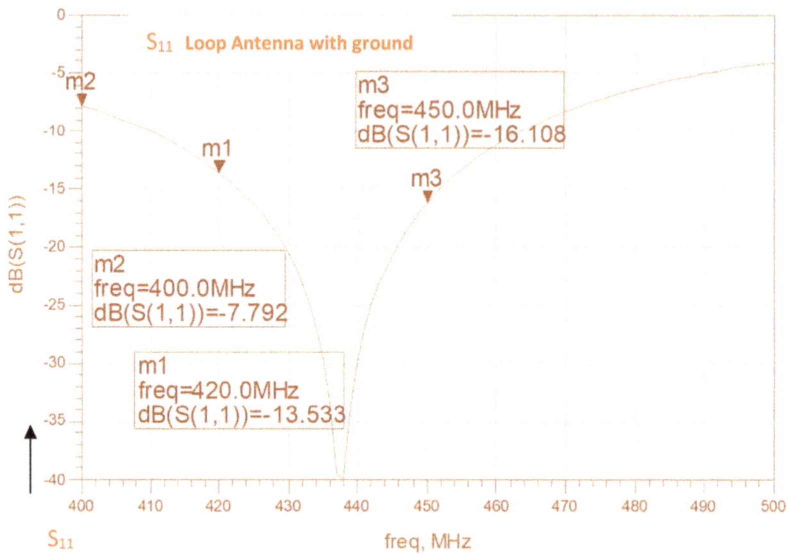

Figure 8.11. Computed S_{11} of a loop antenna with ground.

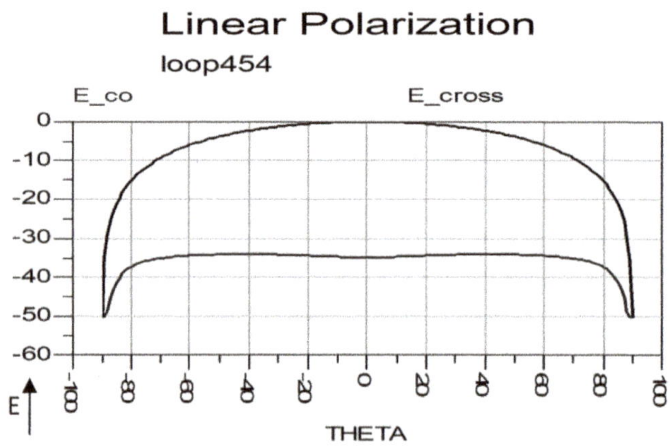

Figure 8.12. Loop antenna's radiation pattern on a human body, presented in figure 8.8(a).

parameters are presented in figure 8.22. Figure 8.23 presents the antenna's measured S_{11} parameters on a human body. The antenna's cross-polarized field strength may be adjusted by varying the slot feed location. The computed 3D radiation pattern of the antenna is shown in figure 8.24, and the computed radiation pattern is shown in figure 8.25.

8.5 Conclusions

This chapter presents several wideband wearable antennas with high efficiency for medical and sports applications. The design considerations, computed and measured

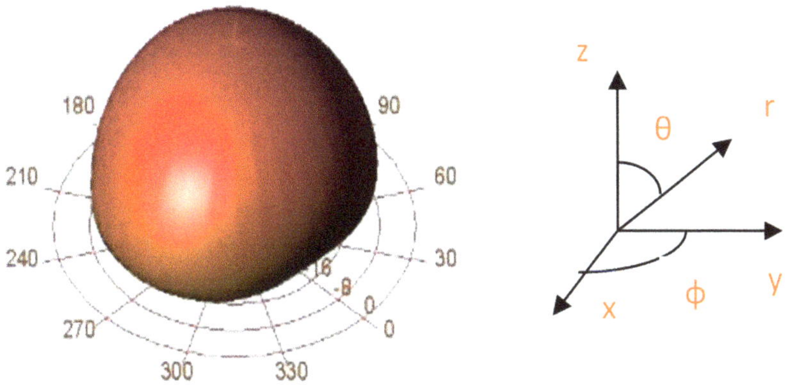

Figure 8.13. Loop antenna with ground, 3D radiation pattern.

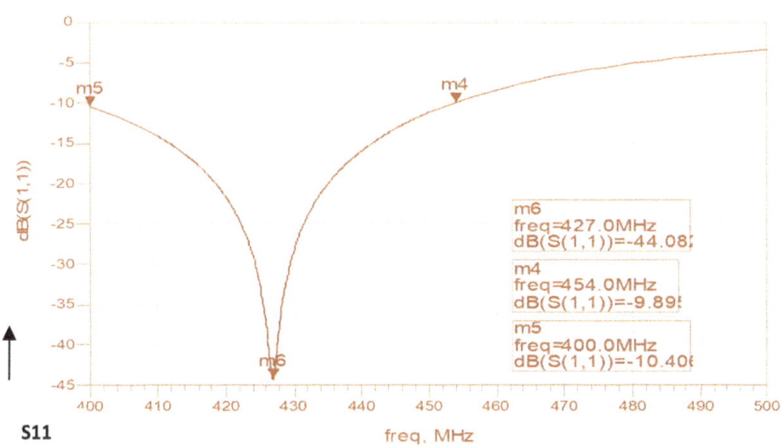

Figure 8.14. Computed S_{11} of a tuned loop antenna, without a ground plane.

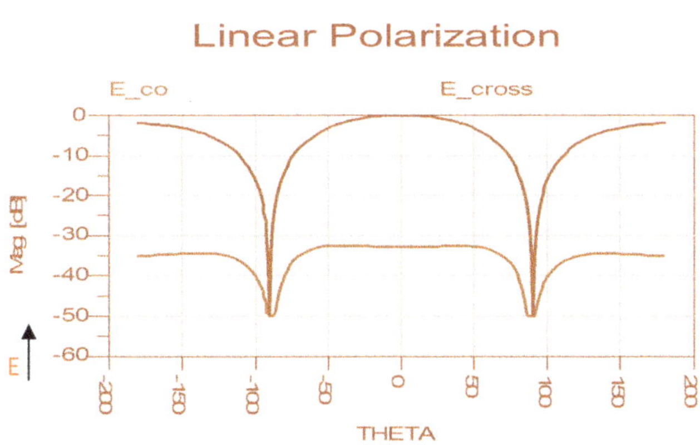

Figure 8.15. Radiation pattern of a printed loop antenna on the human body.

Figure 8.16. Loop antenna with a ground plane printed on a 0.4 mm thick substrate.

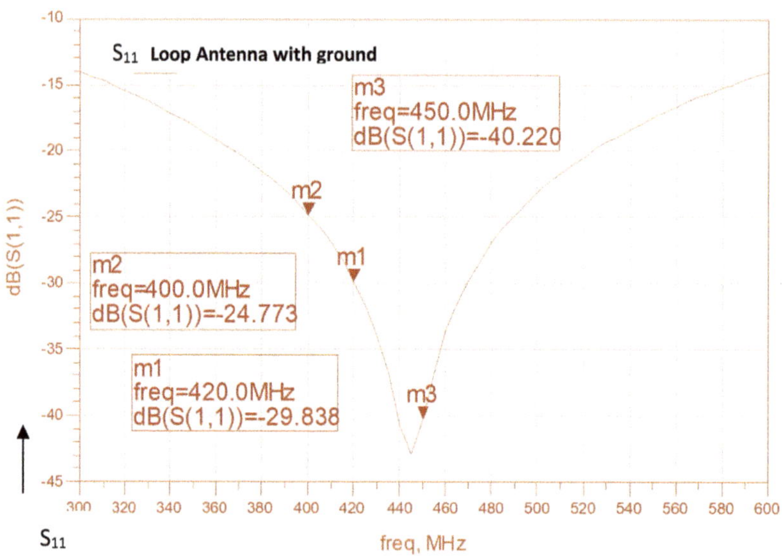

Figure 8.17. Computed S_{11} of a loop antenna printed on a 0.4 mm thick substrate.

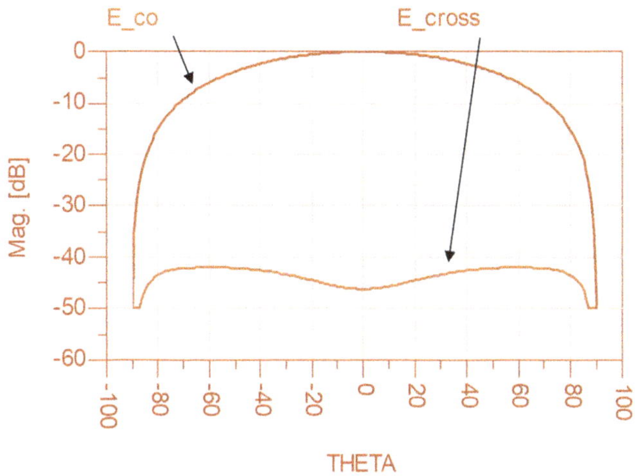

Figure 8.18. Microstrip antennas for medical applications.

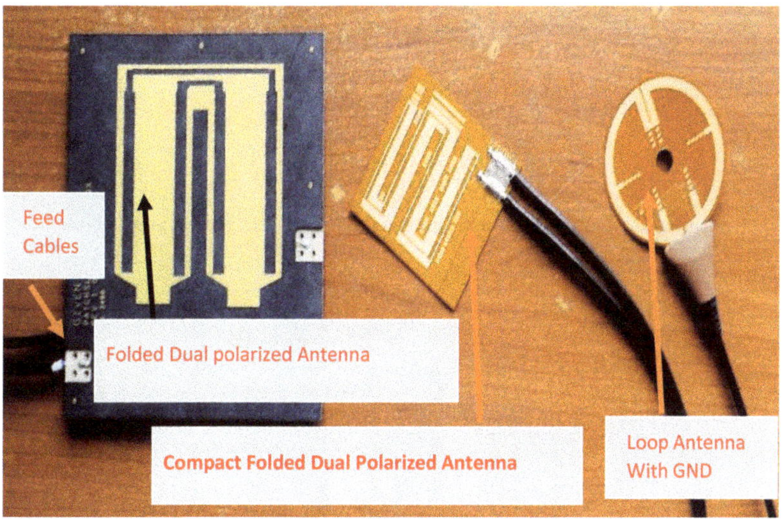

Figure 8.19. Microstrip antennas for medical applications.

results of several wearable printed antennas are presented in this chapter. The antenna's dimensions may vary from 26 × 6 × 0.16 cm to 5 × 5 × 0.05 cm according to the medical system's specification. The bandwidth is around 10% for a VSWR better than 2:1. The beamwidth is around 100°, and the gain varies from 0 to 4 dBi. A wideband tunable microstrip antenna with high efficiency for medical applications has been presented in this chapter. The antenna's dimensions may vary from 26 × 6 × 0.16 cm to 5 × 5 × 0.05 cm according to the medical system's specification. The antenna's bandwidth is

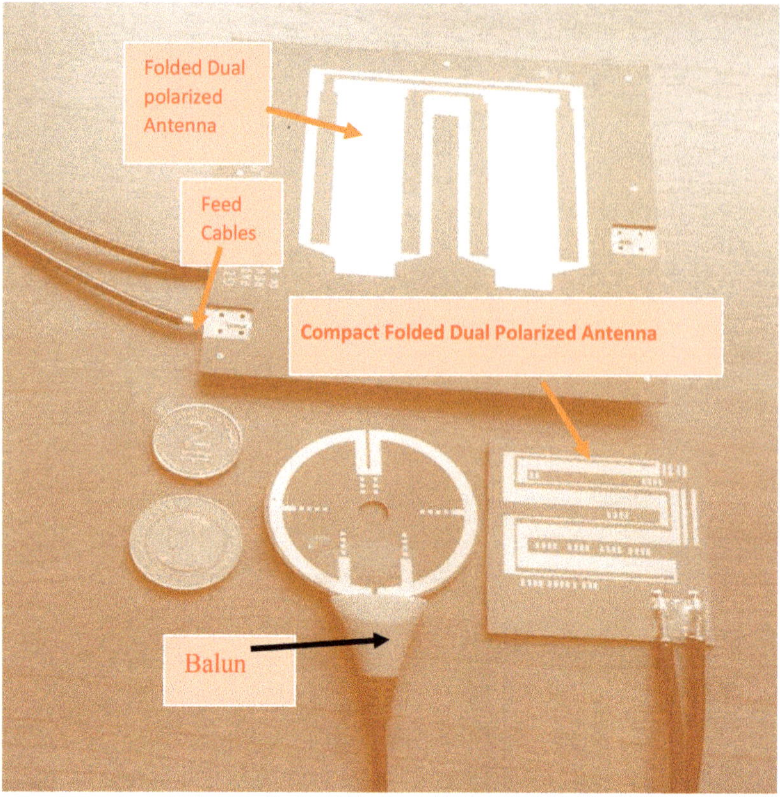

Figure 8.20. Dual polarized antenna and antennas for medical applications.

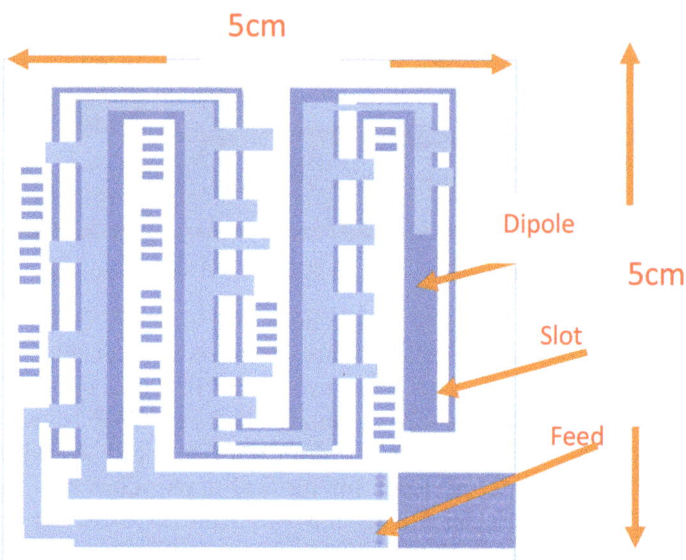

Figure 8.21. Printed compact dual polarized antenna.

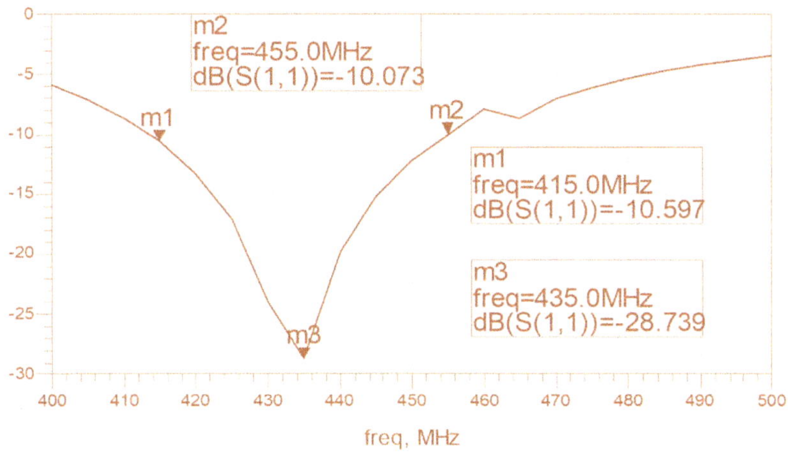

Figure 8.22. Computed S_{11} results of a compact antenna.

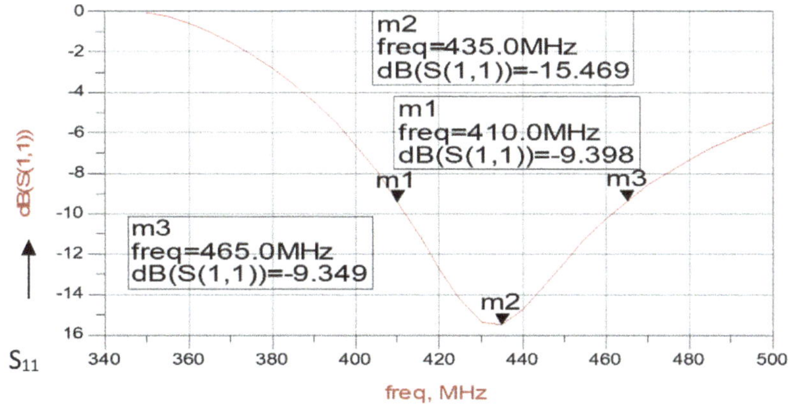

Figure 8.23. Measured S_{11} of a compact antenna on a human body.

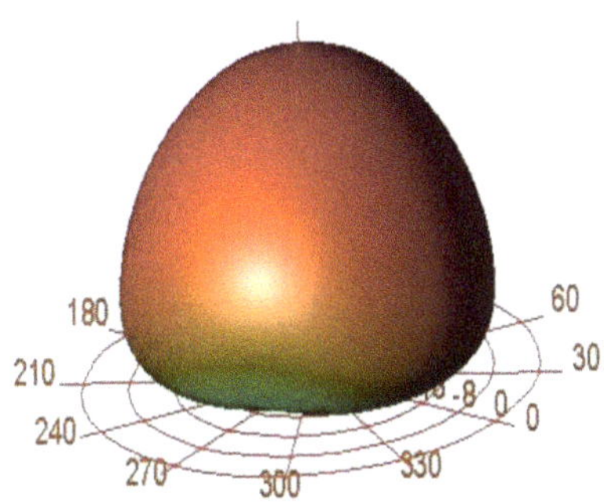

Figure 8.24. Compact antenna's 3D radiation pattern.

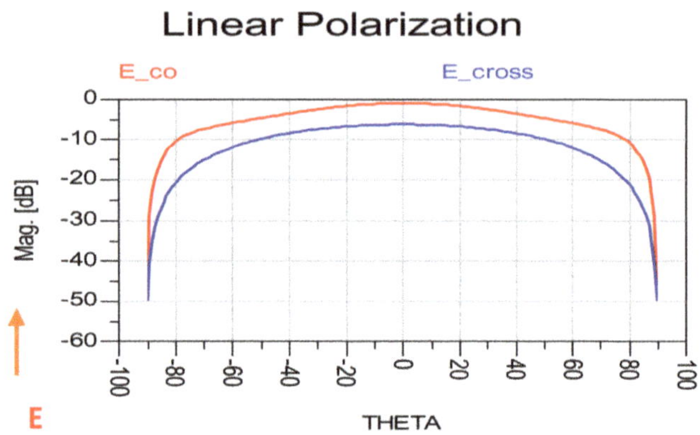

Figure 8.25. Compact dual polarized antenna's radiation pattern.

around 10% for a VSWR better than 2:1. The antenna's beamwidth is around 100°, and the gain varies from 0 to 2 dBi.

A varactor may be used to compensate for variations in the antenna's resonant frequency at different locations on the human body.

Wearable technology provides a powerful new tool for medical and surgical rehabilitation services. The wearable body area network (WBAN) is emerging as an important option for medical centers and patients. Wearable technology provides a convenient platform that may quantify the long-term context and physiological response of individuals. Wearable technology will support the development of individualized treatment systems with real-time feedback to help promote patients' health. Wearable medical systems and sensors can measure body temperature, heart rate, blood pressure, sweat rate, perform gait analysis and other physiological parameters. At present, commercialized wearable sensors have been adopted in various applications of gait analysis.

References

[1] Sabban A 2016 *Wideband RF Technologies and Antenna in Microwave Frequencies* (New York: Wiley)

[2] Sabban A 2017 *Novel Wearable Antennas for Communication and Medical Systems* (London: Taylor and Francis)

[3] Sabban A 2015 *Low Visibility Antennas for Communication Systems* (New York: Taylor and Francis)

[4] Sabban A 2016 Small wearable meta materials antennas for medical systems *Appl. Comput. Electromagn. Soc. J.* **31**

[5] Sabban A 2011 Microstrip antenna arrays *Microstrip Antennas* ed N Nasimuddin (Rijeka: InTech) pp 361–84

[6] Sabban A 2013 New wideband printed antennas for medical applications *IEEE J. Trans. Antennas Propag.* **61** 84–91

[7] Sabban A 2012 *Dual polarized dipole wearable antenna US Patent number:* 8203497

[8] Sabban A 1983 A new wideband stacked microstrip antenna *IEEE Antenna Propagation Symp. (Houston, Texas, USA)*

[9] Sabban A 1981 Wideband microstrip antenna arrays *IEEE Antenna and Propagation Symp. MELCOM (Tel Aviv)*
[10] Chirwa L C, Hammond P A, Roy S and Cumming D R S 2003 Electromagnetic radiation from ingested sources in the human intestine between 150 MHz and 1.2 GHz *IEEE Trans. Biomed. Eng.* **50** 484–92
[11] Werber D, Schwentner A and Biebl E M 2006 Investigation of RF transmission properties of human tissues *Adv. Radio Sci.* **4** 357–60
[12] Fujimoto K and James J R (ed) 1994 *Mobile Antenna Systems Handbook* (Boston, MA: Artech House)
[13] Gupta B, Sankaralingam S and Dhar S 2010 Development of wearable and implantable antennas in the last decade *Microwave Mediterranean Symp. (MMS) (Guzelyurt, Turkey) pp 251–67*
[14] Thalmann T, Popovic Z, Notaros B M and Mosig J R 2009 Investigation and design of a multi-band wearable antenna *3rd European Conf. on Antennas and Propagation, EuCAP (Berlin, Germany) pp 462–5*
[15] Salonen P, Rahmat-Samii Y and Kivikoski M 2004 *Wearable antennas in the vicinity of human body IEEE Antennas and Propagation Society Symp.* **1** *(Monterey, CA) pp 467–70*
[16] Kellomaki T, Heikkinen J and Kivikoski M 2006 Wearable antennas for FM reception *First European Conf. on Antennas and Propagation, EuCAP (The Hague Netherlands) pp 1–6*
[17] Sabban A 2009 Wideband printed antennas for medical applications *APMC 2009 Conf. (Singapore)*
[18] Lee Y 2003 *Antenna Circuit Design for RFID Applications* (AZ, USA: Microchip Technology Inc.) Microchip AN 710c
[19] Sabban A and Gupta K C 1991 Characterization of radiation loss from microstrip discontinuities using a multiport network modeling approach *IEEE Trans. Microwave Theory Tech.* **39** 705–12
[20] Sabban A 1991 Multiport network model for evaluating radiation loss and coupling among discontinuities in microstrip circuits *PhD thesis* University of Colorado at Boulder
[21] Keysight software http://keysight.com/en/pc-1297113/advanced-design-system-ads?c.c.=IL&lc=eng
[22] Mukhopadhyay S C (ed) 2015 *Wearable Electronics Sensors* (Cham: Springer)
[23] Bonfiglio A and De Rossi D (ed) 2011 *Wearable Monitoring Systems* (New York: Springer)
[24] Gao T, Greenspan D, Welsh M, Juang R R and Alm A 2005 Vital signs monitoring and patient tracking over a wireless network *Proc. of IEEE-EMBS 27th Annual Int. Conf. of the Engineering in Medicine and Biology (Shanghai, China, 1–5 September)* pp 102–5
[25] Otto C A, Jovanov E and Milenkovic E A 2006 *WBAN-based system for health monitoring at home Proc. of IEEE/EMBS Int. Summer School, Medical Devices and Biosensors (Boston, MA, USA) 4–6 September pp 20–3*
[26] Zhang G H, Poon C C Y, Li Y and Zhang Y T 2009 *A biometric method to secure telemedicine systems Proc. of the 31st Annual Int. Conf. of the IEEE Engineering in Medicine and Biology Society (Minneapolis, MN, USA) pp 701–4*
[27] Bao S, Zhang Y and Shen L 2005 *Physiological signal based entity authentication for body area sensor networks and mobile healthcare systems Proc. of the 27th Annual Int. Conf. of the IEEE EMBS (Shanghai, China) 1–4 September pp 2455–8*
[28] Ikonen V and Kaasinen E 2008 Ethical assessment in the design of ambient assisted living *Proc. of Assisted Living Systems—Models, Architectures and Engineering Approaches* (Germany: Schloss Dagstuhl) pp 14–7

[29] Srinivasan V, Stankovic J and Whitehouse K 2008 *Protecting your daily in home activity information from a wireless snooping attack Proc. of the 10th Int. Conf. on Ubiquitous Computing (Seoul, Korea) 21–24 September pp 202–11*

[30] Casas R, Blasco M R, Robinet A, Delgado A R, Yarza A R, Mcginn J, Picking R and Grout V 2008 *User modelling in ambient intelligence for elderly and disabled people Proc. of the 11th Int. Conf. on Computers Helping People with Special Needs (Linz, Austria) pp 114–22*

[31] Jasemian Y 2008 *Elderly comfort and compliance to modern telemedicine system at home Proc. of the Second Int. Conf. on Pervasive Computing Technologies for Healthcare (Tampere, Finland) 30 January–1 February pp 60–3*

[32] Atallah L, Lo B, Yang G Z and Siegemund F 2008 *Wirelessly accessible sensor populations (WASP) for elderly care monitoring Proc. of the Second Int. Conf. on Pervasive Computing Technologies for Healthcare (Tampere, Finland) 30 January–1 February pp 2–7*

[33] Hori T, Nishida Y, Suehiro T and Hirai S 2000 *SELF-Network: Design and implementation of network for distributed embedded sensors Proc. of IEEE/RSJ Int. Conf. on Intelligent Robots and Systems (Takamatsu, Japan) 30 October–5 November pp 1373–8*

[34] Mori Y, Yamauchi M and Kaneko K 2000 *Design and implementation of the Vital Sign Box for home healthcare Proc. of IEEE EMBS Int. Conf. on Information Technology Applications in Biomedicine (Arlington, VA, USA) pp 104–9*

[35] Lauterbach C, Strasser M, Jung S and Weber W 2002 *Smart clothes self-powered by body heat Proc. of Avantex Symp. (Frankfurt, Germany) pp 5259–63*

[36] Marinkovic S and Popovici E 2009 *Network coding for efficient error recovery in wireless sensor networks for medical applications Proc. of Int. Conf. on Emerging Network Intelligence (Sliema, Malta) 11–16 October pp 15–20*

[37] Schoellhammer T, Osterweil E, Greenstein B, Wimbrow M and Estrin D 2004 *Lightweight temporal compression of microclimate datasets Proc. of the 29th Annual IEEE Int. Conf. on Local Computer Network (Tampa, FL, USA) 16–18 November pp 516–24*

[38] Barth A T, Hanson M A, Powell H C Jr and Lach J 2009 *Tempo 3.1: A body area sensor network platform for continuous movement assessment Proc. of the 6th Int. Workshop on Wearable and Implantable Body Sensor Networks (Berkeley, CA, USA) pp 71–6*

[39] Gietzelt M, Wolf K H, Marschollek M and Haux R 2008 *Automatic self-calibration of body worn triaxial-accelerometers for application in healthcare Proc. of the Second Int. Conf. on Pervasive Computing Technologies for Healthcare (Tampere, Finland) pp 177–80*

[40] Gao T, Greenspan D, Welsh M, Juang R R and Alm A 2005 *Vital signs monitoring and patient tracking over a wireless network Proc. of the 27th Annual Int. Conf. of the IEEE EMBS (Shanghai, China) 1–4 September pp 102–5 Purwar A, Jeong D U and Chung WY 2007 Activity monitoring from realtime triaxial accelerometer data using sensor network* Proceedings of International Conference on Control, Automation and Systems (Hong Kong, 21–23 March) *pp 2402–6*

[41] Baker C *et al* 2007 *Wireless sensor networks for home health care Proc. of the 21st Int. Conf. on Advanced Information Networking and Applications Workshops (Niagara Falls, Canada) pp 832–7*

[42] Schwiebert L, Gupta S K S and Weinmann J 2001 *Research challenges in wireless networks of biomedical sensors Proc. of the 7th Annual Int. Conf. on Mobile Computing and Networking (Rome, Italy) 16–21 July pp 151–65*

[43] Aziz O, Lo B, King R, Darzi A and Yang G Z 2006 *Pervasive body sensor network: An approach to monitoring the postoperative surgical patient Proc. of Int. Workshop on Wearable and implantable Body Sensor Networks (BSN 2006) (Cambridge, MA, USA) pp 13–18*

[44] Kahn J M, Katz R H and Pister K S J 1999 *Next century challenges: Mobile networking for smart dust Proc. of the ACM MobiCom'99 (Washington, DC, USA) pp 271–8*
[45] Noury N, Herve T, Rialle V, Virone G, Mercier E, Morey G, Moro A and Porcheron T 2000 *Monitoring behavior in home using a smart fall sensor Proc. of IEEE-EMBS Special Topic Conf. on Micro-technologies in Medicine and Biology (Lyon, France) 12–14 October pp 607–10*
[46] Kwon D Y and Gross M 2005 *Combining body sensors and visual sensors for motion training Proc. of the 2005 ACM SIGCHI Int. Conf. on Advances in Computer Entertainment Technology (Valencia, Spain) 15–17 June pp 94–101*
[47] Boulgouris N K, Hatzinakos D and Plataniotis K N 2005 Gait recognition: A challenging signal processing technology for biometric identification *IEEE Signal Process. Mag.* **22** 78–90
[48] Kimmeskamp S and Hennig E M 2001 Heel to toe motion characteristics in Parkinson patients during free walking *Clin. Biomech.* **16** 806–12
[49] Turcot K, Aissaoui R, Boivin K, Pelletier M, Hagemeister N and de Guise J A 2008 New accelerometric method to discriminate between asymptomatic subjects and patients with medial knee osteoarthritis during 3-D gait *IEEE Trans. Biomed. Eng.* **55** 1415–22
[50] Furnée H 1997 Real-time motion capture systems *Three-Dimensional Analysis of Human Locomotion* ed P Allard, A Cappozzo, A Lundberg and C L Vaughan (Chichester: Wiley) pp 85–108
[51] Bamberg S J M, Benbasat A Y, Scarborough D M, Krebs D E and Paradiso J A 2008 Gait analysis using a shoe-integrated wireless sensor system *IEEE Trans. Inf. Technol. Biomed.* **12** 413–23
[52] Choi J H, Cho J, Park J H, Eun J M and Kim M S 2008 *An efficient gait phase detection device based on magnetic sensor array Proc. of the 4th Kuala Lumpur Int. Conf. on Biomedical Engineering***21** *(Kuala Lumpur, Malaysia) 25–28 June pp 778–81*
[53] Hidler J 2004 *Robotic-assessment of walking in individuals with gait disorders Proc. of the 26th Annual Int. Conf. of the IEEE Engineering in Medicine and Biology Society* **7** *(San Francisco, CA, USA) 1–5 September pp 4829–31*
[54] Wahab Y and Bakar N A 2011 *Gait analysis measurement for sport application based on ultrasonic system Proc. of the 2011 IEEE 15th Int. Symp. on Consumer Electronics (Singapore) 14–17 June pp 20–24*
[55] De Silva B, Jan A N, Motani M and Chua K C 2008 *A real-time feedback utility with body sensor networks Proc. of the 5th Int. Workshop on Wearable and Implantable Body Sensor Networks (BSN 08) (Hong Kong) 1–3 June pp 49–53*
[56] Salarian A, Russmann H, Vingerhoets F J G, Dehollain C, Blanc Y,, Burkhard P R and Aminian K 2004 Gait assessment in Parkinson's disease: Toward an ambulatory system for long-term monitoring *IEEE Trans. Biomed. Eng.* **51** 1434–43
[57] Atallah L, Jones G G, Ali R, Leong J J H, Lo B and Yang G Z 2011 *Observing recovery from knee-replacement surgery by using wearable sensors Proc. of the 2011 Int. Conf. on Body Sensor Networks (Dallas, TX, USA) 23–25 May pp 29–34*
[58] ElSayed M, Alsebai A, Salaheldin A, El Gayar N and ElHelw M 2010 *Ambient and wearable sensing for gait classification in pervasive healthcare environments Proc. of the 12th IEEE Int. Conf. on e-Health Networking Applications and Services (Healthcom) (Lyon, France) 1–3 July pp 240–5*

IOP Publishing

Wearable Communication Systems and Antennas (Second Edition)
Design, efficiency, and miniaturization techniques
Albert Sabban

Chapter 9

Analysis and measurements of wearable antennas in the vicinity of the human body

9.1 Introduction

This chapter presents an analysis and measurements of wearable antennas in the vicinity of the human body. Electrical properties of human body tissues have a significant effect on the electrical characteristics of wearable antennas, see [1–21]. The antennas' input impedance variation as a function of distance from the body was computed by employing electromagnetic software [22]. Their radiation characteristics on the human body have been measured by using a phantom. The phantom's electrical characteristics represent those of a human body. It contains a mix of 55% water, 44% sugar, and 1% salt. The antenna under test was placed on the phantom during measurements of the antenna's radiation characteristics. S_{11} and S_{12} parameters were measured directly on a human body by using a network analyzer. The measured results were compared to a known reference antenna.

9.2 Analysis of wearable antennas

The major issue in the design of wearable antennas is the interaction between the RF transmission and the human body. Electrical properties of human body tissues have been used in the design of wearable antennas presented in this book. The dielectric constant and conductivity of human body tissues may be used to calculate the attenuation α of RF transmission through the human body, as given in equations (9.1) and (9.2). The properties of human body tissues are listed in table 9.1 see [8].

$$\gamma = \sqrt{j\omega\mu(\pi + h\omega\varepsilon)} = \alpha + j\beta \tag{9.1}$$

Table 9.1. Properties of human body tissues.

Tissue	Property	434 MHz	800 MHz	1000 MHz
Prostate	σ	0.75	0.90	1.02
	ε	50.53	47.4	46.65
Stomach	σ	0.67	0.79	0.97
	ε	42.9	40.40	39.06
Colon, heart	σ	0.98	1.15	1.28
	ε	63.6	60.74	59.96
Kidney	σ	0.88	0.88	0.88
	ε	117.43	117.43	117.43
Nerve	σ	0.49	0.58	0.63
	ε	35.71	33.68	33.15
Fat	σ	0.045	0.056	0.06
	ε	5.02	4.58	4.52
Lung	σ	0.27	0.27	0.27
	ε	38.4	38.4	38.4

Table 9.2. The attenuation, α, of RF transmission through stomach, skin and pancreas tissues.

Frequency MHz	α Stomach dB cm^{-1}	α Skin dB cm^{-1}	α Pancreas cm^{-1}
150	1.096 431 56	0.993 564 88	1.154 709 08
200	1.209 393 08	1.084 756 96	1.270 231 2
250	1.312 928 12	1.167 876 64	1.374 799 16
300	1.413 919 92	1.249 121 44	1.475 556 6
350	1.486 701 72	1.304 282 84	1.548 581 44
400	1.557 096 52	1.357 282 92	1.618 976 24
450	1.622 474 28	1.406 429 08	1.686 011 88
500	1.678 442 92	1.448 674 64	1.747 639 88
550	1.733 500 16	1.490 112 96	1.808 512 72
600	1.788 062 64	1.531 082 56	1.869 012 32
650	1.838 406 64	1.577 303 56	1.929 650 8
700	1.888 368 72	1.623 298 88	1.990 159 08
750	1.938 096 44	1.669 164	2.050 658 68
800	1.987 711 32	1.714 968 36	2.111 245 08
850	2.048 740 4	1.742 770 4	2.156 684 88
900	2.109 847 6	1.770 303 36	2.202 037 88
950	2.171 067 64	1.797 645 36	2.247 382 2
1000	2.232 426 56	1.824 813 76	2.292 769 92

$$\alpha = \mathrm{Re}(\gamma) \tag{9.2}$$

The attenuation α of RF transmission through stomach, skin and pancreas tissues in dB cm^{-1} is listed in table 9.2. The attenuation α of RF transmission through stomach and pancreas tissues is around 2.2 dB cm^{-1} at 1 GHz.

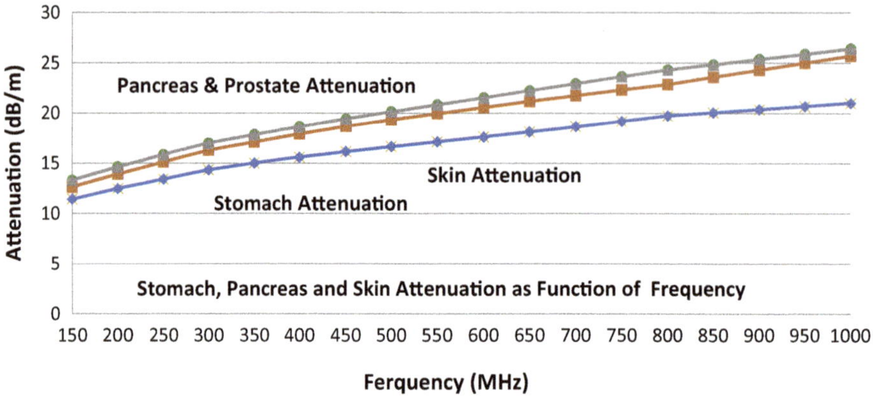

Figure 9.1. The attenuation, α, of RF transmission through stomach, skin and pancreas tissues.

Table 9.3. The attenuation α of RF transmission through blood, fat and small intestine tissues.

Frequency MHz	α Blood dB m^{-1}	α Fat dB m^{-1}	α Small intestine dB m^{-1}
150	26.5193	1.8752	16.7351
200	29.4816	2.300 63	21.4619
250	31.9065	2.592 02	23.9958
300	34.0036	2.850 79	25.4946
350	35.5104	3.102 19	26.4352
400	36.7922	3.355 19	27.0538
450	37.9234	3.493 84	27.4771
500	38.9634	3.629 96	27.7771
550	39.9081	3.771 62	27.996
600	40.7788	3.92648	28.16
650	41.5575	4.081 33	28.2857
700	42.2879	4.236 51	28.384
750	42.9794	4.283 21	28.462
800	43.6395	4.328 99	28.5251
850	44.3448	4.3741	28.5766
900	45.0293	4.418 74	28.6193
950	45.6965	4.4789	28.655
1000	46.3492	4.539 12	28.6851

Figure 9.1 presents attenuation via several human tissues. Stomach tissue attenuation at 500 MHz is around 1.6 dB cm^{-1}. There is good agreement between the measured and computed results. The attenuation α of RF transmission through blood, fat and small intestine tissues in dB m^{-1} is listed in table 9.3. The attenuation α of RF transmission through blood tissues is around 4.6 dB cm^{-1} at 1 GHz. However, the attenuation α of RF transmission through fat tissues is around 0.45 dB cm^{-1} at 1 GHz. Figure 9.2 presents the attenuation of human blood tissues. Blood tissue attenuation at 500 MHz is around 4 dB cm^{-1}.

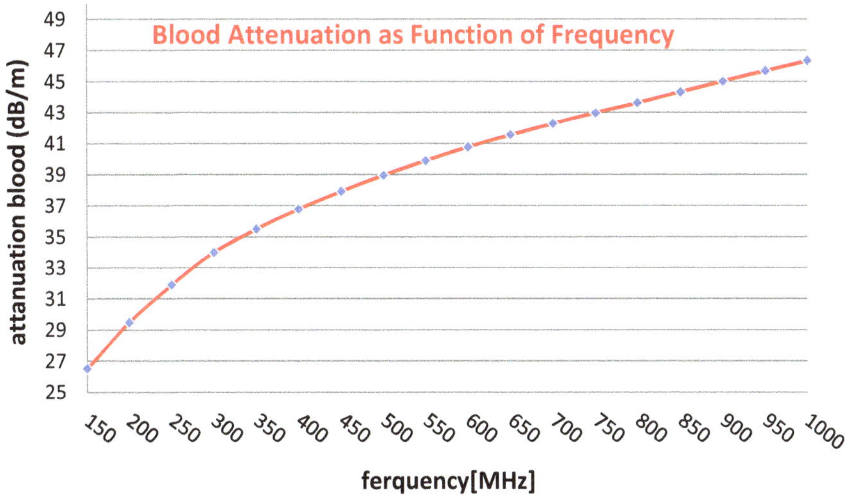

Figure 9.2. The attenuation α of RF transmission through blood.

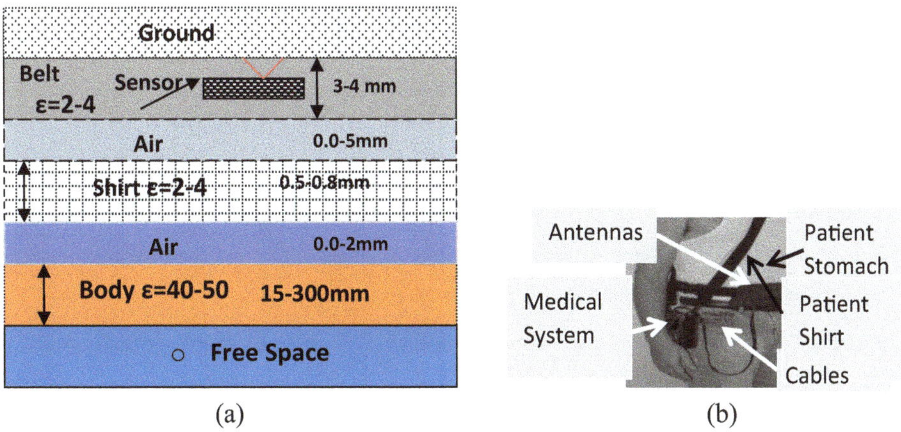

Figure 9.3. (a) Analyzed structure model. (b) Medical system on a patient.

9.3 Design of wearable antennas in the vicinity of the human body

The antennas' input impedance variation as a function of distance from the body has been computed by employing Momentum software. The analyzed structure is presented in figure 9.3(a). These properties were employed in the antenna design. The antenna was placed inside a belt with a thickness between 1 to 4 mm as shown in figure 9.3(b). The patient's body thickness was varied from 15 mm to 300 mm. The dielectric constant of the body was varied from 40 to 50. The antenna was placed inside a belt with a thickness between 2 to 4 mm with a dielectric constant from 2 to 4. The air layer between the belt and the patient's shirt may vary from 0 mm to 8 mm. Shirt thickness varied from 0.5 mm to 1 mm. The dielectric constant of the shirt varied from 2 to 4. Figure 9.4 presents S_{11} results (of the antenna shown

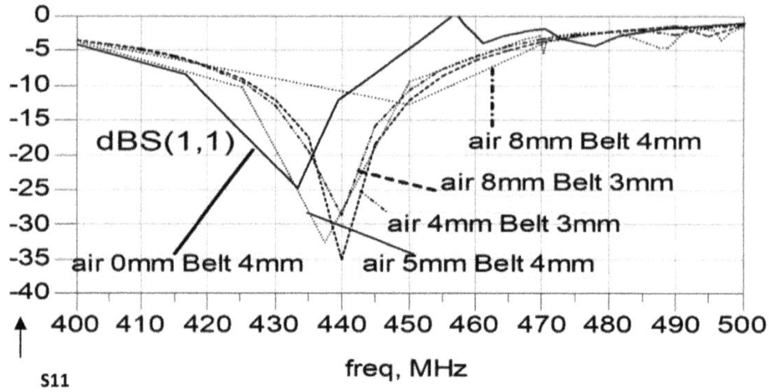

Figure 9.4. S_{11} results for different antenna positions relative to the human body.

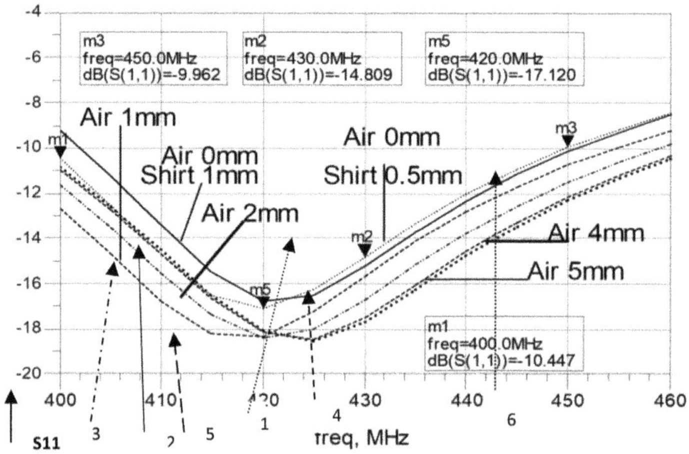

Figure 9.5. Folded antenna S_{11} results for different antenna positions relative to the human body.

in figure 8.1) for different belt thickness, shirt thickness and air spacing between the antennas and the human body. One may conclude from the results shown in figure 9.4 that the antenna has a VSWR better than 2.5:1 for air spacing up to 8 mm between the antennas and the patient's body. For frequencies ranging from 415 MHz to 445 MHz the antenna has a VSWR better than 2:1 when there is no air spacing between the antenna and the patient's body. The results shown in figure 9.5 indicate that the folded antenna (the antenna shown in figure 8.5) has a VSWR better than 2.0:1 for air spacing up to 5 mm between the antennas and the patient's body. Figure 9.6 presents S_{11} results of the folded antenna for different positions relative to the human body. An explanation of figure 9.5 is given in table 9.4. If the air spacing between the sensors and the human body is increased from 0 mm to 5 mm the antenna's resonant frequency is shifted by 5%. The loop antenna (see figure 8.8) has a VSWR better than 2.0:1 for air spacing up to 5 mm between the antennas and patient's body, as presented in figure 9.6. If the air spacing between the sensors and the human body is increased from 0 mm to 5 mm the

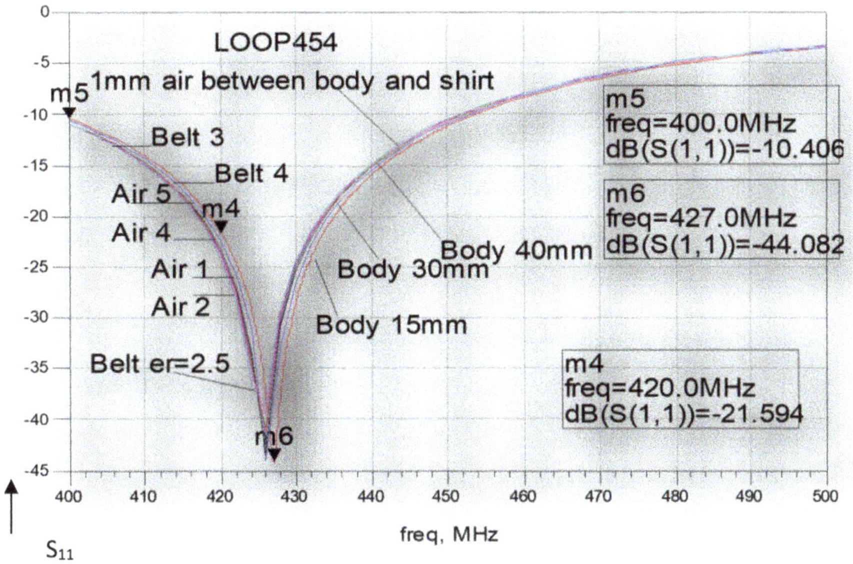

Figure 9.6. Loop antenna S_{11} results for different antenna positions relative to the body.

Table 9.4. Explanation of figure 9.5.

Picture #	Line type	Sensor position
1	Dot	Shirt thickness 0.5 mm
2	Line ———	Shirt thickness 1 mm
3	Dash dot -·---·-·	Air spacing 2 mm
4	Dash -------	Air spacing 4 mm
5	Long dash —— —	Air spacing 1 mm
6	Big dots	Air spacing 5 mm

Table 9.5. Explanation of figure 9.6.

Plot colour	Sensor position
Red	Body 15 mm air spacing 0 mm
Blue	Air spacing 5 mm body 15 mm
Pink	Body 40 mm air spacing 0 mm
Green	Body 30 mm air spacing 0 mm
Sky	Body 15 mm air spacing 2 mm
Purple	Body 15 mm air spacing 4 mm

computed antenna's resonant frequency is shifted by 2%. However, if the air spacing between the sensors and the human body is increased up to 5 mm the measured loop antenna's resonant frequency is shifted by 5%. An explanation of figure 9.6 is given in table 9.5.

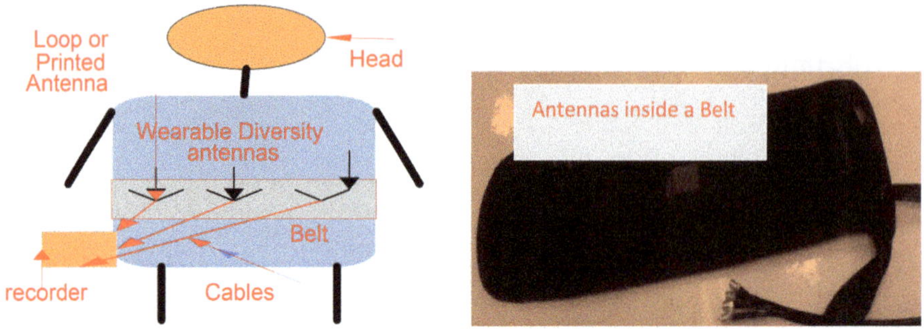

(a) (b)

Figure 9.7. (a) Printed wearable antenna. (b) Wearable antennas inside a belt.

9.4 Wearable antenna arrays

An application of the proposed antenna is shown in figure 9.7(a). Three to four folded dipole or loop antennas may be assembled in a belt and attached to a patient's stomach, see figure 9.7(b). The cable from each antenna is connected to a recorder. The received signal is routed to a switching matrix. The signal with the highest level is selected during the medical test. The antennas receive a signal that is transmitted from various positions in the human body. Folded antennas may be also attached on a patient's back in order to improve the level of the received signal from different locations in the body. Figure 9.8 shows various antenna locations on the back and front of the human body for different medical applications. In several applications the distance separating the transmitting and receiving antennas is less than $2D^2/\lambda$. D is the largest dimension of the radiator. In these applications the amplitude of the electromagnetic field close to the antenna may be quite powerful, but because of rapid fall-off with distance, the antenna does not radiate energy to infinite distances, but instead the radiated power remains trapped in the region near to the antenna. Thus, the near-fields only transfer energy to close distances from the receivers. The receiving and transmitting antennas are magnetically coupled. Change in current flow through one wire induces a voltage across the ends of the other wire through electromagnetic induction. The amount of inductive coupling between two conductors is measured by their mutual inductance. In these applications we have to refer to the near-field and not the far-field radiation.

In figure 9.9(a) several microstrip antennas for medical applications at 434 MHz are shown. The back side of the antennas is presented in figure 9.9(b). The diameter of the loop antenna presented in figure 9.10 is 50 mm. The dimensions of the folded dipole antenna are $7 \times 6 \times 0.16$ cm. The dimensions of the compact folded dipole presented in figure 9.10 are $5 \times 5 \times 0.05$ cm.

9.5 Small wide band dual polarized wearable printed antennas

A small microstrip loaded dipole antenna has been designed. The antenna consists of two layers. The first layer consists of FR4 0.4 mm dielectric substrate. The second

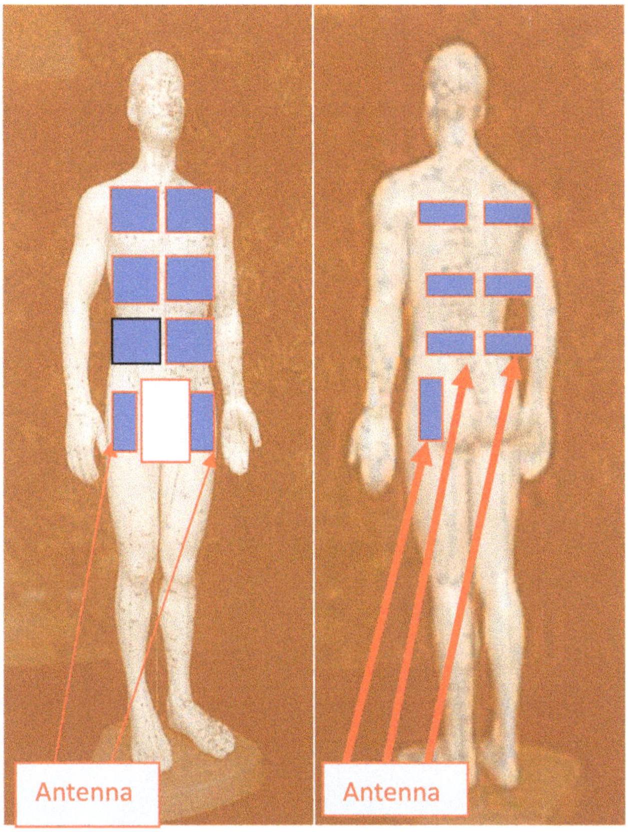

Figure 9.8. Printed patch antenna locations for various medical applications.

layer consists of Kapton 0.4 mm dielectric substrate. The substrate thickness determines the antenna's bandwidth. However, with a thinner substrate we may achieve better flexibility. The proposed antenna is dual polarized. The printed dipole and the slot antenna provide dual orthogonal polarizations. The dual polarized antenna's dimensions are shown in figure 9.11(a); they are $4 \times 4 \times 0.08$ cm. The layout is shown in figure 9.11(b).

The antenna may be attached to a patient's shirt in the stomach or back area. The antenna has been analyzed by using Keysight Momentum software. There is good agreement between the measured and computed results. The antenna's bandwidth is around 15% for a VSWR better than 2:1. The beamwidth is around 100°, and the gain is around 0 dBi. The computed S_{11} parameters are presented in figure 9.12. Figure 9.13 presents the antenna's measured S_{11} parameters on the human body. The antenna's cross-polarized field strength may be adjusted by varying the slot feed location. The computed 3D radiation pattern of the antenna is shown in figure 9.14, and the computed radiation pattern is shown in figure 9.15.

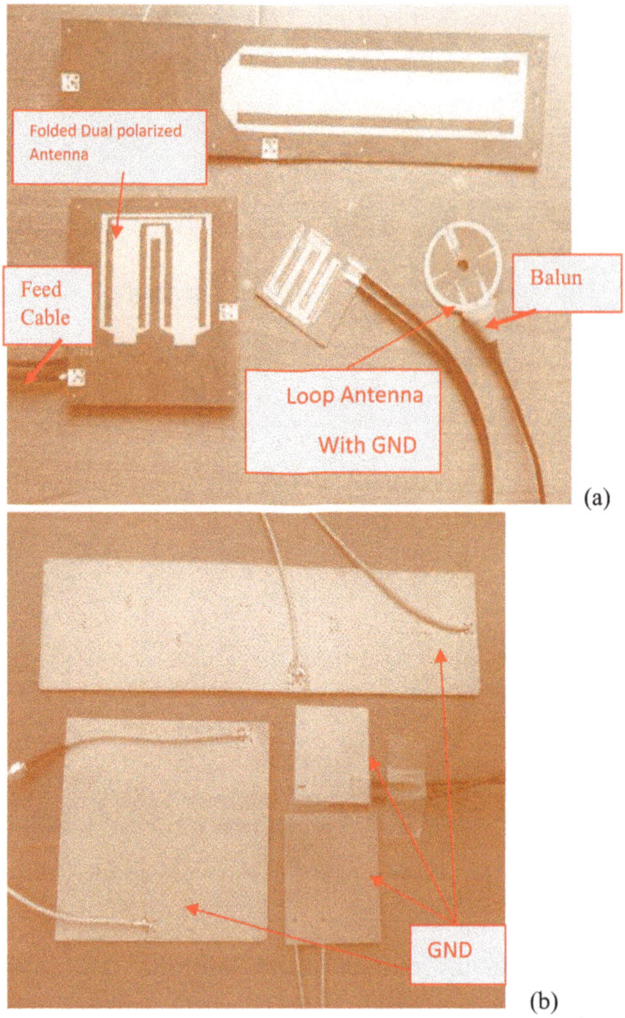

Figure 9.9. (a) Microstrip antennas for medical applications. (b) Back side of the antennas.

9.6 Wearable helix antenna's performance on the human body

In order to compare the variation of the antenna's input impedance as a function of distance from the body to other antennas, a helix antenna has been designed. A helix antenna with nine turns is shown in figure 9.16 and a photo can be seen in figure 9.17. The wearable helix antenna's matching network was printed on a dielectric substrate with dielectric constant of 3.55. The back side of the circuit is copper under the microstrip's matching stubs. However, in the helix antenna area there is no ground plane. The antenna has been designed to operate on the human body. A matching microstrip line network has been designed on RO4003 substrate of 0.8 mm thickness. The helix antenna has a VSWR better than 3:1 at a frequency range from 440 MHz to 460 MHz. The dimensions are $4 \times 4 \times 0.6$ cm. Figure 9.18 presents the

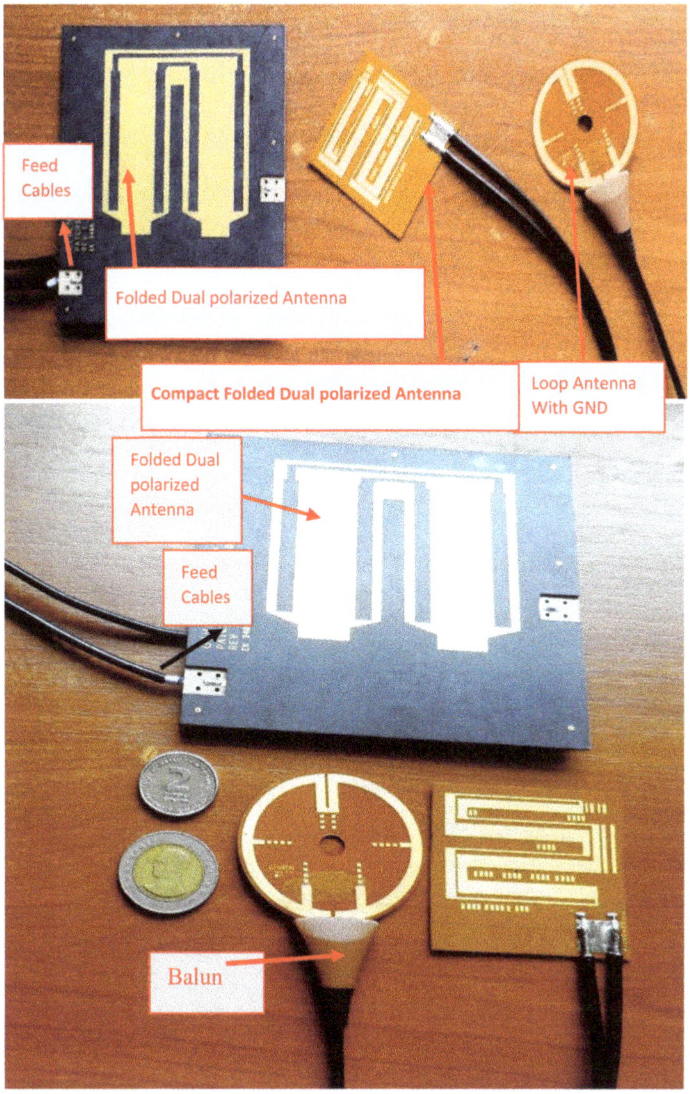

Figure 9.10. Microstrip antennas for medical applications.

measured S_{11} parameters on the human body. The computed E and H radiation plane of the helix antenna is shown in figure 9.19. The input impedance variation as a function of distance from the body is very sensitive. If the air spacing between the helix antenna and the human body is increased from 0 mm to 2 mm the antenna's resonant frequency is shifted by 5%.

However, if the air spacing between the dual polarized antenna and the human body is increased from 0 mm to 5 mm the antenna's resonant frequency is shifted only by 5%.

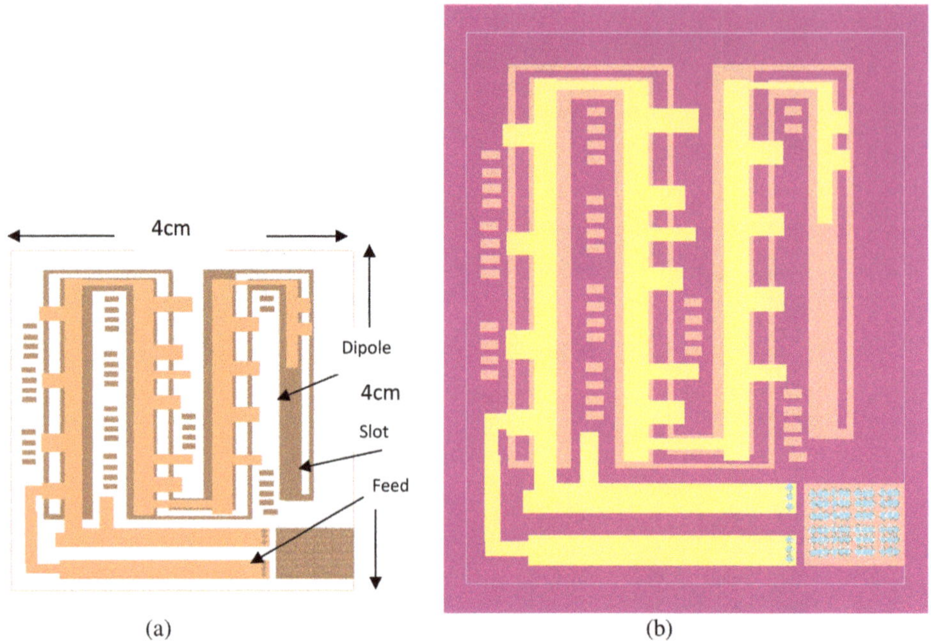

Figure 9.11. Printed compact dual polarized antenna. (a) Antenna dimensions. (b) Layout.

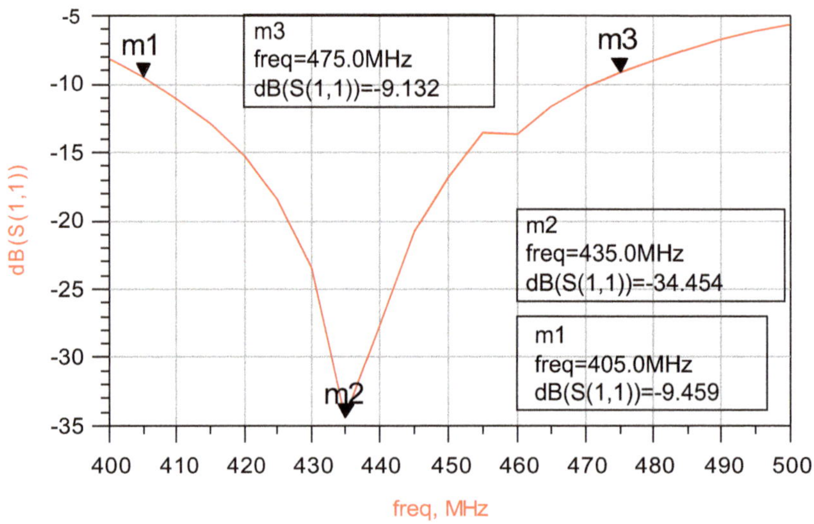

Figure 9.12. Computed S_{11} results of the wide band antenna.

9.7 Wearable antenna measurements in the vicinity of the human body

This section presents measurement techniques of wearable antennas and RF medical systems in the vicinity of the human body. These measurement results are presented

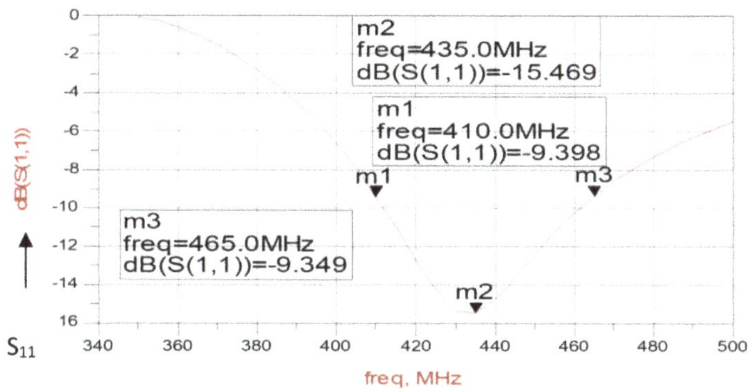

Figure 9.13. Measured S_{11} of the small wide band antenna on the human body.

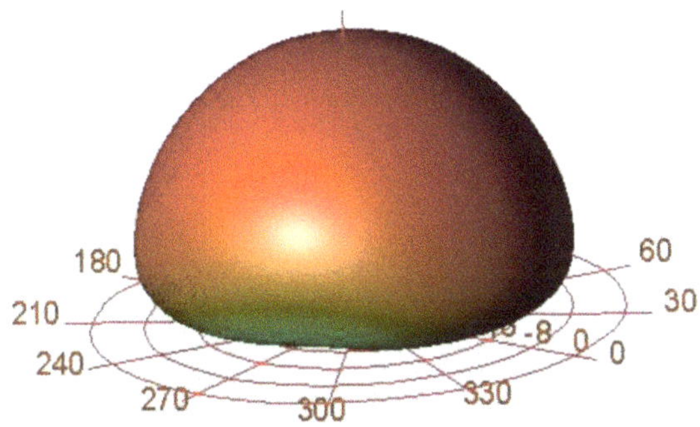

Figure 9.14. Small wide band antenna's 3D radiation pattern.

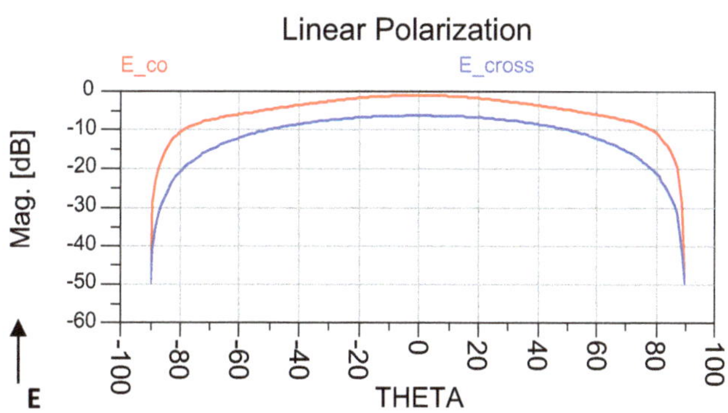

Figure 9.15. Small dual polarized antenna's radiation pattern.

in [1–15]. Wearable antennas and RF medical systems' radiation characteristics on the human body may be measured by using a phantom. The phantom's electrical characteristics represent those of the human body. The phantom has a cylindrical shape with a 40 cm diameter and a length of 1.5 m, and its electrical characteristics are similar to the human body's. The wearable antenna under test was placed on the

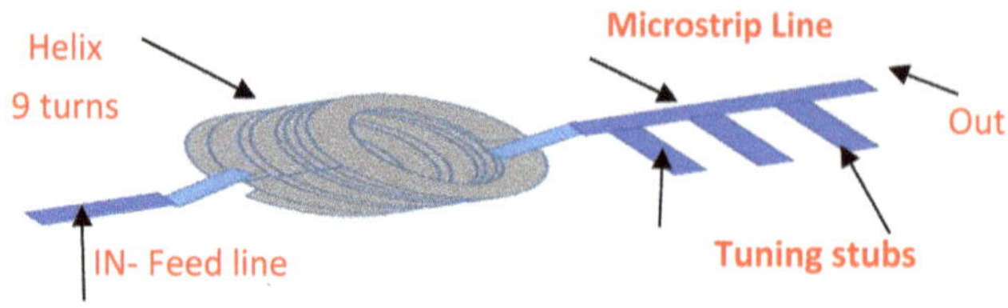

Figure 9.16. Wearable helix antenna layout.

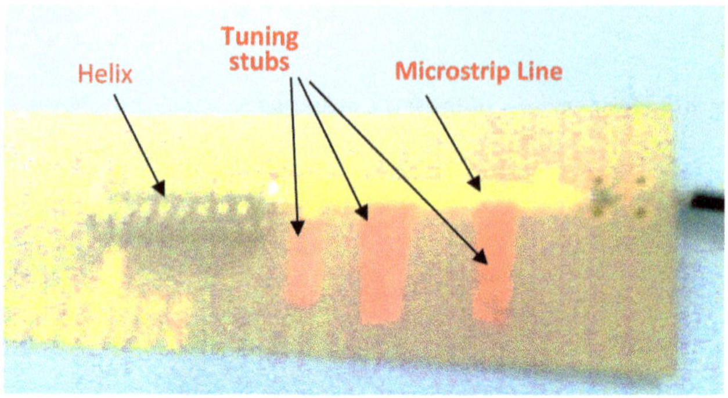

Figure 9.17. Helix antenna for medical applications.

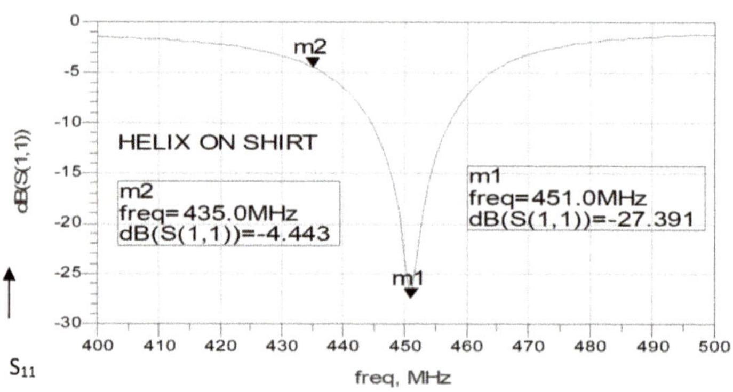

Figure 9.18. Measured S_{11} of the helix antenna on the human body.

phantom during measurements of the antennas' radiation characteristics. The phantom was employed to compare the electrical performance of several new wearable antennas. It was also employed to measure the electrical performance of several antenna belts in the vicinity of the human body.

9.8 Phantom configuration

The phantom represents human body tissues. The phantom contains a mix of water, sugar and salt, and their relative concentration determines the electrical characteristics

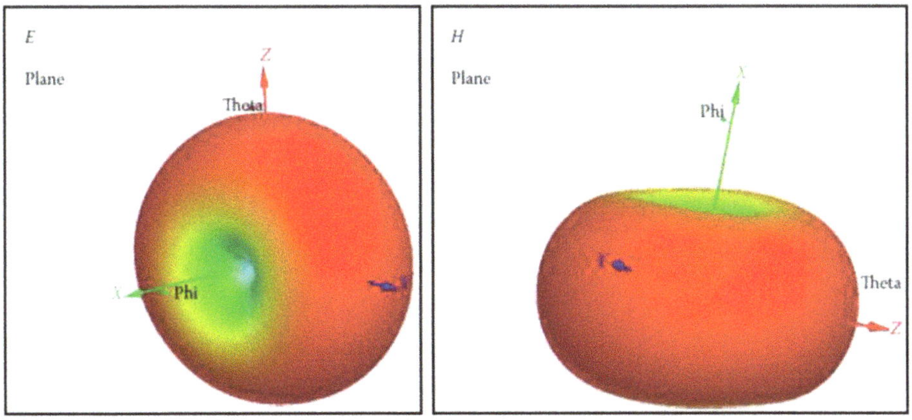

Figure 9.19. *E* and *H* plane radiation pattern of the helix antenna.

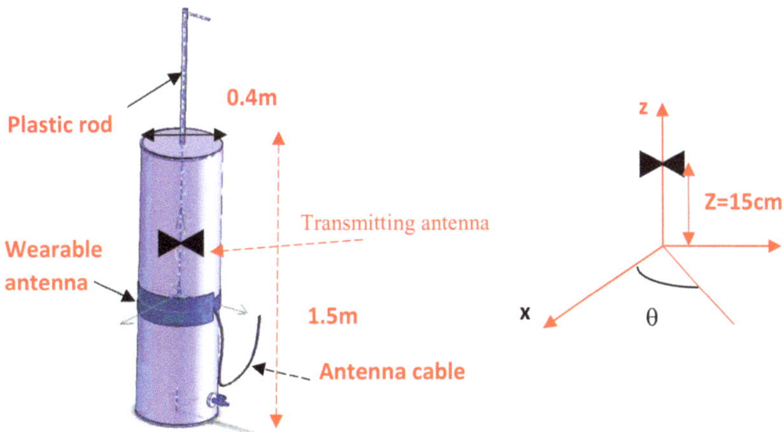

Figure 9.20. Phantom configuration.

of the phantom's environment. A mixture of 55% water, 44% sugar, and 1% salt represents the electrical characteristics of stomach tissues. The phantom may be used to measure electromagnetic radiation from inside or outside the phantom. The phantom is a fiberglass cylinder of 1.5 m height and 0.4 m in diameter as shown in figure 9.20. The thickness of the cylinder's surface is around 2.5 mm. The phantom contains a plastic rod of 5 mm thickness. The position of the plastic rod inside the phantom may be adjusted, and it may be rotated, as shown in figure 9.21. A small transmitting antenna may be attached to the plastic rod at different height positions. The antenna may be rotated in the x–y plane.

9.9 Measurements of wearable antennas using a phantom

The electrical characteristics of several wearable antennas were measured using the phantom. The position of the transmitting antennas along the z axis and x axis was

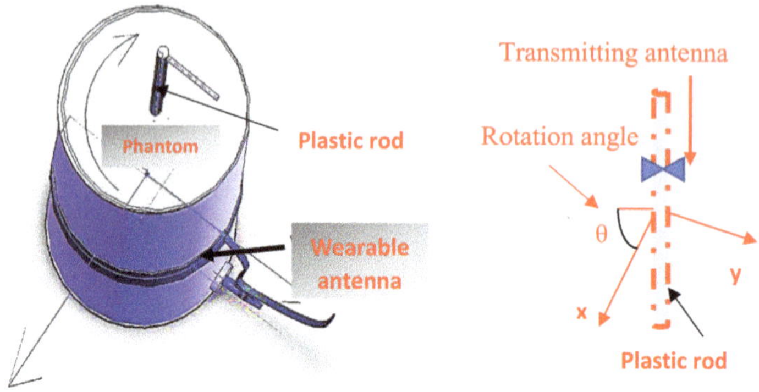

Figure 9.21. Transmitting antenna rotation.

Table 9.6. Measurements of antenna 1.

Antenna 1 X (cm), Z (cm)		Angle θ		
$Z = 0$ cm	0°	90°	180°	270°
Signal level (dB) $X = -5$, $Z = 0$	−60	−63	−81	−65
Signal level (dB) $X = -20$, $Z = 0$	−78	−70	−65	−74
$Z = 15$ cm				
Signal level (dB) $X = -5$, $Z = 15$	−81	−89	−83	−82
Signal level (dB) $X = -20$, $Z = 15$	−86	−89	−81	−85
$Z = -15$ cm				
Signal level (dB) $X = -5$, $Z = -15$	−70	−76	−81	−68
Signal level (dB) $X = -20$, $Z = -15$	−79	−64	−74	−81
Noise test		Angle θ		
$Z = 0$ cm	0°	90°	180°	270°
Signal level (dB)	−72	−74	−74	−74
Noise level (dB)	−95	−95	−95	−95
$Z = 40$ cm				
Signal level (dB)	−72	−74	−74	−74
Noise level (dB)	−88	−88	−88	−88
$Z = -40$ cm				
Signal level (dB)	−72	−74	−74	−74
Noise level (dB)	−94	−93	−94	−95

varied as listed in tables 9.6–9.10, ($Z = 0$, $Z = -15$ cm and $Z = 15$ cm). The angle of the transmitting antenna was varied from 0° to 270°. Signal reception levels and immunity to noise were measured for several types of wearable antenna. We compared the antennas' electrical performance and the best antenna was chosen according to the system's electrical requirements.

Test procedure and process

The test procedure and process is described below.

Table 9.7. Measurements of antenna 2.

Antenna 2 X (cm), Z (cm)			Angle θ	
$Z = 0$ cm	0°	90°	180°	270°
Signal level (dB) $X = -5$, $Z = 0$	−63	−70	−82	−72
Signal level (dB) $X = -20$, $Z = 0$	−84	−66	−66	−79
$Z = 15$ cm				
Signal level (dB) $X = -5$, $Z = 15$	−82	−93	−90	−84
Signal level (dB) $X = -20$, $Z = 15$	−88	−83	−79	−90
$Z = -15$ cm				
Signal level (dB) $X = -5$, $Z = -15$	−74	−82	−85	−92
Signal level (dB) $X = -20$, $Z = -15$	−86	−70	−77	−81
Noise test			Angle θ	
$Z = 0$ cm	0°	90°	180°	270°
Signal level (dB)	−67	−69	−72	−68
Noise level (dB)	−85	−85	−86	−85
$Z = 40$ cm				
Signal level (dB)	−66	−69	−72	−68
Noise level (dB)	−80	−83	−86	−85
$Z = -40$ cm				
Signal level (dB)	−68	−69	−72	−69
Noise level (dB)	−84	−82	−83	−83

Table 9.8. Measurements of antenna 3.

Antenna 3 X (cm), Z (cm)			Angle θ	
$Z = 0$ cm	0°	90°	180°	270°
Signal level (dB) $X = -5$, $Z = 0$	−63	−63	−82	−69
Signal level (dB) $X = -20$, Z = 0	−83	−70	−68	−76
Z = 15 cm				
Signal level (dB) $X = -5$, $Z = 15$	−85	−86	−85	−86
Signal level (dB) $X = -20$, $Z = 15$	−89	−88	−86	−85
Z = −15 cm				
Signal level (dB) $X = -5$, $Z = -15$	−72	−79	−83	−74
Signal level (dB) $X = -20$, $Z = -15$	−85	−69	−77	−82
Noise test			Angle θ	
$Z = 0$ cm	0°	90°	180°	270°
Signal level (dB)	−68	−70	−76	−70
Noise level (dB)	−91	−91	−92	−92
$Z = 40$ cm				
Signal level (dB)	−68	−70	−76	−70
Noise level (dB)	−90	−88	−90	−88
Z = −40 cm				
Signal level (dB)	−68	−70	−76	−70
Noise level (dB)	−89	−87	−90	−87

Table 9.9. Measurements of antenna 4.

Antenna 4 X (cm), Z (cm)		Angle θ		
$Z = 0$ cm	0°	90°	180°	270°
Signal level (dB) $X = -5$, $Z = 0$	−61	−62	−81	−63
Signal level (dB) $X = -20$, $Z = 0$	−79	−69	−67	−73
$Z = 15$ cm				
Signal level (dB) $X = -5$, $Z = 15$	−86	−88	−88	−83
Signal level (dB) $X = -20$, $Z = 15$	−90	−84	−82	−86
$Z = -15$ cm				
Signal level (dB) $X = -5$, $Z = -15$	−70	−81	−82	−67
Signal level (dB) $X = -20$, $Z = -15$	−80	−67	−74	−81
Noise test		Angle θ		
$Z = 0$ cm	0°	90°	180°	270°
Signal level (dB)	−70	−70	−76	−70
Noise level (dB)	−95	−95	−95	−95
$\boldsymbol{Z = 40}$ **cm**				
Signal level (dB)	−70	−70	−76	−70
Noise level (dB)	−92	−92	−91	−92
$\boldsymbol{Z = -40}$ **cm**				
Signal level (dB)	−70	−70	−76	−70
Noise level (dB)	−92	−92	−91	−92

Table 9.10. Measurements of antenna 5.

Antenna 5 X (cm), Z (cm)		Angle θ		
$Z = 0$ cm	0°	90°	180°	270°
Signal level (dB) $X = -5$, $Z = 0$	−67	−52	−58	−60
Signal level (dB) $X = -20$, $Z = 0$	−68	−70	−77	−78
$Z = 15$ cm				
Signal level (dB) $X = -5$, $Z = 15$	−90	−86	−92	−90
Signal level (dB) $X = -20$, $Z = 15$	−84	−86	−90	−86
$Z = -15$ cm				
Signal level (dB) $X = -5$, $Z = -15$	−85	−90	−85	−92
Signal level (dB) $X = -20$, $Z = -15$	−90	−85	−75	−78
Noise test		Angle θ		
$Z = 0$ cm	0°	90°	180°	270°
Signal level (dB)	−70	70	−70	−70
Noise level (dB)	−90	−90	−90	−90
$\boldsymbol{Z = 40}$ **cm**				
Signal level (dB)	−72	−72	−72	−72
Noise level (dB)	−90	−90	−90	−90
$\boldsymbol{Z = -40}$ **cm**				
Signal level (dB)	−73	−73	−73	−73
Noise level (dB)	−90	−90	−90	−90

Test procedure

The test checks two parameters of the antenna array:

- Signal reception levels.
- Immunity to noise.

Measured antennas

The test checks the following antenna arrays:

- A four sensor antenna array in a belt, antennas in orientations of +45°, +45°, +45°,−45°.
- A sensor belt with four loop antennas in orientations of +45°, +45°, +45°, −45°.
- A sensor belt with four antennas in orientations of +45°, −45°, +45°, −45°.
- A thin belt with four sensor arrays in orientations of +45°, −45°, +45°, −45°.
- A four loop antenna array in a sleeve.

Test process

- Place the antenna array on the phantom as shown in figure 9.21 and connect to the recorder.
- Signal reception levels.
- Place the transmitter in the phantom at the following coordinates, 5 min per each location.
- Z values: from −15 cm to +15 cm in 15 cm increments, 0 being the level of the antennas' center.
- X values: from −5 cm to −20 cm in 15 cm increments, 0 being the container's wall where the antennas are attached.
- θ values: from 0° to 270° in 90° increments, 0° being perpendicular to the middle of the antennas' set.

Immunity to noise:

Place the transmitter outside the phantom in the following coordinates, 5 min per each location.

Z values: from −40 cm to +40 cm in 10 cm increments, 0 being the level of the antenna's center.

X values: 100 cm from the container's wall θ values: from 0° to 270° in 90° increments, 0° being perpendicular to the middle of the antenna's set.

9.10 Measurement results of wearable antennas

Measurements of five antenna array configurations were taken. For all the configurations, the lowest measured signal level occurred when the transmitting antenna was located at $z = 15$ cm and $x = -5$ cm. For all the configurations the highest measured signal level occurred when the transmitting antenna was located at $z = 0$ cm and $x = -5$ cm.

9.10.1 Measurements of antenna array 1

The measured antennas consist of four loop antennas with a tuning capacitor as shown in figure 9.22. The loop antennas are printed on FR4 substrate with a

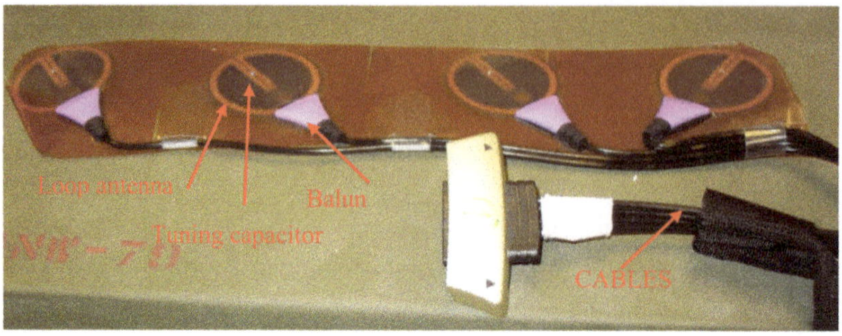

Figure 9.22. Sensor belt with four antennas in orientations of +45°, +45°, +45°, −45°.

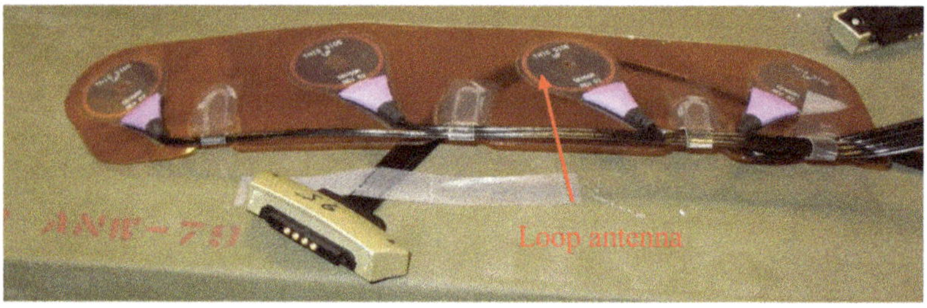

Figure 9.23. Sensor belt with four antennas in orientations of +45°, +45°, +45°, −45° (without a tuning capacitor).

dielectric constant of 4.8 and 0.25 mm thickness. The loop radiators' orientations are, +45°, −45°, −45°, −45°. The antenna was inserted into a thin belt. The measurement results of antenna number 1 are listed in table 9.6.

The highest signal level is at $z = 0$, $X = -5$ cm, $\theta = 0°$. At $\theta = 180°$ the signal level is lower by 21 dB. The lowest signal level is at $z = \pm 15$ cm, $X = -5$ cm, $\theta = 90°$. The noise level is lower by 14 dB to 21 dB than the signal level.

9.10.2 Measurements of antenna array 2

The measured antenna consists of four loop antennas without a tuning capacitor as shown in figure 9.23. The loop radiators' orientations are +45°, −45°, −45°, −45°. The antenna was inserted into a thin belt. The measurement results of antenna number 1 are listed in table 9.7.

The highest signal level is at $z = 0$ cm, $X = -5$ cm, $\theta = 0°$. At $\theta = 180°$ the signal level is lower by 19 dB. The lowest signal level is at $z = 15$ cm, $X = -5$ cm, $\theta = 90°$. The noise level is lower by 14 dB to 18 dB than the signal level.

9.10.3 Measurements of antenna array 3

The measured antenna consists of four tuned loop antennas with a tuning capacitor as shown in figure 9.24. The loop radiators' orientations are +45°, −45°, +45°, −45°. The antenna was inserted into a belt.

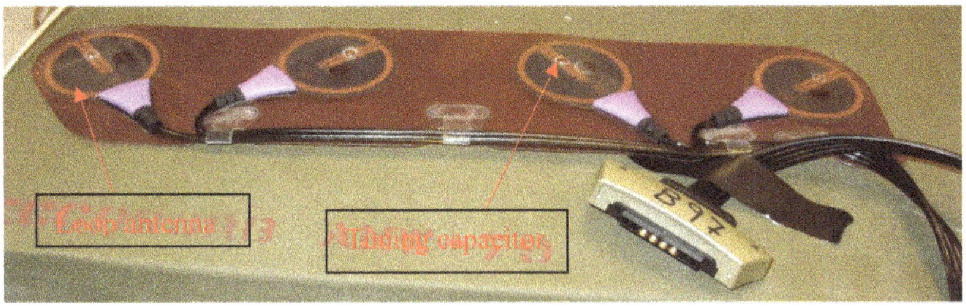

Figure 9.24. Sensor belt with four antennas in orientations of +45°, –45°, +45°, –45°.

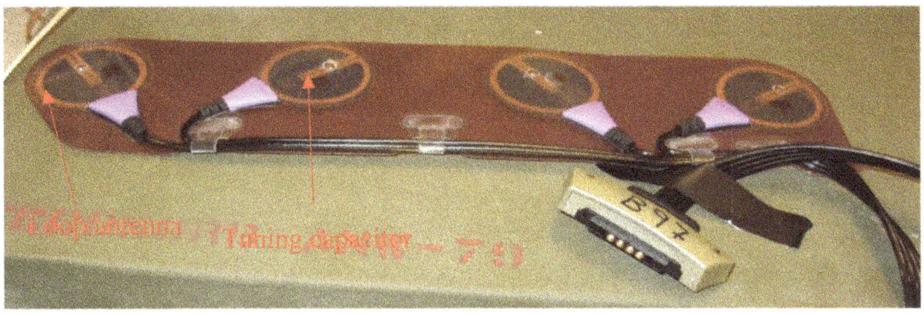

Figure 9.25. Four antennas in orientations of +45°, –45°, +45°, –45°, in a thinner belt.

Measurement results of antenna number 3 are listed in table 9.8.

The highest signal level is at $z = 0$ cm, $X = -5$ cm, $\theta = 0°$. At $\theta = 180°$ the signal level is lower by 19 dB. The lowest signal level is at $z = 15$ cm, $X = -20$ cm, $\theta = 0°$. The noise level is lower by 18 dB to 23 dB than the signal level.

9.10.4 Measurements of antenna array 4 in a thinner belt

The measured antenna consists of four loop antennas with a tuning capacitor as shown in figure 9.25. The loop radiators' orientations are +45°, –45°, +45°, –45°. The antenna was inserted into a thinner belt. The measurements results of antenna number 4 are listed in table 9.9.

The highest signal level is at $z = 0$ cm, $X = -5$ cm, $\theta = 0°$. At $\theta = 180°$ the signal level is lower by 20 dB. The lowest signal level is at $z = 15$ cm, $X = -20$ cm, $\theta = 0°$. The noise level is lower by 18 dB to 25 dB than the signal level.

9.10.5 Measurements of antenna array 5

The measured antenna consists of four loop antennas without a tuning capacitor inserted into a sleeve, as shown in figure 9.26. The sleeve improves the antenna's VSWR from 4:1 to 2:1. The loop radiators' orientations are (90°, 90°, 90°, 90°).

The antennas were inserted into a thin sleeve. The measurement results of antenna number 5 are listed in table 9.10.

The highest signal level is at $z = 0$ cm, $X = -5$ cm, $\theta = 90°$. At $\theta = 180°$ the signal level is lower by 6 dB. At $\theta = 270°$ the signal level is lower by 8 dB. The lowest signal

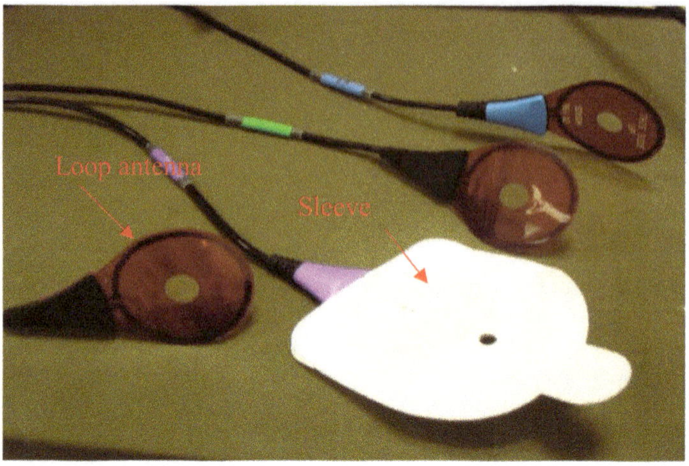

Figure 9.26. Four loop antennas in a sleeve. The radiators' orientations are (90°, 90°, 90°, 90°).

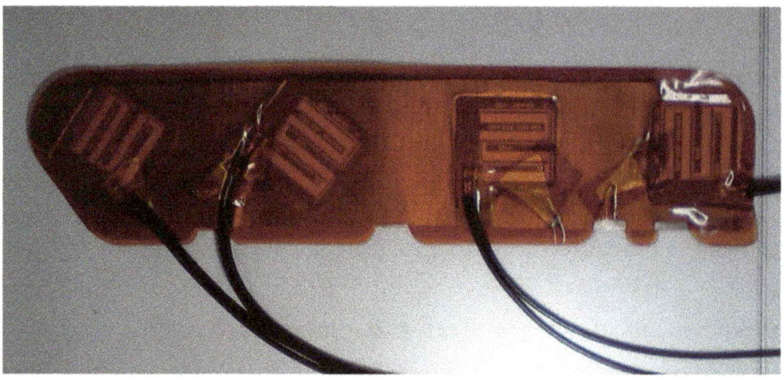

Figure 9.27. Four compact antennas in orientations of 45°, −45°, 0°, 90°, in a thinner belt.

level is at z = 15 cm, X = −20 cm, θ = 180°. The noise level is lower by 18 dB to 22 dB than the signal level.

Four compact antennas in a thinner belt are shown in figure 9.27.

9.11 Conclusions

This chapter presents an analysis and measurements of wearable antennas in the vicinity of the human body. The antennas' S_{11} results for different belt thickness, shirt thickness and air spacing between the antennas and the human body are presented in this chapter. If the air spacing between the dual polarized antenna and the human body is increased from 0 mm to 5 mm the antenna's resonant frequency is shifted by 5%. However, if the air spacing between the helix antenna and the human body is increased only from 0 mm to 2 mm the antenna's resonant frequency is shifted by 5%. The effect of the antenna's location on the human body should be considered in the antenna design process. The proposed antenna may be used in Medicare RF systems.

The antennas' radiation characteristics on the human body have been measured by using a phantom. The phantom's electrical characteristics represent those of the human body. The phantom contains a mix of 55% water, 44% sugar, and 1% salt. The antenna under test was placed on the phantom during measurements of the radiation characteristics. S_{11} and S_{12} parameters were measured directly on the human body by using a network analyzer.

References

[1] Sabban A 2017 () *Novel Wearable Antennas for Communication and Medical Systems* (London: Taylor and Francis)

[2] Sabban A 2016 *Wideband RF Technologies and Antenna in Microwave Frequencies* (New York: Wiley)

[3] Sabban A 2015 *Low Visibility Antennas for Communication Systems* (London: Taylor and Francis)

[4] Sabban A 2016 Small wearable meta materials antennas for medical systems *Appl. Comput. Electromagn. Soc. J.* **31**

[5] Sabban A 2012 Dual polarized dipole wearable antenna US Patent number: *8203497*

[6] Sabban A 2013 New wideband printed antennas for medical applications *IEEE J. Trans. Antennas Propag.* **61** 84–91

[7] Sabban A 2011 Microstrip antenna arrays *Microstrip Antennas* ed N Nasimuddin (Rijeka: InTech) pp 361–84

[8] Chirwa L C, Hammond P A, Roy S and Cumming D R S 2003 Electromagnetic radiation from ingested sources in the human intestine between 150 MHz and 1.2 GHz *IEEE Trans. Biomed. Eng.* **50** 484–92

[9] Werber D, Schwentner A and Biebl E M 2006 Investigation of RF transmission properties of human tissues *Adv. Radio Sci.* **4** 357–60

[10] Fujimoto K and James J R (ed) 1994 *Mobile Antenna Systems Handbook* (Boston, MA: Artech House)

[11] Gupta B, Sankaralingam S and Dhar S 2010 Development of wearable and implantable antennas in the last decade *Microwave Mediterranean Symp. (MMS) (Guzelyurt, Turkey) pp 251–67*

[12] Thalmann T, Popovic Z, Notaros B M and Mosig J R 2009 Investigation and design of a multi-band wearable antenna *3rd European Conf. on Antennas and Propagation, EuCAP* pp 462–65

[13] Salonen P, Rahmat-Samii Y and Kivikoski M 2004 *Wearable antennas in the vicinity of human body IEEE Antennas and Propagation Society Symp.* **1** *(Monterey, CA) pp 467–70*

[14] Kellomaki T, Heikkinen J and Kivikoski M 2006 Wearable antennas for FM reception *First European Conf. on Antennas and Propagation, EuCAP (The Hague, Netherlands) pp 1–6*

[15] Sabban A 2009 Wideband printed antennas for medical applications *APMC Conf. (Singapore)*

[16] Lee Y 2003 *Antenna Circuit Design for RFID Applications* (AZ, USA: Microchip Technology Inc.) Microchip AN 710c

[17] Sabban A 1983 A new wideband stacked microstrip antenna *IEEE Antenna and Propagation Symp. (Houston, TX, USA)*

[18] Sabban A 1981 Wideband microstrip antenna arrays *IEEE Antenna and Propagation Symp. MELCOM (Tel Aviv)*

[19] James J R, Hall P S and Wood C 1981 *Microstrip Antenna Theory and Design* (London: Peter Peregrinus)

[20] Sabban A and Gupta K C 1991 Characterization of radiation loss from microstrip discontinuities using a multiport network modeling approach *IEEE Trans. Microwave Theory Tech.* **39** 705–12

[21] Sabban A 1991 *Multiport network model for evaluating radiation loss and coupling among discontinuities in microstrip circuits PhD Thesis* University of Colorado at Boulder

[22] Keysight software, http://www.keysight.com/en/pc-1297113/advanced-design-system-ads?c.c.=IL&lc=eng

IOP Publishing

Wearable Communication Systems and Antennas (Second Edition)

Design, efficiency, and miniaturization techniques

Albert Sabban

Chapter 10

Wearable RFID technology and antennas

10.1 Introduction

RFID, **R**adio **F**requency **Id**entification, is an electronic method of exchanging data via radio frequency waves. RFID uses electromagnetic fields to automatically identify and track tags attached to people, animals and objects. There are three major components in a RFID system: the transponder, tag, antenna and controller. The tags contain electronically-stored information. Passive tags collect energy from a nearby RFID reader by employing radio waves. Active tags have a local power source (such as a battery) and may operate hundreds of meters from the RFID reader. Unlike a barcode, the tag does not need to be within the line of sight of the reader, so it may be embedded in the tracked person or object. The RFID tag, antenna and controller may be assembled on the same board. Printed small antennas are the best choice for RFID devices. Microstrip antennas have been widely studied in books and papers in the last decade [1–18]. However, compact wearable printed antennas are not widely used at 13.5 MHz in UHF RIFD systems. RFID loop antennas are commonly used, and several are presented in [16]. RFID loop antennas have low efficiency and narrow bandwidth. Wideband compact wearable printed antennas for RFID applications are presented in this chapter. RF transmission properties of human tissues have been investigated in papers [8, 9]. The effect of the human body on antenna performance for RFID applications is investigated in this chapter. The proposed antennas may be used as wearable antennas on people or animals. The proposed antennas may be also attached to cars, trucks, containers and various other objects. HF tags work better at close range but are more effective at penetrating non-metal objects, especially objects with high water content.

10.2 RFID technology

RFID technology has been available for more than 50 years. However, progress in electronic, communication and materials technologies has enhanced the development

doi:10.1088/978-0-7503-5224-6ch10

and production of low-cost RFID tags. A lack of standards in the RFID industry caused a delay in the development and production of multipurpose international RFID tags. RFID technology is an electronic method of exchanging data between people, objects and products over radio frequency waves. RFID tags track vehicles, airline passengers, Alzheimer's patients, employees and animals. Data stored on RFID tags can be changed, updated and locked (table 10.1).

A RFID system consists of three parts.

- An antenna. Usually a printed antenna. The antenna receives and transmits the signal.
- A transceiver with a decoder to interpret the data. The transceiver is an integrated circuit for storing and processing information that modulates and demodulates radio frequency signals.
- A transponder and a DC power unit that collects power from the incident reader signal.

The antenna radiates radio frequency fields in a relatively short range. When a RFID tag passes through the field the antenna radiates and the tag detects the activation signal. The activation signal 'starts-up' the RFID chip, and it transmits the information on its microchip to be picked up by the antenna.

Tags may either be read-only, having a factory assigned serial number that is used as a key into a database. Tags can read and write. Object specific data may be written into the tag by the system user. Field programmable tags may be write-once and read-multiple.

The tag information is stored in a nonvolatile memory. The RFID tag includes either fixed or programmable logic for processing the transmission and sensor data, respectively. A RFID reader transmits an encoded radio signal to interrogate the tag. The RFID tag receives the message and then responds with its identification and other information. This can be just a unique tag serial number, or may be product information such as a stock number, lot, batch number, production date and other specific information. Tags have individual serial numbers. The RFID system can scan several tags that might be within the range of the RFID reader and read them simultaneously.

A passive RFID tag gets power from the reader through the inductive coupling method. The reader consists of a coil connected to an AC supply such that a magnetic field is formed around it. The tag coil is placed in the vicinity of the reader coil and an electromotive force is induced, based on Faraday's law of induction. The EMF causes a flow of current in the coil, thus producing a magnetic field around it. By employing Lenz's law, the magnetic field of the tag coil opposes the reader's magnetic field and there will be a subsequent increase in the current through the reader coil. The reader intercepts this as the load information. This system may operate for very short distance communication. The tag AC voltage appearing across the tag coil is converted to DC by using a rectifier.

In an active RFID system, the reader transmits a signal to the tag via the antenna. The tag receives this information and resends this information along with the information in its memory. The reader receives this signal and transmits to the

Table 10.1. RFID frequency bands and applications.

Band	Regulations	Range	Data speed	ISO/IEC 18000 section	Applications
120–150 kHz (LF)	Unregulated	10 cm	Low	Part 2	Animal identification, factory data collection
13.56 MHz (HF)	ISM band worldwide	10 cm–1 m	Low to moderate	Part 3	Smart cards (ISO/IEC 15693, ISO/IEC 14443 A, B). Non-fully ISO compatible memory cards Microprocessor ISO compatible cards
433 MHz (UHF)	Short range devices	1–100 m	Moderate	Part 7	Defense applications, with active tags
865–868 MHz (Europe)	ISM band	1–12 m	Moderate to high	Part 6	EAN, various standards; used by railroads
902–928 MHz (North America) UHF					
2450–5800 MHz (microwave)	ISM band	1–2 m	High	Part 4	802.11 WLAN, Bluetooth standards
3.1–10 GHz (microwave)	Ultra-wideband	Max. 200 m	High	Not defined	Requires semi-active or active tags

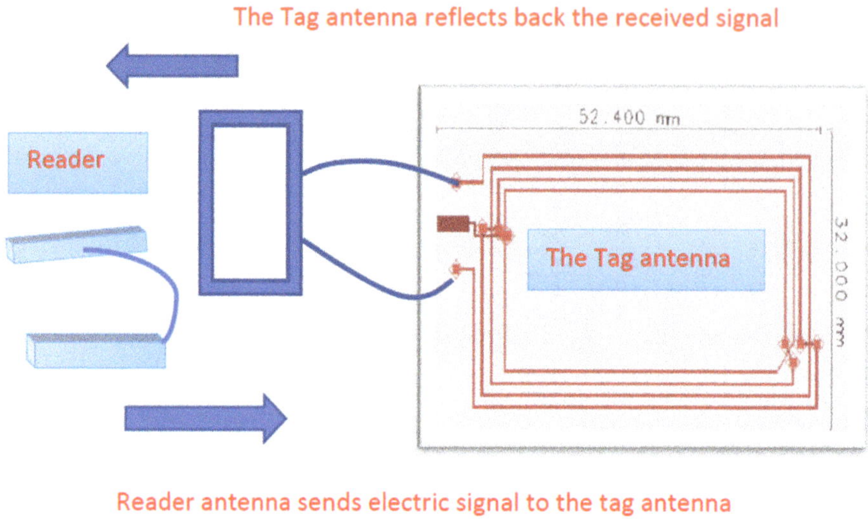

Figure 10.1. RFID concept.

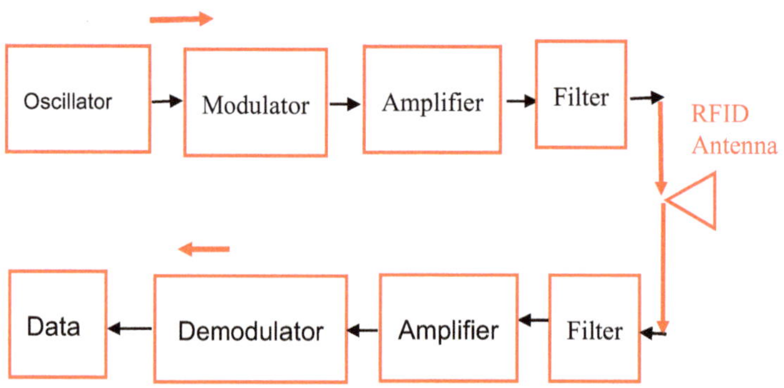

Figure 10.2. Block diagram of RFID tag reader.

processor for further processing. A RFID concept tag concept is shown in figure 10.1.

Figure 10.2 presents a block diagram for one application of a RFID tag reader. As shown, the block diagram consists of three main subsystems: a transceiver, a decoder, and a display module. The RFID tag will send its data as an amplitude modulated (AM) signal, so the reader will filter the signal to select and amplify the signal. After this filtering, the reader will demodulate the signal to obtain the signal transmitted by the tag. The decoder will decode the signal transmitted by the demodulator and perform the necessary transformations to get the identification data. Finally, this data is sent to a video display module to generate the necessary signals to output the data to a display.

10.3 RFID standards

The Food and Drug Administration gave final approval (17.10.2004) to apply digital solutions to sell their VeriChip RFID tags for implantation into patients in hospitals. The intent was to provide immediate positive identification of patients both in hospitals and in emergency services. Doctors, emergency-room personnel and ambulance crews can immediately identify patients without the need to search for wallets and purses for identification. If, for example, you had a pre-existing medical condition or allergy, this could be taken into account immediately.

Since RFID tags can be attached to cash, clothing, and possessions, or implanted in people and animals, the possibility of reading personally-linked information without consent has raised serious privacy concerns. These concerns resulted in standard specifications development addressing privacy and security issues. ISO/IEC 18000 and ISO/IEC 29167 use on-chip cryptography methods for untraceability, tag and reader authentication, and over-the-air privacy. ISO/IEC 20248 specifies a digital signature data structure for RFID and barcodes providing data, source and read method authenticity. This work is done within ISO/IEC JTC 1/SC 31 Automatic identification and data capture techniques. Tags can also be used in shops to expedite checkout, and to prevent theft by customers and employees.

At 2011 VeriMed's VeriChip was the only RFID tag that has been cleared by the FDA for human implants. The RFID tags may be implanted in the fatty tissue under the skin in the back of the upper arm. Each VeriMed microchip contains a unique identification number that emergency personnel may scan to immediately identify a patient and access their personal health information. This process helps a medical team to treat a patient without delay. This is especially important for patients that cannot communicate. Although the FDA approved the use of the device for anyone 12 years of age or older, it is very important for patients with diabetes, stroke, seizure disorders, dementia, Alzheimer's and organ transplants. The estimated life of the tags is 20 years. The subdermal RFID tag location is invisible to the naked eye. A unique verification number is transmitted to a suitable reader when the person is within range. In July 2017 a Wisconsin company was the first in the United States to offer implanted chips for opening doors and logging in to computers. The company employees may agree to implant an RFID chip in their hand. The tiny chip uses Near-Field Communication (NFC) technology to allow employees to unlock doors, make vending machine purchases, log in to computers and access office tools like photocopiers, all with a wave of the hand.

10.4 Dual polarized 13.5 MHz compact printed antenna

One of the most critical elements of any RFID system is the electrical performance of its antenna. The antenna is the main component for transferring energy from the transmitter to the passive RFID tags, receiving the transponder's signal reply and avoiding in-band interference from electrical noise and other nearby RFID components. Low profile compact printed antennas are crucial in the development of RIFD systems.

A compact microstrip loaded dipole antenna has been designed at 13.5 MHz to provide horizontal polarization. The antenna consists of two layers. The first layer consists of FR4 0.8 mm dielectric substrate. The second layer consists of Kapton 0.8 mm dielectric substrate. The substrate thickness determines the antenna's bandwidth. A printed slot antenna provides a vertical polarization. The proposed antenna is dual polarized.

The printed dipole and the slot antenna provide dual orthogonal polarizations. The dual polarized RFID antenna is shown in figure 10.3. The dimensions are 6.4 × 6.4 × 0.16 cm. It may be attached to a person's shirt in their stomach or back area. The antenna has been analyzed by using Keysight ADS software, see [19].

The antenna's S_{11} parameters are better than −21 dB at 13.5 MHz. The gain is around −10 dBi, and the beamwidth is around 160°. The computed S_{11} parameters are presented in figure 10.4. There is good agreement between the measured and computed results. Figure 10.5 presents the antenna's measured S_{11} parameters on a human body. The antenna's cross-polarized field strength may be adjusted by varying the slot feed location. The computed radiation pattern is shown in figure 10.6, and the computed 3D radiation pattern of the antenna is shown in figure 10.7.

10.5 Varying the antenna feed network

Several designs with different feed networks have been developed. A compact antenna with different feed networks is shown in figure 10.8. The dimensions are 8.4 × 6.4 × 0.16 cm. Figure 10.9 presents the antenna's computed S_{11} resuts on a human body. There is good agreement between the measured and computed results. The computed radiation pattern is shown in figure 10.10. Table 10.2 compares the electrical performance of a loop antenna with a compact dual polarized antenna.

10.6 Wearable loop antennas for RFID applications

Several RFID loop antennas are presented in [15]. The disadvantages of loop antennas with a number of turns are low efficiency and narrow bandwidth. The real part of the loop antenna's impedance approaches 0.5 Ω. The image part of the loop antenna's impedance may be represented as high inductance. A matching network

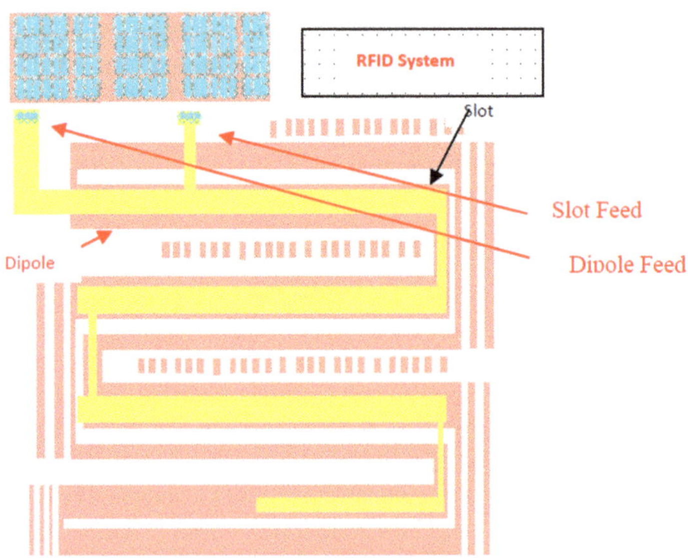

Figure 10.3. Printed compact dual polarized antenna, 64 × 64 × 1.6 mm.

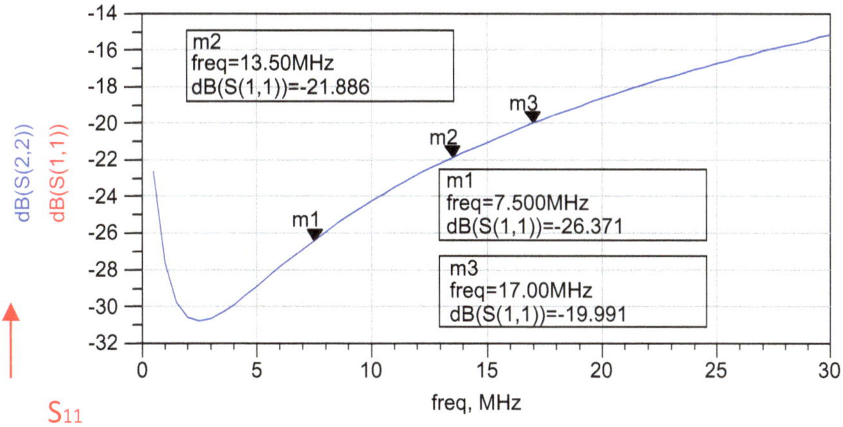

Figure 10.4. Computed S_{11} results.

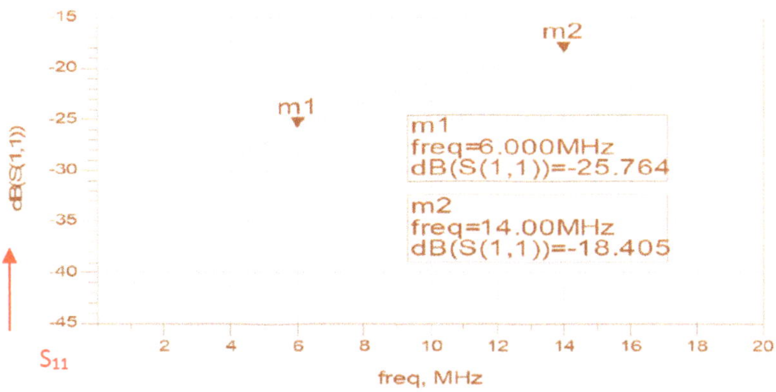

Figure 10.5. Measured S_{11} of the 13.5 MHz antenna on a human body.

may be used to match the antenna to 50 Ω. The matching network consists of an RLC matching network. This matching network has a narrow bandwidth. The loop antenna's efficiency is lower than 1%.

A square four-turn loop antenna has been designed at 13.5 MHz by using Keysight ADS software. The antenna is printed on FR4 substrate. The dimensions are 32 × 52.4 × 0.25 mm. The antenna's layout is shown in figure 10.11(a), and a photo is shown in figure 10.11(b). S_{11} results of the printed loop antenna are shown in figure 10.12. The antenna's S_{11} parameters are better than –9.5 dB without an external matching network. The computed radiation pattern is shown in figure 10.13. The computed radiation pattern takes into account an infinite ground plane.

The microstrip antenna's input impedance variation as a function of distance from the body has been computed by employing momentum software. The analyzed structure is presented in figure 6.19. Properties of human body tissues are listed in table 9.1 see [8]. These properties were used in the antenna design. S_{11} parameters for different human body thicknesses have been computed. We may note that the

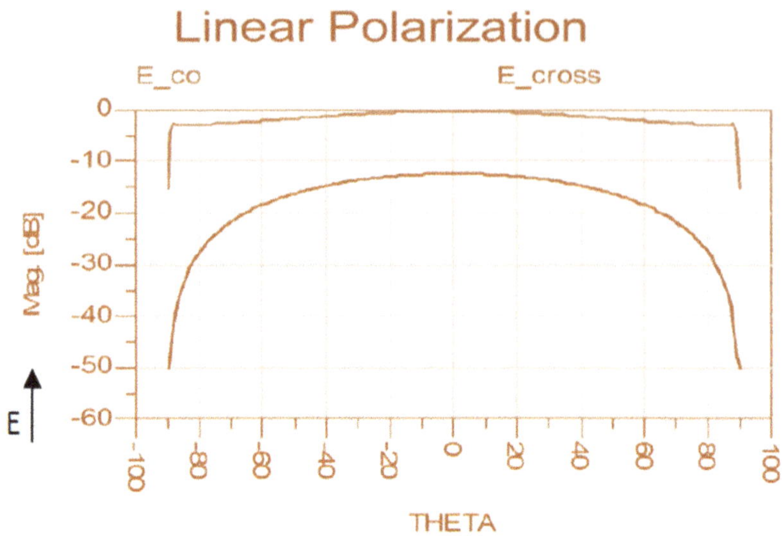

Figure 10.6. Radiation pattern of the 13.5 MHz antenna.

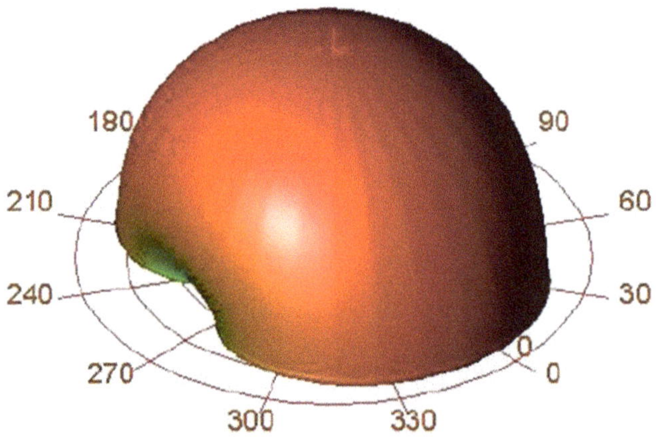

Figure 10.7. 3D antenna radiation pattern.

differences in the results for body thickness of 15 mm to 100 mm are negligible. S_{11} parameters for different positions relative to the human body have also been computed. If the air spacing between the antenna and the human body is increased from 0 mm to 10 mm the antenna's S_{11} parameters may change by less than 1%. The VSWR is better than 1.5:1.

10.7 Proposed antenna applications

An application of the proposed antenna is shown in figure 10.14. The RFID antennas may be assembled in a belt and attached to the a person's stomach. The antennas may be employed as transmitting or receiving antennas. They may receive or transmit information to medical systems.

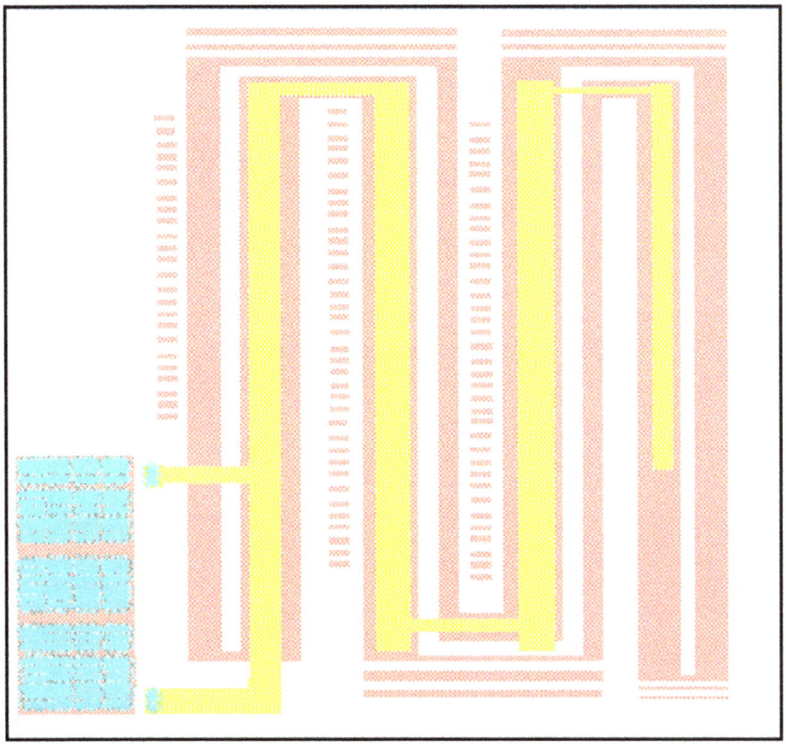

Figure 10.8. RFID printed antenna, 8.4 × 6.4 × 0.16 cm.

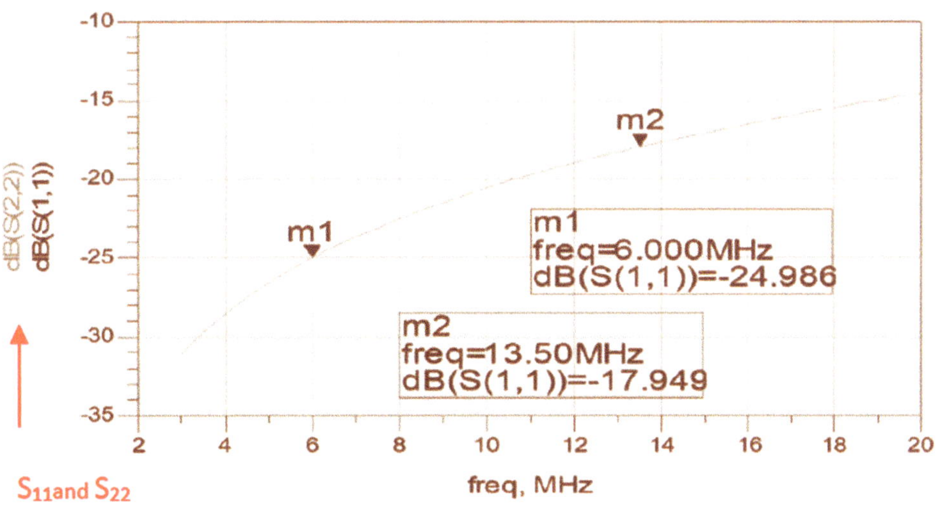

Figure 10.9. RFID antenna's computed S_{11} and S_{22} results.

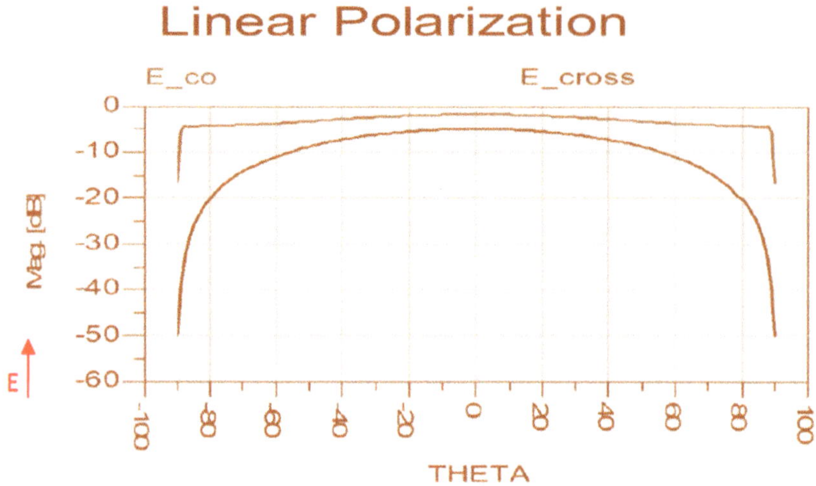

Figure 10.10. Compact antenna's (8.4 × 6.4 × 0.16 cm) radiation pattern.

Table 10.2. Comparison of a loop antenna's and microstrip antenna's parameters.

Antenna	Beamwidth° 3 dB	Gain dBi	VSWR
Loop antenna	140	−25	2:1
Microstrip antenna	160	−10	1.2:1

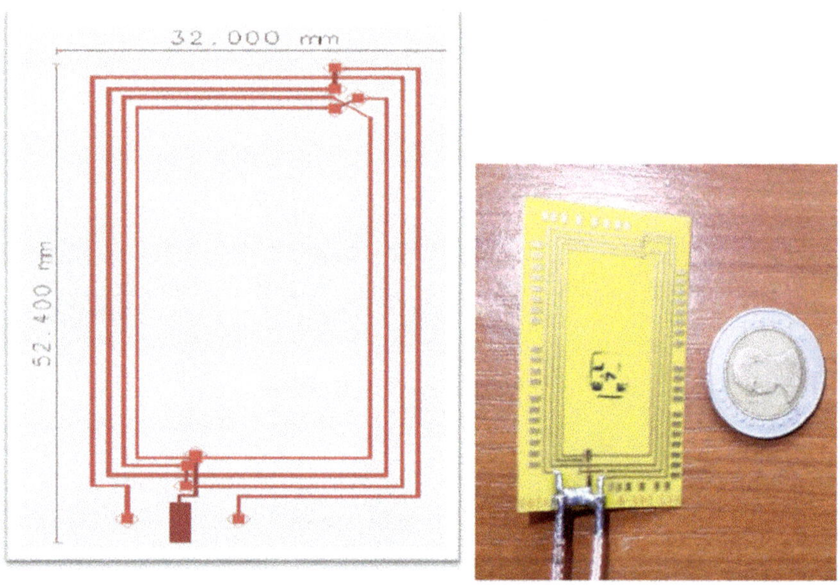

(a) (b)

Figure 10.11. A square four-turn loop antenna. (a) Layout. (b) Photo.

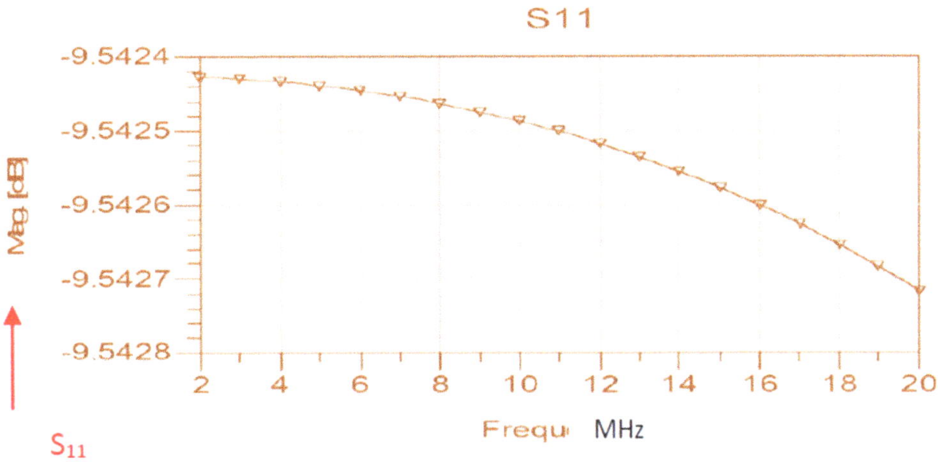

Figure 10.12. Loop antenna's computed S_{11} results.

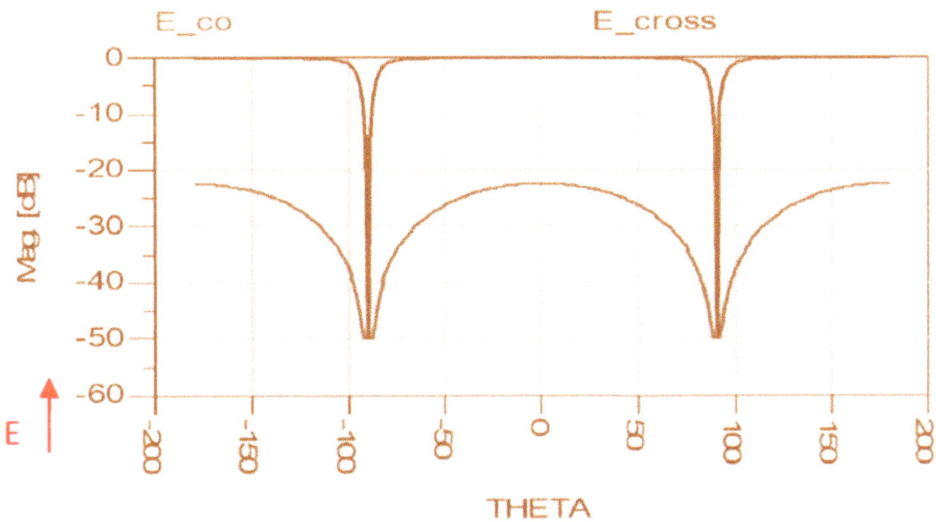

Figure 10.13. Loop antenna's radiation patterns for an infinite ground plane.

In RFID systems the distance between the transmitting and receiving antennas is less than $2D^2/\lambda$, where D is the largest dimension of the antenna. The receiving and transmitting antennas are magnetically coupled. In these applications we refer to the near-field and not to the far-field radiation pattern. Figures 10.15 and 10.16 present a compact printed antenna for RFID applications. The presented antenna may be assembled in a belt and attached to a patient's stomach or back.

10.8 Conclusions

This chapter presents wideband microstrip antennas with high efficiency for medical applications. The antenna's dimensions may vary from 26 × 6 × 0.16 cm to 5 × 5 × 0.05 cm

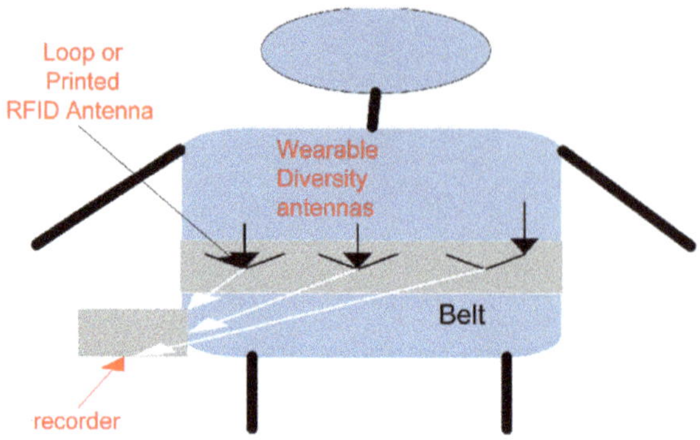

Figure 10.14. Wearable RFID antenna.

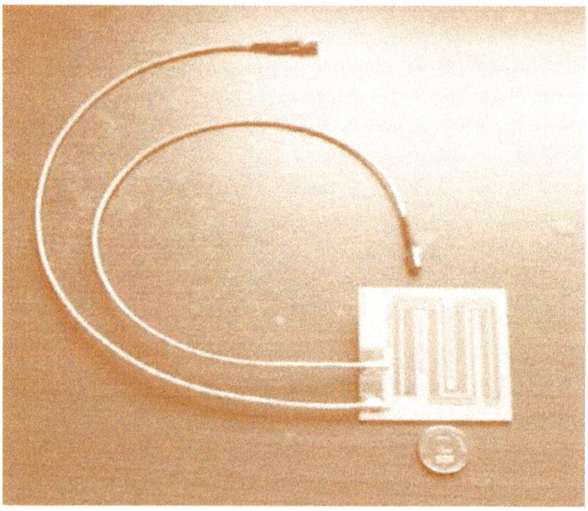

Figure 10.15. A microstrip antenna for RFID applications.

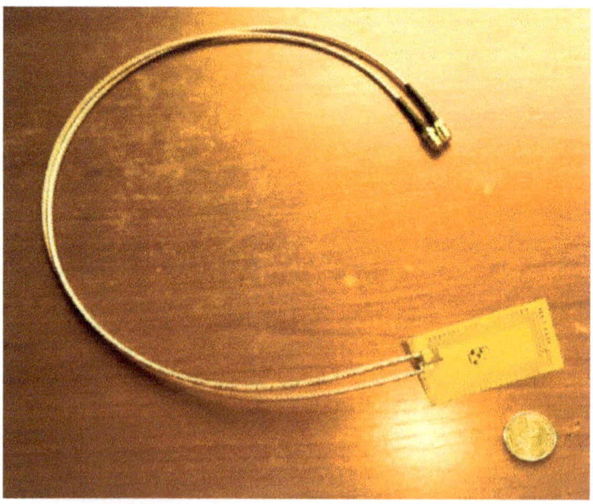

Figure 10.16. Loop antenna for RFID applications.

according to the medical system's specification. The bandwidth is around 10% for a VSWR better than 2:1. The beamwidth is around 100°, and the gain varies from 0 to 4 dBi. The antenna's S_{11} results for different belt thickness, shirt thickness and air spacing between the antenna and the human body are presented in this chapter. If the air spacing between the dual polarized antenna and the human body is increased from 0 mm to 5 mm the antenna's resonant frequency is shifted by 5%. However, if the air spacing between the helix antenna and the human body is increased only from 0 mm to 2 mm the antenna's resonant frequency is shifted by 5%. The effect of the antenna's location on the human body should be considered in the design process. The proposed antenna may be used in Medicare RF systems.

A wideband tunable microstrip antenna with high efficiency for medical applications has been presented in this chapter. The dimensions may vary from 26 × 6 × 0.16 cm to 5 × 5 × 0.05 cm according to the medical system's specification. The antenna's bandwidth is around 10% for a VSWR better than 2:1. The beamwidth is around 100°, and the gain varies from 0 to 2 dBi. If the air spacing between the dual polarized antenna and the human body is increased from 0 mm to 5 mm the antenna's resonant frequency is shifted by 5%. A varactor is employed to compensate for variations in the antenna's resonant frequency at different locations on the human body.

This chapter also presents wideband compact printed antennas, microstrip and loop antennas, for RFID applications. The beamwidth is around 160°, and the gain is around −10 dBi. The proposed antennas may be used as wearable antennas by people or animals. They may be attached to cars, trucks and other various objects. If the air spacing between the antenna and the human body is increased from 0 mm to 10 mm the antenna's S_{11} parameters may change by less than 1%. The antenna's VSWR is better than 1.5:1 for all of the tested environments.

References

[1] Sabban A 2016 *Wideband RF Technologies and Antenna in Microwave Frequencies* (New York: Wiley)

[2] James J R, Hall P S and Wood C 1981 *Microstrip Antenna Theory and Design* (London: Peter Peregrinus)

[3] Sabban A and Gupta K C 1991 Characterization of radiation loss from microstrip discontinuities using a multiport network modeling approach *IEEE Trans. Microwave Theory Tech.* **39** 705–12

[4] Sabban A 1983 A new wideband stacked microstrip antenna *IEEE Antenna and Propagation Symp. (Houston, TX, USA)*

[5] Sabban A 2012 *Dual polarized dipole wearable antenna US Patent number:* 8203497

[6] Sabban A 1981 Wideband microstrip antenna arrays *IEEE Antenna and Propagation Symp. MELCOM (Tel Aviv)*

[7] Sabban A 2011 Microstrip antenna arrays *Microstrip Antennas* ed N Nasimuddin (Rijeka: InTech) pp 361–84

[8] Chirwa L C, Hammond P A, Roy S and Cumming D R S 2003 Electromagnetic radiation from ingested sources in the human intestine between 150 MHz and 1.2 GHz *IEEE Trans. Biomed. Eng.* **50** 484–92

[9] Werber D, Schwentner A and Biebl E M 2006 Investigation of RF transmission properties of human tissues *Adv. Radio Sci.* **4** 357–60
[10] Fujimoto K and James J R (ed) 1994 *Mobile Antenna Systems Handbook* (Boston, MA: Artech House)
[11] Gupta B, Sankaralingam S and Dhar S 2010 Development of wearable and implantable antennas in the last decade *Microwave Mediterranean Symp. (MMS) (Guzelyurt, Turkey) pp 251–67*
[12] Thalmann T, Popovic Z, Notaros B M and Mosig J R 2009 Investigation and design of a multi-band wearable antenna *3rd European Conf. on Antennas and Propagation, EuCAP (Berlin, Germany) pp 462–5*
[13] Salonen P, Rahmat-Samii Y and Kivikoski M 2004 *Wearable antennas in the vicinity of human body IEEE Antennas and Propagation Society Symp.***1** *(Monterey, CA) pp 467–70*
[14] Kellomaki T, Heikkinen J and Kivikoski M 2006 Wearable antennas for FM reception *First European Conf. on Antennas and Propagation, EuCAP (The Hague, Netherlands) pp 1–6*
[15] Sabban A 2009 Wideband printed antennas for medical applications *PMC 2009 Conf. (Singapore)*
[16] Lee Y 2003 *Antenna Circuit Design for RFID Applications* (AZ, USA: Microchip Technology Inc.) Microchip AN 710c.
[17] Sabban A 2015 *Low Visibility Antennas for Communication Systems* (New York: Taylor and Francis)
[18] Sabban A 1991 Multiport network model for evaluating radiation loss and coupling among discontinuities in microstrip circuits *PhD thesis* University of Colorado at Boulder
[19] Keysight software, http://keysight.com/en/pc-1297113/advanced-design-system-ads?c.c.=IL&lc=eng

IOP Publishing

Wearable Communication Systems and Antennas
(Second Edition)
Design, efficiency, and miniaturization techniques
Albert Sabban

Chapter 11

Novel wearable printed antennas for wireless communication and medical systems

Communications and the biomedical industry have rapidly grown in the last decade. Low profile small antennas are crucial in the development of commercial compact systems. Small printed antennas, however, suffer from low efficiency.

11.1 Wideband wearable metamaterial antennas for wireless communication applications

Metamaterial technology is used to design small wideband wearable antennas with high efficiency. The design considerations and computed and measured results of printed metamaterial antennas with high efficiency are presented in this chapter. The proposed antenna may be used in communications and Medicare systems. The antenna's S_{11} results for different positions on the human body are presented. The gain and directivity of the patch antenna with split ring resonators (SRR) is higher by 2.5 dB than the patch antenna without SRR. The resonant frequency of the antenna with SRR on the human body is shifted by 3%.

11.1.1 Introduction

Microstrip antennas are widely used in communication systems. Microstrip antennas have several advantages such as being low profile, flexible, lightweight, small volume and with low production costs. Compact printed antennas have been presented in journals and books, as referred to in [1–4]. However, small printed antennas suffer from low efficiency. Metamaterial technology is used to design small printed antennas with high efficiency. Printed wearable antennas were presented in [5]. Artificial media with negative dielectric permittivity were presented in [6]. Periodic SRR and metallic post structures may be used to design materials with

doi:10.1088/978-0-7503-5224-6ch11

dielectric constant and permeability less than 1 as presented in [6–14]. In this chapter, metamaterial technology is used to develop small antennas with high efficiency. RF transmission properties of human tissues have been investigated in several papers such as [15, 16], and several wearable antennas have been presented in papers in the last few years (see [17–24]). New wearable printed metamaterial antennas with high efficiency are discussed in this chapter. The bandwidth of metamaterial antennas with SRR and metallic strips is around 50% for a VSWR better than 2.3:1. The computed and measured results of metamaterial antennas on the human body are also discussed.

11.1.2 Printed antennas with split ring resonators

A microstrip dipole antenna with SRR is shown in figure 11.1. This antenna provides horizontal polarization, and the slot antenna provides vertical polarization. The resonant frequency of the antenna with SRR is 400 MHz. The resonant frequency of the antenna without SRR is 10% higher. The antenna shown in figure 11.1 consists of two layers. The dipole feed network is printed on the first layer. The radiating dipole with SRR is printed on the second layer. The thickness of each layer is 0.8 mm. The dipole and the slot antenna create a dual polarized antenna. The computed S_{11} parameters are presented in figure 11.2.

The length of the dual polarized antenna with SRR shown in figure 11.1 is 19.8 cm. The length of the dual polarized antenna without SRR shown in figure 11.3

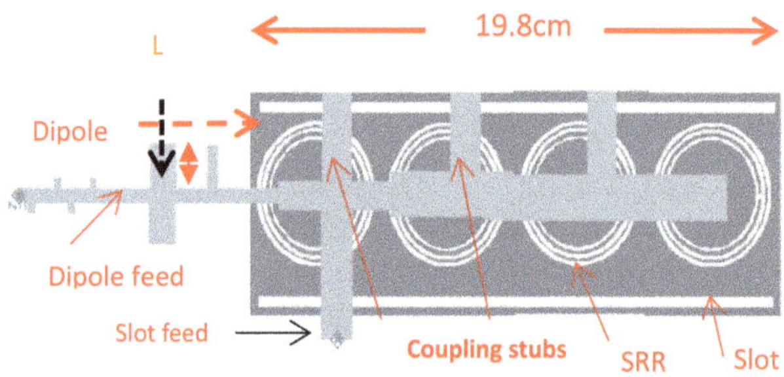

Figure 11.1. Printed antenna with split ring resonators.

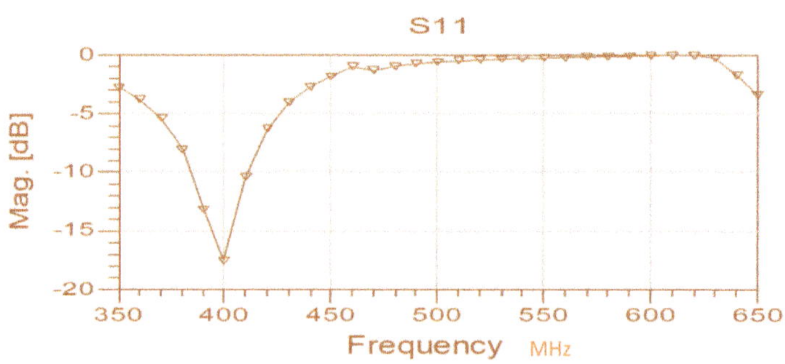

Figure 11.2. Antenna with split ring resonators, computed S_{11}.

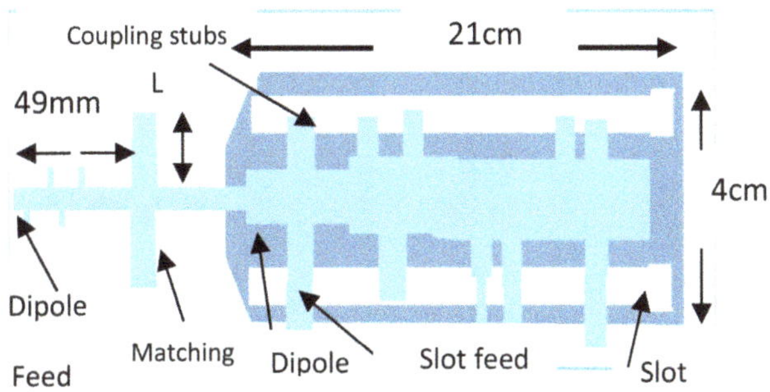

Figure 11.3. Dual polarized microstrip antenna.

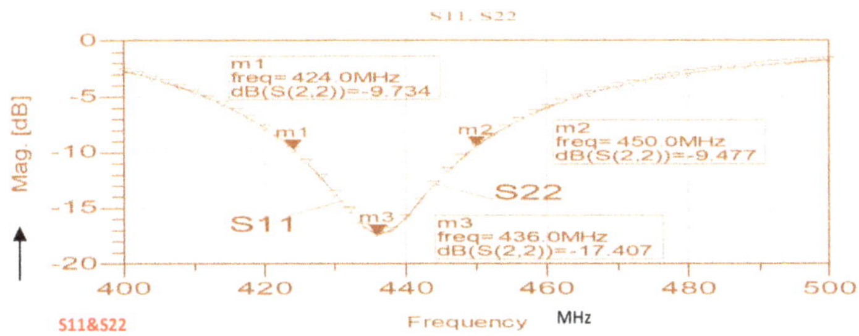

Figure 11.4. Computed S_{11} and S_{22} results, antenna without SRR.

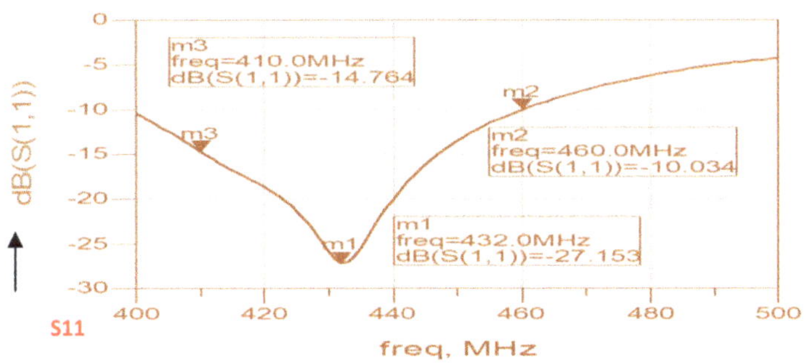

Figure 11.5. Measured S_{11} of the antenna without SRR.

is 21 cm. The ring width is 1.4 mm and the spacing between the rings is 1.4 mm. The antennas have been analyzed by using Agilent ADS software. The matching stubs' locations and dimensions have been optimized to get the best VSWR results. The length of the stub L in figures 11.1 and 11.3 is 10 mm. The locations and number of the coupling stubs may vary the antenna's axial ratio from 0 dB to 30 dB. The number of coupling stubs may be minimized. The number of coupling stubs in figure 11.1 is three. The antenna's axial ratio value may be also adjusted by varying the slot feed location. The dimensions of the antenna shown in figure 11.3 are presented in [5]. The bandwidth shown in figure 11.3 is around 10% for a VSWR better than 2:1. The antenna's beamwidth is 100°, and the gain is around 2 dBi. The computed S_{11} parameters are presented in figure 11.4. Figure 11.5 presents the antenna's measured

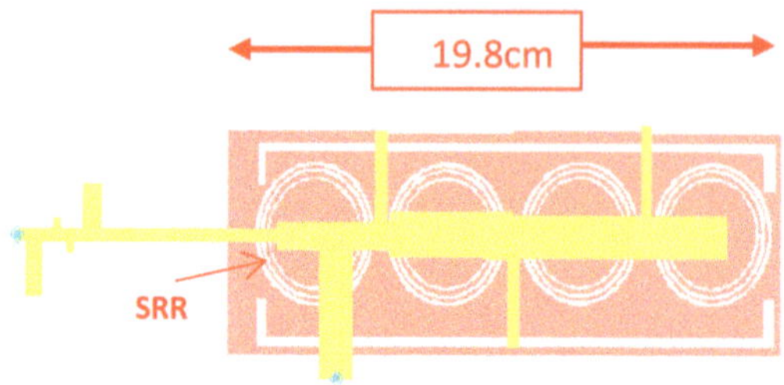

Figure 11.6. Antenna with SRR with two resonant frequencies.

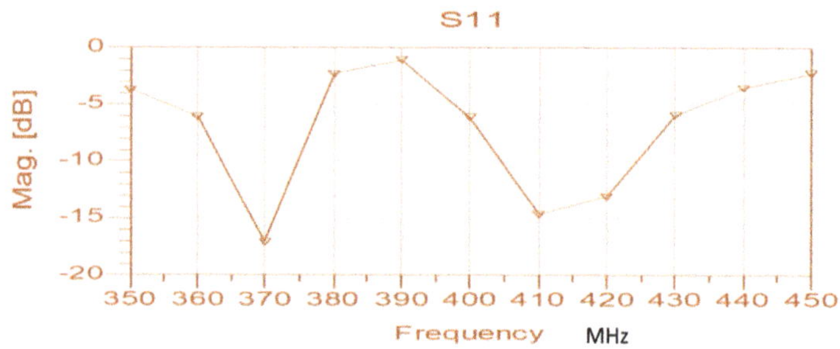

Figure 11.7. S_{11} for antenna with two resonant frequencies.

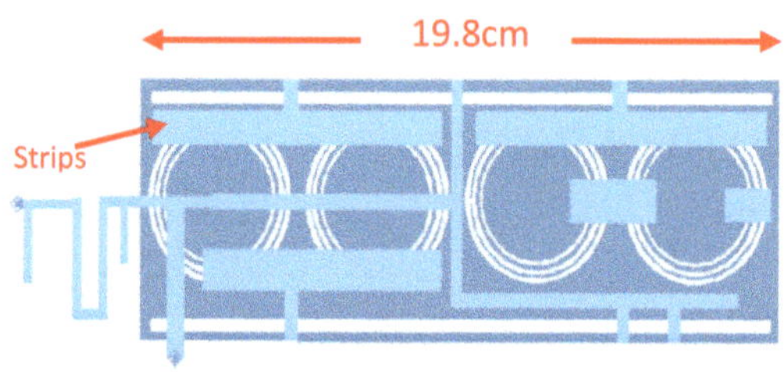

Figure 11.8. Antenna with SRR and metallic strips.

S_{11} parameters. There is good agreement between the measured and computed results. The antenna presented in figure 11.1 has been modified as shown in figure 11.6. The location and the dimension of the coupling stubs have been modified to get two resonant frequencies. The first resonant frequency is 370 MHz and is lower by 20% than the resonant frequency of the antenna without SRR (figure 11.7).

Metallic strips have been added to the antenna with SRR as presented in figure 11.8. The computed S_{11} parameters of the antenna with metallic strips are presented in figure 11.9. The antenna's bandwidth is around 50% for a VSWR better than 3:1. The computed radiation pattern is shown in figure 11.10. The 3D computed radiation pattern is shown in figure 11.11. The directivity and gain of the antenna with SRR is around 5 dBi, see figure 11.12. The directivity of the antenna without SRR is around 2 dBi. The length of the antenna with SRR is smaller by 5% than the antenna without SRR. Moreover, the resonant frequency of the antenna with SRR is lower by 5%–10%.

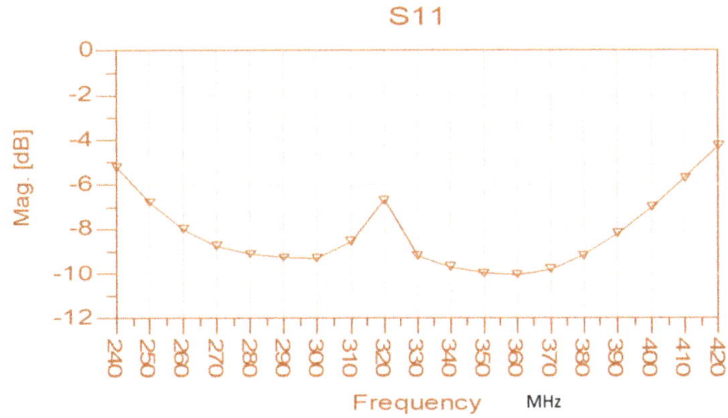

Figure 11.9. S_{11} for antenna with SRR and metallic strips.

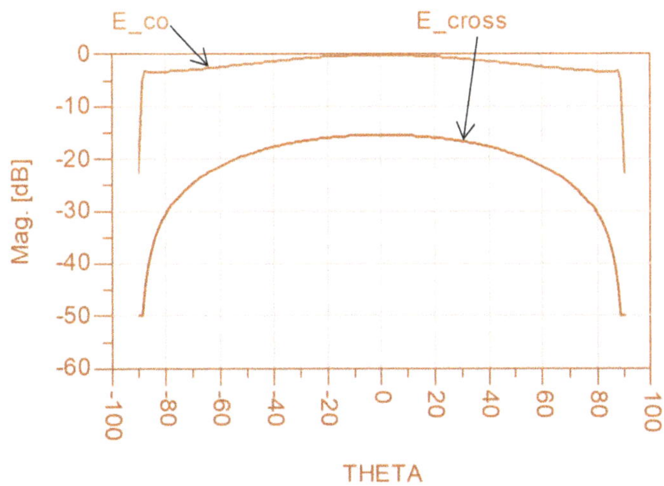

Figure 11.10. Radiation pattern for antenna with SRR.

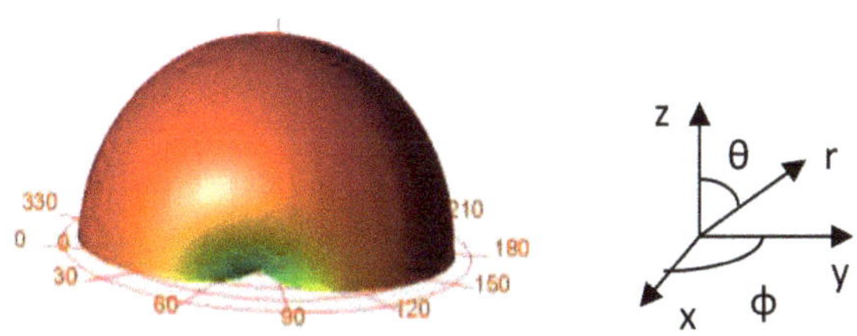

Figure 11.11. 3D radiation pattern for antenna with SRR.

The feed network of the antenna presented in figure 11.8 has been optimized to yield a VSWR better than 2:1 in the frequency range of 250 MHz to 440 MHz. Optimization of the number of coupling stubs and the distance between the coupling stubs may be used to tune the antenna's resonant frequency. An optimized antenna

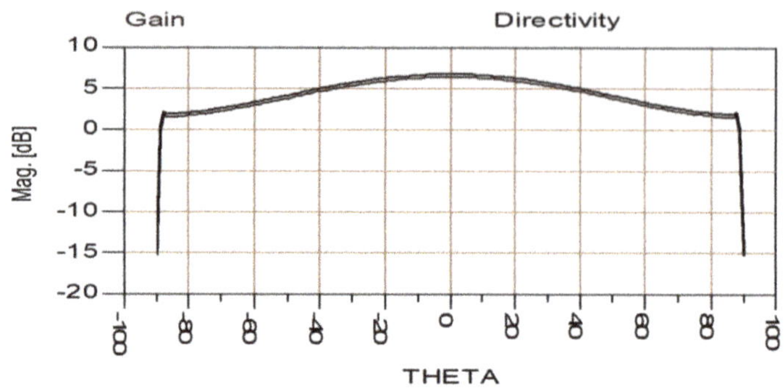

Figure 11.12. Directivity of the antenna with SRR.

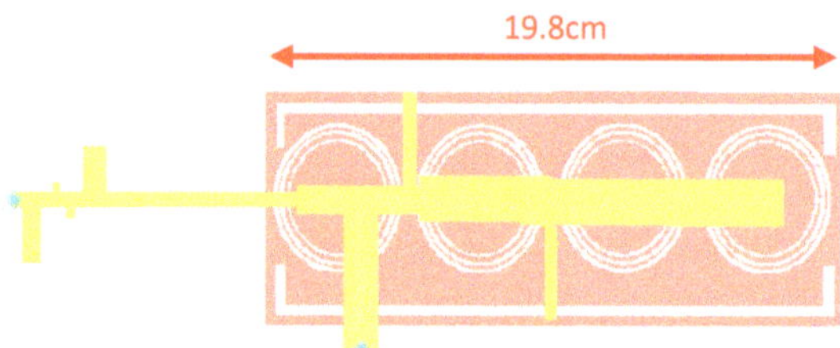

Figure 11.13. Antenna with SRR and two coupling stubs.

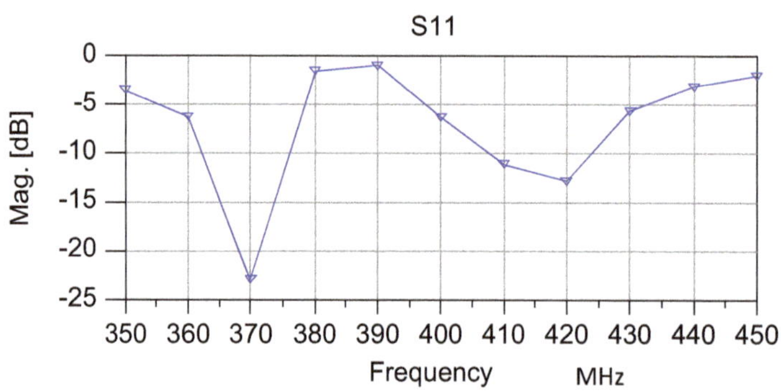

Figure 11.14. S_{11} for an antenna with SRR and two coupling stubs.

with two coupling stubs has two resonant frequencies. The first resonant frequency is 370 MHz and the second resonant frequency is 420 MHz. An antenna with SRR and two coupling stubs is presented in figure 11.13.

The computed S_{11} parameters of the antenna with two coupling stubs are presented in figure 11.14. The 3D radiation pattern for the antenna with SRR and two coupling stubs is shown in figure 11.15.

The antenna with metallic strips has been optimized to yield a wider bandwidth as shown in figure 11.16. The computed S_{11} parameters of the modified antenna with metallic strips are presented in figure 11.17. The antenna's bandwidth is around 50% for a VSWR better than 2.3:1.

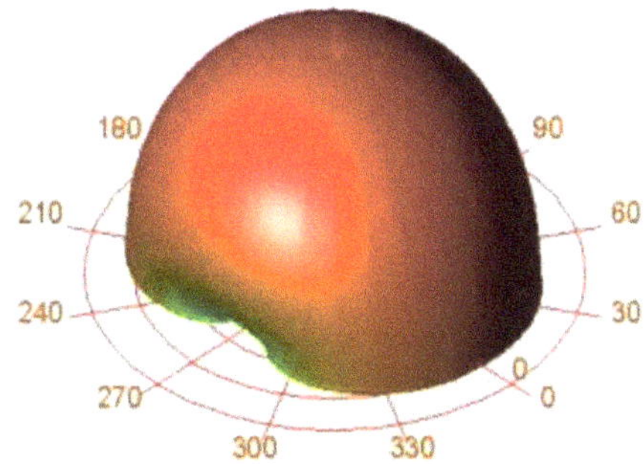

Figure 11.15. 3D radiation pattern for an antenna with SRR and two coupling stubs.

Figure 11.16. Wideband antenna with SRR and metallic strips.

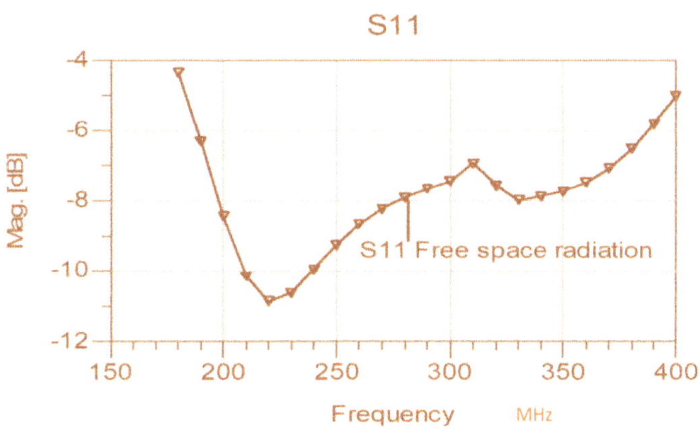

Figure 11.17. S_{11} for an antenna with SRR and metallic strips.

11.1.3 Folded dipole metamaterial antenna with SRR

The length of the antenna shown in figure 11.3 may be reduced from 21 cm to 7 cm by folding the printed dipole as shown in figure 11.18. Tuning bars are located along the feed line to tune the antenna to the desired frequency. The antenna's bandwidth is around 10% for a VSWR better than 2:1 as shown in figure 11.19. The beamwidth is around 100°, and the gain is around 2 dBi. The size of the antenna with SRR shown in figure 11.6 may be reduced by folding the printed dipole as shown in figure 11.20. The dimensions of the folded dual polarized antenna with SRR presented in figure 11.20 are 11 × 11 × 0.16 cm. Figure 11.21 presents the antenna's computed

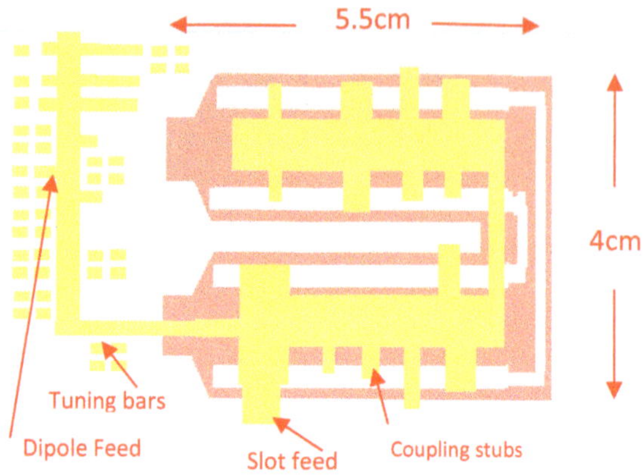

Figure 11.18. Folded dipole antenna, 7 × 5 × 0.16 cm.

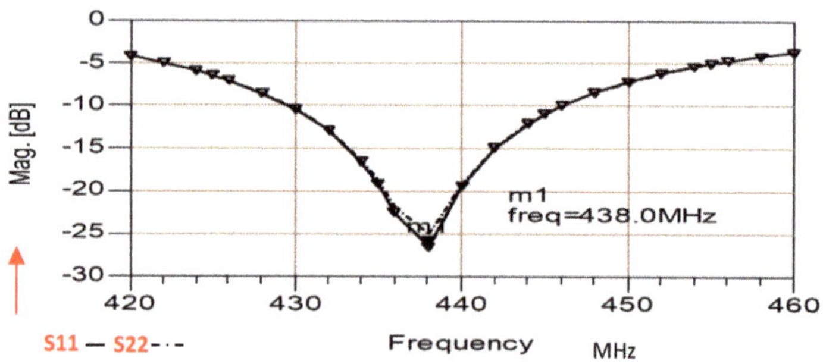

Figure 11.19. Folded antenna's computed S_{11} and S_{22} results.

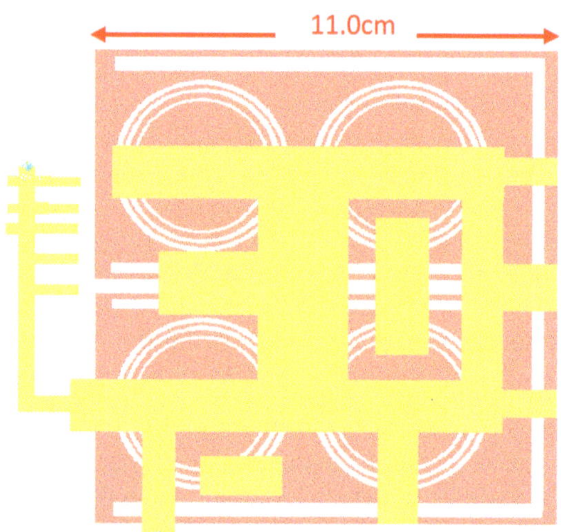

Figure 11.20. Folded dual polarized antenna with SRR.

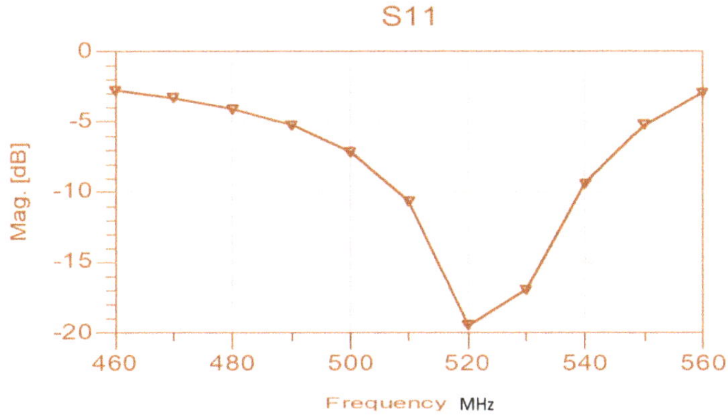

Figure 11.21. Folded antenna with SRR, computed S_{11}.

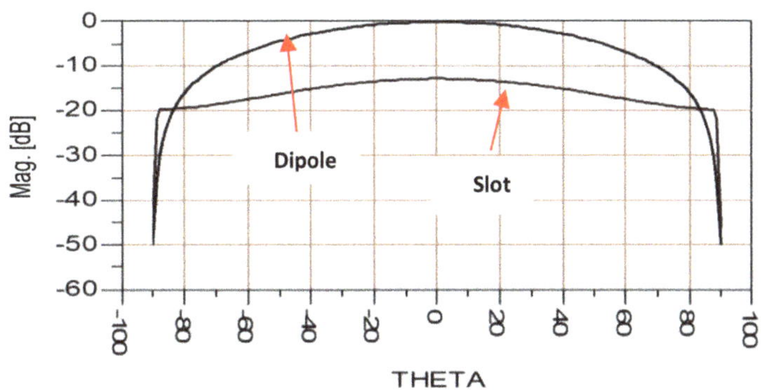

Figure 11.22. Radiation pattern of the folded antenna with SRR.

S_{11} parameters. The bandwidth is 10% for a VSWR better than 2:1. The computed radiation pattern of the folded antenna with SRR is shown in figure 11.22.

11.2 Stacked patch antenna loaded with SRR

First, a microstrip stacked patch antenna [1–3] was designed. The second step was to design the same antenna with SRR. The antenna consists of two layers. The first layer consists of FR4 dielectric substrate with dielectric constant of 4 and 1.6 mm thick. The second layer consists of RT-Duroid 5880 dielectric substrate with a dielectric constant of 2.2 and 1.6 mm thick. The dimensions of the microstrip stacked patch antenna shown in figure 11.23 are 33 × 20 × 3.2 mm. The antenna has been analyzed by using Agilent ADS software. The antenna's bandwidth is around 5% for a VSWR better than 2.5:1. The beamwidth is around 72°, and the gain is around 7 dBi. The computed S_{11} parameters are presented in figure 11.24. The radiation pattern of the microstrip stacked patch is shown in figure 11.25. The antenna with SRR is shown in figure 11.26. This antenna has the same structure as the antenna shown in figure 11.23. The ring width is 0.2 mm and the spacing between the rings is 0.25 mm. Twenty-eight SRR are placed on the radiating

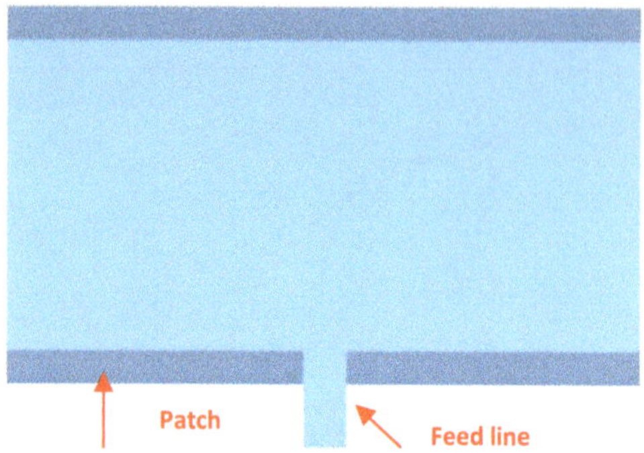

Figure 11.23. A microstrip stacked patch antenna.

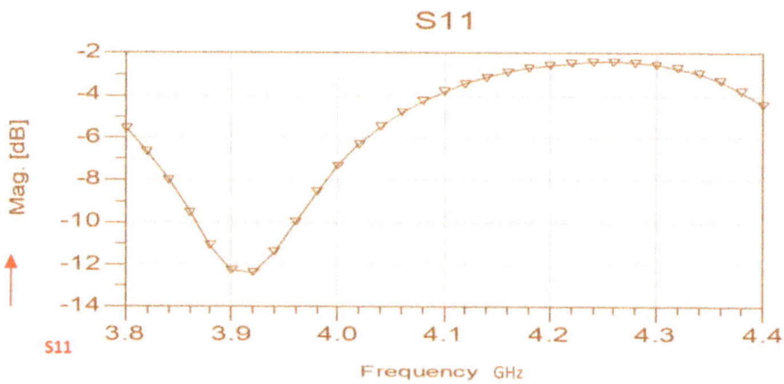

Figure 11.24. Computed S_{11} of the microstrip stacked patch.

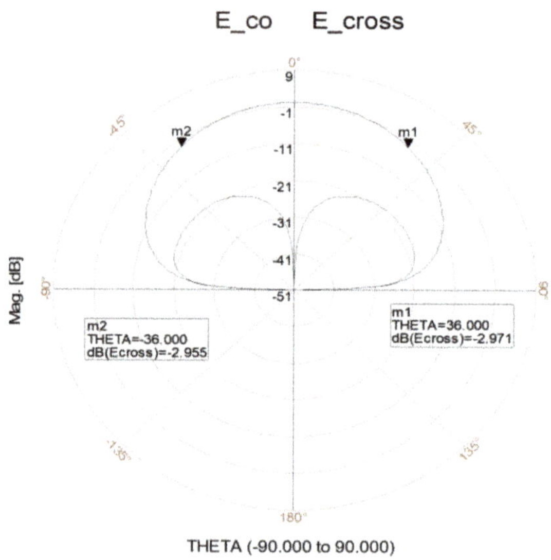

Figure 11.25. Radiation pattern of the microstrip stacked patch.

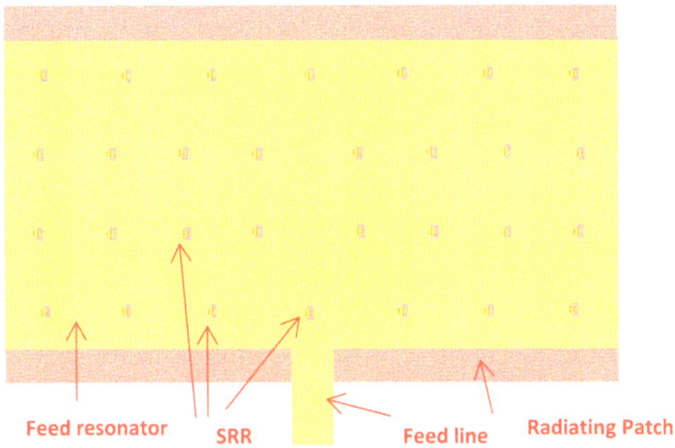

Figure 11.26. Printed antenna with split ring resonators.

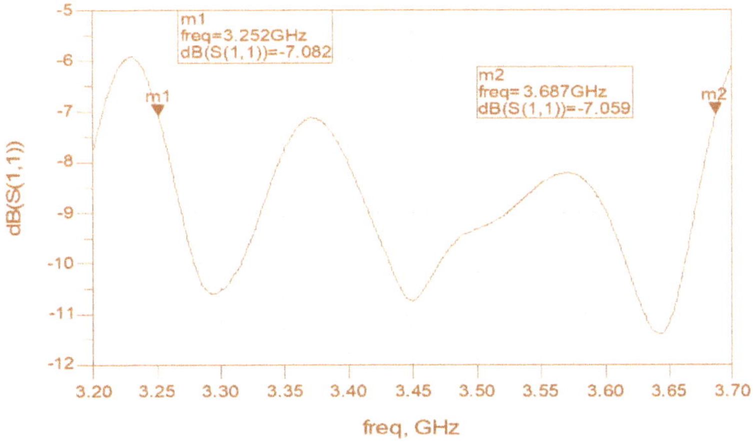

Figure 11.27. Patch with split ring resonators, measured S_{11}.

element. There is good agreement between the measured and computed results. The measured S_{11} parameters of the antenna with SRR are presented in figure 11.27. The antenna's bandwidth is around 12% for a VSWR better than 2.5:1. By adding an air space of 4 mm between the antenna layers the VSWR was improved to 2:1. The gain is around 9–10 dBi, and the efficiency is around 95%. The antenna's computed radiation pattern is shown in figure 11.28. The patch antenna with SRR performs as a loaded patch antenna. The effective area of a patch antenna with SRR is higher than the effective area of a patch antenna without SRR. The resonant frequency of a patch antenna with SRR is lower by 10% than the resonant frequency of a patch antenna without SRR.

The antenna's beamwidth is around 70°. The gain and directivity of the stacked patch antenna with SRR is higher by 2 dB to 3 dB than the patch antenna without SRR.

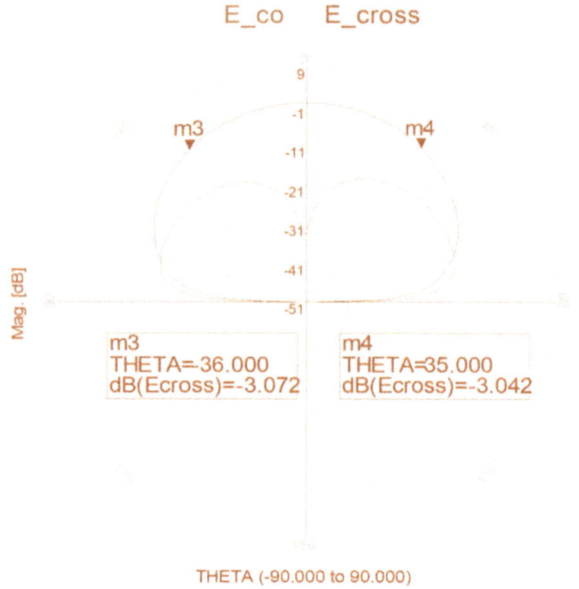

Figure 11.28. Radiation pattern for patch with SRR.

11.3 Patch antenna loaded with split ring resonators

A patch antenna with SSR has been designed. The antenna is printed on RT-Duroid 5880 dielectric substrate with a dielectric constant of 2.2 and 1.6 mm thick. The dimensions of the microstrip patch antenna shown in figure 11.29 are 36 × 20 × 1.6 mm. The antenna's bandwidth is around 5% for S_{11} lower than –9.5 dB. However, the antenna's bandwidth is around 10% for a VSWR better than 3:1. The beamwidth is around 72°, and the gain is around 7.8 dBi. The directivity of the antenna is 8. The antenna's gain is 6.03, and the efficiency is 77.25%. The measured S_{11} parameters are presented in figure 11.30. The gain and directivity of the patch antenna with SRR is higher by 2.5 dB than the patch antenna without SRR.

11.4 Metamaterial antenna characteristics in the vicinity of the human body

The antenna's input impedance variation as a function of distance from the human body has been computed by using the structure presented in figure 11.31. The electrical properties of human body tissues are listed in table 11.1 (see [15]). The antenna's location on the human body may be considered by calculating S_{11} for different dielectric constants of the body. The variation of the dielectric constant of the body from 43 at the stomach to 63 at the colon zone shifts the antenna's resonant frequency by 2%. The antenna was placed inside a belt with a thickness between 1 to 4 mm and a dielectric constant from 2 to 4.

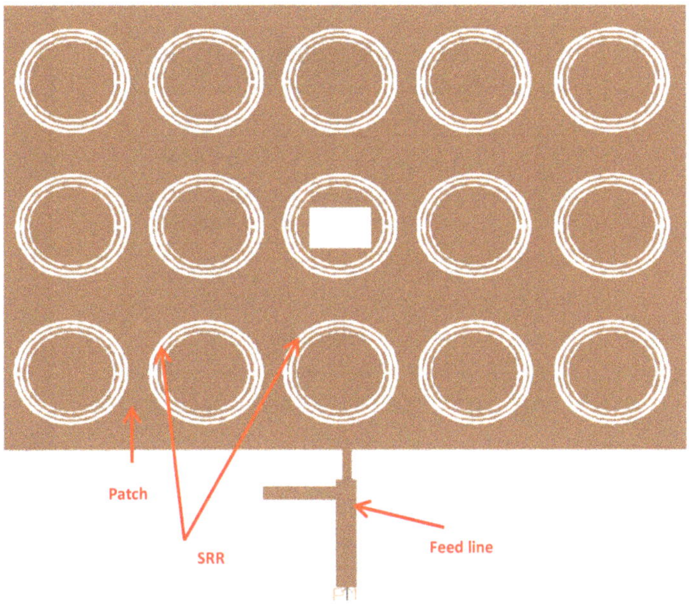

Figure 11.29. Patch antenna with split ring resonators.

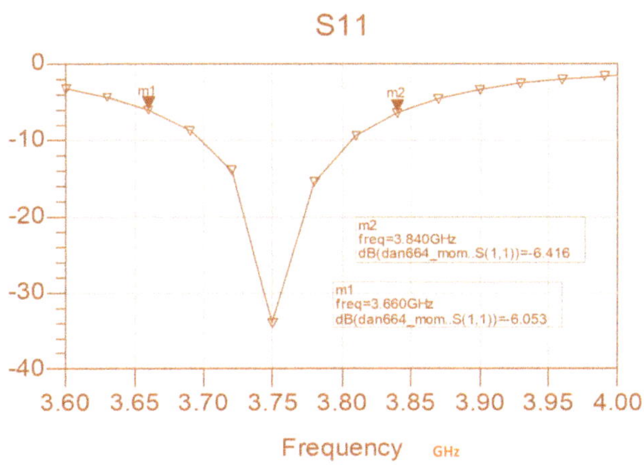

Figure 11.30. Patch with split ring resonators, computed S_{11}.

The air spacing between the belt and a patient's shirt is varied from 0 mm to 8 mm. The dielectric constant of a patient's shirt was varied from 2 to 4.

Figure 11.32 presents S_{11} results of the antenna with SRR shown in figure 11.13 on the human body. The antenna's resonant frequency is shifted by 3%. Figure 11.33 presents S_{11} results of the antenna with SRR and metallic strips, shown in figure 11.16, on the human body. The antenna's resonant frequency is shifted by 1%.

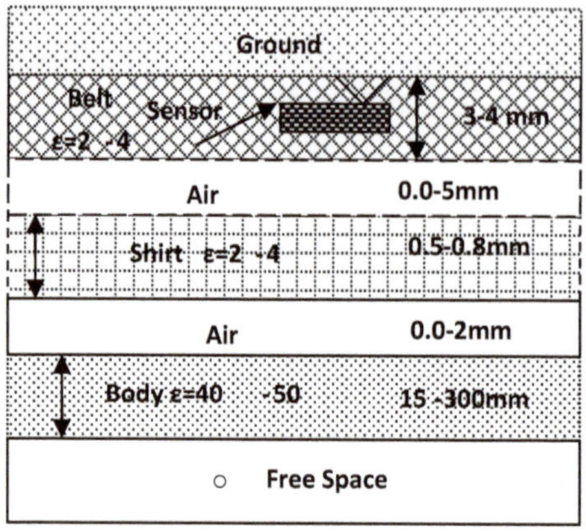

Figure 11.31. Wearable antenna environment.

Table 11.1. Summary of electrical properties of human body tissues.

Tissue	Property	434 MHz	600 MHz
Prostate	σ	0.75	0.90
	ε	50.53	47.4
Skin	σ	0.57	0.6
	ε	41.6	40.43
Stomach	σ	0.67	0.73
	ε	42.9	41.41
Colon, muscle	σ	0.98	1.06
	ε	63.6	61.9
Lung	σ	0.27	0.27
	ε	38.4	38.4

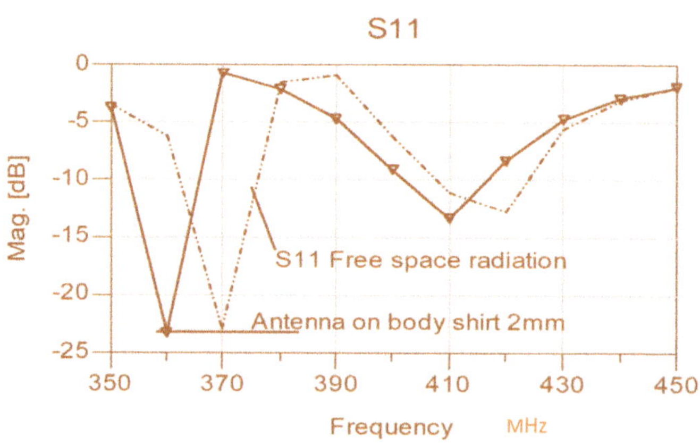

Figure 11.32. S_{11} of the antenna with SRR on the human body.

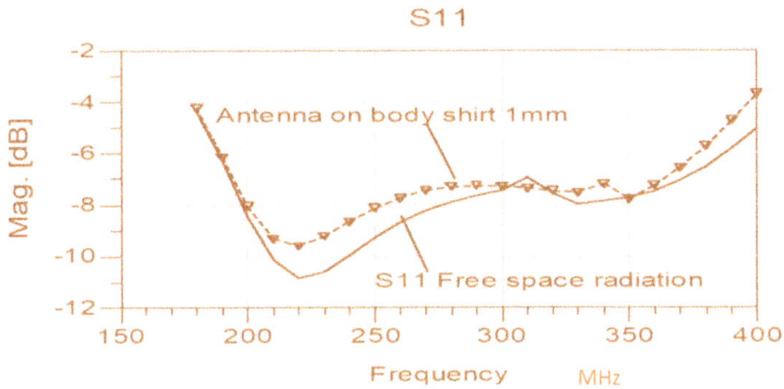

Figure 11.33. Antenna with SRR S_{11} results on the human body.

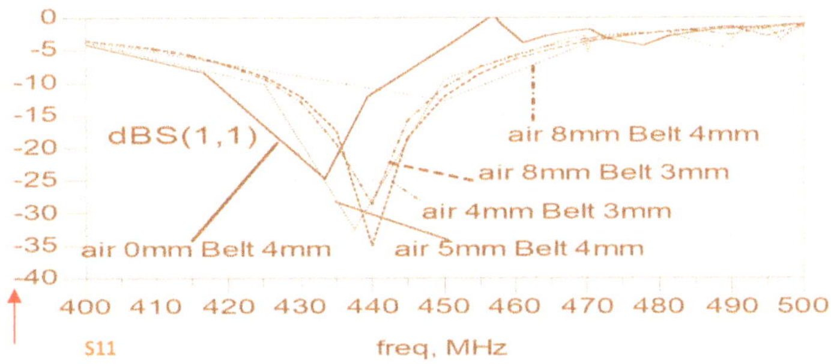

Figure 11.34. S_{11} results of the antenna shown in figure 11.3 on the human body.

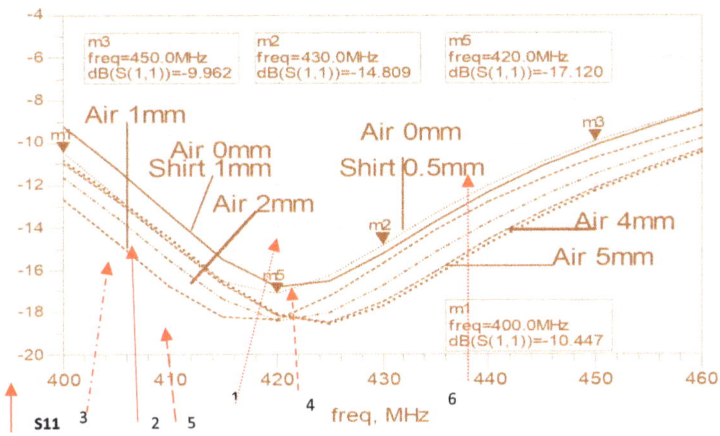

Figure 11.35. S_{11} results for different locations relative to the human body for the antenna shown in figure 11.6.

Figure 11.34 presents S_{11} results (of the antenna shown in figure 11.3) for different air spacing between an antenna and the human body, belt thickness and shirt thickness. The results presented in figure 11.33 indicate that the antenna has a VSWR better than 2.5:1 for air spacing up to 8 mm between the antenna and the body. Figure 11.35 presents S_{11} results for different positions relative to the human

Table 11.2. Explanation of figure 11.35.

Picture #	Line type	Sensor position
1	Dot ··················	Shirt thickness 0.5 mm
2	Line ———	Shirt thickness 1 mm
3	Dash dot -·-·-·-·	Air spacing 2 mm
4	Dash -------	Air spacing 4 mm
5	Long dash — — —	Air spacing 1 mm
6	Big dots ··········	Air spacing 5 mm

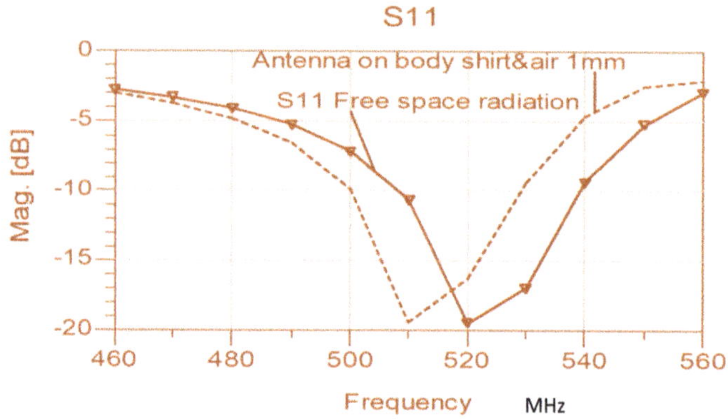

Figure 11.36. Folded antenna with SRR, S_{11} results on the body.

body of the folded antenna shown in figure 11.6. An explanation of figure 11.35 is given in table 11.2. If the air spacing between the antenna and the human body is increased from 0 mm to 5 mm the antenna's resonant frequency is shifted by 5%. A tunable wearable antenna may be used to control the antenna's resonant frequency at different positions on the human body, see [25].

Figure 11.36 presents S_{11} results of the folded antenna with SRR, shown in figure 11.8, on the human body. The antenna's resonant frequency is shifted by 2%. The radiation pattern of the folded antenna with SRR on the human body is presented in figure 11.37.

11.5 Metamaterial wearable antennas

The proposed wearable metamaterial antennas may be placed inside a belt as shown in figure 11.38. Three to four antennas may be placed in a belt and attached to a patient's stomach. More antennas may be attached to the patient's back to improve the level of the received signal from different locations in the human body. The cable from each antenna is connected to a recorder. The received signal is transferred via a SP8T switch to the receiver. The antennas receive a signal that is transmitted from various positions in the human body. The medical system selects the signal with the highest power.

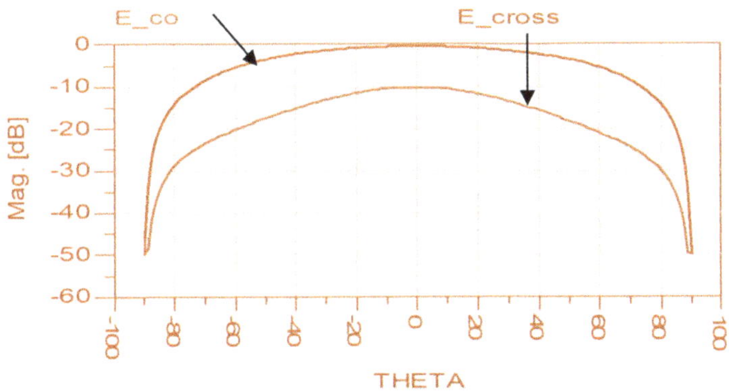

Figure 11.37. Radiation pattern of the folded antenna with SRR on the human body.

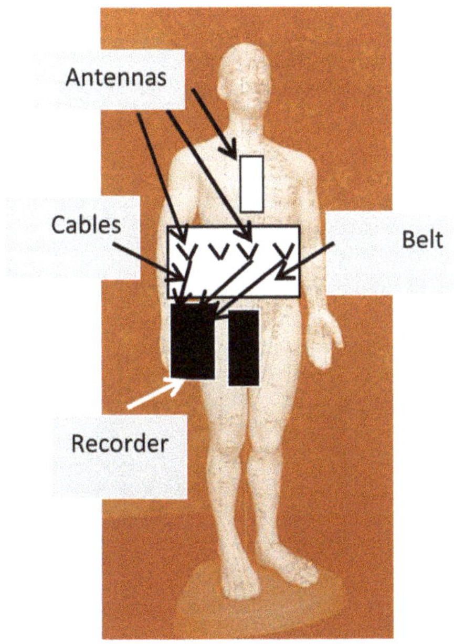

Figure 11.38. Medical system with a printed wearable antenna.

In several systems the distance separating the transmitting and receiving antennas is in the near-field zone. In these cases, the electric field intensity decays rapidly with distance. The near-fields only transfer energy to close distances from the antenna and do not radiate energy to far distances. The radiated power is trapped in the region near to the antenna. In the near-field zone the receiving and transmitting antennas are magnetically coupled. The inductive coupling value between two antennas is measured by their mutual inductance. In these systems we must consider only the near-field electromagnetic coupling.

In figures 11.39–11.42 several photos of printed antennas for medical applications are shown. The dimensions of the folded dipole antenna are $7 \times 6 \times 0.16$ cm. The

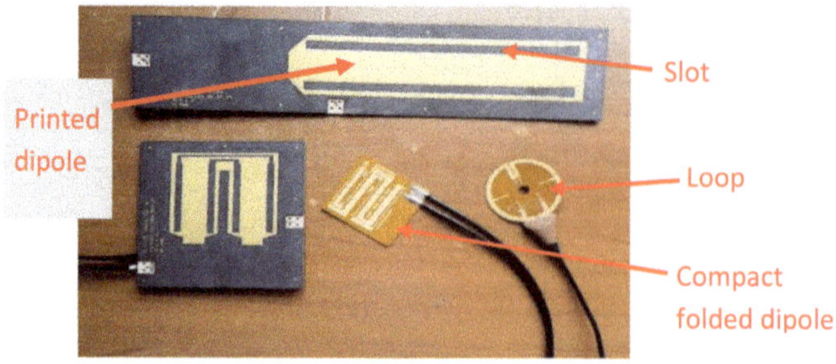

Figure 11.39. Microstrip antennas for medical applications.

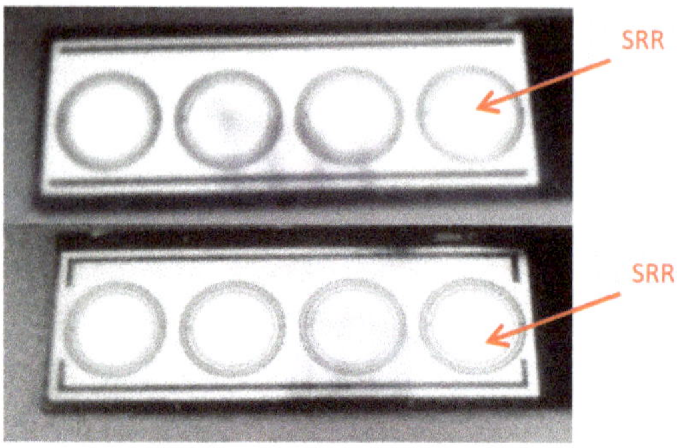

Figure 11.40. Metamaterial antennas for medical applications.

Figure 11.41. Metamaterial patch antenna with SRR.

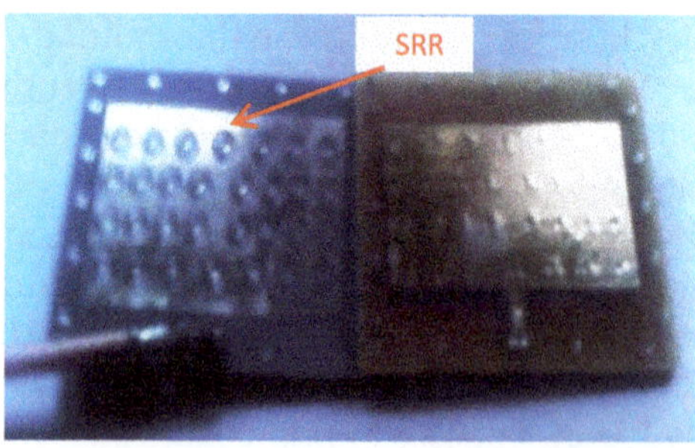

Figure 11.42. Metamaterial stacked patch antenna with SRR.

dimensions of the compact folded dipole presented in [5] and shown in figure 11.39 are 5 × 5 × 0.5 cm. The antennas' electrical characteristics on the human body have been measured by using a phantom. The phantom has been designed to represent the human body's electrical properties as presented in [5]. The tested antenna was attached to the phantom during the measurements of the antennas' electrical parameters.

11.6 Wideband stacked patch with SRR

A wideband microstrip stacked patch antenna with air spacing [1–3] has been designed. The antenna has been designed with SRR. The antenna consists of two layers. The first layer consists of FR4 dielectric substrate with a dielectric constant of 4 and 1.6 mm thick. The second layer consists of RT-Duroid 5880 dielectric substrate with a dielectric constant of 2.2 and 1.6 mm thick. The layers are separated by air spacing. The dimensions of the microstrip stacked patch antenna shown in figure 11.43 are 33 × 20 × 3.2 mm. The antenna has been analyzed by using Agilent ADS software. The antenna's bandwidth is around 10% for a VSWR better than 2.0:1. The beamwidth is around 72°, the gain is around 90 dBi to 10 dBi, and the efficiency is around 95%. The computed S_{11} parameters are presented in figure 11.44. A radiation pattern of the stacked patch is shown in figure 11.45. There is good agreement between the measured and computed results.

Figure 11.32 presents S_{11} results of the antenna with SRR shown in figure 11.13 on the human body. The antenna's resonant frequency is shifted by 3%. Figure 11.33 presents S_{11} results of the antenna with SRR and metallic strips, shown in figure 11.16, on the human body. The antenna's resonant frequency is shifted by 1%.

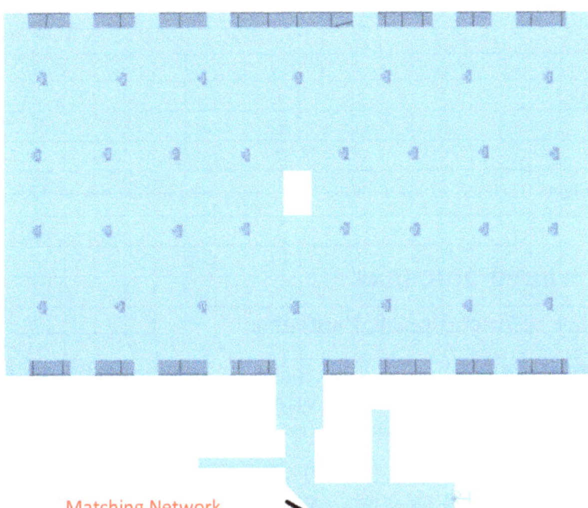

Figure 11.43. Wideband stacked patch antenna with SRR.

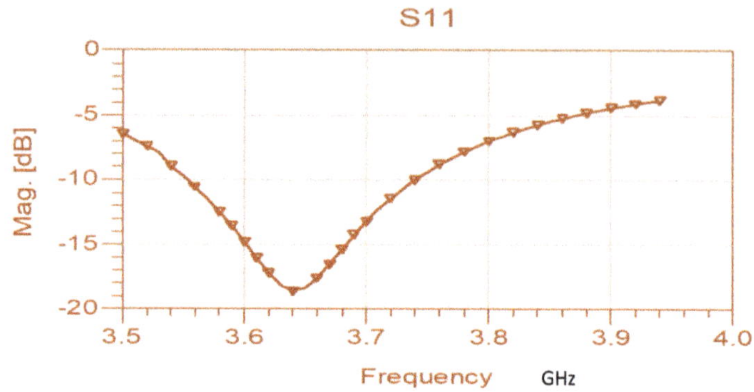

Figure 11.44. Wideband stacked antenna with SRR, S_{11} results.

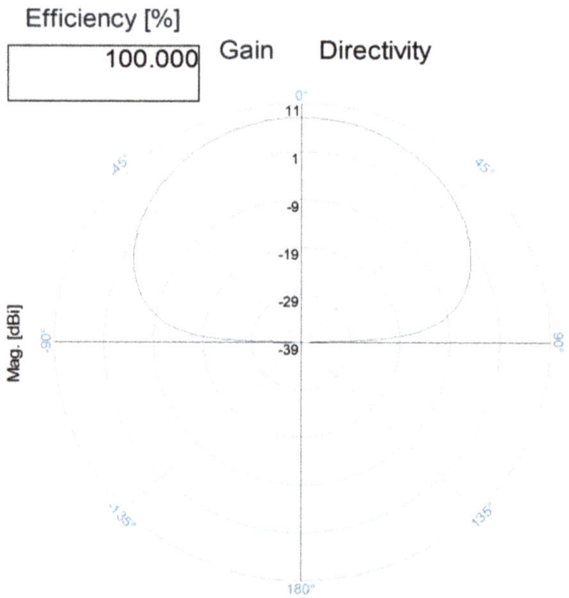

Figure 11.45. Radiation pattern of the stacked antenna with SRR.

11.7 Fractal printed antennas

11.7.1 Introduction to fractal printed antennas

A fractal antenna is an antenna that uses antenna design with similar fractal segments to maximize the antenna's effective area. Fractal antennas are also referred to as multilevel structures with space-filling curves. The key aspect lies in the repetition of a motif over two or more scale sizes or 'iterations'. Fractal antennas are very compact, multiband or wideband, and have useful applications in cellular telephone and microwave communications. Several fractal antennas have been presented in books, papers and patents, see [26–41].

11.7.2 Fractal structures

A curve, with endpoints, is represented by a continuous function whose domain is the unit interval [0, 1]. The curve may lie in a plane or in a 3D space. A fractal curve is a densely self-intersecting curve that passes through every point of the unit square. A fractal curve is a continuous mapping from the unit interval to the unit square.

In mathematics, a space-filling curve is a curve whose range contains the entire 2D unit square. Most space-filling curves are constructed iteratively as a limit of a sequence of piecewise linear continuous curves, each one closely approximating the space-filling limit. Space-filling curves, where two sub curves intersect (in the technical sense), there is self-contact without self-crossing. A space-filling curve can be (everywhere) self-crossing if its approximation curves are self-crossing. A space-filling curve's approximations can be self-avoiding, as presented in figure 11.46. In three dimensions, self-avoiding approximation curves can even contain joined ends. Space-filling curves are special cases of fractal constructions. No differentiable space-filling curve can exist.

The term 'fractal curve' was introduced by B Mandelbrot [26, 27] to describe a family of geometrical objects that are not defined in standard Euclidean geometry. Fractals are geometric shapes that repeat themselves over a variety of scale sizes. One of the key properties of a fractal curve is self-similarity. A self-similar object is unchanged after increasing or shrinking its size. An example of a repetitive geometry is the Koch curve, presented in 1904 by Helge von Koch and shown in figure 11.46(b). Koch generated the geometry by using a segment of a straight line and raised an equilateral triangle over its middle third. Repeating once more the process of erecting equilateral triangles over the middle third of a straight line results in the structures shown in figure 11.47(a). Iterating the process infinitely many times results in a 'curve' of infinite length. This geometry is continuous everywhere but is nowhere

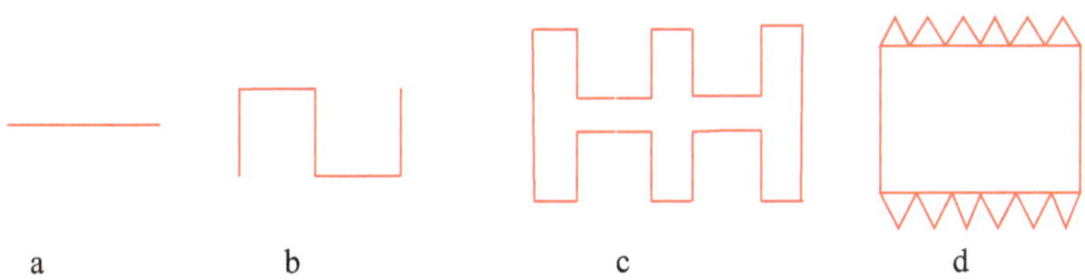

Figure 11.46. (a) Line. (b) Motif of bended line. (c) Bended line fractal structures. (d) Fractal structure.

Figure 11.47. (a) Koch Fractal structures. (b) Koch snowflakes.

Figure 11.48. Folded fractal structures.

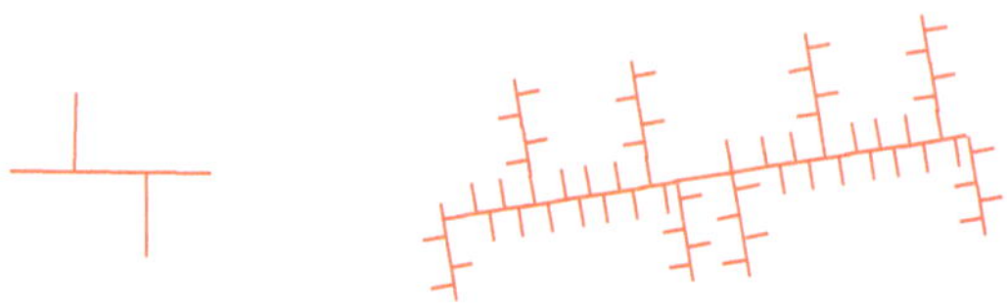

Figure 11.49. Variations of Koch fractal structures.

differentiable. If one applies the Koch process to an equilateral triangle, after many iterations, it converges to a Koch snowflake, shown in figure 11.47. This process can be applied to several geometries as shown in figures 11.48 and 11.49. Variations of these geometries have been presented in many papers [26–28].

11.7.3 Fractal antennas

Fractal geometries may be applied to the design of antennas and antenna arrays. The advantages of printed circuit technology and printed antennas can be used enhance the design of fractal printed antennas and microwave components. The effective area of a fractal antenna is significantly higher than the effective area of a regular printed antenna. Fractal antennas may operate with good performance at several different frequencies simultaneously. Fractal antennas are compact multi-band antennas. The directivity of fractal antennas is usually higher than the directivity of a regular printed antenna. The number of elements in a fractal antenna array may be reduced by around a quarter of the number of elements in a regular array. A fractal antenna could be considered as a non-uniform distribution of radiating elements. Each of the elements contributes to the total radiated power density at a given point with a given amplitude and phase. By spatially superposing these line radiators we can study the properties of a fractal antenna array.

Small antenna features are:

- A large input reactance (either capacitive or inductive) that usually must be compensated for with an external matching network.
- A small radiating resistance.
- Small bandwidth and low efficiency.
- This means that it is highly challenging to design a resonant antenna in a small space in terms of the wavelength at resonance.

The use of microstrip antennas is well known in mobile telephony handsets [30]. The PIFA (Planar Inverted F Antennas) configuration is popular in mobile communication systems. The advantages of PIFA antennas are their low profile, low fabrication costs and easy integration within the system structure. One of the miniaturization techniques used in this antenna system is based on space-filling curves. In some particular cases of antenna configuration systems, the antenna shape may be described as a multilevel structure. The multilevel technique has been already proposed to reduce the physical dimensions of microstrip antennas. The present integrated multiservice antenna system for a communication system comprises the following parts and features.

The antenna includes a conducting strip or wire shaped by a space-filling curve, composed of at least 200 connected segments forming a substantial right angle with each adjacent segment smaller than a hundredth of the free-space operating wavelength. The important reduction size of such an antenna system is obtained by using space-filling geometries. A space-filling curve can be described as a curve that is large in terms of physical length but small in terms of the area in which the curve can be included. A space-filling curve can be fitted over a flat or curved surface, and due to the angles between segments, the physical length of the curve is always larger than that of any straight line that can be fitted in the same area (surface).

Additionally, to properly shape the structure of a miniature antenna, the segments of the space-filling curves must be shorter than a tenth of the free-space operating wavelength. The antenna is fed with a two conductor structure such as a coaxial cable, with one of the conductors connected to the lower tip of the multilevel structure and the other conductor connected to the metallic structure of the system, which acts as a ground plane. This type of antenna features a size reduction below 20% of the typical size of a conventional external quarter-wave whip antenna. This feature, together with the small profile of the antenna, which can be printed in a low-cost dielectric substrate, allows a simple and compact integration of the antenna structure.

Reducing the size of the radiating elements can be achieved by using a PIFA configuration, consisting on connecting two parallel conducting sheets, separated either by air or a dielectric, magnetic or magneto dielectric material. The sheets are connected through a conducting strip near one of the sheet's corners and orthogonally mounted to both sheets.

The antenna is fed through a coaxial cable, having its outer conductor connected to the first sheet, the second sheet being coupled either by direct contact or capacitive to the inner conductor of the coaxial cable.

In figures 11.50(a) and (b), two examples of space-filling perimeters of the conducting sheet to achieve an optimized miniaturization of the antenna are presented.

11.8 Anti-radar fractals and/or multilevel chaff dispersers

Definition of chaff

Chaff is one of the forms of countermeasure employed against radar. It usually consists of a large number of electromagnetic dispersers and reflectors, normally arranged in the form of strips of metal foil packed in a bundle. Chaff is usually employed to foil or confuse surveillance and tracking radar.

Figure 11.50. (a) Patch with a space-filling perimeter of the conducting sheet. (b) Microstrip patch with a space-filling perimeter of the conducting sheet.

Geometry of dispersers

Here we present a new geometry of dispersers or reflectors that improves the properties of radar chaff [31]. Some of the geometries presented here are related to some forms for antennas. Multilevel and fractal structure antennas are distinguished by being of reduced size and having multiband behavior, as has been expounded already in patent publications [32].

The main electrical characteristic of a radar chaff disperser is:
Its radar cross section (RCS), which is related to the reflective capability of the disperser.

A fractal curve for a chaff disperser is defined as a curve comprising at least ten segments that are connected so that each element forms an angle with its neighbors; no pair of these segments defines a longer straight segment, these segments being smaller than a tenth of the resonant wavelength in free space of the entire structure of the dispenser.
In many of the configurations presented, the size of the entire disperser is smaller than a quarter of the lowest operating wavelength.

The space-filling curves (or fractal curves) can be characterized by:

1. They are long in terms of physical length, but small in terms of the area in which the curve can be included. The dispersers with a fractal form are long electrically but can be included in a very small surface area. This means it is possible to obtain smaller packaging and a denser chaff cloud using this technique.
2. Frequency response: Their complex geometry provides a spectrally richer signature when compared with rectilinear dispersers known in the state-of-the-art.

The fractal structure properties of the disperser not only introduce an advantage in terms of reflected radar signal response, but also in terms of the aerodynamic profile of dispersers. It is known that a surface offers greater resistance to air than a line or a 1D form.

Therefore, giving a fractal form to the dispersers with a dimension greater than unity ($D > 1$) increases resistance to the air and improves the time of suspension.

11.9 Definition of a multilevel fractal structure

Multilevel structures are a geometry related to fractal structures. In the case of radar chaff, a multilevel structure is defined as a structure that includes a set of polygons, which are characterized by having the same number of sides, wherein these polygons are electromagnetically coupled either by means of capacitive coupling, or by means of an ohmic contact. The region of contact between the directly connected polygons is smaller than 50% of the perimeter of the polygons mentioned in at least 75% of the polygons that constitute the defined multilevel structure.

A multilevel structure provides both:

- A reduction in the size of dispensers and an enhancement of their frequency response.
- Can resonate in a non-harmonic way and can even cover simultaneously and with the same relative bandwidth at least a portion of numerous bands.

  The fractal structure (SFC) is preferred when a reduction in size is required, while multilevel structures are preferred when it is required that the most important considerations be given to the spectral response of radar chaff.

The main advantages for configuring the form of the chaff dispersers are:

1. The dispersers are small; consequently, more disperser can be encapsulated in the same cartridge, rocket or launch vehicle.
2. The dispersers are also lighter. Therefore, they can remain floating in the air longer than conventional chaff.
3. Due to the smaller size of the chaff dispersers, the launching devices (cartridges, rockets, etc) can be smaller regarding chaff systems in the state-of-the-art providing the same RCS.
4. Due to the lighter weight of the chaff dispersers, the launching devices can shoot the packages of chaff farther from the launching devices and locations.
5. Chaff constituted by multilevel and fractal structures provides larger RCS at longer wavelengths than conventional chaff dispersers of the same size.
6. The dispersers with long wavelengths can be configured and printed on light dielectric supports having a non-aerodynamic form and opposing a greater resistance to the air and thereby having a longer suspension time.
7. The dispersers provide a better frequency response regarding dispersers of the state-of-the-art. In the following images such size compression structures based on fractal curves are presented (figure 11.51).

Figure 11.52 shows several examples of Hilbert fractal curves (with increasing iteration order), which can be used to configure the chaff disperser.

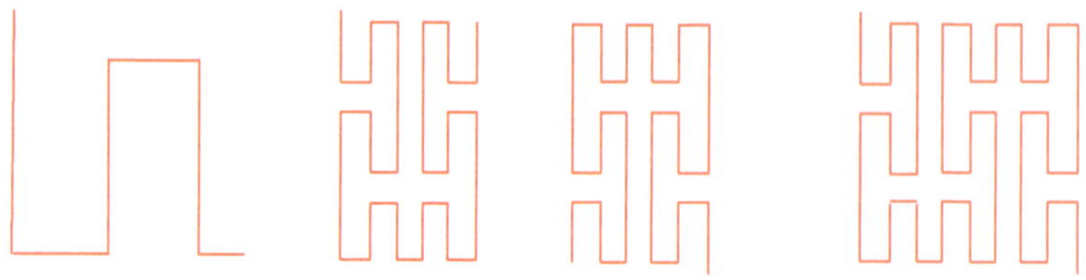

Figure 11.51. Fractal curves that can be used to configure a chaff disperser.

Figure 11.52. Hilbert fractal curves.

11.10 Advanced antenna system

The main advantage of the advanced antenna system lies in the multiband and multiservice performance of the antenna. This enables convenient and easy connection of a simple antenna for most communication systems and applications. The main advantages addressed by advanced antennas featured similar parameters (input impedance, radiation pattern) at several bands maintaining their performance, compared with conventional antennas. Fractal shapes permit one to obtain a compact antenna of reduced dimensions compared to other conventional antennas. Multilevel antennas introduced a higher flexibility to the design of multiservice antennas for real applications, extending the theoretical capabilities of ideal fractal antennas to practical, commercial antennas.

11.10.1 Comparison between Euclidean antennas and fractal antennas

Most conventional antennas are of Euclidean design/geometry, where the closed antenna's area is directly proportional to the antenna's perimeter. Thus, for example, when the length of a Euclidean square is increased by a factor of three, the enclosed area of the antenna is increased by a factor of nine. Gain, directivity, impedance and efficiency of Euclidean antennas are a function of the antenna's size to wavelength ratio.

Euclidean antennas are typically required to operate within a narrow range (e.g. 10%–40%) around a central frequency (fc), which in turn dictates the size of the antenna (e.g. half- or quarter-wavelength). When the size of a Euclidean antenna is made to be much smaller than the operating wavelength (λ), it becomes very inefficient because the antenna's radiation resistance decreases and becomes less than its ohmic resistance (i.e. it does not couple electromagnetic excitations efficiently to free space). Instead, it stores energy reactively within its vicinity (reactive impedance X_c). These aspects of Euclidean antennas work together to make it difficult for small Euclidean antennas to couple or match to feeding or excitation circuitry and causes them to have a high Q factor (lower bandwidth). The Q (Quality) factor may be defined as approximately the ratio of input reactance X_{in} to radiation resistance R_r, $Q = X_{in}/R_r$.

The Q factor may also be defined as the ratio of average stored electric energies (or magnetic energies stored) to the average radiated power. Q can be shown to be inversely proportional to bandwidth.

Thus, small Euclidean antennas have a very small bandwidth, which is of course undesirable (a matching network may be needed). Many known Euclidean antennas are based upon closed-loop shapes.

Unfortunately, when small in size, such loop-shaped antennas are undesirable because, as discussed above, the radiation resistance decreases significantly when the antenna's size is decreased. This is because the physical area (A) contained within the loop-shaped antenna's contour is related to the loop perimeter.

The radiation resistance of a circular loop-shaped Euclidean antenna is defined by R_r, as given in equation (11.1), k is a constant.

$$R_r = \eta\pi(2/3)(KA/\lambda)^2 = 20\pi^2(C/\lambda) \tag{11.1}$$

Since the resistance R_c is only proportional to the perimeter (C), then for $C < 1$, the resistance R_c is greater than the radiation resistance R_r and the antenna is highly inefficient. This is generally true for any small circular Euclidean antenna. A small-sized antenna will exhibit a relatively large ohmic resistance and a relatively small radiation resistance R_r. This low efficiency limits the use of small antennas.

Fractal geometry is a non-Euclidean geometry which can be used to overcome the problems with small Euclidean antennas. Radiation resistance, R_r, of a fractal antenna decreases as a small power of the perimeter (C) compression, in which a fractal loop or island always has a substantially higher radiation resistance than a small Euclidean loop antenna of equal size. Fractal geometry may be grouped into:

- Random fractals, which may be called chaotic or Brownian fractals.
- Deterministic or exact fractals. In deterministic fractal geometry, a self-similar structure results from the repetition of a design or motif (generator) with self-similarity and structure at all scales. In deterministic or exact self-similarity, fractal antennas may be constructed through recursive or iterative means. In other words, fractal structures are often composed of many copies of themselves at different scales, thereby allowing them to defy the classical antenna's performance constraint which is size to wavelength ratio.

11.10.2 Multilevel and space-filling ground planes for miniature and multiband antennas

There is a new family of antenna ground planes of reduced size and enhanced performance based on an innovative set of geometries.

These new geometries are known as multilevel and space-filling structures, which have been previously used in the design of multiband and miniature antennas.

One of the key issues of the present antenna system is considering the ground plane of an antenna as an integral part that mainly contributes to its radiation and impedance performance (impedance level, resonant frequency, and bandwidth).

The multilevel and space-filling structures are used in the ground plane of the antenna obtaining a better return loss or VSWR, better bandwidth, multiband behavior or a combination of all these effects. The technique can be seen as a way to reduce the size of the ground plane and therefore the size of the overall antenna. The key point of the present antenna system is shaping the ground plane of an antenna in such a way that the combined effect of the ground plane and the radiating element enhances the performance and characteristics of the whole antenna device, either in terms of bandwidth, VSWR, multiband, efficiency, size, or gain.

Multilevel geometry

The resulting geometry is no longer a solid, conventional ground plane, but a ground plane with a multilevel or space-filling geometry, at least in a portion of the ground plane.

A multilevel geometry for a ground plane consists of a conducting structure including a set of polygons, featuring the same number of sides, electromagnetically coupled either by means of a capacitive coupling or ohmic contact. The contact region between directly connected polygons is narrower than 50% of the perimeter of polygons in at least 75% of polygons defining the conducting ground plane. In this definition of multilevel geometry, circles and ellipses are included as well, since they can be understood as polygons with an infinite number of sides.

Space-filling curve

A space-filling curve (hereafter, SFC) is a curve that is large in terms of physical length but small in terms of the area in which the curve can be included.

A curve is composed of at least ten segments that are connected in such a way that each segment forms an angle with their neighbors; that is, no pair of adjacent segments defines a larger straight segment, and wherein the curve can be optionally periodic along a fixed straight direction of space if, and only if, the period is defined by a non-periodic curve composed of at least ten connected segments and no pair of adjacent and connected segments defines a straight longer segment.

A SFC can be fitted over a flat or curved surface, and due to the angles between segments, the physical length of the curve is always larger than that of any straight line that can be fitted in the same area (surface).

Additionally, to properly shape the ground plane, the segments of the SFCs included in the ground plane must be shorter than a tenth of the free-space operating wavelength.

Figure 11.53 shows several examples of fractal geometries that can be used as SFCs.

Figure 11.54 shows several examples of Hilbert fractal curves that can be used as SFCs.

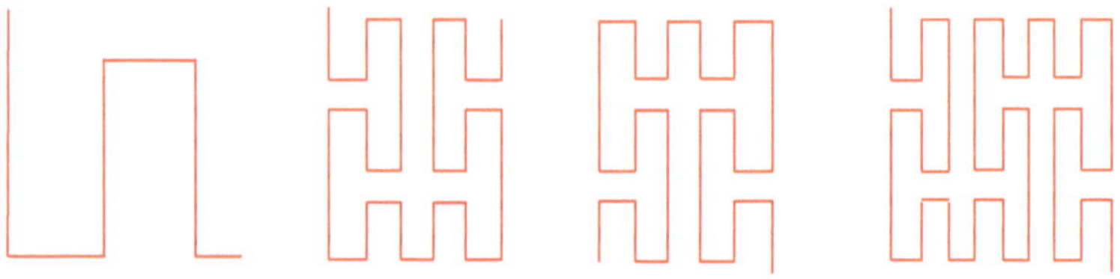

Figure 11.53. Fractal curves that can be used as space-filling curves.

The curves shown in figure 11.53 are some examples of such SFCs. Due to the special geometry of multilevel and space-filling structures, the current distributes over the ground plane in such a way that it enhances the antenna's performance and features in terms of:

- Reduced size compared to antennas with a solid ground plane.
- Enhanced bandwidth compared to antennas with a solid ground plane.
- Multi-frequency performance.
- Better VSWR feature at the operating band or bands.
- Better radiation efficiency.
- Enhanced gain.

Figure 11.55(a) shows a patch antenna above an example of a new ground plane structure formed by both multilevel and space-filling geometries. Figure 11.55(b) shows a monopole antenna above a ground plane structure formed by both multilevel and space-filling geometries.

Figure 11.56 shows several examples of different contour-shaped multilevel ground planes, such as rectangular (figures 11.56(a), (b)) and circular ground planes (figure 11.56(c)).

Figure 11.54. Hilbert fractal curves that can be used as space-filling curves.

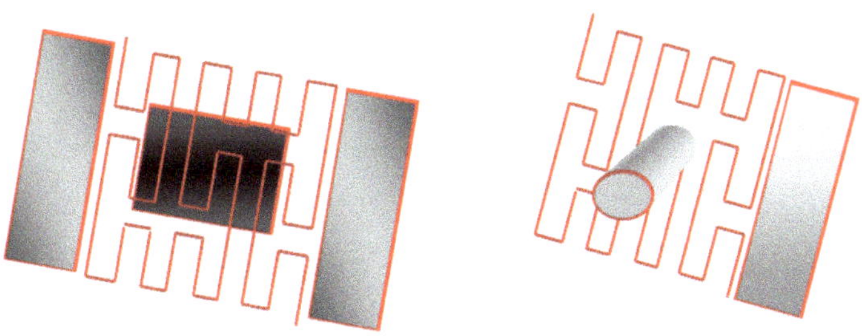

Figure 11.55. (a) Patch antenna above a new ground plane structure. (b) Monopole antenna above a ground plane structure formed by both multilevel and space-filling geometries.

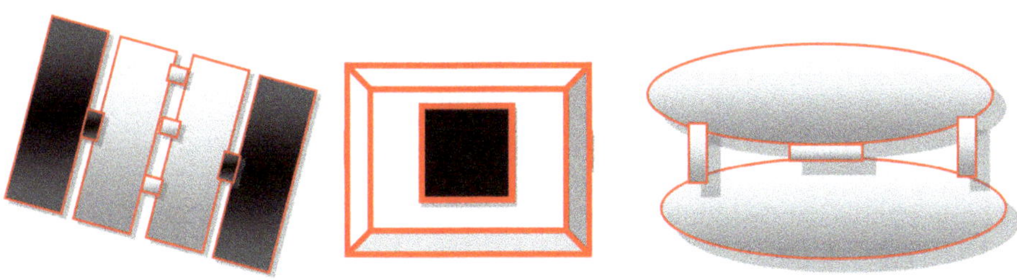

a b c

Figure 11.56. Examples of different contour-shaped multilevel ground planes. (a) Rectangular ground plane. (b) Multilevel rectangular ground plane. (c) Circular ground plane.

Figure 11.57. Fractal antenna resonators.

11.11 Applications of fractal printed antennas

In this chapter several designs of fractal printed antennas are presented for communication applications. These fractal antennas are compact and efficient. The antenna gain is around 8 dBi with 90% efficiency.

11.11.1 New 2.5 GHz fractal antenna with a space-filling perimeter on the radiator

A new fractal microstrip antenna was designed as presented in figure 11.57. The antenna was printed on Duroid substrate 0.8 mm thick with 2.2 dielectric constant. The antenna's dimensions are 5.2 × 48.8 × 0.08 cm. It was designed using ADS software.

The antenna's bandwidth is around 2%, around 2.5 GHz for a VSWR better than 3:1. The bandwidth may be improved to 5%, for a VSWR better than 2:1, by adding a second layer above the resonator. A patch radiator is printed on the second layer as presented in figure 11.58. The radiator was printed on FR4 substrate 0.8 mm thick

Figure 11.58. Fractal antenna patch radiator.

with 4.5 dielectric constant. The electromagnetic fields radiated by the resonator are electromagnetic coupled to the patch radiator. The patch radiator's dimensions are 45.2 × 48.8 × 0.08 cm. The stacked fractal antenna's structure is shown in figure 11.59. The spacing between the two layers may be varied to get a wider bandwidth. The stacked fractal antenna's S_{11} parameters with 8 mm air spacing between the layers is presented in figure 11.60. The S_{11} parameters of the fractal stacked patch antenna with 10 mm air spacing are given in figure 11.61. The antenna's bandwidth is improved to 5% for a VSWR better than 2:1: The fractal stacked patch antenna's radiation pattern is shown in figure 11.62. The beamwidth is around 76°, with 8 dBi gain and 91% efficiency. A photo of the stacked fractal patch antenna is shown in figure 11.63. The antenna's resonators are shown in figure 11.63(a), and the antenna's radiator is shown in figure 11.63(b). A modified version of the antenna is shown in figure 11.64. The S_{11} parameters of the modified fractal stacked patch antenna with 8 mm air spacing are given in figure 11.65. The antenna's bandwidth is around 10% for a VSWR better than 3:1. The fractal stacked patch antenna's radiation pattern is shown in figure 11.66. The beamwidth is around 76°, with 8 dBi gain and 91.82% efficiency.

Figure 11.59. Fractal stacked patch antenna structure.

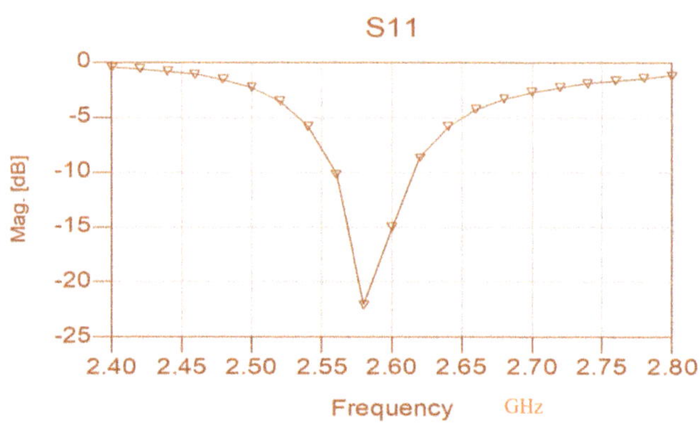

Figure 11.60. S_{11} parameters of the fractal stacked patch antenna with 8 mm air spacing.

11.11.2 New stacked patch 2.5 GHz fractal printed antennas

A new fractal microstrip antenna was designed as presented in figure 11.67. The antenna was printed on Duroid substrate 0.8 mm thick with 2.2 dielectric constant. The antenna's dimensions are 45.8 × 39.1 × 0.08 cm. It was designed using ADS software.

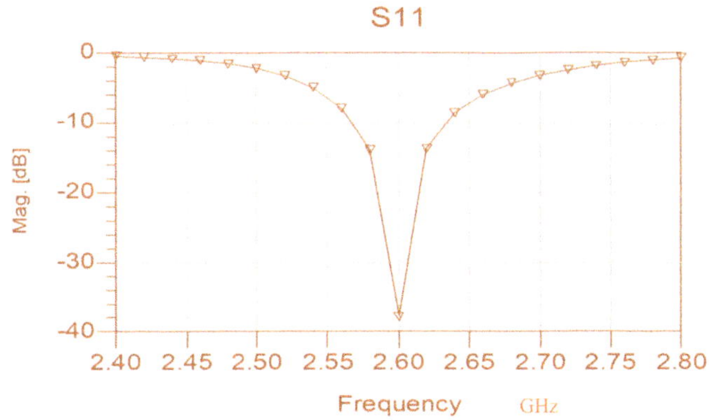

Figure 11.61. S_{11} parameters of the fractal stacked patch antenna with 10 mm air spacing.

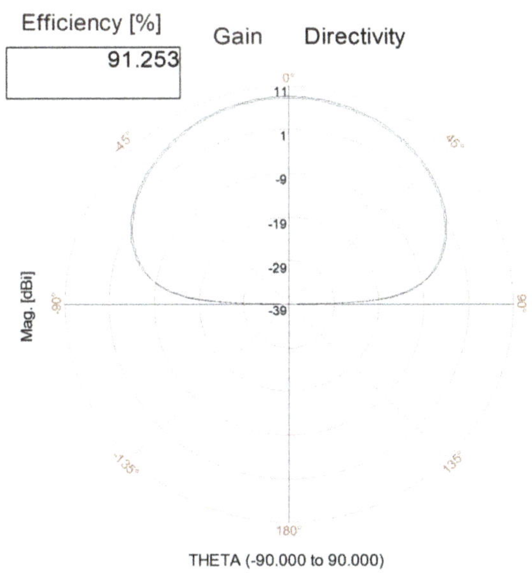

Figure 11.62. Fractal stacked patch antenna radiation pattern with 10 mm air spacing.

The antenna's resonator bandwidth is around 2%, around 2.52 GHz for a VSWR better than 3:1. The bandwidth may be improved to 6%, for a VSWR better than 3:1, by adding a second layer above the resonator. A patch radiator is printed on the second layer as presented in figure 11.68. The radiator was printed on FR4 substrate 0.8 mm thick with 4.5 dielectric constant. The electromagnetic fields radiated by the resonator are electromagnetically coupled to the patch radiator. The patch radiator's dimensions are 45.8 × 39.1 × 0.08 cm. The stacked fractal antenna structure is shown in figure 11.69. The spacing between the two layers may be varied to get a

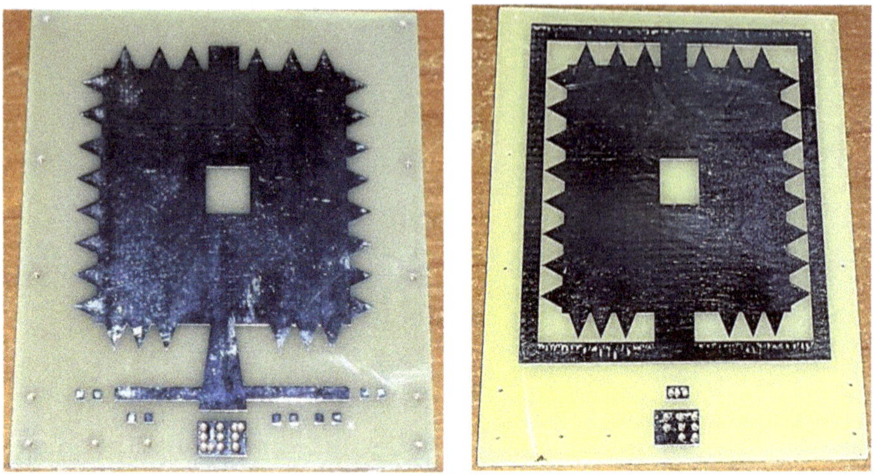

a b

Figure 11.63. Fractal stacked patch antenna. (a) Resonator. (b) Radiator.

Figure 11.64. A modified fractal stacked patch antenna structure.

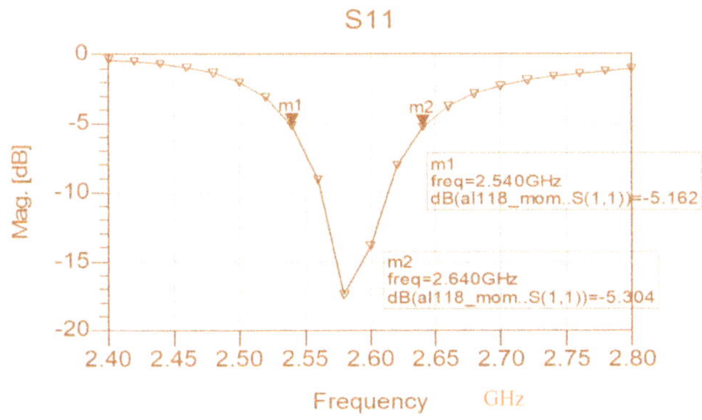

Figure 11.65. S_{11} parameters of the modified fractal stacked antenna with 8 mm air spacing.

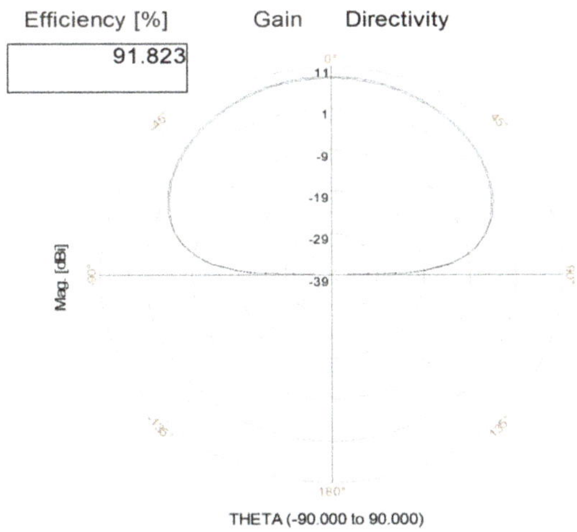

Figure 11.66. Fractal stacked patch antenna radiation pattern with 8 mm air spacing.

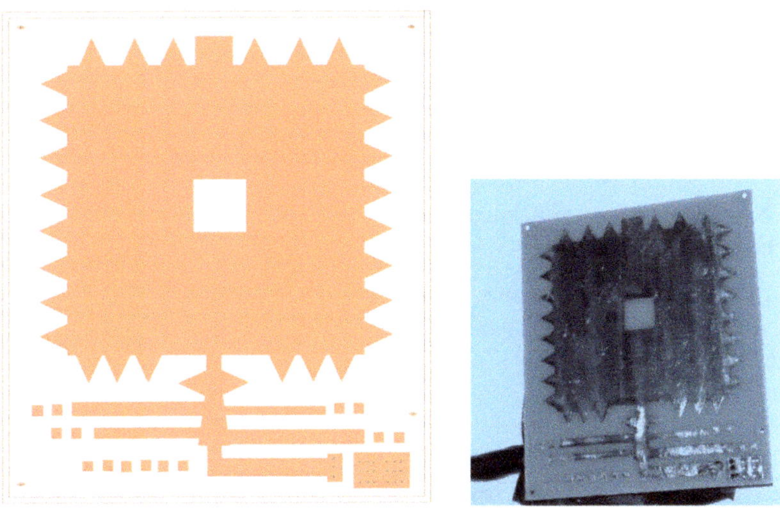

a b

Figure 11.67. Resonator of a fractal stacked patch antenna. (a) Layout. (b) Resonator photo.

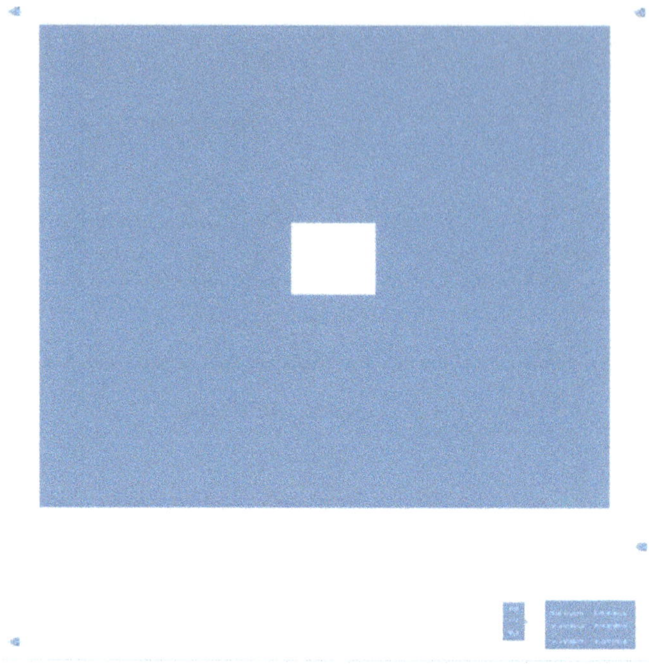

Figure 11.68. Radiator of a fractal stacked patch antenna.

wider bandwidth. The single layer fractal antenna's S_{11} parameters are presented in figure 11.70.

A comparison of the computed and measured S_{11} parameters of the fractal stacked patch antenna with no air spacing is given in figure 11.71. There is good agreement between the measured and computed results. The fractal stacked patch antenna's radiation pattern is shown in figure 11.72. The beamwidth is around 82°, with 7.5 dBi gain and 97.2% efficiency. A photo of the fractal stacked patch antenna is shown in figure 11.73. The antenna's resonator is shown in figure 11.74(a), and the radiator is shown in figure 11.74(b).

11.11.3 New 8 GHz fractal printed antennas with a space-filling perimeter of the conducting sheet

A new fractal microstrip antenna was designed as presented in figure 11.74. The antenna was printed on Duroid substrate 0.8 mm thick with 2.2 dielectric constant. The antenna's dimensions are 17.2 × 21.8 × 0.08 cm. It was designed using ADS software.

The antenna's resonator bandwidth is around 3%, around 7.2 GHz for a VSWR better than 2:1. The bandwidth is improved to 22%, for a VSWR better than 3:1, by adding a second layer above the resonator. A patch radiator is printed on the second layer as presented in figure 11.75. The radiator was printed on FR4 substrate 0.8 mm thick with 4.5 dielectric constant. The electromagnetic fields radiated by the

Figure 11.69. Layout of the fractal stacked patch antenna.

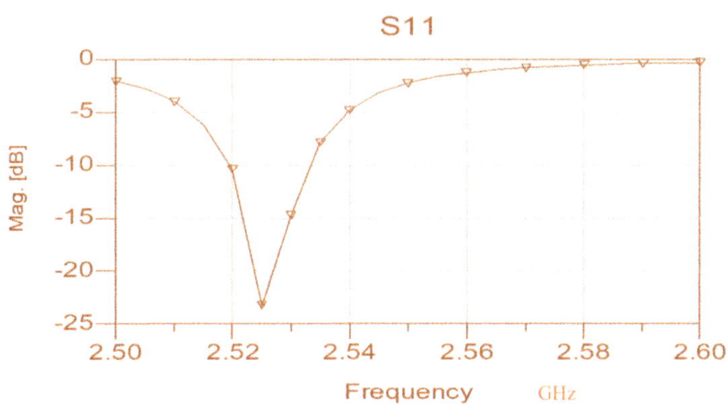

Figure 11.70. Computed S_{11} parameters of the single layer fractal antenna.

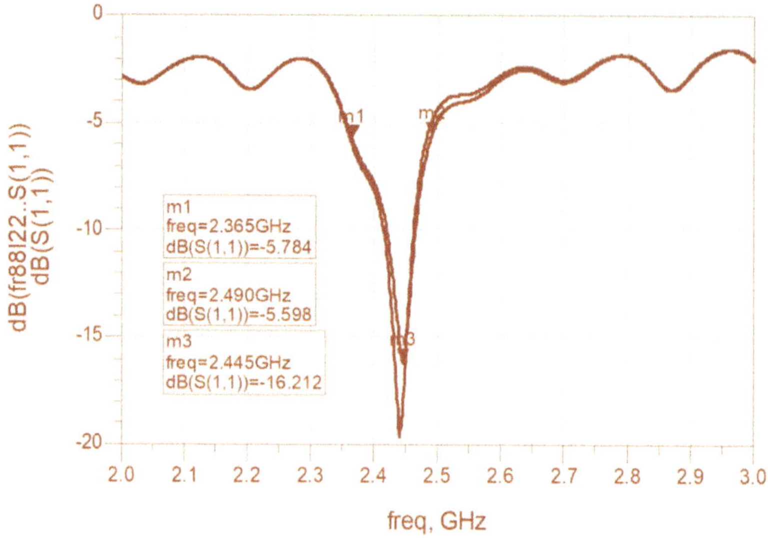

Figure 11.71. Measured and computed S_{11} parameters of the fractal stacked patch antenna with no air spacing between the layers.

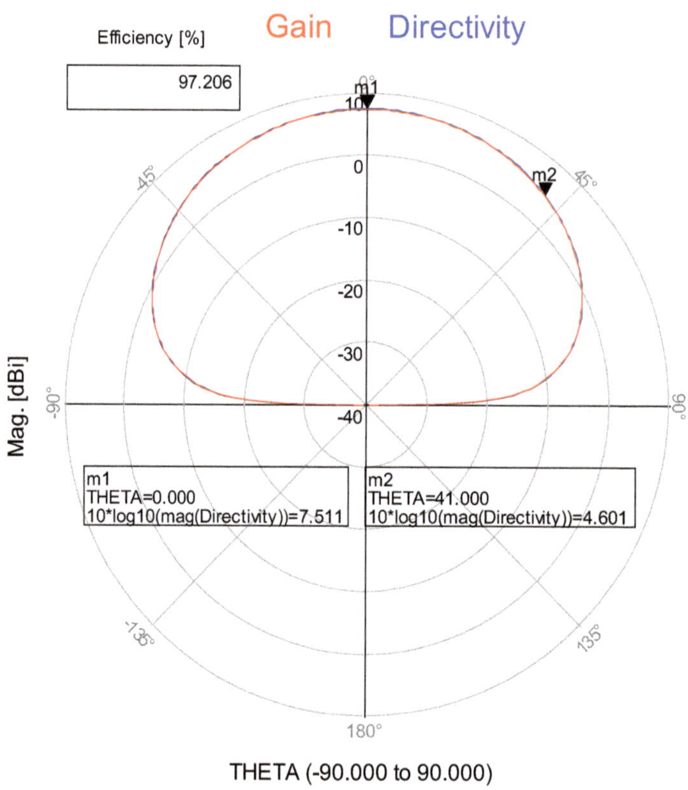

Figure 11.72. Fractal stacked patch antenna radiation pattern with 8 mm air spacing.

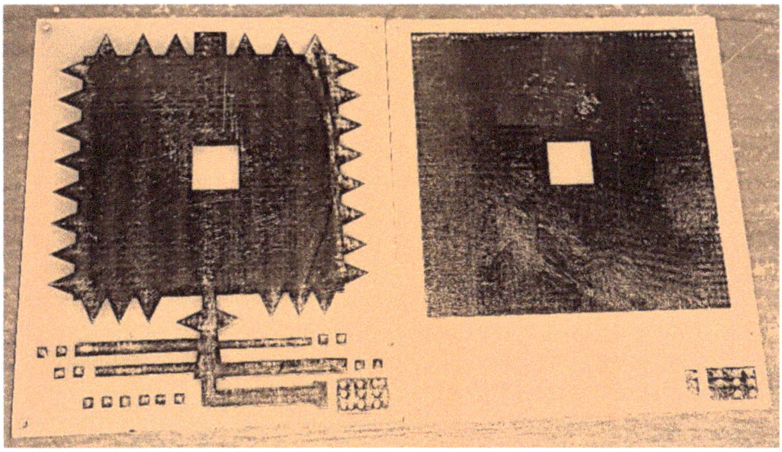

a b

Figure 11.73. Fractal stacked patch antenna. (a) Resonator. (b) Radiator.

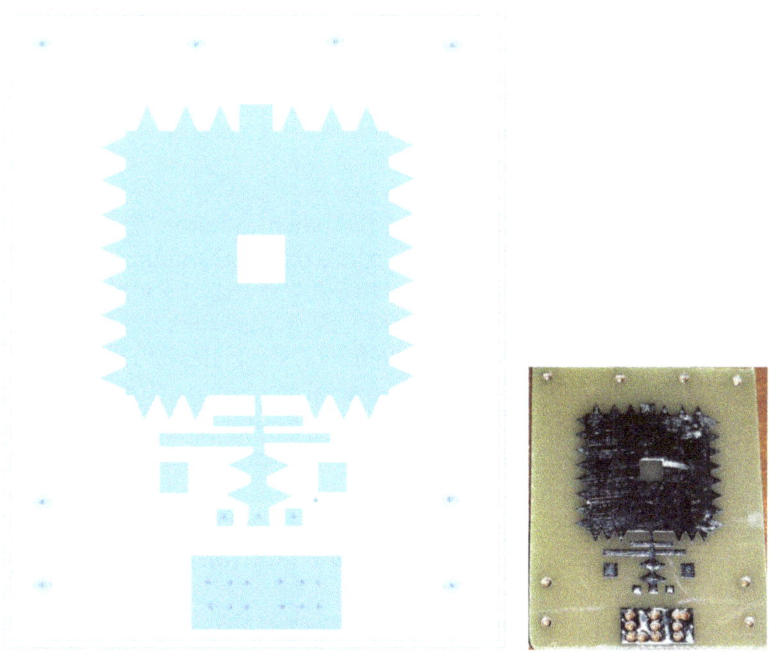

Figure 11.74. (a) Resonator of the 8 GHz fractal stacked patch antenna. (b) 8 GHz fractal resonator photo.

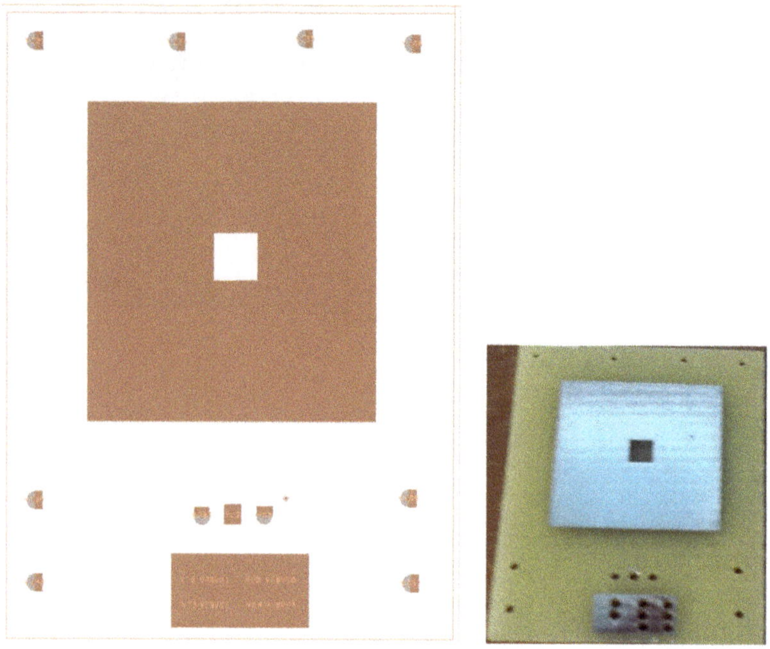

Figure 11.75. (a) Radiator of the 8 GHz fractal stacked patch antenna. (b) Radiator photo.

resonator are electromagnetic coupled to the patch radiator. The patch radiator's dimensions are 17.2 × 21.8 × 0.08 cm. The stacked fractal antenna structure is shown in figure 11.76. The spacing between the two layers may be varied to get a wider bandwidth. The single layer fractal antenna's S_{11} parameters are presented in figure 11.77. The computed S_{11} parameters of the fractal stacked patch antenna with 2 mm air spacing are given in figure 11.78. The fractal stacked patch antenna's radiation pattern at 7.5 GHz is shown in figure 11.79. The beamwidth is around 82°, with 7.8 dBi gain and 82.2% efficiency. The fractal stacked patch antenna's radiation pattern at 8 GHz is shown in figure 11.80. The beamwidth is around 82°, with 7.5 dBi gain and 95.3% efficiency. A photo of the fractal stacked patch antenna is shown in figure 11.81. The antenna's resonator is shown in figure 11.81(a), and the radiator is shown in figure 11.81(b).

11.11.4 New stacked patch 7.4 GHz fractal printed antennas

A new fractal microstrip antenna was designed as presented in figure 11.82. The antenna was printed on Duroid substrate 0.8 mm thick with 2.2 dielectric constant. The dimensions are 18 × 12 × 0.08 cm. It was designed using ADS software.

The antenna's resonator bandwidth is around 2%, around 7.4 GHz for a VSWR better than 2:1. The antenna's bandwidth is improved to 10%, for a VSWR better

Figure 11.76. Layout of the 8 GHz fractal stacked patch antenna.

than 3:1, by adding a second layer above the resonator. A patch radiator is printed on the second layer as presented in figure 11.83. The radiator was printed on FR4 substrate 0.8 mm thick with 4.5 dielectric constant. The electromagnetic fields radiated by the resonator are electromagnetically coupled to the patch radiator. The fractal stacked patch dimensions are $18 \times 12 \times 0.08$ cm. The stacked fractal antenna structure is shown in figure 11.84. The spacing between the two layers may be varied to get a wider bandwidth. The computed S_{11} parameters of the fractal stacked patch antenna with 3 mm air spacing are given in figure 11.85. The fractal stacked patch antenna's radiation pattern at 7.5 GHz is shown in figure 11.86. The beamwidth is around 86°, with 7.9 dBi gain and 89.7% efficiency.

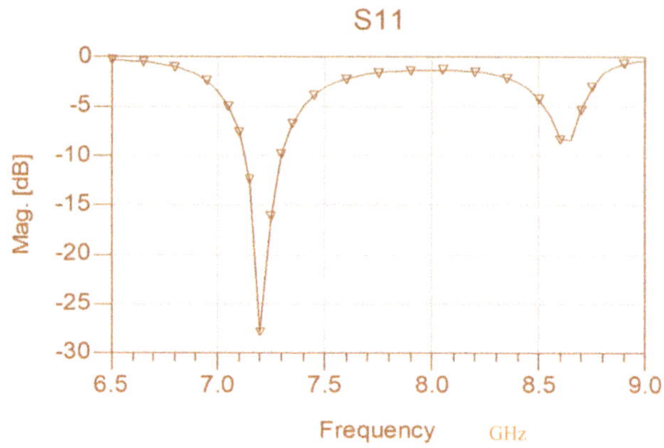

Figure 11.77. Computed S_{11} of the 8 GHz fractal stacked patch antenna.

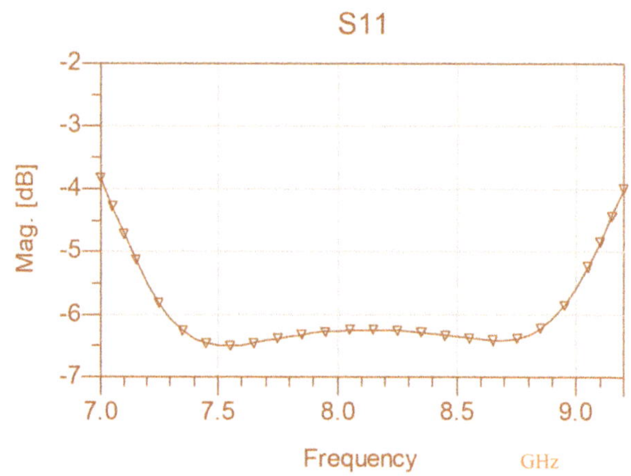

Figure 11.78. Computed S_{11} of the 8 GHz fractal stacked patch antenna with 2 mm air spacing.

A modified antenna structure is presented in figure 11.87. The antenna's matching network has been modified and S_{11} at 7.45 GHz is −23.5 dB as shown in figure 11.88. The computed S_{11} parameters of the fractal stacked patch antenna with 3 mm air spacing are given in figure 11.88. The fractal stacked patch antenna's bandwidth is around 9% for a VSWR better than 3:1. The radiation pattern at 7.5 GHz is shown in figure 11.89. The beamwidth is around 85°, with 7.8 dBi gain and 86.2% efficiency.

11.12 Conclusion

Metamaterial technology is used to develop small antennas with high efficiency. A new class of printed metamaterial antennas with high efficiency is presented in this

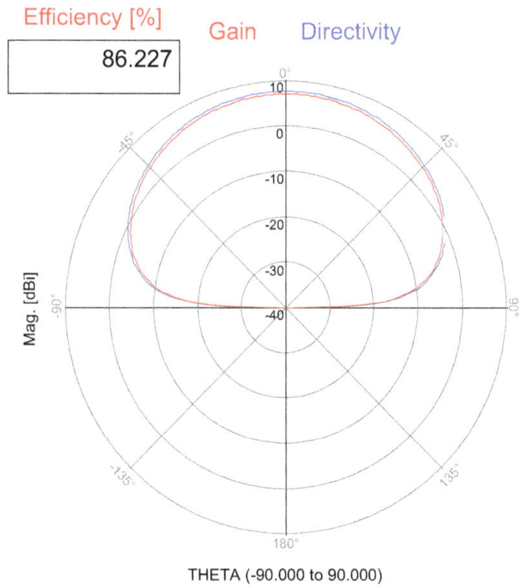

Figure 11.79. Fractal stacked patch antenna radiation pattern with 2 mm air spacing at 7.5 GHz.

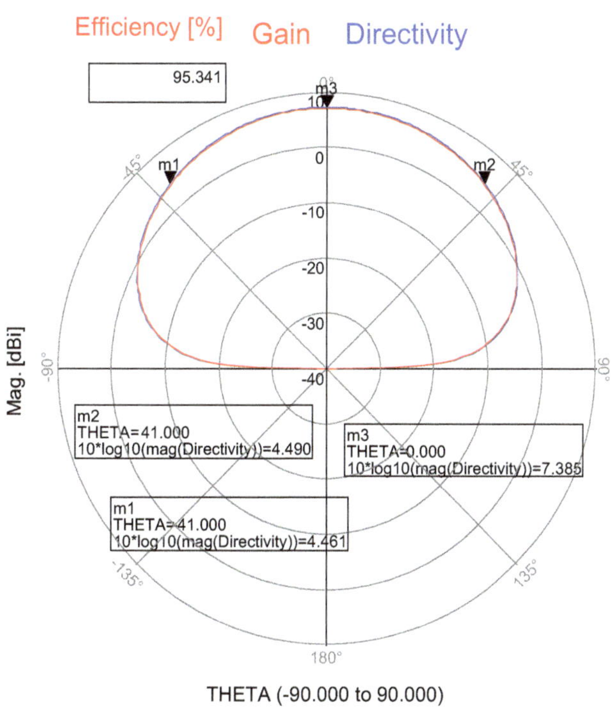

Figure 11.80. Fractal stacked patch antenna radiation pattern with 2 mm air spacing at 8 GHz.

a b

Figure 11.81. A photo of the fractal stacked patch antenna. (a) Resonator. (b) Radiator.

Figure 11.82. Layout of the 7.4 GHz fractal resonator.

Figure 11.83. Radiator of a 7.4 GHz fractal stacked patch antenna.

Figure 11.84. Layout of the 8 GHz fractal stacked patch antenna.

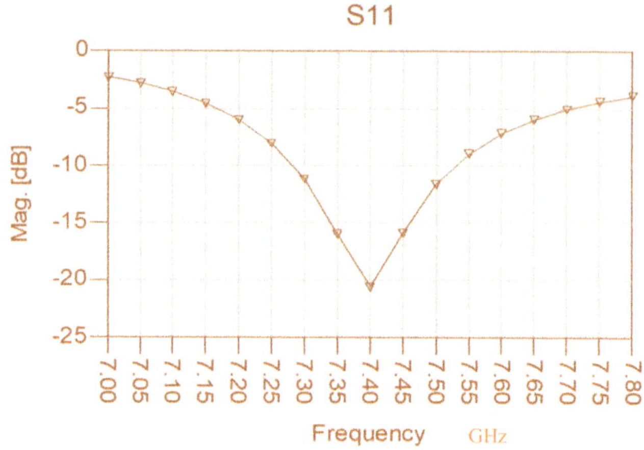

Figure 11.85. Computed S_{11} of the 7.4 GHz modified fractal antenna with 3 mm air spacing.

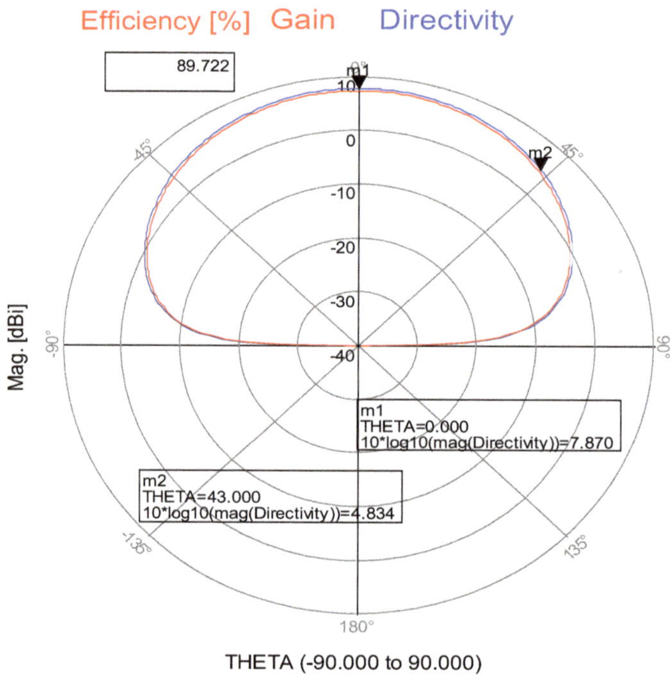

Figure 11.86. Fractal stacked patch antenna radiation pattern with 3 mm air spacing at 7.5 GHz.

chapter. The bandwidth of the antenna with SRR and metallic strips is around 50% for a VSWR better than 2.3:1. Optimization of the number of coupling stubs and the distance between them may be used to tune the antenna's resonant frequency and number of resonant frequencies. The length of the antennas with SRR is smaller by

Figure 11.87. Layout of the modified 7.4 GHz fractal stacked patch antenna.

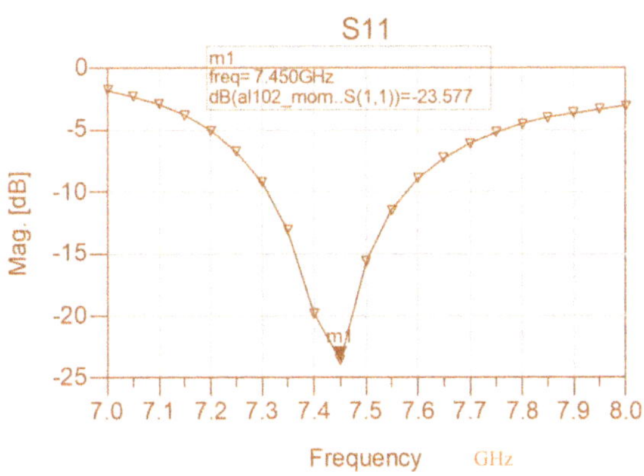

Figure 11.88. Computed S_{11} of the 7.4 GHz modified fractal antenna with 3 mm air spacing.

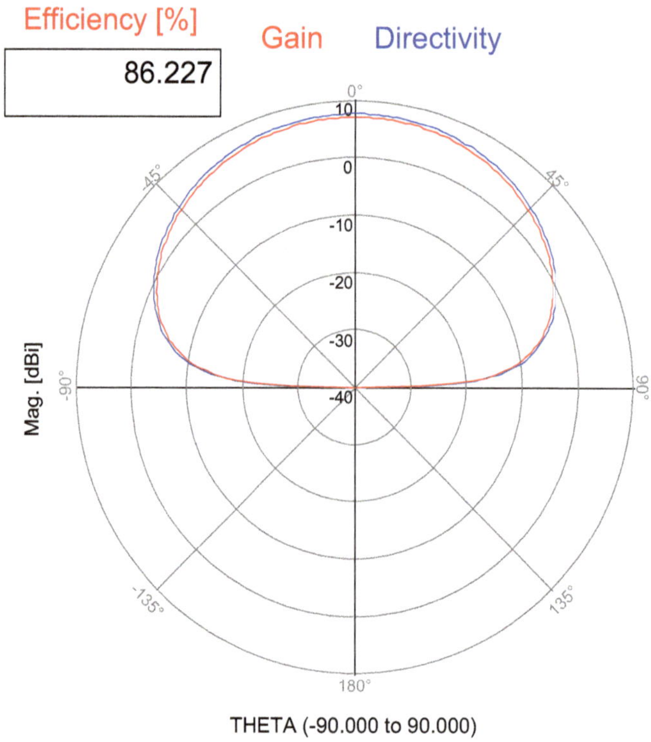

Figure 11.89. Modified fractal patch antenna radiation pattern with 3 mm air spacing at 7.5 GHz.

5% than the antennas without SRR. Moreover, the resonant frequency of the antennas with SRR is lower by 5%–10% than the antennas without SRR. The gain and directivity of the patch antenna with SRR is higher by 2–3 dB than the patch antenna without SRR. The resonant frequency of the antenna with SRR on the human body is shifted by 3%.

Several designs of fractal printed antennas have been presented for communication applications. These fractal antennas are compact and efficient. The space-filling technique and Hilbert curves were employed to design the fractal antennas. The antenna's bandwidth is around 10% with a VSWR better than 3:1. The gain is around 8 dBi with 90% efficiency.

References

[1] James J R, Hall P S and Wood C 1981 *Microstrip Antenna Theory and Design* (London: Peter Peregrinus)

[2] Sabban J A and Gupta K C 1991 Characterization of radiation loss from microstrip discontinuities using a multiport network modeling approach *IEEE Trans. Microwave Theory Tech.* **39** 705–12

[3] Sabban A 1983 A new wideband stacked microstrip antenna *IEEE Antenna Propagation Symp. (Houston, Texas, USA)*

[4] Sabban A 2011 Microstrip antenna arrays *Microstrip Antennas* ed N Nasimuddin (Rijeka: InTech) pp 361–84
[5] Sabban A 2013 New wideband printed antennas for medical applications *IEEE J. Trans. Antennas Propag.* **61** 84–91
[6] Pendry J B, Holden A J, Stewart W J and Youngs I 1996 Extremely low frequency plasmons in metallic mesostructures *Phys. Rev. Lett.* **76** 4773–6
[7] Pendry J B, Holden A J, Robbins D J and Stewart W J 1999 Magnetism from conductors and enhanced nonlinear phenomena *IEEE Trans. Microwave Theory Tech.* **47** 2075–84
[8] Marques R, Mesa F, Martel J and Medina F 2003 Comparative analysis of edge and broadside coupled split ring resonators for metamaterial design. Theory and experiment *IEEE Trans. Antennas Propag.* **51** 2572–81
[9] Marques R, Baena J D, Martel J, Medina F, Falcone F, Sorolla M and Martin F 2003 *Novel small resonant electromagnetic particles for metamaterial and filter design Proc. ICEAA'03 (Torino, Italy) pp 439–42*
[10] Marques R, Martel J, Mesa F and Medina F 2002 Left-handed-media simulation and transmission of EM waves in resonator-subwavelength split-ring-loaded metallic waveguides *Phys. Rev. Lett.* **89** 183901
[11] Baena J D, Marques R, Martel J and Medina F 2003 *Experimental results on metamaterial simulation using SRR-loaded waveguides Proc. IEEE- AP/S Int. Symp. on Antennas and Propagation (Ohio, USA) pp 106–109*
[12] Marques R, Martel J, Mesa F and Medina F 2002 A new 2-D isotropic left- handed metamaterial design: theory and experiment *Microwave Opt. Tech. Lett.* **35** 405–8
[13] Shelby R A, Smith D R, Nemat-Nasser S C and Schultz S 2001 Microwave transmission through a two-dimensional, isotropic, left-handed metamaterial *Appl. Phys. Lett.* **78** 489–91
[14] Zhu J and Eleftheriades G V 2009 A compact transmission-line metamaterial antenna with extended bandwidth *IEEE Antennas Wirel. Propag. Lett.* **8** 295–8
[15] Chirwa L C, Hammond P A, Roy S and Cumming D R S 2003 Electromagnetic radiation from ingested sources in the human intestine between 150 MHz and 1.2 GHz *IEEE Trans. Biomed. Eng.* **50** 484–92
[16] Werber D, Schwentner A and Biebl E M 2006 Investigation of RF transmission properties of human tissues *Adv. Radio Sci.* **4** 357–60
[17] Gupta B, Sankaralingam S and Dhar S 2010 Development of wearable and implantable antennas in the last decade *2010 Mediterranean Microwave Symp. (MMS) (Cyprus) pp 251–67*
[18] Thalmann T, Popovic Z, Notaros B M and Mosig J R 2009 Investigation and design of a multi-band wearable antenna *3rd European Conf. on Antennas and Propagation, EuCAP (Berlin, Germany) pp 462–5*
[19] Salonen P, Rahmat-Samii Y and Kivikoski M 2004 *Wearable antennas in the vicinity of human body IEEE Antennas and Propagation Society Int. Symp., 2004***1** *(Monterey, CA) pp 467–70*
[20] Kellomaki T, Heikkinen J and Kivikoski M 2006 Wearable antennas for FM reception *First European Conf. on Antennas and Propagation, EuCAP 2006 (Nice, France) pp 1–6*
[21] Sabban A 2009 Wideband printed antennas for medical applications *APMC 2009 Conf. (Singapore)*
[22] Alomainy A and Sani A *et al* 2009 Transient characteristics of wearable antennas and radio propagation channels for ultrawideband body-centric wireless communication *IEEE Trans. Antennas Propag.* **57** 875–84

[23] Klemm M and Troester G 2006 Textile UWB antenna for wireless body area networks *IEEE Trans. Antennas Propag.* **54** 3192–7

[24] Izdebski P M, Rajagoplan H and Rahmat-Sami Y 2009 Conformal ingestible capsule antenna: a novel Chandelier meandered design *IEEE Trans. Antennas Propag.* **57** 900–9

[25] Sabban A 2012 Wideband tunable printed antennas for medical applications *IEEE Antenna and Propagation Symp. (Chicago, IL, USA)*

[26] Mandelbrot B B 1983 *The Fractal Geometry of Nature* (NY, USA: Freeman)

[27] Mandelbrot B B 1967 How long is the coast of Britain? Statistical self-similarity and fractional dimension *Science* **156** 636–8

[28] Falkoner F J 1990 *The Geometry of Fractal Sets* (Cambridge: Cambridge Univ. Press)

[29] Balanis C A 1997 *Antenna Theory Analysis and Design* 2nd edn (New York: Wiley)

[30] Virga K and Rahmat-Samii Y 1997 Low-profile enhanced-bandwidth PIFA antennas for wireless communications packaging *IEEE Trans. Microwave Theory Tech.* **45** 1879–88

[31] Skolnik M I 1981 *Introduction to Radar Systems* (London: Mc. Graw Hill)

[32] *Patent US* 5087515, *Patent US* 4976828, *Patent US* 4763127, *Patent US* 4600642, *Patent US* 3952307, *Patent US* 3725927.

[33] 2002 *European Patent Application* EP 1317018 A2/27.11.2002

[34] Chiou T and Wong K 2001 Design of compact microstrip antennas with a slotted ground plane *IEEE-APS Symp. (Boston, USA) 8–12 July*

[35] Hansen R C 1981 Fundamental limitations on antennas *Proc. IEEE* **69** 170–82

[36] Pozar D 1995 *The Analysis and Design of Microstrip Antennas and Arrays* (Piscataway, NJ: IEEE Press)

[37] Zurcher J F and Gardiol F E 1995 *Broadband Patch Antennas* (Boston, USA: Artech House)

[38] Minin I 2010 Fractal antenna applications *Microwave and Millimeter Wave Technologies from Photonic Bandgap Devices to Antenna and Applications* ed M V Rusu and R Baican (Rijeka: InTech) pp 351–82

[39] Rusu M V, Hirvonen M, Rahimi H, Enoksson P, Rusu C, Pesonen N, Vermesan O and Rustad H 2008 Minkowski fractal microstrip antenna for RFID tags *Proc. EuMW2008 Symp. (Amsterdam)*

[40] Rahimi H, Rusu M, Enoksson P, Sandström D and Rusu C 2007 Small patch antenna based on fractal design for wireless sensors *MME07 18th Workshop on Micromachining, Micromechanics, and Microsystems (Portugal) pp 16–18 September*

[41] Sabban A 2015 *Low Visibility Antennas for Communication Systems* (USA: Taylor and Francis)

Chapter 12

Active wearable printed antennas for medical applications

Low profile, compact active and tunable antennas are needed in several communication systems. Printed antennas possess attractive features such as being low profile, flexible, lightweight, with a small volume and low production costs. Microstrip antenna's resonant frequency is altered due to environmental conditions, different antenna locations and system modes of operation. Wearable printed active and tunable antennas are not widely presented in the literature. However, in the last decade, microstrip antennas have been widely presented in books and papers [1–7]. A new class of wideband active and tunable wearable antennas for medical applications is presented in this section. For example, the antenna's VSWR is better than 2:1 at 434 MHz ± 5%. The beamwidth is around 100°, and the gain is around 0–2 dBi. A voltage-controlled varactor is used to control the antenna's resonant frequency at different locations on the human body. Amplifiers may be connected to the wearable antenna feed line to increase the system's dynamic range.

12.1 Tunable printed antennas

Communications and the biomedical industry have continuously grown in the last decade. Low profile, compact tunable antennas are crucial in the development of wearable human biomedical systems. Tunable antennas consist of a radiating element and a voltage-controlled diode, varactor. Varactor diodes are semiconductor devices that are voltage-controlled with variable capacitance. The radiating element may be a microstrip patch antenna, dipole or loop antenna. The antenna's resonant frequency may be tuned by using a varactor to compensate for variations in the antenna's resonant frequency at different locations.

doi:10.1088/978-0-7503-5224-6ch12

12.2 Varactors: theory

Varactor diodes are semiconductor devices that are used in many microwave systems where a voltage-controlled, variable capacitance is required.

PN junction diodes exhibit a variable capacitance effect and PN diodes may be used as a voltage-controlled, variable capacitance. However, special PN diodes are optimized and fabricated to give the required capacitance values. Varactor diodes normally enable much higher ranges of capacitance change to be achieved as a result of the optimized design of PN diodes. Varactor diodes are widely used in RF devices. The circuit capacitance is varied by applying a controlled voltage. Varactor diodes are used in voltage-controlled oscillators (VCOs). They are also used in tunable filters and antennas.

Varactor diode basics

The varactor diode consists of a standard PN junction, see figure 12.1. The diode is operated under reverse bias conditions and this gives rise to three regions and there is no conduction. The left and right ends of the diode are the P and N regions, where the current can be conducted. However, around the junction is the depletion region where no current carriers are available. As a result, current can be carried in the P and N regions, but the depletion region is an insulator. This is similar to a capacitor structure. It has conductive plates separated by an insulating dielectric. The capacitance of a capacitor depends on the plate area, the dielectric constant of the insulator between the plates, and the distance between the two plates. In the case of the varactor diode, it is possible to increase and decrease the width of the depletion region by changing the level of the reverse bias. This has the effect of changing the distance between the plates of the capacitor. However, to be able to use varactor

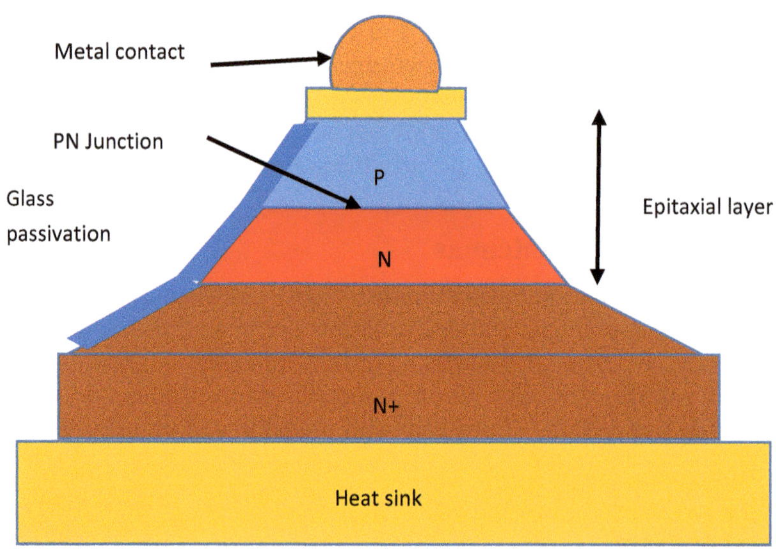

Figure 12.1. Internal structure of a varactor.

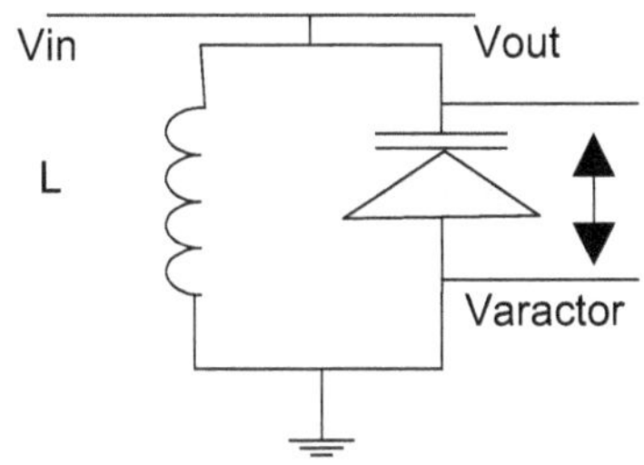

Figure 12.2. Varactor model.

diodes to their best advantage it is necessary to understand the features of varactor diodes including the capacitance ratio, Q, gamma, reverse voltage, etc.

Varactors provide an electrically controllable capacitance, which can be used in tuned circuits. It is small and inexpensive. Its disadvantages compared to a manually controlled variable capacitor are a lower Q, nonlinearity, lower voltage rating and a more limited range. A varactor model is shown in figure 12.2.

Any PN junction has a junction capacitance that is a function of the voltage across the junction. The electric field in the depletion layer that is set up by the ionized donors and acceptors is responsible for the voltage difference that balances the applied voltage. A higher reverse bias widens the depletion layer, uncovering more fixed charge and raising the junction potential. The capacitance of the junction is $C = Q\,(\mathrm{V})/V$, and the *incremental capacitance* is $c = \Delta Q\,(\mathrm{V})/\Delta V$. The capacitance to be used in the formula for the resonant frequency is the incremental capacitance, where it is assumed that the incremental voltage ΔV is small compared to V. Finite voltages give rise to nonlinearities. The capacitance decreases as the reverse bias increases, according to the relation $C = C_o/(1 + V/V_o)^n$, where C_o and V_o are constants. The diode's forward voltage is approximately V_o. The exponent n depends on how the doping density of the semiconductors depends on distance away from the junction. For a graded junction (linear variation), $n = 0.33$. For an abrupt junction (constant doping density), $n = 0.5$. If the density jumps abruptly at the junction, then decreases (called hyperabrupt), n can be made as high as $n = 2$. The varactor capacitance is given in equation (12.1). The circuit frequency, f_r, may be calculated by using equation (12.2).

$$C = \frac{A\varepsilon}{d} \tag{12.1}$$

where C is the capacitance, A is the plate area, and d is the diode thickness.

$$f_r = \frac{1}{2\pi\sqrt{LC}} \tag{12.2}$$

Types of varactors

Abrupt and hyperabrupt type: When the changeover p–n junction is abrupt then it is called an abrupt varactor. When the change is very abrupt, it is called a hyperabrupt varactor. Varactors are used in oscillators to sweep for different frequencies.

Gallium arsenide varactor diodes: The semiconductor material used is gallium arsenide. These diodes are used for frequencies from 18 GHz up to and beyond 600 GHz.

12.3 Dually polarized tunable printed antenna

A compact tunable microstrip dipole antenna has been designed to provide horizontal polarization. The antenna consists of two layers. The first layer consists of RO3035 0.8 mm dielectric substrate. The second layer consists of RT-Duroid 5880 0.8 mm dielectric substrate. The substrate thickness affects the antenna's bandwidth. The printed slot antenna provides a vertical polarization. The printed dipole and the slot antenna provide dual orthogonal polarizations. The dimensions of the dual polarized antenna are 26 × 6 × 0.16 cm. Tunable compact folded dual polarized antennas have also been designed. The dimensions of the compact antennas are 5 × 5 × 0.05 cm. Varactors are connected to the antenna feed lines as shown in figure 12.3. The voltage-controlled varactors are used to control the antenna's resonant frequency. The varactor bias voltage may be varied automatically to set the antenna's resonant frequency at different locations on the human body. The antenna may be used as a wearable antenna on a human body. It may be attached to a patient's shirt in the stomach or back area. The antenna has been analyzed using Agilent ADS software. There is good agreement between the measured and computed results. The antenna's bandwidth is around 10% for a VSWR better than 2:1. The beamwidth is around 100°, and the gain is around 2 dBi.

Figure 12.4 presents the measured S_{11} parameters without a varactor. Figure 12.5 presents the antenna's S_{11} parameters as a function of different varactor capacitances. Figure 12.6 presents the tunable antenna's resonant frequency as a function of the varactor capacitance. The antenna's resonant frequency varies around 5% for capacitances up to 2.5 pF. The beamwidth is 100°. The antenna's cross-polarized field strength may be adjusted by varying the slot feed location.

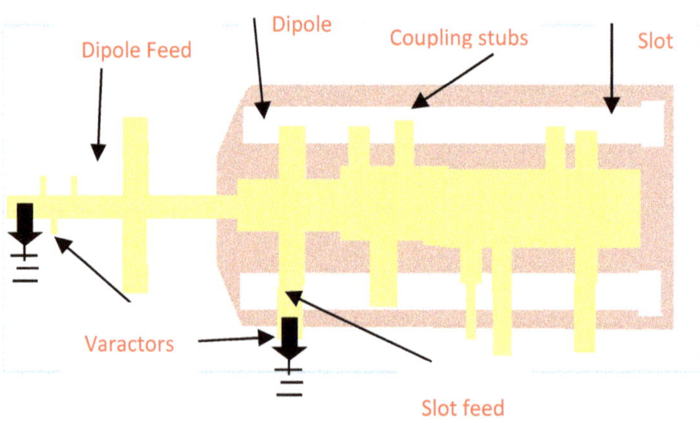

Figure 12.3. Dual polarized tunable antenna, 26 × 6 × 0.16 cm.

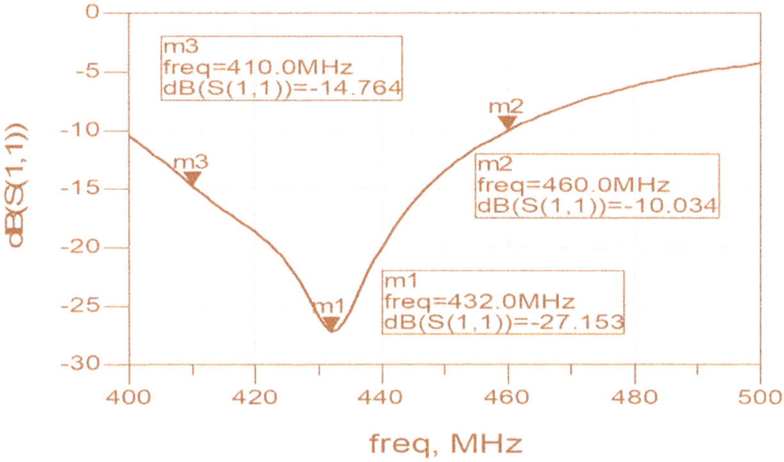

Figure 12.4. Measured S_{11} on the human body.

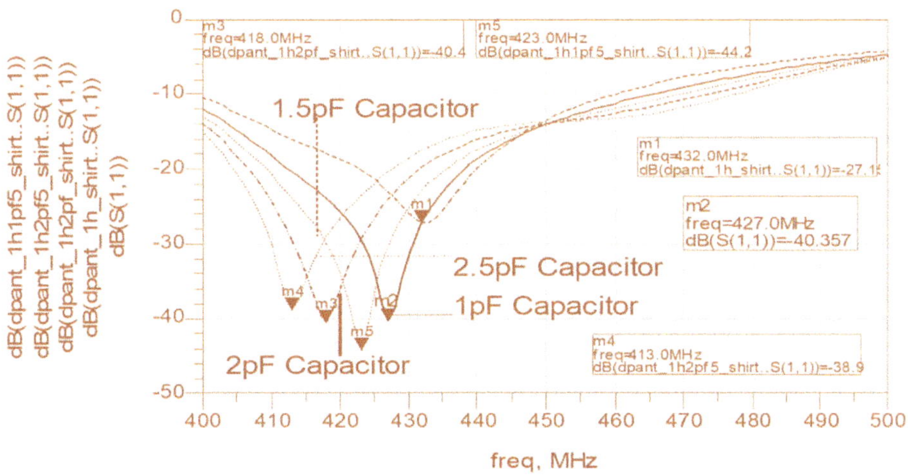

Figure 12.5. The tunable S_{11} parameter as a function of varactor capacitance.

12.4 Wearable tunable antennas

As presented in chapter 6, the antennas' input impedance varies as a function of distance from the body. The properties of human body tissues are listed in tables 9.1 and 11.1 see [8, 9]. Wearable antennas have been presented in books and papers in the last decade [10–17]. The electrical performance of wearable antennas was computed using ADS software [18]. The analyzed structure is presented in figure 12.7. Wearable tunable antennas are not widely presented Therefore, in this section several wearable tunable antennas are discussed. Figure 12.8 presents S_{11} results for different belt and shirt thickness, and air spacing between the antennas and the

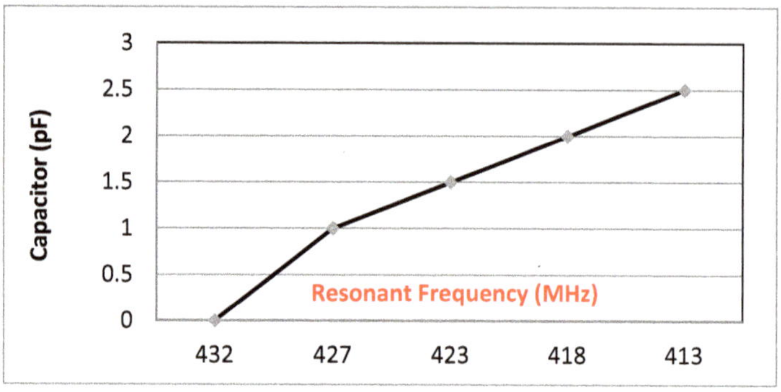

Figure 12.6. Resonant frequency as a function of varactor capacitance.

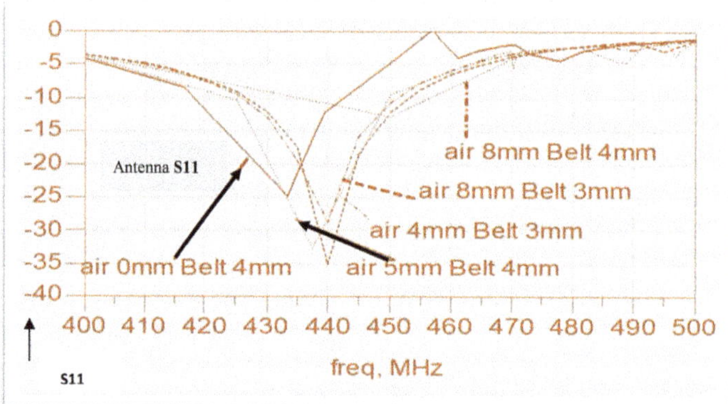

Figure 12.7. S_{11} of the antenna for different spacing relative to the human body.

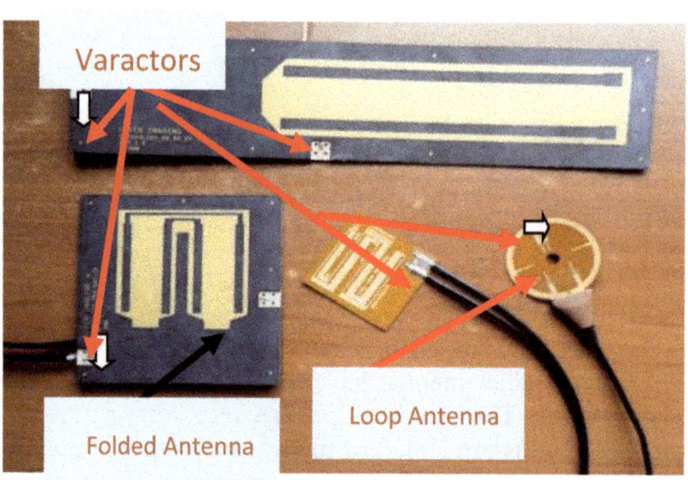

Figure 12.8. Tunable antennas for medical applications.

human body. When the antenna's resonant frequency is shifted, the voltage across the varactor is varied to tune the antenna's resonant frequency.

If the air spacing between the sensors and the human body is increased from 0 mm to 5 mm the antenna's resonant frequency is shifted by 5%. A voltage-controlled varactor is used to tune the antenna's resonant frequency due to different antenna locations on a human body. Figure 12.8 presents several compact tunable antennas for medical applications. A voltage-controlled varactor may be also used to tune the loop antenna's resonant frequency at different antenna locations on the body.

12.5 Varactors: electrical characteristics

Tuning varactors are voltage variable capacitors designed to provide electronic tuning of microwave components. Varactors are manufactured on silicon and gallium arsenide substrates. Gallium arsenide varactors offer higher Q and may be used at higher frequencies than silicon varactors. Hyperabrupt varactors provide nearly linear variation of frequency with applied control voltage. However, abrupt varactors provide inverse fourth root frequency dependence. MACOM offers several gallium arsenide hyperabrupt varactors such as the MA46 series. Figure 12.9 presents the C–V curves of varactors MA46505 to MA46506.

Figure 12.10 presents the C–V curves of varactors MA46H070 to MA46H074.

12.6 Measurements of wearable tunable antennas

Figure 12.11 presents a compact tunable antenna with a varactor. A varactor was connected to the antenna's feed line. The varactor bias voltage was varied from 0 V to 9 V. Figure 12.12 presents measured S_{11} as a function of varactor bias voltage. The antenna's resonant frequency was shifted by 5% for a bias voltage between 7 V to 9 V. We may conclude that varactors may be used to compensate for

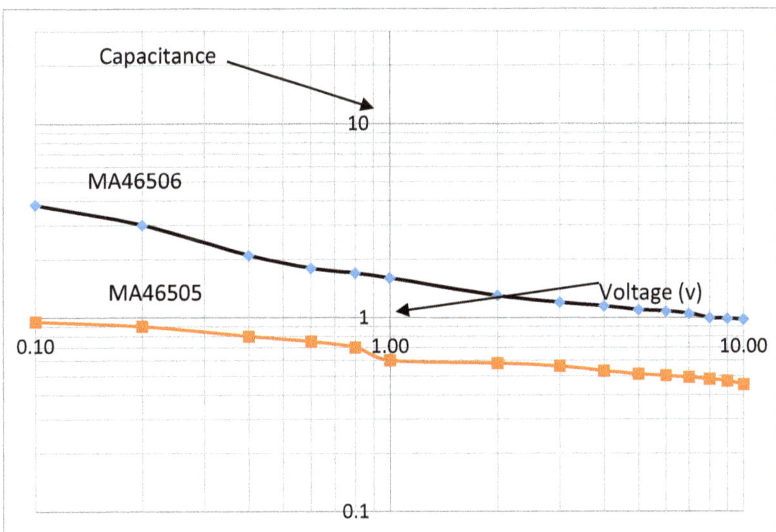

Figure 12.9. Varactor capacitance as a function of bias voltage.

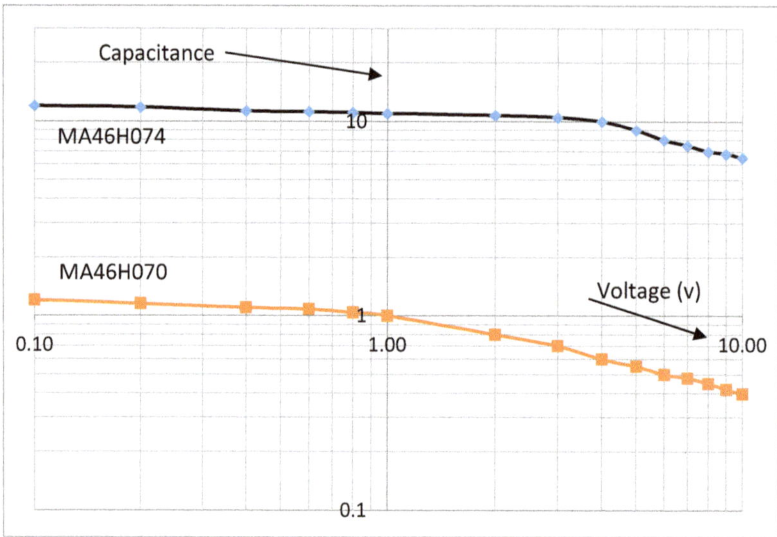

Figure 12.10. *C–V* curves of varactors MA46H070 to MA46H074.

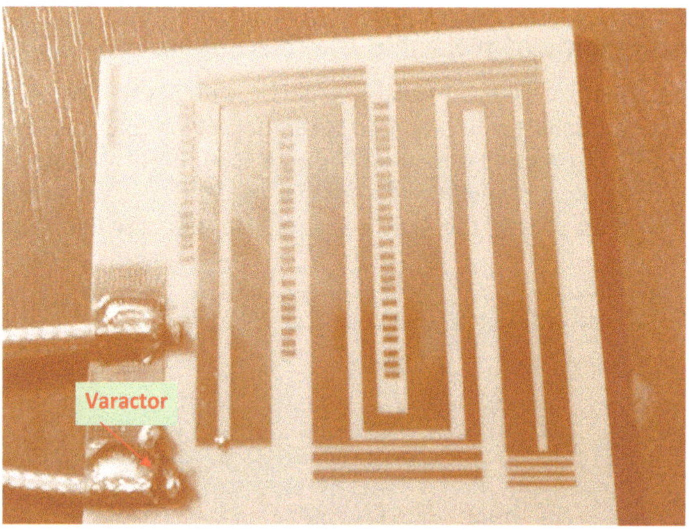

Figure 12.11. Tunable antenna with a varactor.

variations in the antenna's resonant frequency at different locations on the human body.

12.7 Folded wearable dual polarized tunable antenna

The dimensions of the folded dual polarized antenna presented in figure 12.13 are 7 × 5 × 0.16 cm. The length and width of the coupling stubs in figure 12.13 are

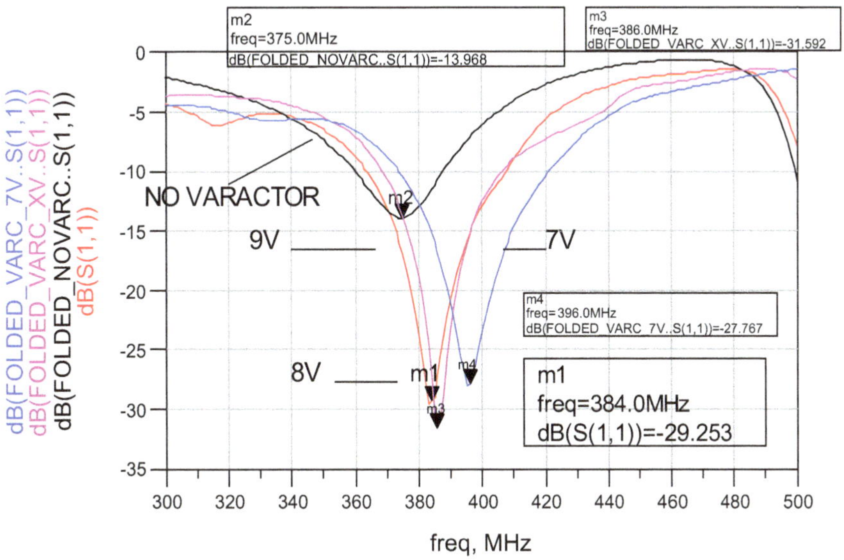

Figure 12.12. Measured S_{11} as a function of varactor bias voltage.

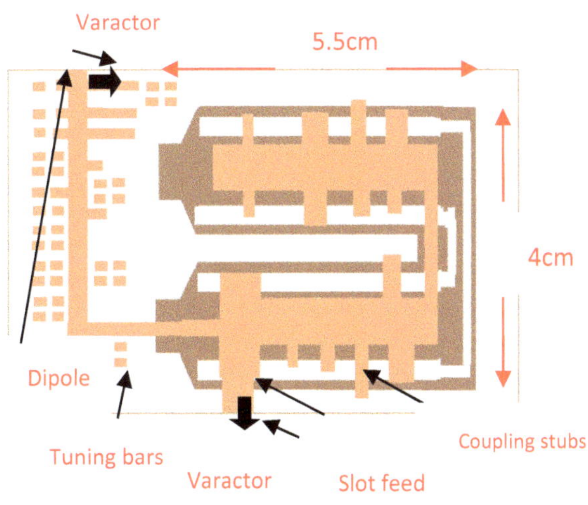

Figure 12.13. Tunable folded dual polarized antenna.

12 mm by 9 mm. Small tuning bars are located along the feed line to tune the antenna to the desired resonant frequency.

Figure 12.14 presents the antenna's computed S_{11} and S_{22} parameters. The computed radiation pattern of the folded dipole is shown in figure 12.15.

12.8 Medical applications for wearable tunable antennas

Three to four tunable folded dipole or tunable loop antennas may be assembled in a belt and attached to a patient's stomach or back, as shown in figure 12.16. The bias

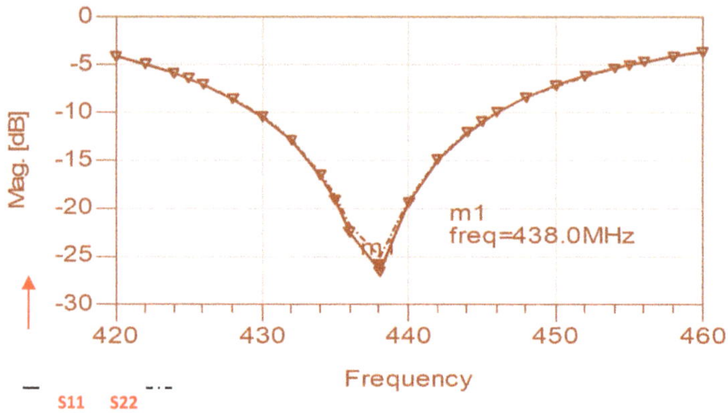

Figure 12.14. Folded antenna's computed S_{11} and S_{22} results.

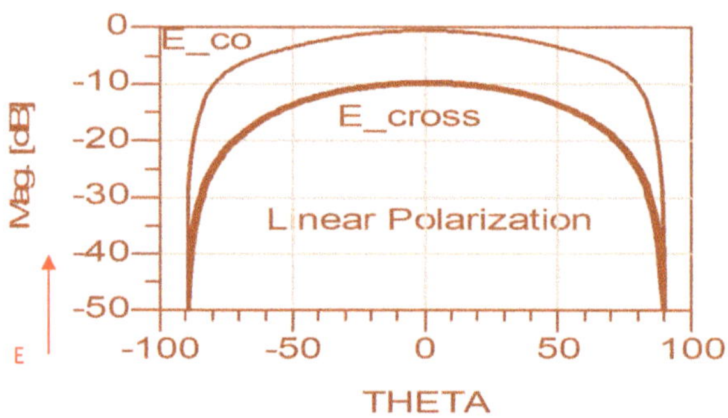

Figure 12.15. Folded antenna's radiation pattern.

voltage to the varactors is supplied by a recorder battery. The RF and DC cables from each antenna are connected to a recorder. The received signal is routed to a switching matrix. The signal with the highest level is selected during the medical test. The varactors' bias voltage may be varied to tune the antenna's resonant frequency. The antenna's receive a signal that is transmitted from various positions in the human body. A tunable antenna may be attached to a patient's back in order to improve the level of the received signal from different locations in the human body. In several applications, the distance separating the transmitting and receiving antennas is less than the far-field distance, $2D^2/\lambda$, where D is the largest dimension of the source of the radiation. λ is the wavelength. In these applications the amplitude of the electromagnetic field close to the antenna falls-off rapidly with the distance from the antenna. The electromagnetic fields do not radiate energy to infinite distances, but instead their energies remain trapped in the antenna's near zone. The near-fields transfer energy only to close distances from the receivers. In these applications we have to refer to the near-field and not to the far-field radiation. The receiving and transmitting antennas are magnetically coupled. Change in current flow through one wire induces a voltage across the ends of the other wire through electromagnetic induction. The proposed tunable wearable antennas may

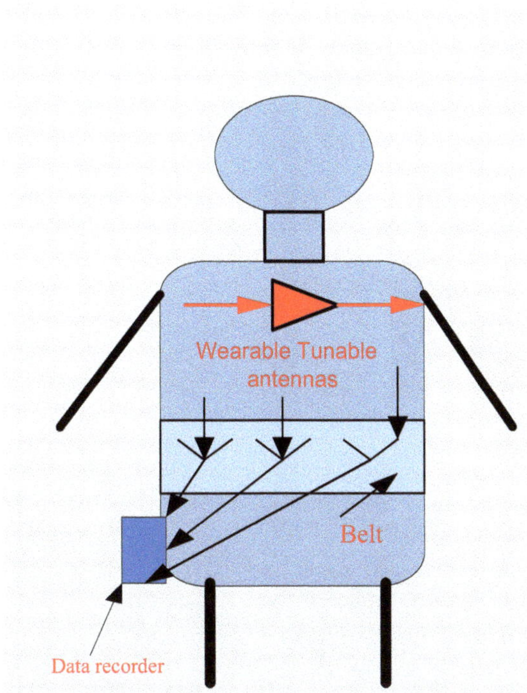

Figure 12.16. Tunable wearable antenna.

be placed on a patient's body as shown in figure 12.17(a). The patient in figure 12.17(b) is wearing a wearable antenna. The antenna's belt is attached to the patient's front or back. Figure 12.18 presents several compact tunable antennas for medical applications. A voltage-controlled varactor may be also used to tune the wearable antenna's resonant frequency at different antenna locations on the body.

12.9 Active wearable antennas

Active wearable antennas may be used in receiving or transmitting channels. In transmitting channels a power amplifier is connected to the antenna. In receiving channels a low noise amplifier is connected to the receiving antenna.

12.9.1 Basic concept of the active antenna

Active antennas (AAs) are devices combining a radiating element with active components. The radiating element is designed to provide the optimal load to the active elements. The integration of the antenna and the active components drastically reduces the complexity of the matching network. In the last decade, active antennas have been employed in wireless and medical communication systems, [19–27]. The current major applications of active antennas are large electronically scanned arrays, phased arrays. Indeed, arrays of active antennas, or active arrays, are well suited for mobile terminals requiring dynamic satellite tracking. The most common approach toward achieving fast-beam scanning is

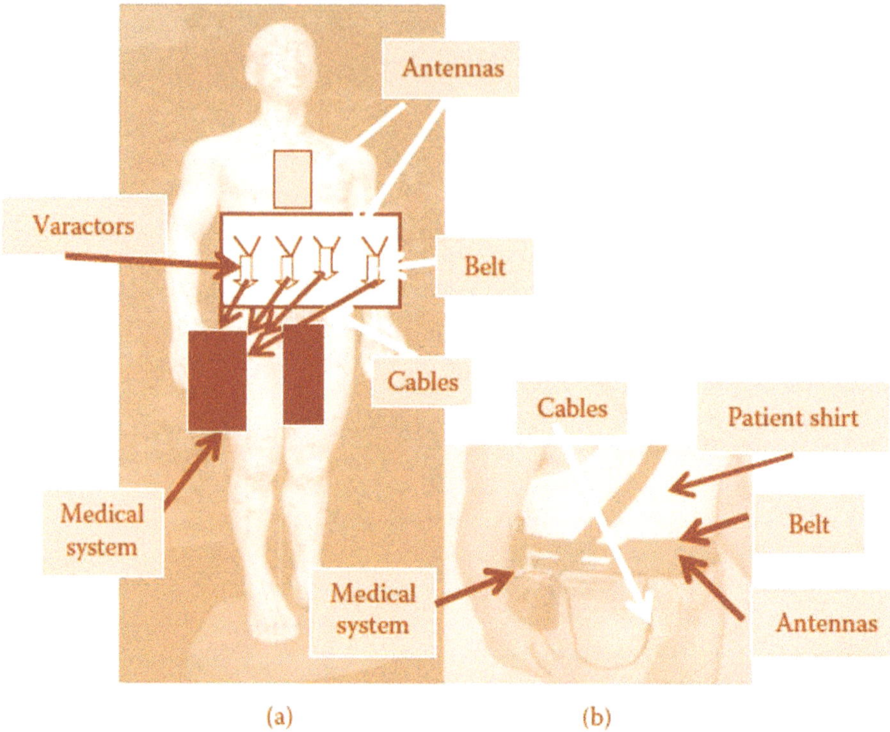

Figure 12.17. (a) Medical system with printed wearable antennas. (b) Patient with a printed wearable antenna.

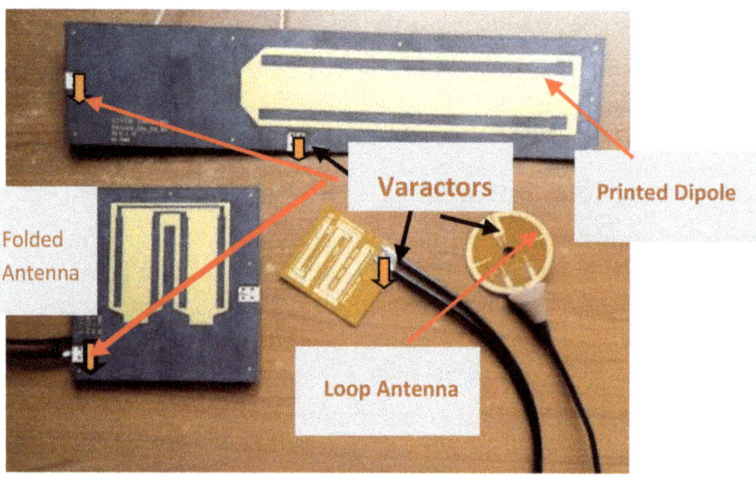

Figure 12.18. Tunable antennas for medical applications.

through the integration of monolithic microwave integrated circuit (MMIC) phase shifters, low noise amplifiers, and solid state power amplifiers with antenna elements. In some cases, hybrid electro/mechanical arrays combining mechanical steering with electrical steering/shaping are considered. This architecture is often used to reduce the number of active control elements by limiting the electrical scanning in only one plane. This is often the case for mobile user terminals where azimuth scanning is performed by mechanical rotation and elevation agility is realized by a linear phased array. In the last decade, printed transmission lines and antennas replaced coaxial transmission lines and metallic radiators in phased array systems. Developments in MMIC technology and other fabrication processes enabled an automated low-cost production process of phased arrays with a high integration level.

Phased arrays emerged as a new promising technology for radar and communication systems around 1970. Phased arrays replace mechanically scanned antennas. Phased arrays are much faster for beam switching than mechanically scanned antennas. Phased arrays significantly reduce the size, weight and power associated with a gimbal. Early phased array antennas were passive antennas. The front end of the antenna was composed of array elements with phase shifters. A passive manifold was employed for RF combining to form a single beam. At the output of the manifold was a switch with an LNA (receiving channel) and power amplifier (transmitting channel). Solid state electronic device capability was not developed enough to include active amplifiers at the front end of the array for each array element. With the LNA and amplifier behind the manifold the amount of RF loss was quite large, causing inefficiency of transmission (power aperture) and limitations on the sensitivity of the receiver. With the great progress of GaAs MMIC technology in the last twenty years, solid state device dimensions have been minimized to the size of the array elements, enabling distributed phased arrays architectures. High power amplifiers and LNAs could be placed close to the front end and connected to each radiating element. This resulted in drastic power efficiency improvement and much higher receiver sensitivity, since the only loss before the first LNA was the radiating element and a radome. The amplifiers were packaged in transmit/receive (T/R) modules with a phase shifter and attenuator. This phased array architecture was called 'active' phased arrays.

12.9.2 Active wearable receiving loop antenna

Figure 12.19 presents a basic receiver block diagram with an active antenna.

In figure 12.19 the low noise amplifier (LNA) is an integral part of the antenna.

An E PHEMT LNA was connected to a loop antenna. The active antenna layout is shown in figure 12.20.

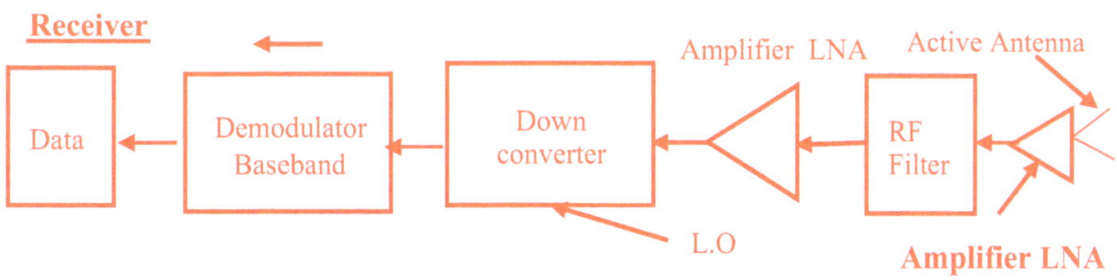

Figure 12.19. Receiver block diagram with an active receiving antenna.

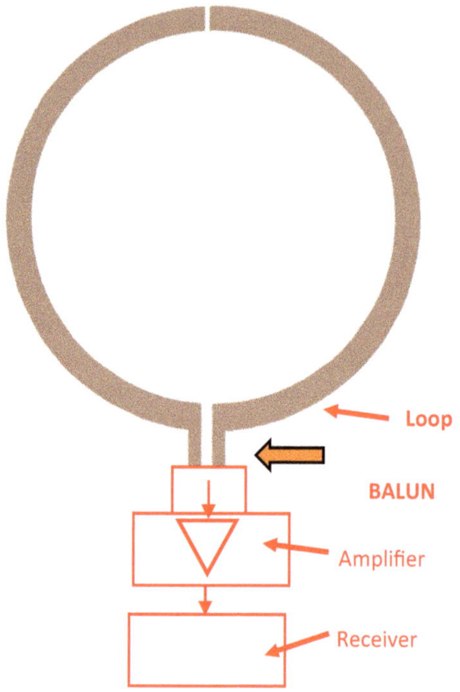

Figure 12.20. Active printed loop antenna layout.

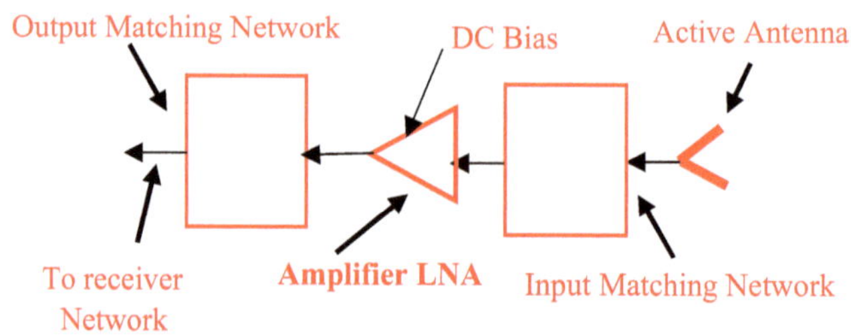

Figure 12.21. Receiving active antenna block diagram.

A receiving active antenna block diagram is presented in figure 12.21. The radiating element is connected to the LNA via an input matching network. An output matching network connects the amplifier port to the receiver. A DC bias network supplies the required voltages to the amplifiers. The amplifier specification is listed in table 12.1. The amplifier's complex S parameters are listed in table 12.2, and the noise parameters are listed in table 12.3. The loop antenna's S_{11} parameters on human body are presented in figure 12.22. A textile sleeve covers the loop antenna to match the loop to the antenna's environment. The radiating loop antenna and the textile sleeve are attached to the human body.

Table 12.1. LNA amplifier specification.

Parameter	Specification	Remarks
Frequency range	0.4–3 GHz	
Gain	26 dB @0.4 GHz	V_{ds} = 3 V; I_{ds} = 60 mA
	18 dB @2 GHz	
NF	0.4 dB @0.4 GHz	V_{ds} =3 V; I_{ds} = 60 mA
	0.5 dB @2 GHz	
P1dB	18.9 dBm @0.4 GHz	V_{ds} = 3 V; I_{ds} = 60 mA
	19.1 dBm @2 GHz	
OIP3	32.1 dBm @0.4 GHz	V_{ds} = 3 V; I_{ds} = 60 mA
	33.6 dBm @2 GHz	
Max. input power	17 dBm	
V_{gs}	0.48 V	V_{ds} =3 V; I_{ds} = 60 mA
V_{ds}	3 V	
I_{ds}	60 mA	
Supply voltage	±5 V	
Package	Surface mount	
Operating temperature	−40 °C to 80 °C	
Storage temperature	−50 °C to 8100 °C	

Table 12.2. LNA amplifier's *S* parameters.

F GHz	S_{11}	S_{11}°	S_{21}	S_{21}°	S_{12}	S_{12}°	S_{22}	S_{22}°
0.10	0.986	−17.17	25.43	168.9	0.008	88.22	0.55	−14.38
0.19	−31.76	0.964	24.13	158.9	0.016	74.88	0.54	−22.98
0.279	0.93	−45.77	22.97	149.5	0.021	65.77	0.51	−33.65
0.323	0.92	−53.39	22.45	145.3	0.026	62.38	0.49	−39.2
0.413	0.89	−65.72	20.98	137.27	0.03	57.9	0.46	−49.3
0.50	0.87	−77.1	19.54	130.3	0.034	53.03	0.43	−57.5
0.59	0.83	−87.12	18.08	124.14	0.038	48.18	0.40	−64.12
0.726	0.8	−100.8	16.22	115.7	0.042	42.06	0.36	−74.86
0.816	0.77	−108.8	15.07	110.75	0.044	39.53	0.34	−80.87
1.04	0.74	−126.2	12.74	100.13	0.049	33.69	0.29	−94.96
1.21	0.71	−137.6	11.25	92.91	0.051	30.05	0.26	−104
1.53	0.687	−154.2	9.29	82.06	0.055	26.08	0.22	−119
1.75	0.67	−164.1	8.24	75.31	0.058	23.14	0.20	−128.4
2.02	0.67	−174.6	7.27	67.82	0.06	20.88	0.18	−138.8

The antenna's bandwidth is around 20% for a VSWR better than 3:1. The active loop antenna's S_{21} parameters and gain, on the human body, are presented in figure 12.23. The active antenna's gain is 25 ± 2.5 dB for frequencies ranging from 350

Table 12.3. Noise parameters.

F GHz	NFMIN	N11X	N11Y	*rn*
0.5	0.079	0.3284	24.56	0.056
0.7	0.112	0.334	36.08	0.05
0.9	0.144	0.3396	47.4	0.045
1	0.16	0.3424	52.98	0.042
1.9	0.306	0.3682	100.93	0.029
2	0.322	0.3711	106.01	0.029
2.4	0.387	0.3829	125.79	0.029
3	0.484	0.401	153.93	0.036
3.9	0.629	0.429	−167.3	0.059
5	0.808	0.4645	−125.53	0.11
5.8	0.937	0.4912	−99.03	0.162
6	0.969	0.498	−92.92	0.177

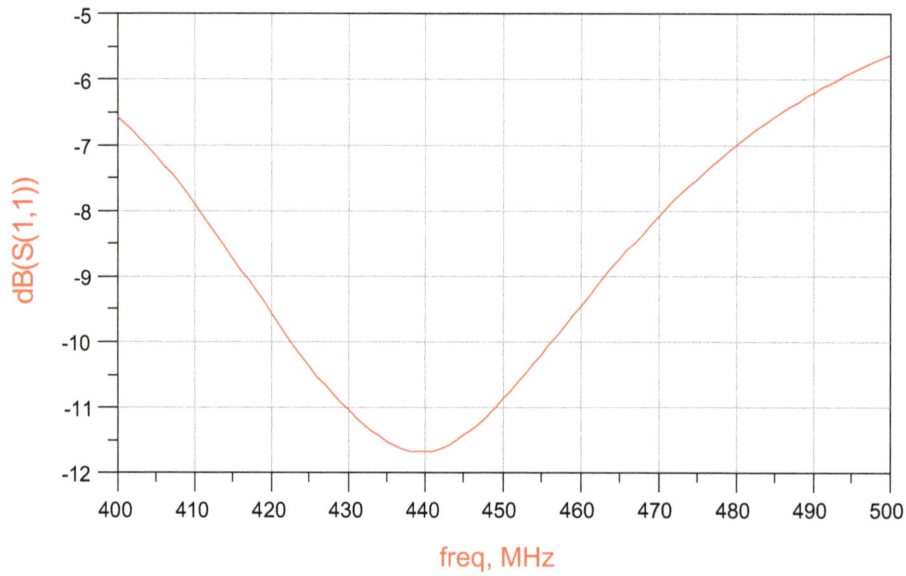

Figure 12.22. Loop antenna's S_{11} parameters on the human body.

MHz to 580 MHz. The active loop antenna's noise figure is presented in figure 12.24. It is 0.7 ± 0.2 dB for frequencies ranging from 400 MHz to 900 MHz.

12.9.3 Compact dual polarized receiving active antenna

A printed compact dual polarized antenna is shown in figure 12.25. The compact antenna consists of two layers. The first layer consists of FR4 0.25 mm dielectric substrate. The second layer consists of Kapton 0.25 mm dielectric substrate. In figure 12.26 the LNA, presented in section 12.9.2, is an integral part of the dual

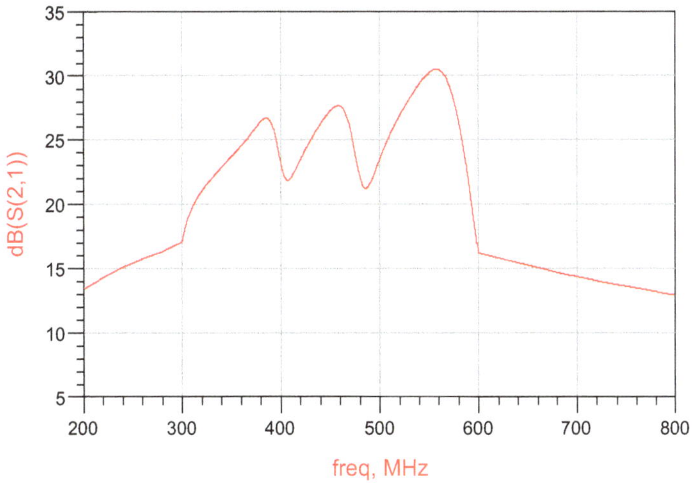

Figure 12.23. Active loop antenna's S_{21} parameters, and gain, on the human body.

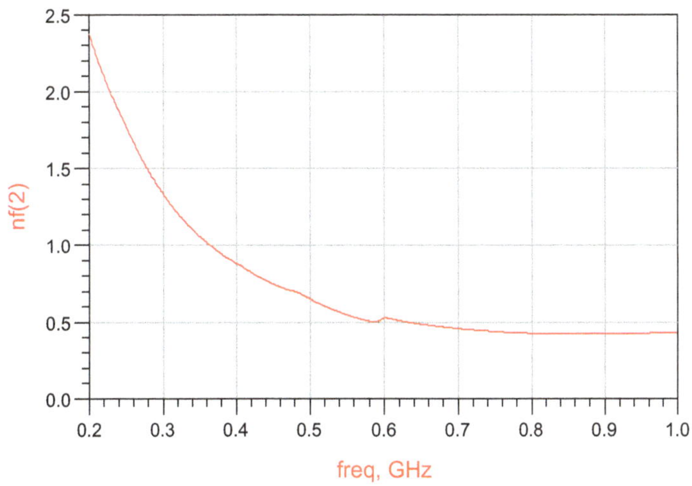

Figure 12.24. Active loop antenna's noise figure.

polarized receiving antenna. The active dual polarized antenna's S_{21} parameters, and gain, on the human body is presented in figure 12.27. The active antenna's gain is 25 ± 3 dB for frequencies ranging from 400 MHz to 650 MHz. There is a good match between the gain of the vertical and horizontal antennas. The gain difference between the gain of the vertical and horizontal antenna is around ±0.5 dB. The active dual polarized antenna noise figure is presented in figure 12.28. It is 0.8 ± 0.4 dB for frequencies ranging from 400 MHz to 900 MHz.

12.10 Active transmitting antenna

Figure 12.29 presents a basic transmitter block diagram with an active antenna. The high power amplifier (HPA) is an integral part of the antenna.

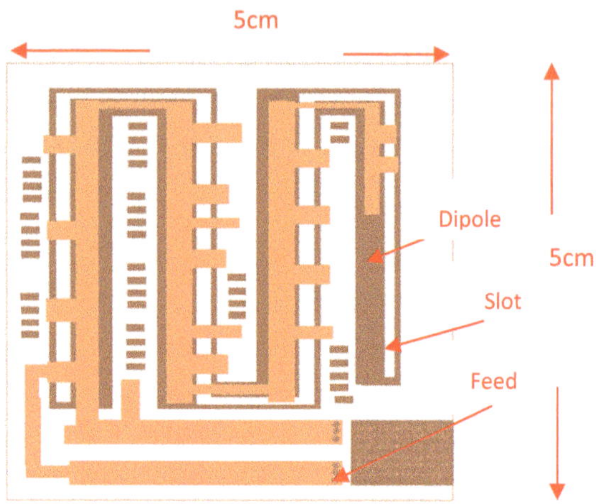

Figure 12.25. Printed compact dual polarized antenna.

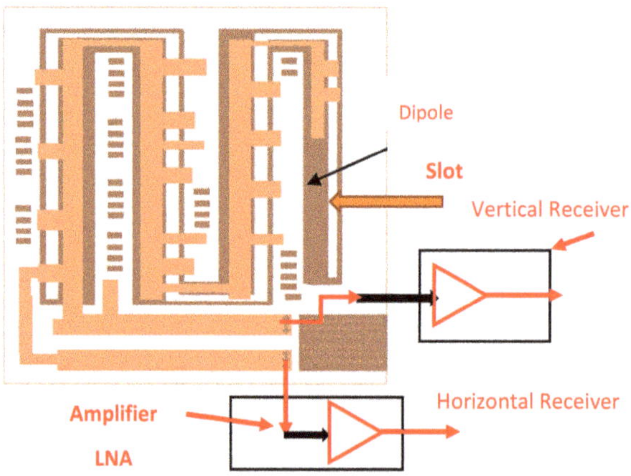

Figure 12.26. Active printed compact dual polarized receiving antenna layout.

12.10.1 Compact dual polarized active transmitting antenna

Printed compact dual polarized transmitting antenna is shown in figure 12.30. The dimensions are $5 \times 5 \times 0.05$ cm.

The active transmitting dual polarized antenna layout is shown in figure 12.31. In figure 12.32 the HPA is an integral part of the antenna and is connected to the transmitting antenna. The HPA is a MMIC GaAs MESFET. A transmitting active antenna block diagram is presented in figure 12.32. The radiating element is connected to the HPA via an output HPA matching network. An HPA input matching network connects the amplifier port to the transmitter.

The amplifier's specification is listed in table 12.4. The amplifier's pin description is given and presented in figure 12.33.

The HPA complex S parameters are listed in table 12.5.

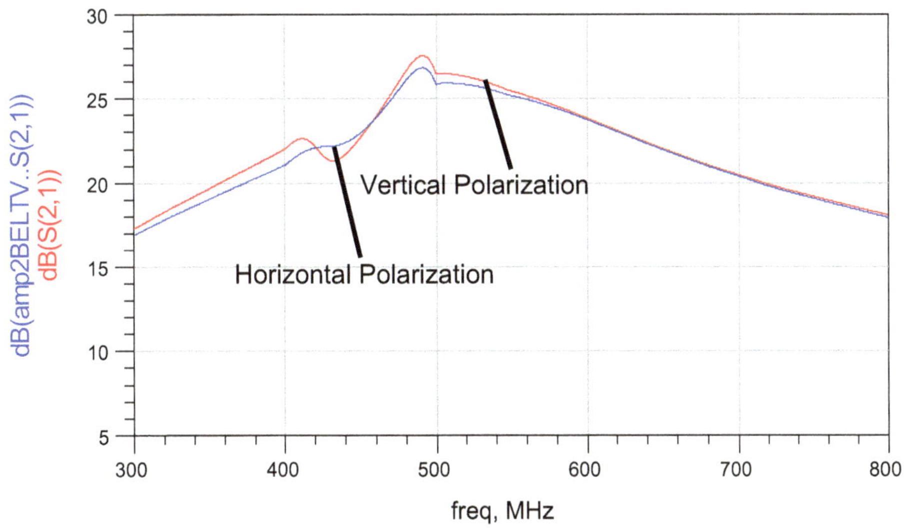

Figure 12.27. Active dual polarized antenna's S_{21} parameters, and gain, on the human body.

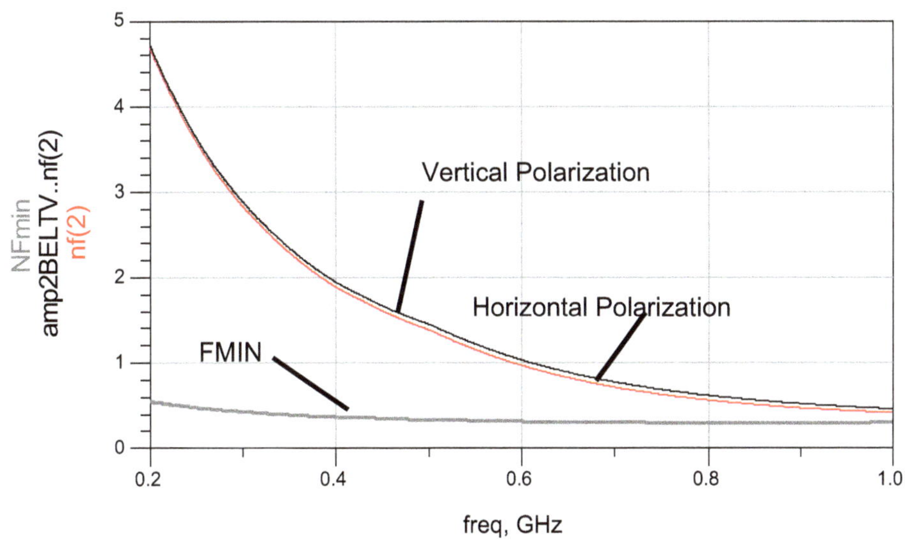

Figure 12.28. Active dual polarize antenna's noise figure.

The active transmitting dual polarized antenna's S_{11} parameters on the human body is presented in figure 12.34. The active transmitting dual polarized antenna's S_{21} parameters, and gain, on the human body is presented in figure 12.35. The active dual polarized antenna's gain is 14 ± 3 dB for frequencies ranging from 380 MHz to 600 MHz. The active transmitting dual polarized antenna's output power is around 18 dBm.

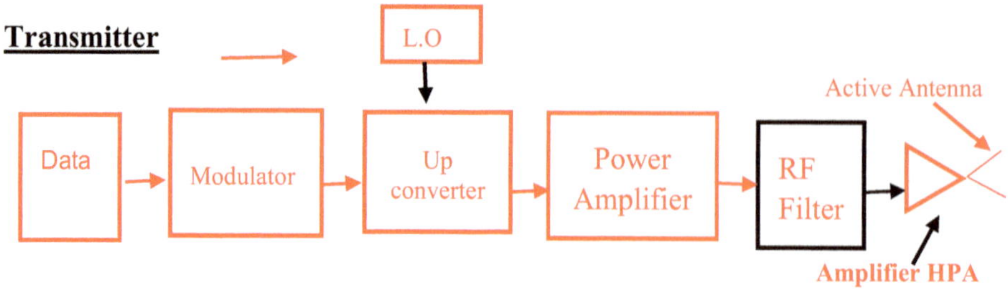

Figure 12.29. Transmitter block diagram with an active transmitting antenna.

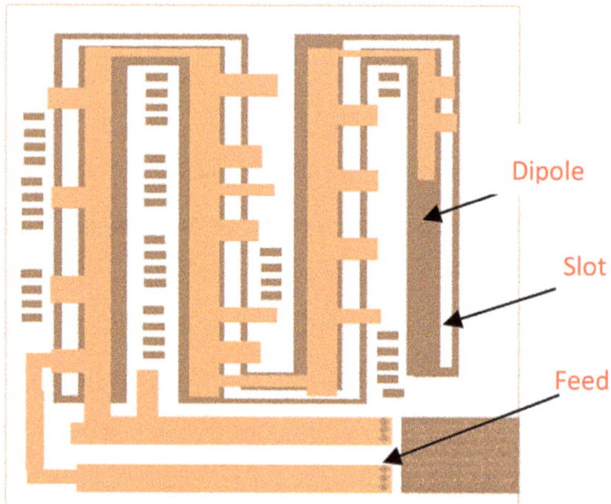

Figure 12.30. Transmitting printed compact dual polarized antenna.

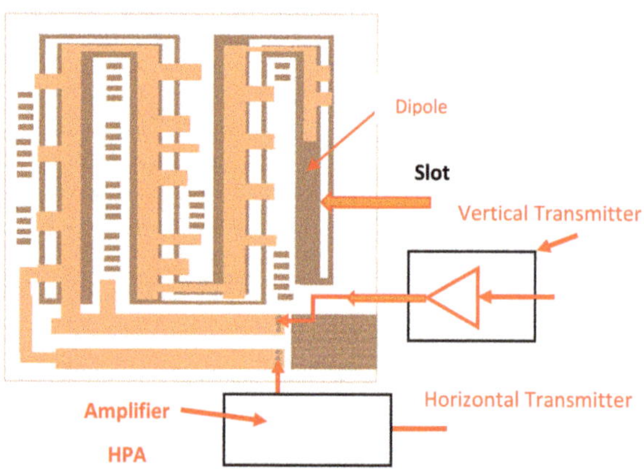

Figure 12.31. Transmitting active printed dual polarized antenna layout.

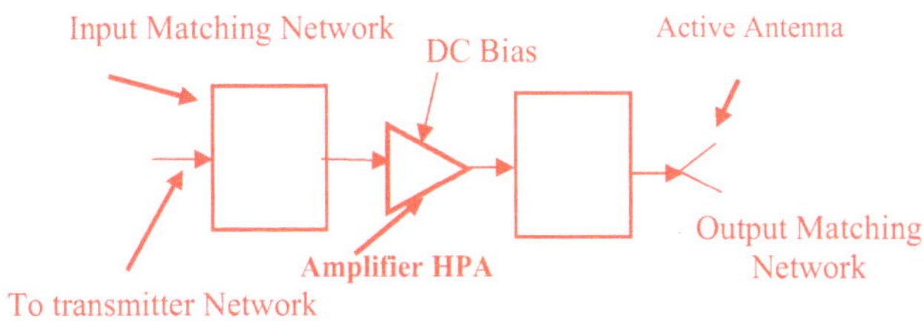

Figure 12.32. Transmitting active antenna block diagram.

Table 12.4. HPA amplifier's specification.

Parameter	Specification	Remarks
Frequency range	0.4–2.5 GHz	
Gain	15 dB @0.4 GHz	$V_{ds} = 5$ V; $I_{ds} = 85$ mA
	17.8 dB @2 GHz	
N.F	5.5 dB @0.4 GHz	$V_{ds} = 5$ V; $I_{ds} = 85$ mA
	5.5 dB @2 GHz	
P1dB	18.0 dBm @0.4 GHz	$V_{ds} = 5$ V; $I_{ds} = 85$ mA
	18.0 dBm @2 GHz	
OIP3	29 dBm @0.4 GHz	$V_{ds} = 5$ V; $I_{ds} = 85$ mA
	29 dBm @2 GHz	
Max. Input power	10 dBm	
V_{gs}	0.48 V	$V_{ds} = 5$ V; $I_{ds} = 85$ mA
V_{ds}	5 V	
I_{ds}	85 mA	
Supply voltage	±5 V	
Package	Surface Mount	
Operating temperature	−40 °C to 80 °C	
Storage temperature	−50 °C to 100 °C	

Pin	Function
1	DC bias
3	RF in
6	RF out
2,4,5,7,8	Ground. Via Holes.

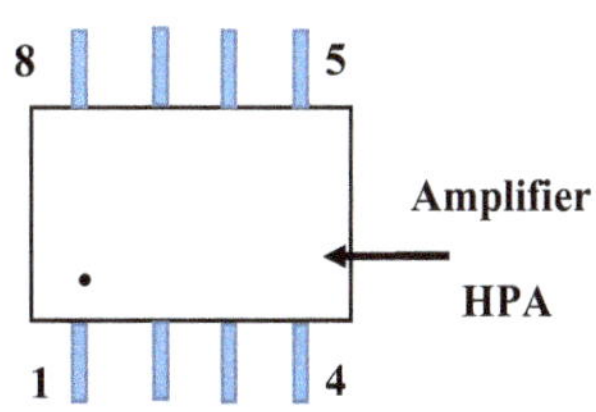

Figure 12.33. Amplifier pin description.

12.10.2 Active transmitting loop antenna

The printed loop antenna is shown in figure 12.36. The dimensions are 5 × 5 × 0.05 cm. In figure 12.36 the HPA is an integral part of the antenna. The high power amplifier is a MMIC GaAs MESFET.

Table 12.5. High power amplifier *S* parameters.

F GHz	S_{11} dB	$S_{11}{}^{\circ}$	S_{21} dB	$S_{21}{}^{\circ}$	S_{12} dB	$S_{12}{}^{\circ}$	S_{22} dB	$S_{22}{}^{\circ}$
0.20	0.065	−38.75	−3.09	−139.2	−47.56	157.63	−1.03	−74.66
0.28	−0.14	−60.8	7.46	163.7	−40.45	114	−3.6	−109.3
0.344	−1.1	−77	11.8	118.7	−37.9	78.4	−7.2	−131.6
0.4	−2.2	−88.5	13.8	85.24	−36.9	52.76	−11	−143.5
0.48	−3.7	−101.8	15.35	46.5	−36.6	25.4	−17	−143.1
0.52	−4.44	−107.5	15.8	30.2	−36.7	14.1	−19.4	−132.5
0.56	−5.1	−112.7	16.2	15.3	−36.8	3.64	−20.5	−118.9
0.64	−6.4	−122	16.8	−11.4	−37.2	−12.6	−19.2	−100.3
0.712	−7.4	−100.8	17.13	−32.7	−37.5	−25.4	−17.7	−100.6
0.8	−8.45	−137.6	17.5	−56.8	−38.1	−37.8	−16.3	−108.8
0.88	−9.2	−144.6	17.72	−77.1	−38.5	−49.4	−15.7	−119.7
1.04	−10.4	−158.6	18.1	−115.1	−39.6	−67.5	−15.3	−144.5
1.12	−10.8	−166.2	18.23	−133.3	−40.3	−75.8	−15.35	−157.5
1.24	−11.3	−178.7	18.37	−159.7	−41.3	−86.9	−15.9	−178.7
1.36	−11.8	167.4	18.4	174.4	−42.4	−91.4	−16.5	159.2
1.48	−12.2	151.2	18.4	149	−43.6	−94.9	−17.5	136.8
1.6	−12.8	134.3	18.3	123.3	−44.2	−93.4	−18.9	113.7
1.8	−14.3	101.2	17.9	83	−43	−86.3	−22	69.5
2	−16.5	61.8	17.3	43.5	−40.4	94.6	−27	6.42
2.16	−18.5	22.1	16.8	12.9	−38.	−105.5	−27.8	−70.2
2.28	−19.6	−14.9	16.3	−9.6	−37.2	−116.2	−25.1	−113.2
2.4	−19.4	−53.9	15.7	−31.8	−36	−128	−22.2	−147.2
2.56	−17.7	99.7	15	−60	−34.6	−145.6	−19.3	−179.4
2.7	−15.7	131	14.3	−84.3	−33.8	−160.3	−17.5	158.1
2.86	−13.7	159	13.5	−111.1	−33	−177.7	−16	134.7
3	−12.2	179.1	12.7	−134.1	−32.4	167.4	15.2	116.3

The active transmitting loop antenna's S_{21} parameters, and gain, on the human body is presented in figure 12.37. The active antenna's gain is 14 ± 3 dB for frequencies ranging from 400 MHz to 600 MHz. The active transmitting loop antenna's S_{11} parameters on the human body is presented in figure 12.38. The active transmitting loop antenna's S22 parameters on the human body is presented in figure 12.39. The output power is around 18 dBm.

12.11 Conclusions

This chapter presents wideband active printed antennas with high efficiency for commercial and medical applications. The dimensions may vary from 26 cm by 6 cm by 0.16 cm to 5 cm by 5 cm by 0.05 cm according to the medical system's

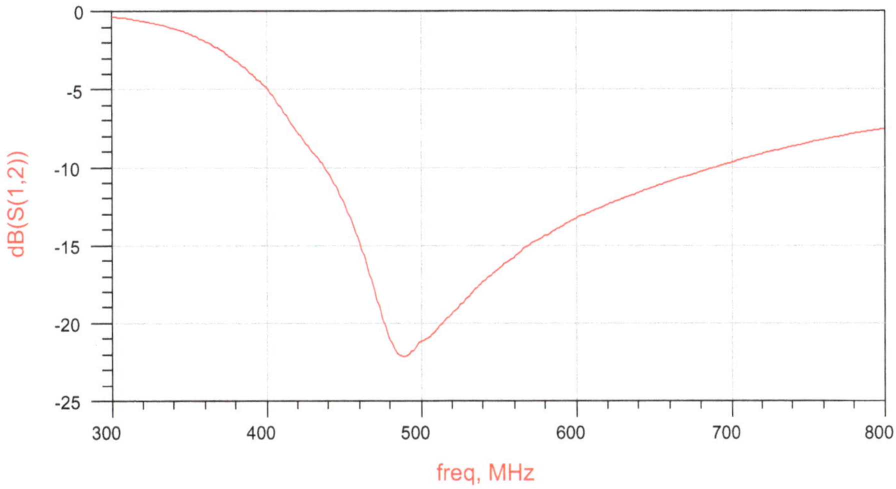

Figure 12.34. Active dual polarized antenna's S_{11} parameters.

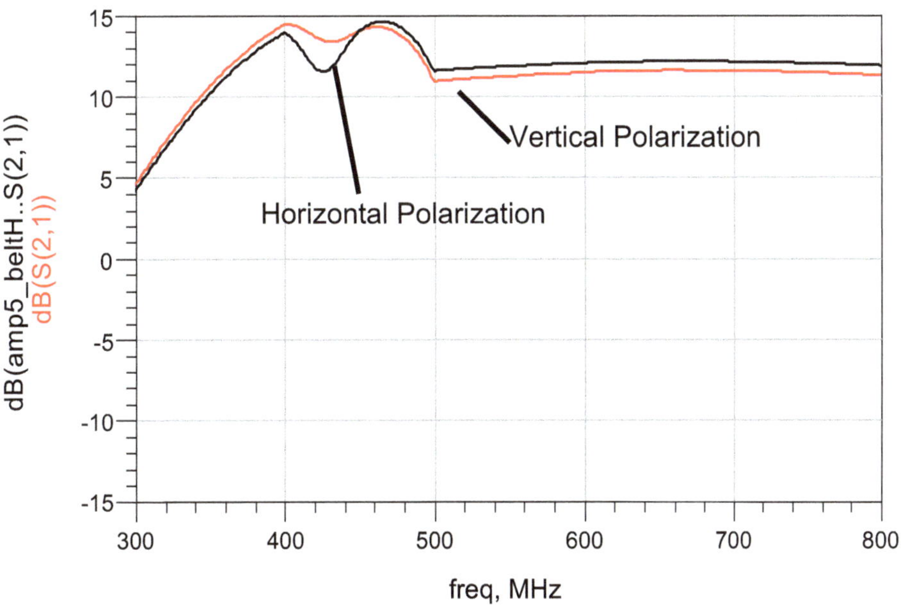

Figure 12.35. Active dual polarized antenna's S_{21} parameters, and gain, on the human body.

specification. The antenna's bandwidth is around 10% for a VSWR better than 2:1. The antenna's beamwidth is around 100°. The tunable antenna's gain varies from 0 to 2 dBi. If the air spacing between the dual polarized antenna and the human body is increased from 0 mm to 5 mm the antenna's resonant frequency is shifted by 5%. A varactor is employed to compensate for variations in the antenna's resonant frequency at different locations on the human body.

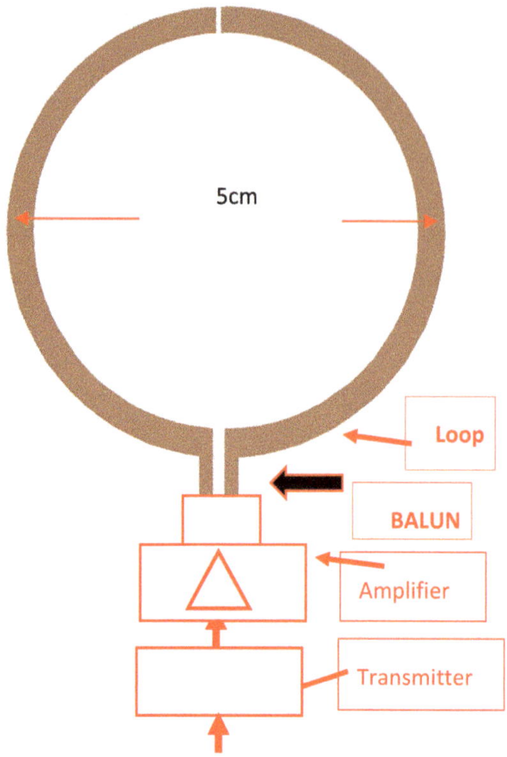

Figure 12.36. Active transmitting printed loop antenna layout.

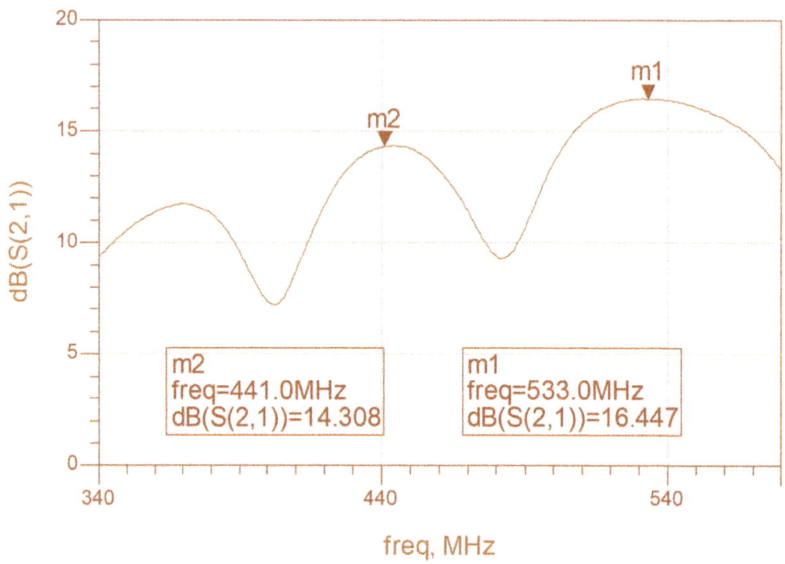

Figure 12.37. Active transmitting loop antenna's S_{21} parameters.

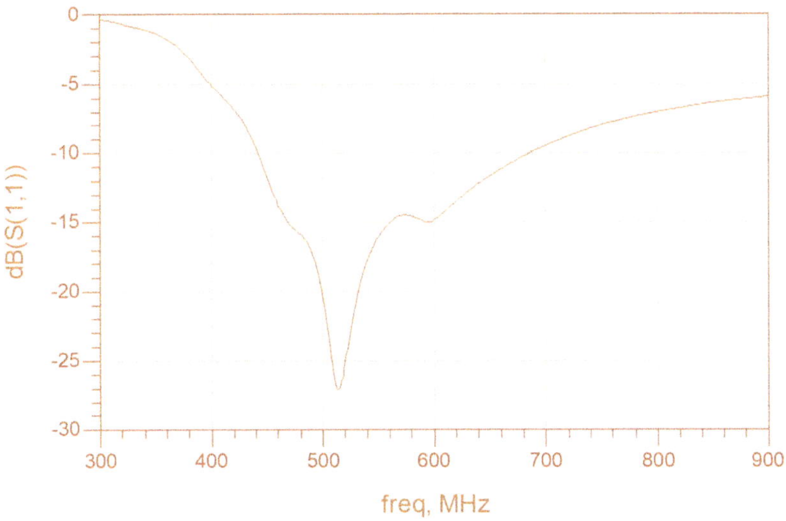

Figure 12.38. Active transmitting loop antenna's S_{11} parameters.

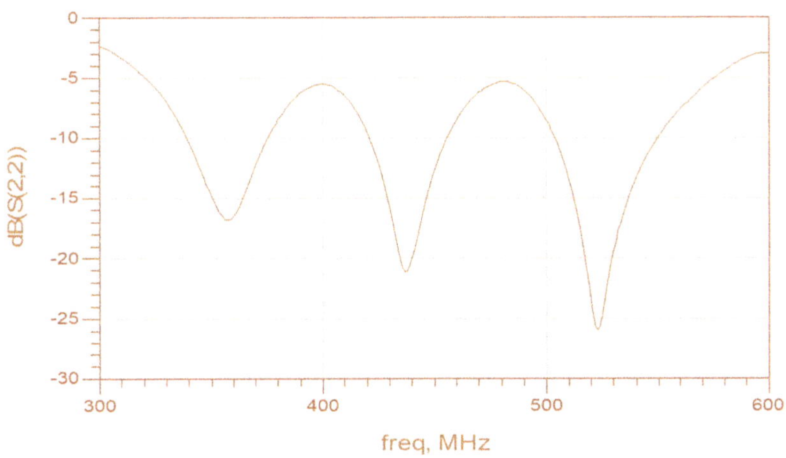

Figure 12.39. Active transmitting loop antenna's S_{22} parameters.

Active wearable antennas may be used in receiving or transmitting channels. In transmitting channels a power amplifier is connected to the antenna. In receiving channels a low noise amplifier is connected to the receiving antenna. The active loop antenna's gain is 25 ± 2.5 dB for frequencies ranging from 350 MHz to 580 MHz. The active loop antenna's noise figure is 0.7 ± 0.2 dB for frequencies ranging from 400 MHz to 900 MHz. The active dual polarized antenna's gain is 25 ± 3 dB for frequencies ranging from 400 MHz to 650 MHz. The gain difference between the gain of the vertical and horizontal antenna is around ±0.5 dB. The active dual polarized antenna's noise figure is 0.8 ± 0.4 dB for frequencies ranging from 400 MHz to 900 MHz. The active transmitting dual polarized antenna's gain is

14 ± 3 dB for frequencies ranging from 380 MHz to 600 MHz. The output power is around 18 dBm.

References

[1] James J R, Hall P S and Wood C 1981 *Microstrip Antenna Theory and Design* (London, UK: Peter Peregrinus)

[2] Sabban A and Gupta K C 1991 Characterization of radiation loss from microstrip discontinuities using a multiport network modeling approach *IEEE Trans. Microwave Theory Tech.* **39** 705–12

[3] Sabban A 1983 A new wideband stacked microstrip antenna *IEEE Antenna and Propagation Symp. (Houston, Texas, USA)*

[4] Sabban A 2015 *Low Visibility Antennas for Communication Systems* (USA: Taylor and Francis)

[5] Kastner R, Heyman E and Sabban A 1988 Spectral domain iterative analysis of single and double-layered microstrip antennas using the conjugate gradient algorithm *IEEE Trans. Antennas Propag.* **36** 1204–12

[6] Sabban A 1981 Wideband microstrip antenna arrays *IEEE Antenna and Propagation Symp. MELCOM (Tel Aviv)*

[7] Sabban A 2011 Microstrip antenna arrays *Microstrip Antennas* ed N Nasimuddin (Rijeka: InTech) pp 361–84

[8] Chirwa L C, Hammond P A, Roy S and Cumming D R S 2003 Electromagnetic radiation from ingested sources in the human intestine between 150 MHz and 1.2 GHz *IEEE Trans. Biomed. Eng.* **50** 484–92

[9] Werber D, Schwentner A and Biebl E M 2006 Investigation of RF transmission properties of human tissues *Adv. Radio Sci.* **4** 357–60

[10] Gupta B, Sankaralingam S and Dhar S 2010 Development of wearable and implantable antennas in the last decade *Mediterranean Microwave Symp. (MMS) (Guzelyurt, Turkey)* pp 251–67

[11] Thalmann T, Popovic Z, Notaros B M and Mosig J R 2009 Investigation and design of a multi-band wearable antenna *3rd European Conf. on Antennas and Propagation, EuCAP (Berlin, Germany)* pp 462–5

[12] Salonen P, Rahmat-Samii Y and Kivikoski M 2004 *Wearable antennas in the vicinity of human body IEEE Antennas and Propagation Society Int. Symp.* **1** *(Monterey, CA)* pp 467–70

[13] Kellomaki T, Heikkinen J and Kivikoski M 2006 Wearable antennas for FM reception *First European Conf. on Antennas and Propagation, EuCAP (Nice, France)* pp 1–6

[14] Sabban A 2009 Wideband printed antennas for medical applications *APMC 2009 Conf. (Singapore)*

[15] Lee Y 2003 *Antenna Circuit Design for RFID Applications* (AZ, USA: Microchip Technology Inc.) Microchip AN 710c

[16] Sabban A 1986 *Microstrip antenna arrays US Patent* US 1986/4,623,893

[17] Sabban A 2012 *Dual polarized dipole wearable antenna US Patent number:* 8203497

[18] ADS software, Agilent http://home.agilent.com/agilent/product.jspx?c.c.=IL&lc=eng&ckey=1297113&nid=-34346.0.00&id=1297113

[19] Wheeler H A 1975 Small antennas *IEEE Trans. Antennas Propag.* **23** 462–9

[20] Lin J and Itoh T 1994 Active integrated antennas *IEEE Trans. Microwave Theory Tech.* **42** 2186–94

[21] Mortazwi A, Itoh T and Harvey J 1998 *Active Antennas and Quasi-Optical Arrays* (New York: Wiley)

[22] Jacobsen S and Klemetsen Ø 2008 Improved detectability in medical microwave radio-thermometers as obtained by active antennas *IEEE Trans. Biomed. Eng.* **55** 2778–85

[23] Jacobsen S and Klemetsen Ø 2007 Active antennas in medical microwave radiometry *Electron. Lett.* **43** 606–8

[24] Ellingson S W, Simonetti J H and Patterson C D 2007 Design and evaluation of an active antenna for a 29–47 MHz radio telescope array *IEEE Trans. Antennas Propag.* **55** 826–31

[25] Segovia-Vargas D, Castro-Galan D, Garcia-Munoz L E and Gonzalez-Posadas V 2008 Broadband active receiving patch with resistive equalization *IEEE Trans. Microwave Theory Tech.* **56** 56–64

[26] Rizzoli V, Costanzo A and Spadoni P 2008 Computer-aided design of ultra-wideband active antennas by means of a new figure of merit *IEEE Microwave Wirel. Compon. Lett.* **18** 290–2.

[27] Yun G 2008 Compact active integrated microstrip antennas with circular polarisation diversity *IET Microwaves Antennas Propag.* **2** 82–7

Chapter 13

New wideband passive and active wearable slot and notch antennas for wireless and medical communication systems

Wideband wearable antennas are crucial in novel wearable communication systems. Slot antennas are low profile and low cost and may be employed in wearable communication systems. The dynamic range and the efficiency of communication systems may be improved by using active wearable antennas.

13.1 Slot antennas

Slot antennas may be printed on a dielectric substrate or cut out of the surface they are to be mounted on [1–16]. Slot antennas may be excited by connecting a transmission line to the slot edges as shown in figure 13.1(a), and by a microstrip line as shown in figure 13.1(b). The radiation pattern of slot antennas is determined by the size and shape of the slot in a radiating surface. Slot antennas may be used at frequencies between 200 MHz and up to 40 GHz.

The feed transmission line excites electric field distribution within the slot, and currents that travel around the slot perimeter. If we replace the slot with metal and the ground plane with air we get a dipole antenna. The slot antenna is dual to the dipole antenna as shown in figure 13.1(a). Babinet's principle [1] relates the electromagnetic fields of a slot antenna to its dual antenna, and the slot antenna to the dipole antenna. Babinet's principle states that the impedance of the slot antenna Z_s is related to the impedance of its dual antenna Z_d by the equation (13.1). Where η is the intrinsic impedance of free space and is equal to $120\pi\ \Omega$.

$$Z_s Z_d = \frac{\pi\eta^2}{4} \tag{13.1}$$

doi:10.1088/978-0-7503-5224-6ch13

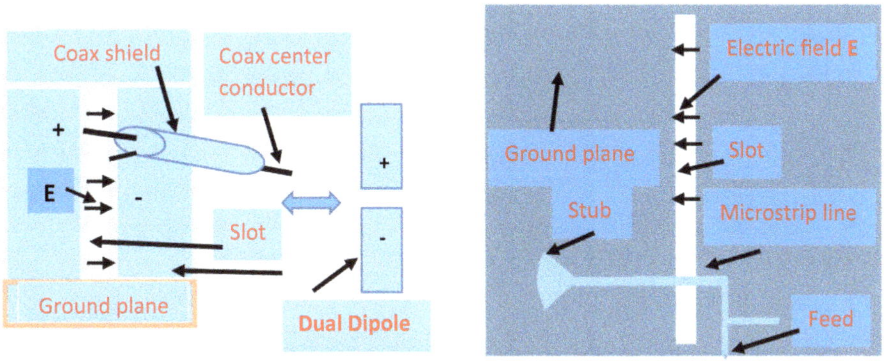

Figure 13.1. (a) Slot antenna and its dual dipole. (b) Printed slot antenna and microstrip line feed.

The impedance of a 0.5λ long dipole is 73 Ω. By using equation (13.1) the impedance of a 0.5λ long slot antenna is around 486 Ω. The second major result of Babinet's principle is that the fields of the dual antenna are almost the same as the slot antenna (the fields' components are interchanged, and called 'duals'). That is, the fields of the slot antenna (given with a subscript *s*) are related to the fields of its complement (given with a subscript *d*) as written in equation (13.2).

$$E_{\theta s} = H_{\theta d}$$

$$E_{\phi s} = H_{\phi d}$$

$$H_{\theta s} = \frac{-E_{\theta d}}{\eta^2} \tag{13.2}$$

$$H_{\theta s} = \frac{-E_{\phi d}}{\eta^2}$$

The polarization of the dual antennas is reversed. The dipole antenna in figure 13.1 is vertically polarized and the slot antenna will be horizontally polarized. In the far-fields the dipole electromagnetic fields vary as $\frac{1}{r}$ and $\sin\theta$ as written in equation (13.3).

$$\begin{aligned} E_r &= 0 \\ E_\theta &= j\eta_0 \frac{l\beta I_0 \sin\theta}{4\pi r} e^{j(\omega t - \beta r)} \\ H_\varphi &= j\frac{l\beta I_0 \sin\theta}{4\pi r} e^{j(\omega t - \beta r)} \end{aligned} \tag{13.3}$$

In the far-fields the slot electromagnetic fields vary as $\frac{1}{r}$ and $\sin\theta$ as written in equation (13.4).

$$\begin{aligned} E_r &= 0 \\ H_{\theta s} &= -j\frac{l\beta I_0 \sin\theta}{4\pi r\eta_0}e^{j(\omega t-\beta r)} \\ E_{\varphi s} &= j\frac{l\beta I_0 \sin\theta}{4\pi r}e^{j(\omega t-\beta r)} \end{aligned} \tag{13.4}$$

13.2 Slot radiation pattern

The antenna's radiation pattern represents the radiated fields in space at a point $P(r, \theta, \varphi)$ as function of θ, φ. The pattern is three-dimensional. When φ is constant and θ varies we get the E plane radiation pattern. When φ varies and θ is constant, usually $\theta = \pi/2$, we get the H plane radiation pattern.

Slot *E* plane radiation pattern

The slot E plane radiation pattern is given in equation (13.5) and presented in figure 13.2.

$$|E_{\varphi s}| = j\frac{l\beta I_0 \sin\theta}{4\pi r} = A\sin\theta \tag{13.5}$$

At a given point $P(r, \theta, \varphi)$ the slot E plane radiation pattern is given in equation (13.6).

$$\begin{aligned} &|E_{\varphi s}| = j\frac{l\beta I_0 |\sin\theta|}{4\pi r} = A|\sin\theta| \\ &\text{Choose } A = 1 \\ &|E_{\varphi s}| = |\sin\theta| \end{aligned} \tag{13.6}$$

A slot E plane radiation pattern in a spherical coordinate system is shown in figure 13.3.

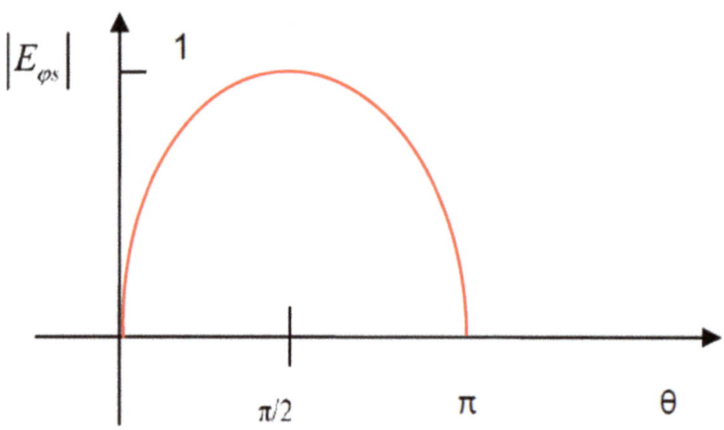

Figure 13.2. Slot E plane radiation pattern.

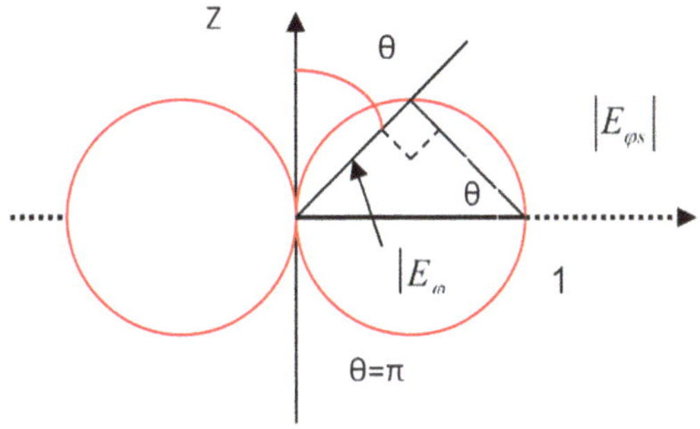

Figure 13.3. Slot *E* plane radiation pattern in a spherical coordinate system.

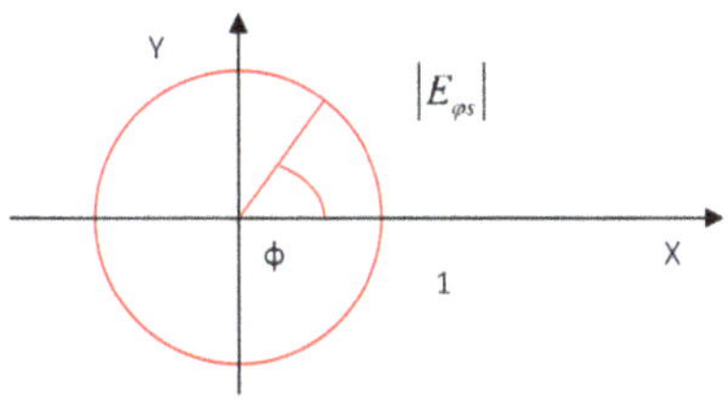

Figure 13.4. Slot *H* plane radiation pattern for $\theta = \pi/2$.

Slot *H* plane radiation pattern

For $\theta = \pi/2$ the slot *H* plane radiation pattern is given in equation (13.7) and presented in figure 13.4. At a given point $P(r, \theta, \varphi)$ the slot *H* plane radiation pattern is given in equation (13.7).

$$\begin{aligned} &\left| E_{\varphi s} \right| = j\frac{l\beta I_0}{4\pi r} = A \\ &\text{Choose } A = 1 \\ &\left| E_{\varphi s} \right| = 1 \end{aligned} \tag{13.7}$$

The slot *H* plane radiation pattern in the *xy* plane is a circle with $r = 1$.
The radiation pattern of a vertical slot is omnidirectional. It radiates equal power in all azimuthal directions perpendicular to the axis of the antenna. A slot *H* plane radiation pattern in a spherical coordinate system is shown in figure 13.4.

13.3 Slot antenna impedance

Antenna impedance determines the efficiency of the transmitting and receiving energy in antennas. The dipole impedance is given in equation (13.8). The slot impedance is given in equation (13.9).

$$\begin{aligned} R_{rad} &= \frac{2W_T}{I_0^2} \\ \text{For a Dipole: } R_{rad} &= \frac{80\pi^2 l^2}{\lambda^2} \end{aligned} \tag{13.8}$$

$$Z_s = \frac{\pi\eta^2}{4Z_d} = \frac{\eta^2\lambda^2}{320\pi l^2} \tag{13.9}$$

By using equation (13.9) the impedance of a 0.5λ long slot antenna is around 565 Ω.

13.4 A wideband wearable printed slot antenna

A wideband wearable printed slot antenna is shown in figure 13.5. The slot antenna is printed on RT-Duroid 5880 dielectric substrate with a dielectric constant of 2.2 and 1.2 mm thick. The antenna's electrical parameters were calculated and optimized using ADS software. The dimensions of the slot antenna shown in figure 13.5 are 66 × 60 × 1.2 mm. The slot antenna's center frequency is 2.5 GHz. The computed S_{11} parameters are presented in figure 13.6. The antenna's bandwidth is around 50% for a VSWR better than 2:1, and around 70% for VSWR better than 3:1. The radiation pattern of the slot antenna is shown in figure 13.7. The beamwidth

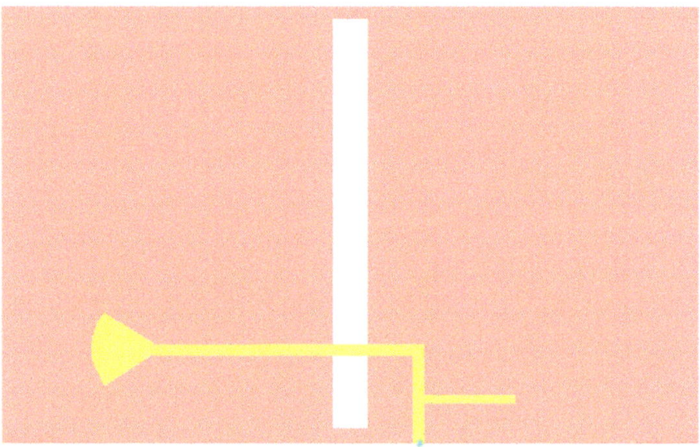

Figure 13.5. A wideband wearable printed slot antenna.

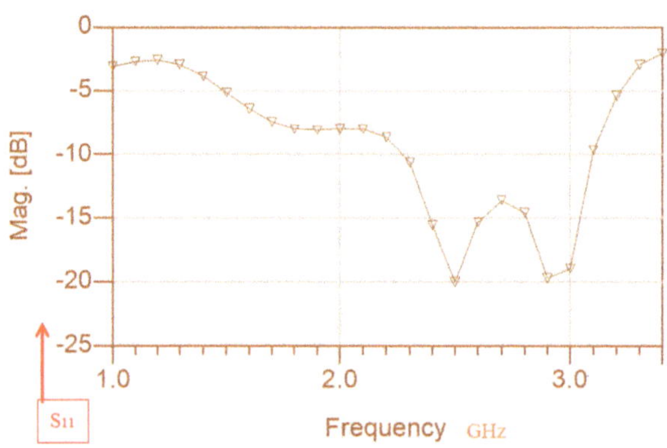

Figure 13.6. S_{11} of a wideband wearable printed slot antenna.

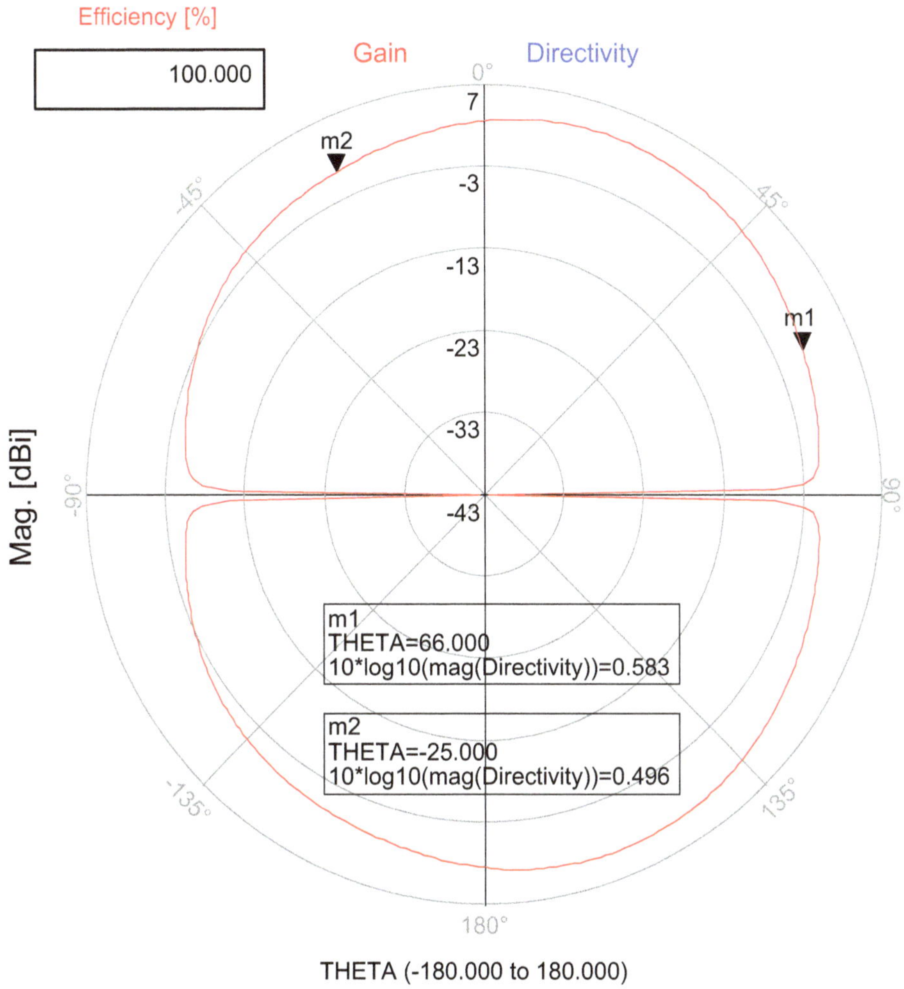

Figure 13.7. Radiation pattern of a wideband wearable printed slot antenna at 2 GHz.

is around 90° at 2 GHz, as shown in figure 13.7, and the gain is around 3 dBi. The radiation pattern at 2.5 GHz is shown in figure 13.8.

13.5 A wideband T shape wearable printed slot antenna

A wideband T shape wearable printed slot antenna is shown in figure 13.9. The slot antenna is printed on RT-Duroid 5880 dielectric substrate with a dielectric constant of 2.2 and 1. 2 mm thick. The antenna's electrical parameters were calculated and optimized by using ADS software. The dimensions of the slot antenna shown in figure 13.9 are 66 × 60 × 1.2 mm. The slot antenna's center frequency is around 2.25 GHz. The computed S_{11} parameters are presented in figure 13.10. The antenna's bandwidth is around 57% for a VSWR better than 2:1, and around 90% for a VSWR better than 3:1. The radiation pattern of the T shape slot antenna

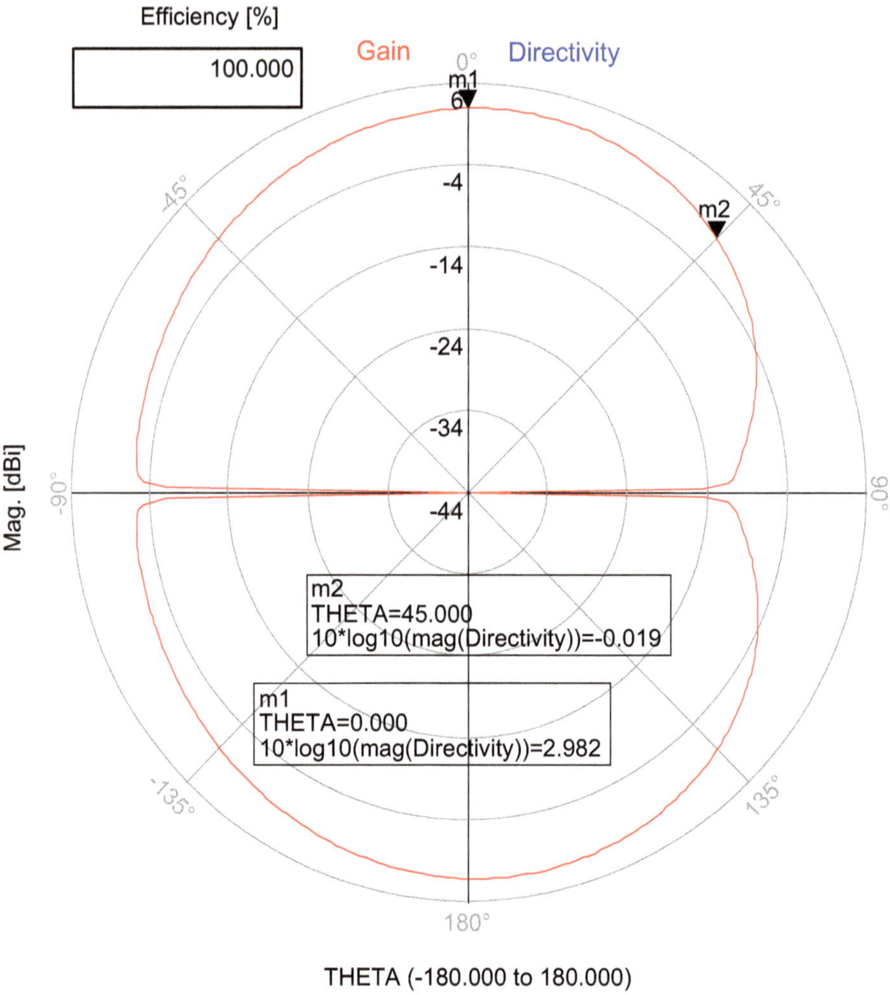

Figure 13.8. Radiation pattern of a wideband wearable printed slot antenna at 2.5 GHz.

is shown in figure 13.11. The antenna's beamwidth is around 82° at 1.5 GHz, as shown in figure 13.11, and the gain is around 3 dBi.

The computed S_{11} parameters of the T shape slot on a human body are presented in figure 13.12. The dielectric constant of human body tissue was taken as 45. The antenna was attached to a shirt with a dielectric constant of 2.2 1 mm thick. The antenna's bandwidth is around 50% for a VSWR better than 2:1, and around 57% for a VSWR better than 3:1. The antenna's center frequency is shifted by 10%. The feed network of the antenna shown in figure 13.9 was optimized to match the antenna to the human body's environment, see figure 13.13. The computed S_{11} parameters of the modified T shape slot on a human body are presented in figure 13.14. The modified antenna's VSWR is better than 3:1 for frequencies ranging from 0.8 GHz to 3.9 GHz. The antenna's gain at 1.5 GHz of the modified antenna is

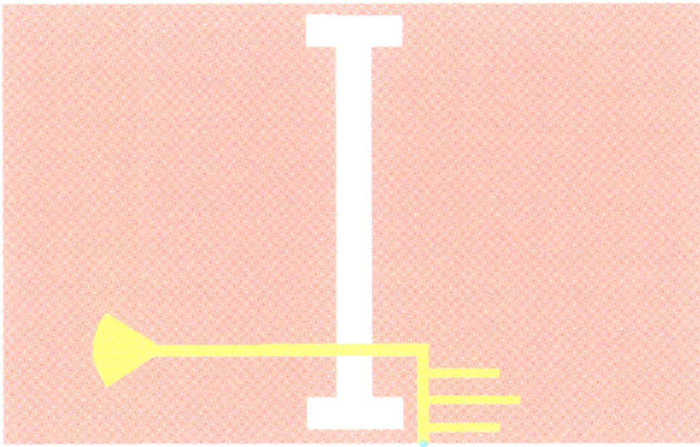

Figure 13.9. A wideband T shape wearable printed slot antenna.

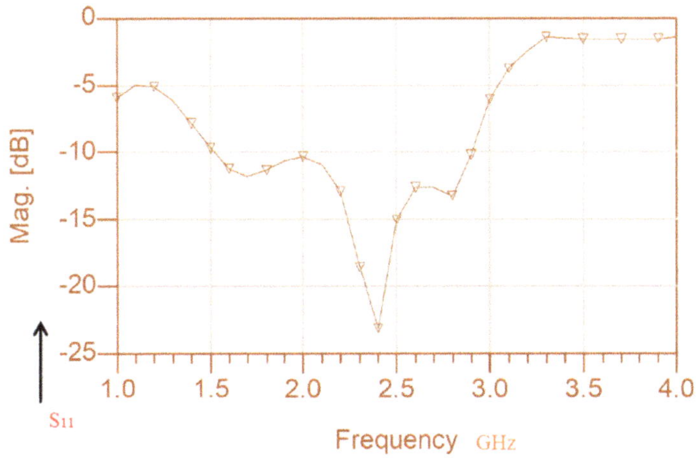

Figure 13.10. S_{11} of a wideband wearable printed slot antenna.

around 3 dBi (figure 13.15). The radiation pattern of the modified T shape slot antenna at 1.5 GHz is shown in figure 13.16.

13.6 Wideband wearable notch antenna for wireless communication systems

The wireless communication industry has rapidly grown in the last few years. Due to huge progress in the development of communication systems in the last decade, development of wideband communication systems has continuously grown. However, development of wideband efficient antennas is one of the major challenges in the development of wideband wireless communication systems. Low-cost, compact antennas are crucial in the development of communication systems.

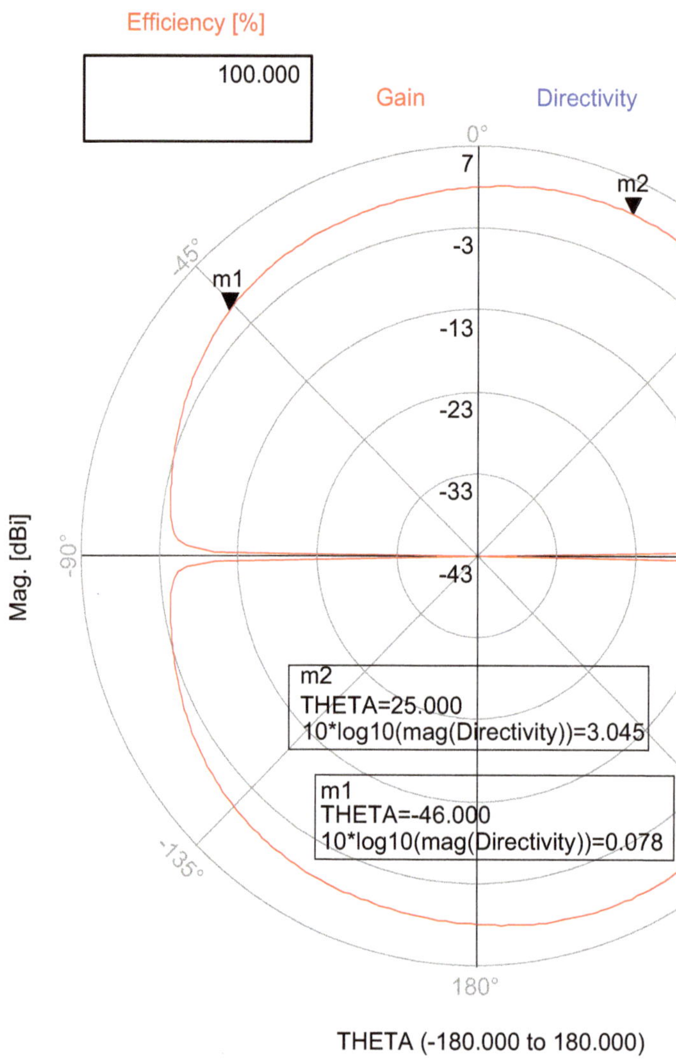

Figure 13.11. Radiation pattern of a wideband wearable printed slot antenna at 1.5 GHz.

Printed notch antennas and miniaturization techniques are employed to develop efficient compact notch antennas.

Wideband notch antenna 2.1 GHz to 7.8 GHz

A wideband notch antenna has been designed. The antenna is printed on RT-Duroid 5880 dielectric substrate with a dielectric constant of 2.2 and 1.2 mm thick. The notch antenna is shown in figure 13.17. The notch antenna's dimensions are 116.4 × 71.4 mm. The antenna's center frequency is 5 GHz, and the bandwidth is around 100% for S_{11} lower than −6.5 dB, as presented in figure 13.18. The notch antenna's VSWR is better than 3:1 for frequencies from 2.1 GHz to 7.8 GHz. The antenna's beamwidth is around 84°, and the gain is around 2.5 dBi. Figure 13.19

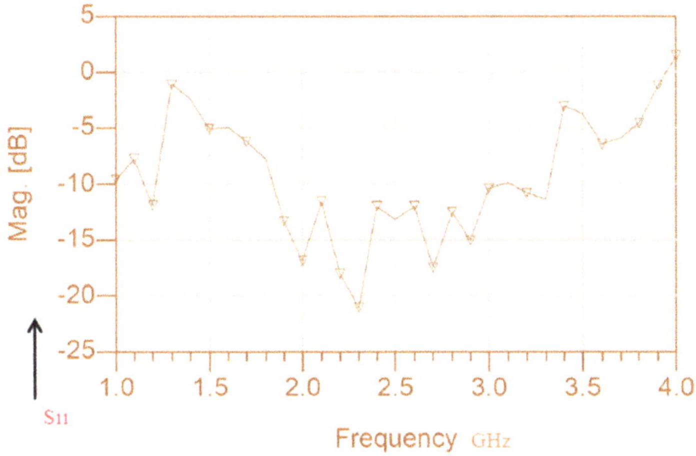

Figure 13.12. S_{11} of a wideband wearable printed slot antenna on a human body.

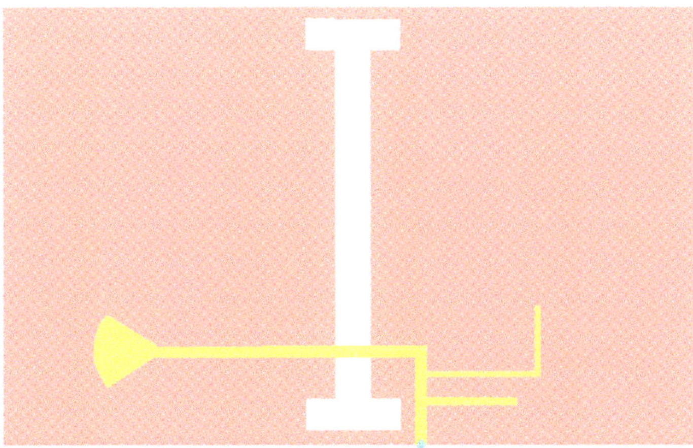

Figure 13.13. A modified wideband T shape wearable printed slot antenna.

presents the radiation pattern of the wideband notch antenna at 3.5 GHz. Figure 13.20 presents the radiation pattern of the wideband notch antenna at 3 GHz.

13.7 Wearable tunable slot antennas for wireless communication systems

A wideband wearable tunable slot antenna is shown in figure 13.21. A tunable slot antenna consists of a slot antenna and a voltage-controlled diode, varactor, [17]. The antenna's resonant frequency may be tuned by using a varactor to compensate for variations in the antenna's resonant frequency at different locations. The slot antenna is printed on RT-Duroid 5880 dielectric substrate with a dielectric constant of 2.2 and 1.2 mm thick. The antenna's electrical parameters were calculated and

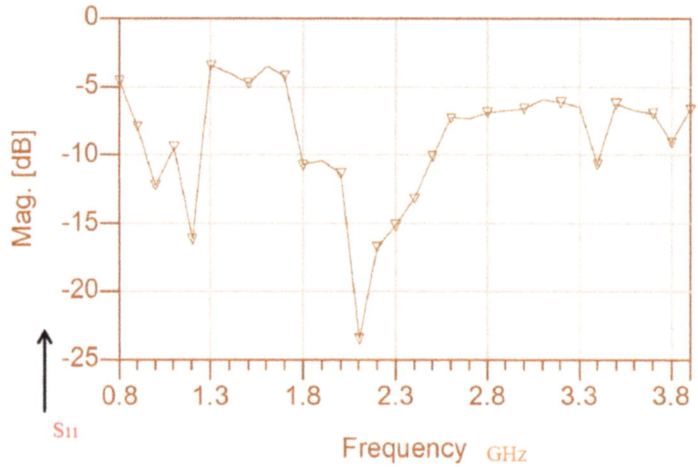

Figure 13.14. S_{11} of the modified T shape wearable slot antenna on a human body.

optimized using ADS software. The dimensions of the slot antenna shown in figure 13.21 are 66 × 60 × 1.2 mm. A varactor is connected to the slot feed line. The varactor's bias voltage may be varied automatically to set the antenna's resonant frequency at different locations and environments. The slot antenna's center frequency is 2.5 GHz. The S_{11} parameters for varactor capacitances ranging from 0.1 pF to 1 pF are presented in figure 13.22. The antenna's bandwidth is around 40% for a VSWR better than 2:1, and 60% for a VSWR better than 3:1.

13.8 A wideband T shape tunable wearable printed slot antenna

A wideband T shape wearable printed slot antenna is shown in figure 13.23. The slot antenna is printed on RT-Duroid 5880 dielectric substrate with a dielectric constant of 2.2 and 1.2 mm thick. The antenna's electrical parameters were calculated and optimized by using ADS software. The dimensions of the slot antenna shown in figure 13.22 are 66 × 60 × 1.2 mm. The slot antenna's center frequency is around 2.25 GHz. The computed S_{11} parameters are presented in figure 13.24. The antenna's bandwidth is around 57% for a VSWR better than 2:1, and around 90% for a VSWR better than 3:1. A varactor is connected to the slot feed line. The varactor's bias voltage may be varied automatically to set the antenna's resonant frequency at different locations and environments. The S_{11} parameters for varactor capacitances ranging from 0.1 pF to 1 pF are presented in figure 13.25.

13.9 Wearable active slot antennas for wireless communication systems

A wideband active wearable receiving slot antenna is shown in figure 13.26. Active slot antennas are devices combining radiating elements with active components such as amplifiers and diodes. The radiating element is designed to provide the optimal

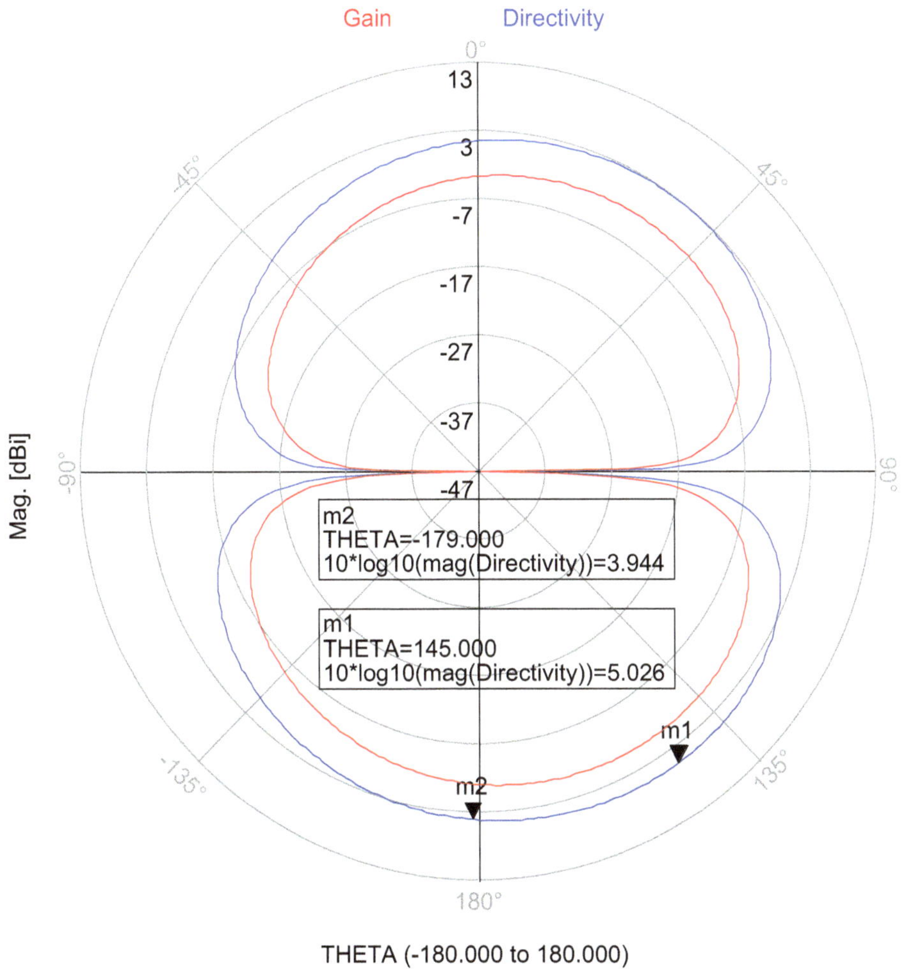

Figure 13.15. Radiation pattern of the modified wideband wearable slot antenna at 1 GHz.

load to the active elements. The slot antenna is printed on RT-Duroid 5880 dielectric substrate with a dielectric constant of 2.2 and 1.2 mm thick. The antenna's electrical parameters were calculated and optimized using ADS software. The dimensions of the slot antenna shown in figure 13.26 are 66 × 60 × 1.2 mm. An E PHEMT LNA, low noise amplifier, was connected to a slot antenna. The radiating element is connected to the LNA via an input matching network. An output matching network connects the amplifier port to the receiver. A DC bias network supplies the required voltages to the amplifiers. The amplifier's specification is listed in table 12.1. The amplifier's complex S parameters are listed in table 12.2, and the noise parameters are listed in table 12.3. The active slot antenna's S_{11} parameters are presented in figure 13.27, and the S_{22} parameters are presented in figure 13.28. The antenna's bandwidth is around 40% for a VSWR better than 3:1. The active slot antenna's S_{21} parameters (gain) are presented in figure 13.29. The active antenna's gain is 18 ± 2.5

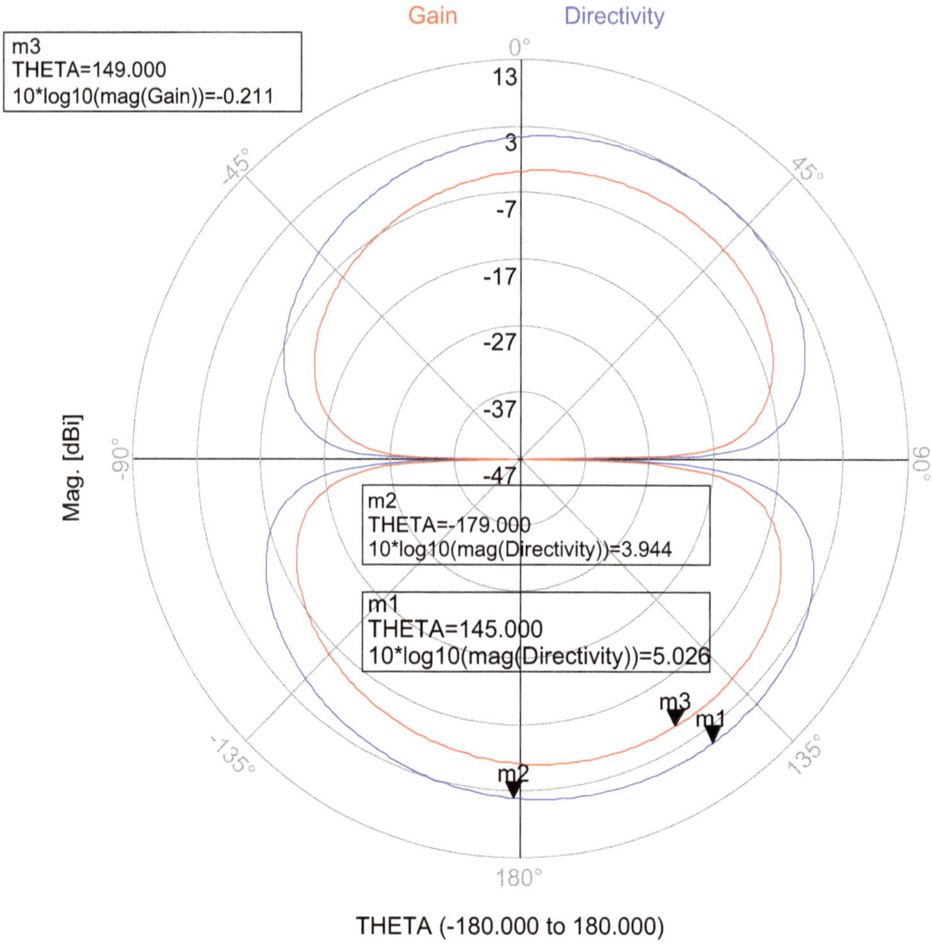

Figure 13.16. Radiation pattern of the modified wideband wearable slot antenna at 1.5 GHz.

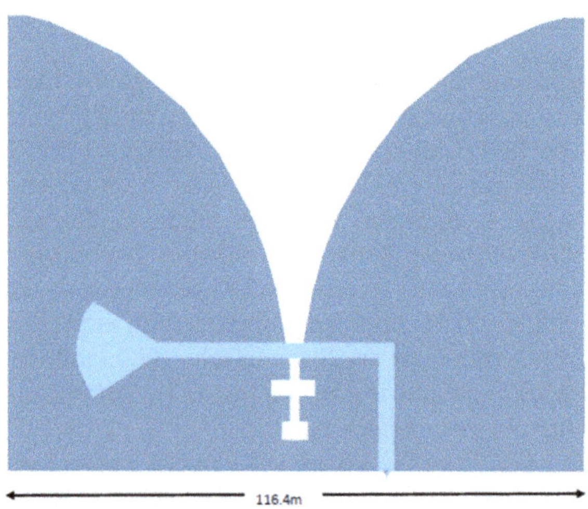

Figure 13.17. A wideband 2.1 GHz to 7.8 GHz notch antenna.

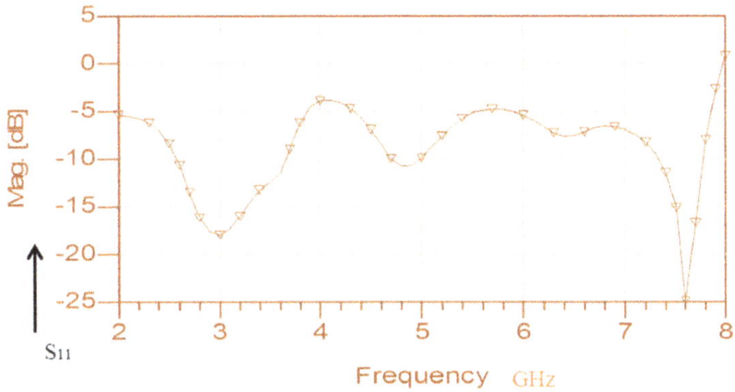

Figure 13.18. A wideband 2.1 GHz to 7.8 GHz notch antenna, computed S_{11}.

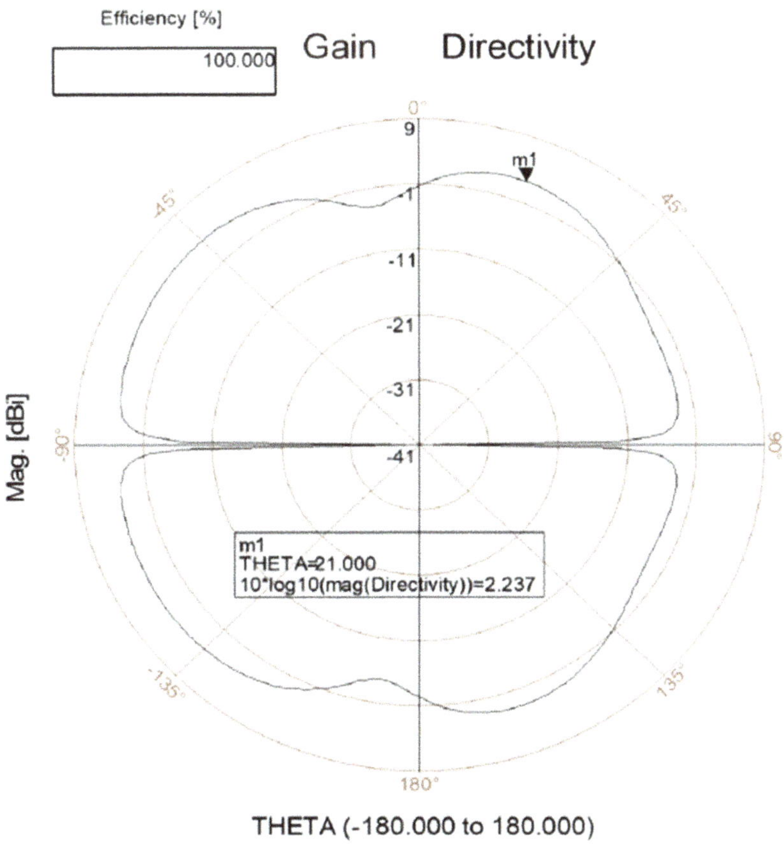

Figure 13.19. Radiation pattern of the wideband notch antenna at 3.5 GHz.

dB for frequencies ranging from 200 MHz to 580 MHz, and 12 ± 2 dB for frequencies ranging from 1.3 GHz to 3.3 GHz. The active slot antenna's noise figure is presented in figure 13.30. It is 0.5 ± 0.3 dB for frequencies ranging from 200 MHz to 3.3 GHz.

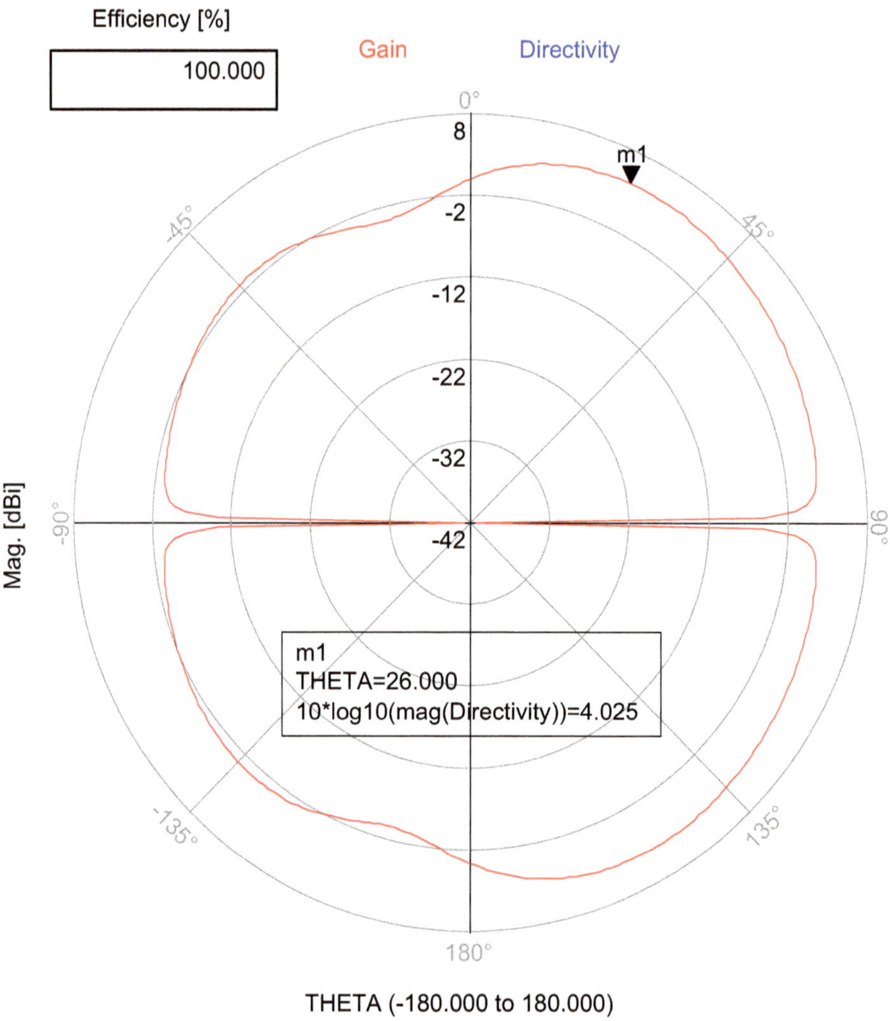

Figure 13.20. Radiation pattern of the wideband notch antenna at 3 GHz.

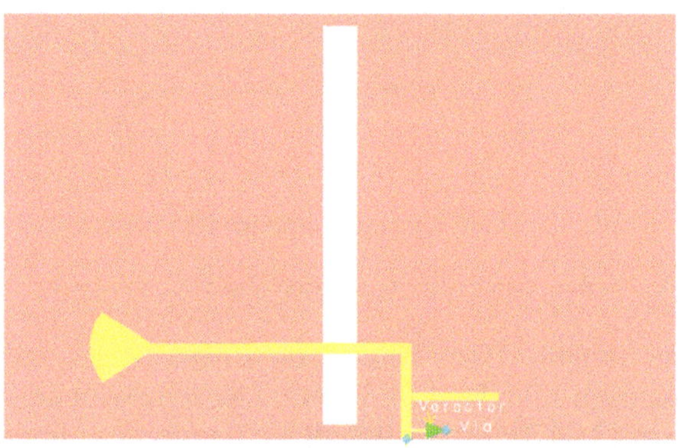

Figure 13.21. A wideband tunable wearable slot antenna.

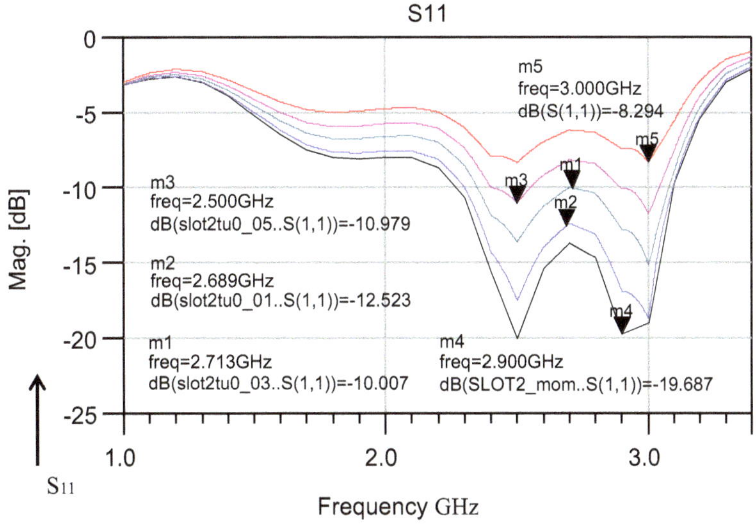

Figure 13.22. S_{11} of a wideband tunable wearable printed slot antenna.

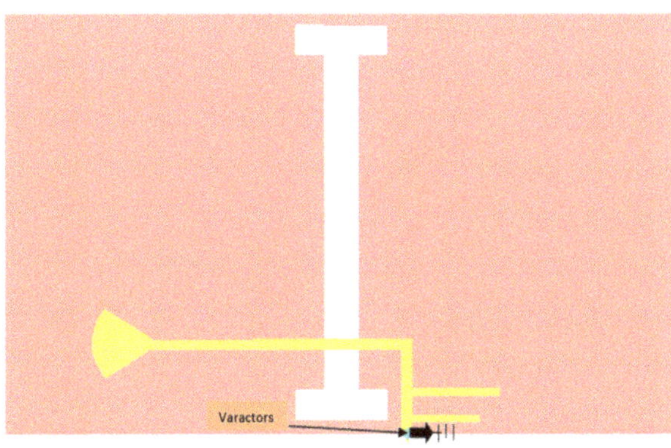

Figure 13.23. A wideband tunable wearable T shape slot antenna.

13.10 Wearable active T shape slot antennas for wireless communication systems

A wideband active wearable receiving T shape slot antenna is shown in figure 13.31. The radiating element is designed to provide the optimal load to the active elements. The slot antenna is printed on RT-Duroid 5880 dielectric substrate with a dielectric constant of 2.2 and 1.2 mm thick. The antenna's electrical parameters were calculated and optimized using ADS software. The dimensions of the slot antenna shown in figure 13.31 are 66 × 60 × 1.2 mm. An E PHEMT LNA, low noise amplifier, was connected to a slot antenna. The radiating element is connected to the LNA via an input matching network. An output matching network connects the amplifier port to the receiver. A DC bias network supplies the required voltage

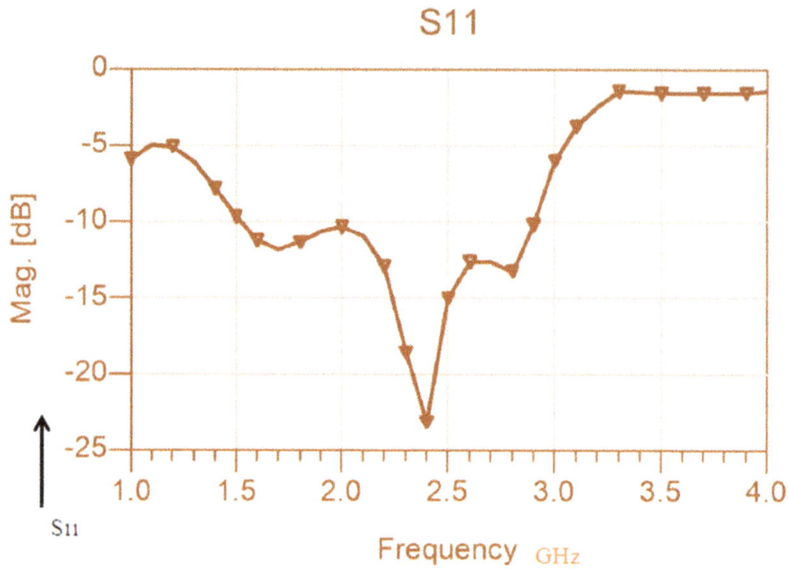

Figure 13.24. S_{11} of a wideband wearable T shape slot antenna without varactor.

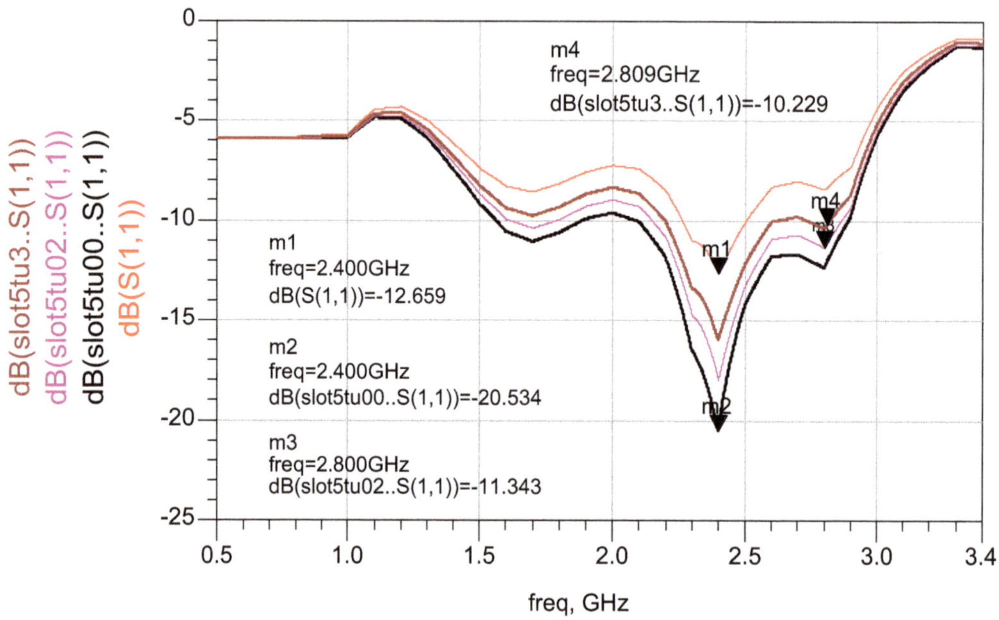

Figure 13.25. S_{11} of a wideband tunable wearable T shape slot antenna.

to the amplifiers. The amplifier's specification is listed in table 12.1. The amplifier's complex *S* parameters are listed in table 12.2, and the noise parameters are listed in table 12.3. An active slot antenna's S_{11} parameters are presented in figure 13.32. The antenna's bandwidth is around 40% for a VSWR better than 2:1. The active slot

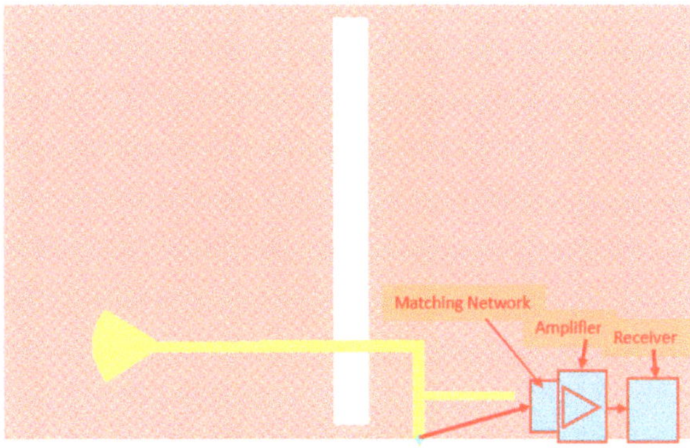

Figure 13.26. A wideband active receiving wearable slot antenna.

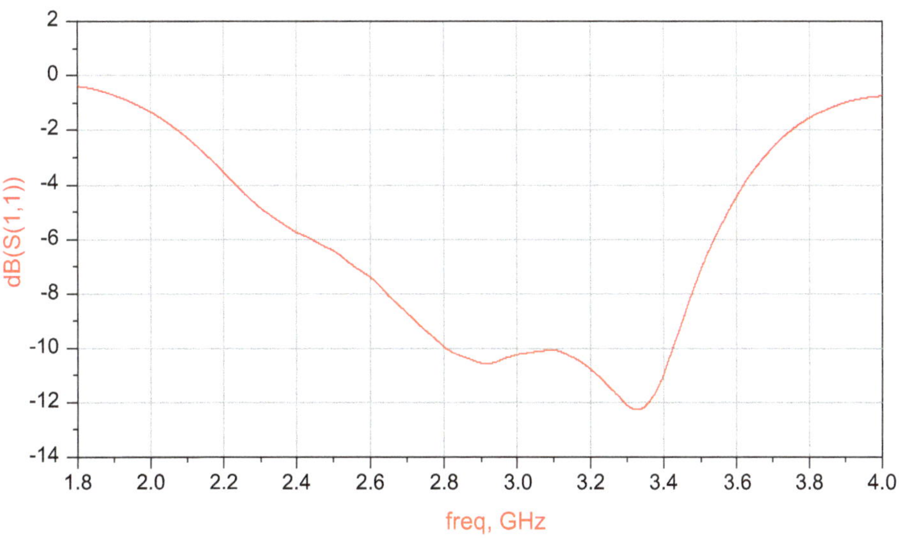

Figure 13.27. Active slot antenna's S_{11} parameters.

antenna's S_{21} parameters (gain) are presented in figure 13.33. The active antenna's gain is 18 ± 2.5 dB for frequencies ranging from 200 MHz to 580 MHz. The active antenna gain is 12.5 ± 2.5 dB for frequencies ranging from 1 GHz to 3 GHz. The active slot antenna noise figure is presented in figure 13.34. The noise figure is 0.5 ± 0.3 dB for frequencies ranging from 300 MHz to 3.2 GHz. The active slot antenna's S_{22} parameters are presented in figure 13.35.

13.11 New fractal compact ultra-wideband, 1 GHz to 6 GHz, notch antenna

A wideband notch antenna with a fractal structure has been designed. The antenna is printed on RT-Duroid 5880 dielectric substrate with a dielectric constant of 2.2 and 1.2 mm thick. The notch antenna is shown in figure 13.36. The notch antenna's

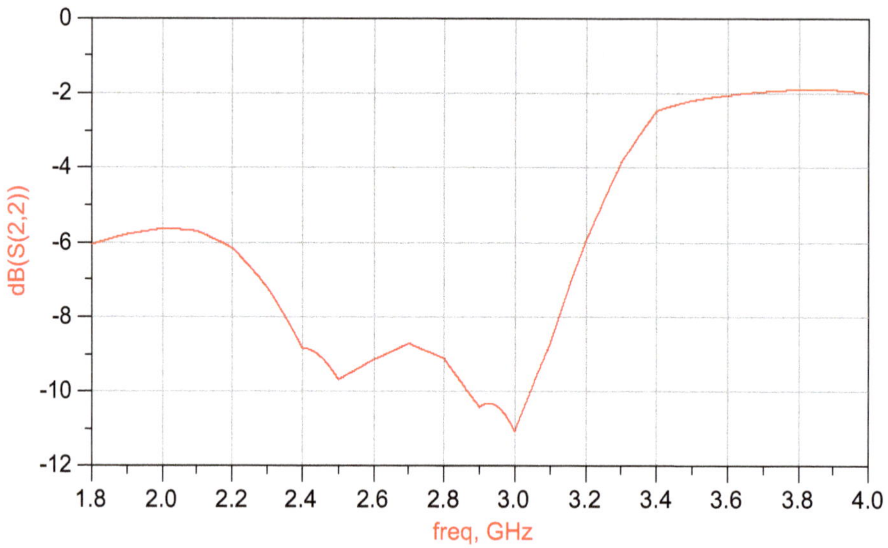

Figure 13.28. Active slot antenna's S_{22} parameters.

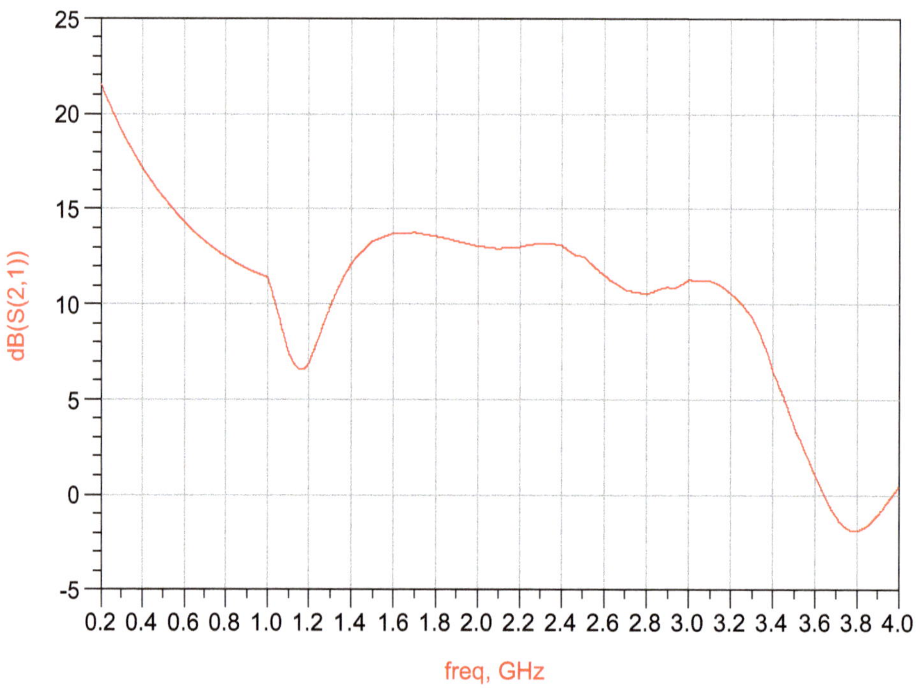

Figure 13.29. Active slot antenna's S_{21} parameters (gain).

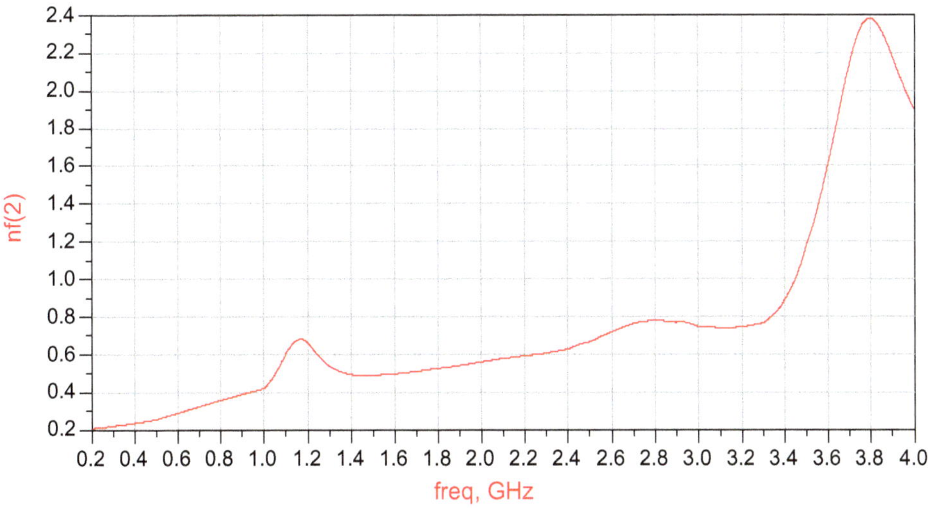

Figure 13.30. Active slot antenna's noise figure.

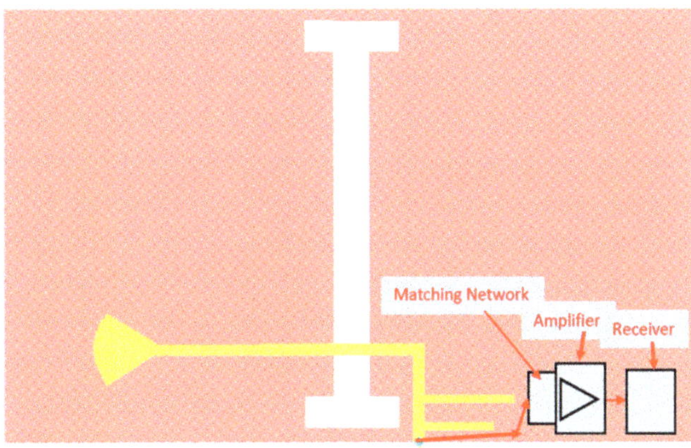

Figure 13.31. A wideband active receiving wearable slot antenna.

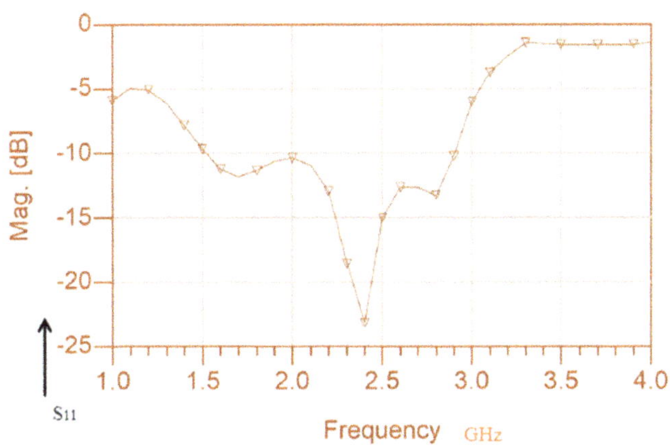

Figure 13.32. Active T shape slot antenna's S_{11} parameters.

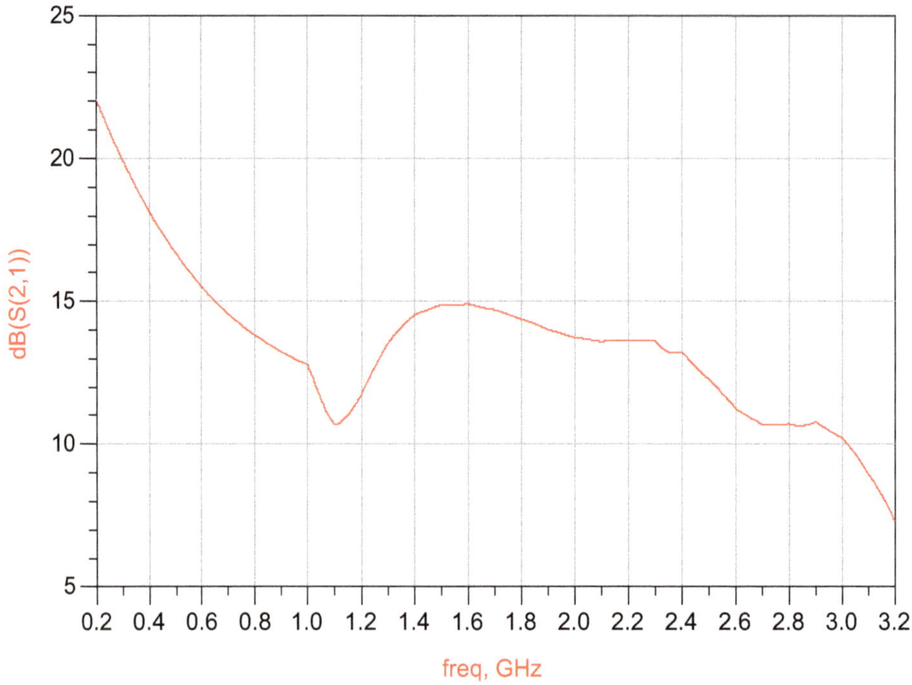

Figure 13.33. Active T shape slot antenna's S_{21} parameters.

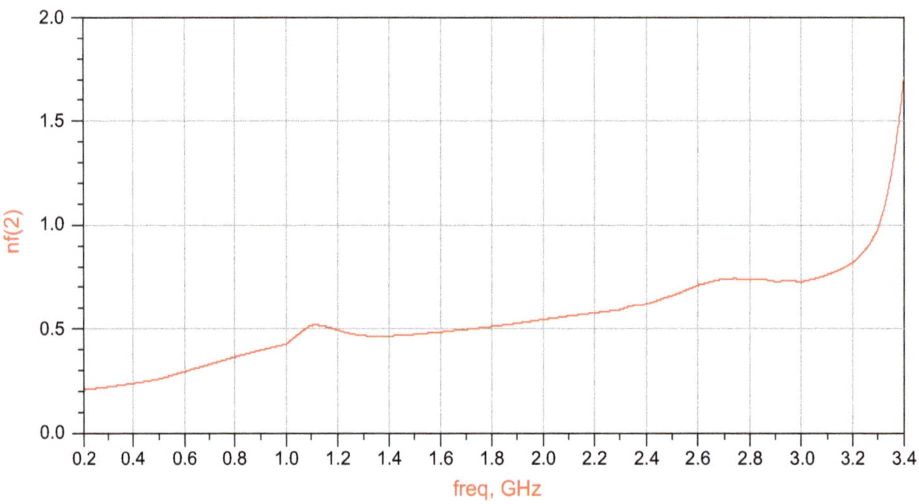

Figure 13.34. Active slot antenna's noise figure.

dimensions are 74.5 × 57.1 mm. The antenna's center frequency is 2.75 GHz, and the bandwidth is around 200% for S_{11} lower than −6.5 dB, as presented in figure 13.37. The notch antenna's VSWR is better than 3:1 for frequencies from 1 GHz to 5.5 GHz. The antenna's beamwidth is around 84°, and the gain is around 3.5 dBi, as

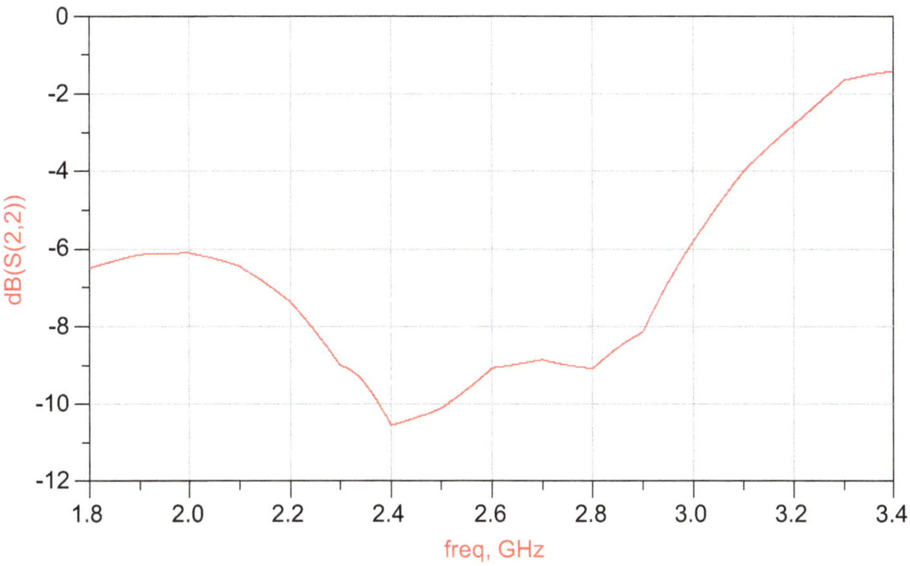

Figure 13.35. Active T shape slot antenna's S_{22} parameters.

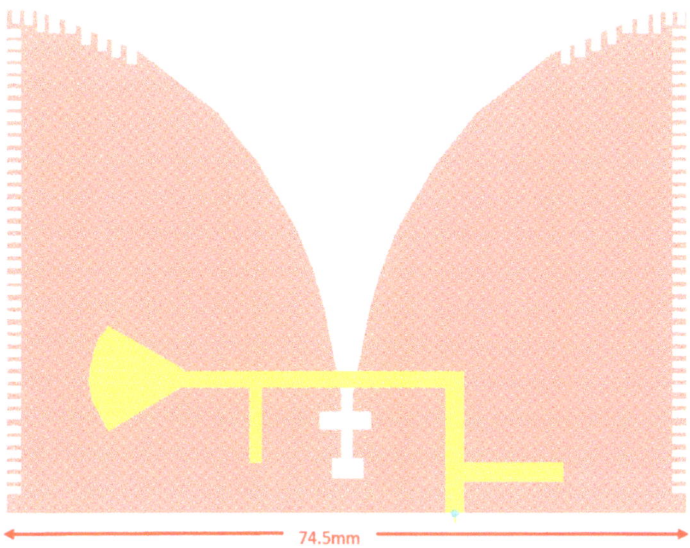

Figure 13.36. A wideband notch antenna with a fractal structure.

presented in figure 13.38. An *H* plane radiation pattern of the wideband notch antenna with a fractal structure is presented in figure 13.39.

13.12 New compact ultra-wideband notch antenna 1.3 GHz to 3.9 GHz

A wideband notch antenna with a fractal structure has been designed. The antenna is printed on RT-Duroid 5880 dielectric substrate with a dielectric constant of 2.2 and

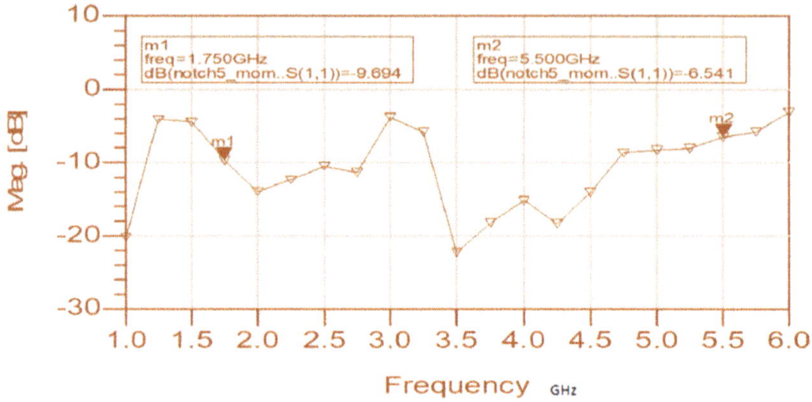

Figure 13.37. A wideband notch antenna with a fractal structure, computed S_{11}.

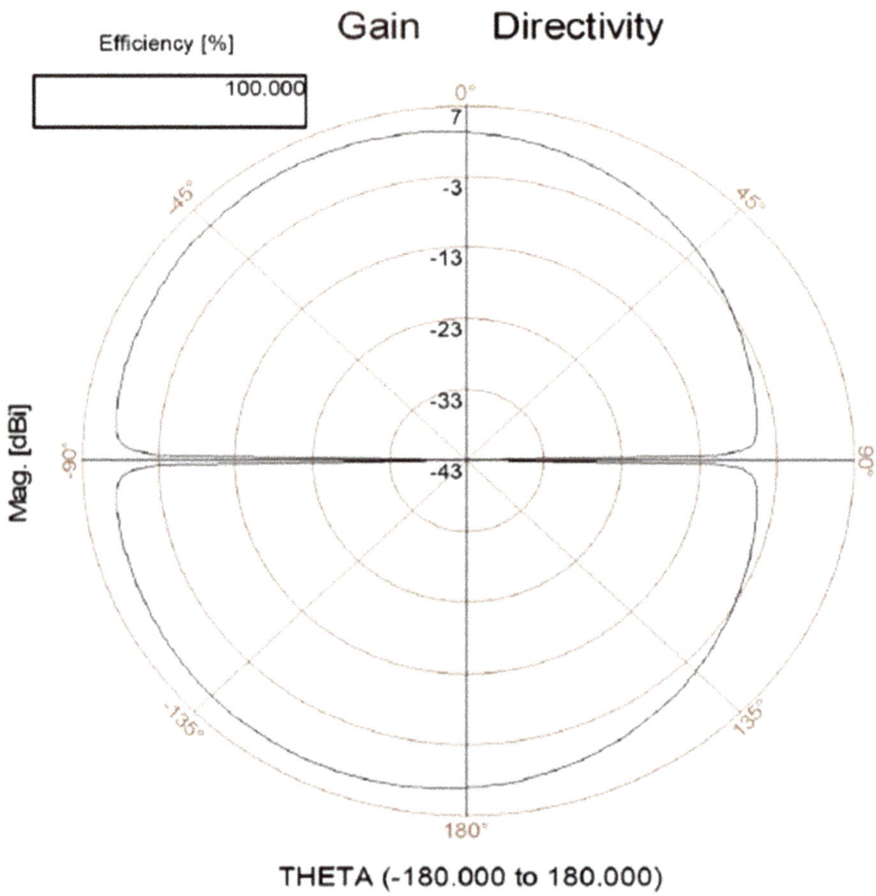

Figure 13.38. *E* plane radiation pattern of the wideband notch antenna with a fractal structure.

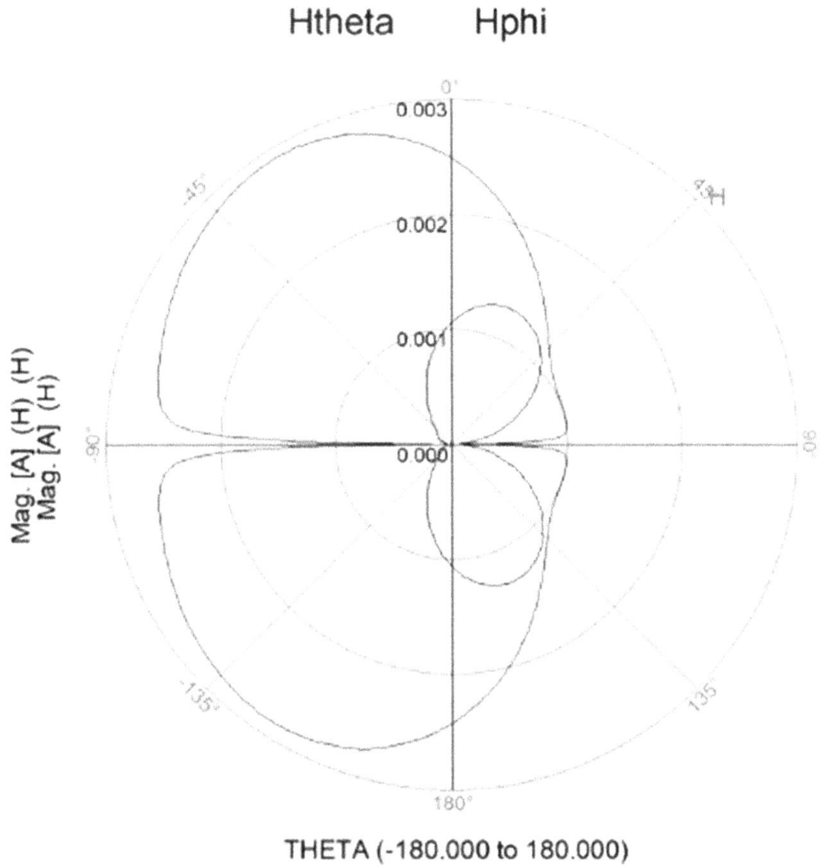

Figure 13.39. *H* plane radiation pattern of the wideband notch antenna with a fractal structure.

1.2 mm thick. The notch antenna is shown in figure 13.40. The notch antenna's dimensions are 52.2 × 36.8 mm. The antenna's center frequency is 2.7 GHz. The antenna's bandwidth is around 100% for S_{11} lower than −6.5 dB, as presented in figure 13.41. The notch antenna's VSWR is better than 3:1 for frequencies from 1.3 GHz to 3.9 GHz. The antenna's beamwidth is around 84°, and the gain is around 3.5 dBi.

By using a fractal structure the notch antenna's length and width was reduced by around 50%.

13.13 New compact ultra-wideband notch antenna 5.8 GHz to 18 GHz

A wideband notch antenna with a fractal structure has been designed. The antenna is printed on RT-Duroid 5880 dielectric substrate with a dielectric constant of 2.2 and 1.2 mm thick. The notch antenna is shown in figure 13.42. The dimensions are 11 × 7.7 mm. The antenna's center frequency is 12 GHz, and the bandwidth is around 100% for S_{11} lower than −5 dB, as presented in figure 13.43. The notch antenna's VSWR is better than 3:1 for more than 90% of the frequency range from

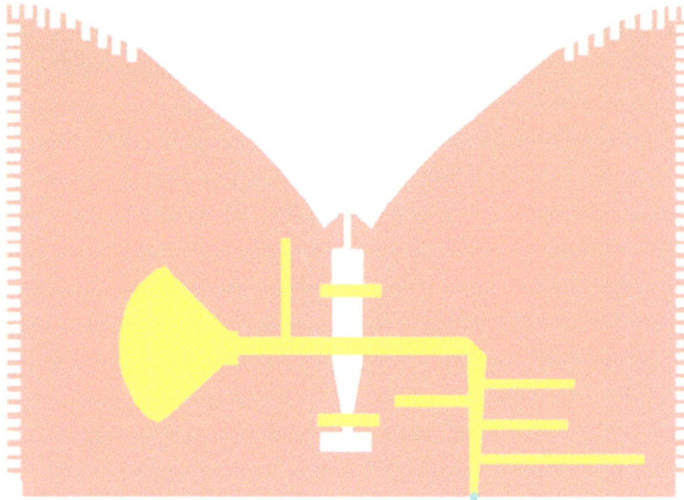

Figure 13.40. A wideband 1.3 GHz to 3.9 GHz notch antenna with a fractal structure.

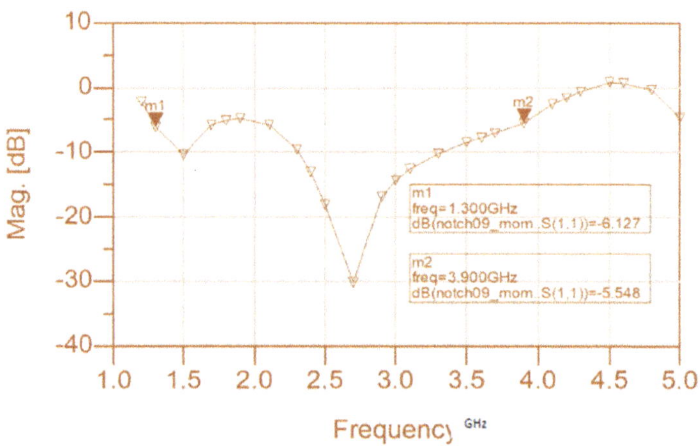

Figure 13.41. A wideband 1.3 GHz to 3.9 GHz notch antenna with a fractal structure, S_{11} results.

5.8 GHz to 18 GHz. The antenna's beamwidth is around 84°, and the gain is around 3.5 dBi. Figure 13.44 presents the radiation pattern of the wideband notch antenna with a fractal structure at 8 GHz.

The antenna's matching network was optimized to get better S_{11} results at 16 GHz to 18 GHz. The length and width of the stubs were tuned to get better S_{11} results at 16 GHz to 18 GHz.

13.14 New fractal active compact ultra-wideband, 0.5 GHz to 3 GHz, notch antenna

A wideband active notch antenna with a fractal structure has been designed. The antenna is printed on RT-Duroid 5880 dielectric substrate with a dielectric constant

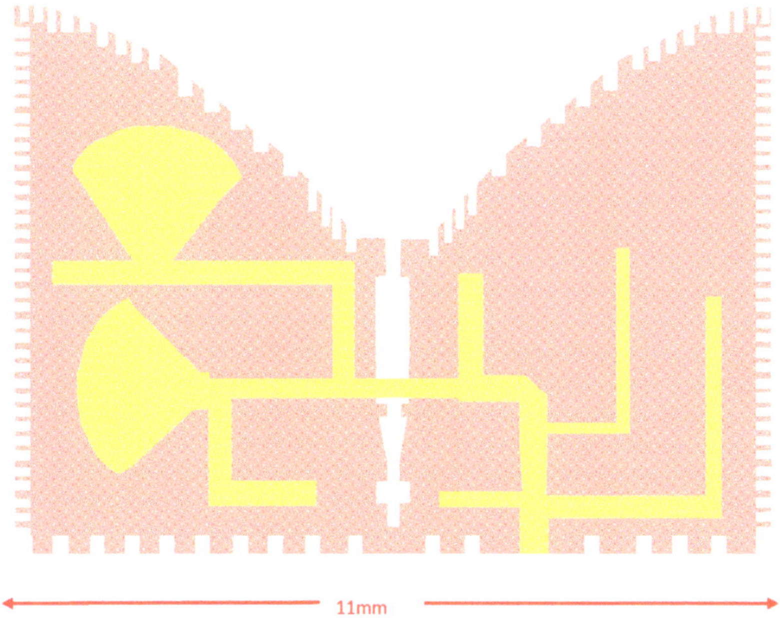

Figure 13.42. A wideband 5.8 GHz to 18 GHz notch antenna with a fractal structure.

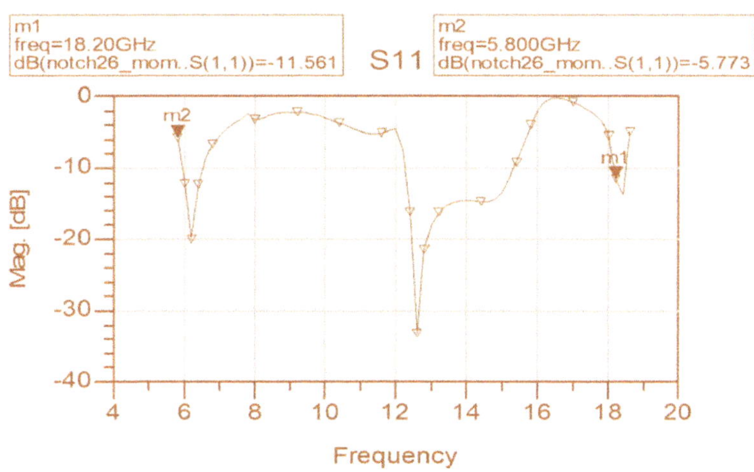

Figure 13.43. A wideband 5.8 GHz to 18 GHz notch antenna with a fractal structure, S_{11} results.

of 2.2 and 1.2 mm thick. The active notch antenna is shown in figure 13.45. The dimensions are 74.5 × 57.1 mm. The antenna's center frequency is 1.75 GHz, and the bandwidth is around 200% for S_{11} lower than −5 dB. The active notch antenna's VSWR is better than 3:1 for frequencies from 0.5 GHz to 3 GHz, and the beamwidth is around 84°.

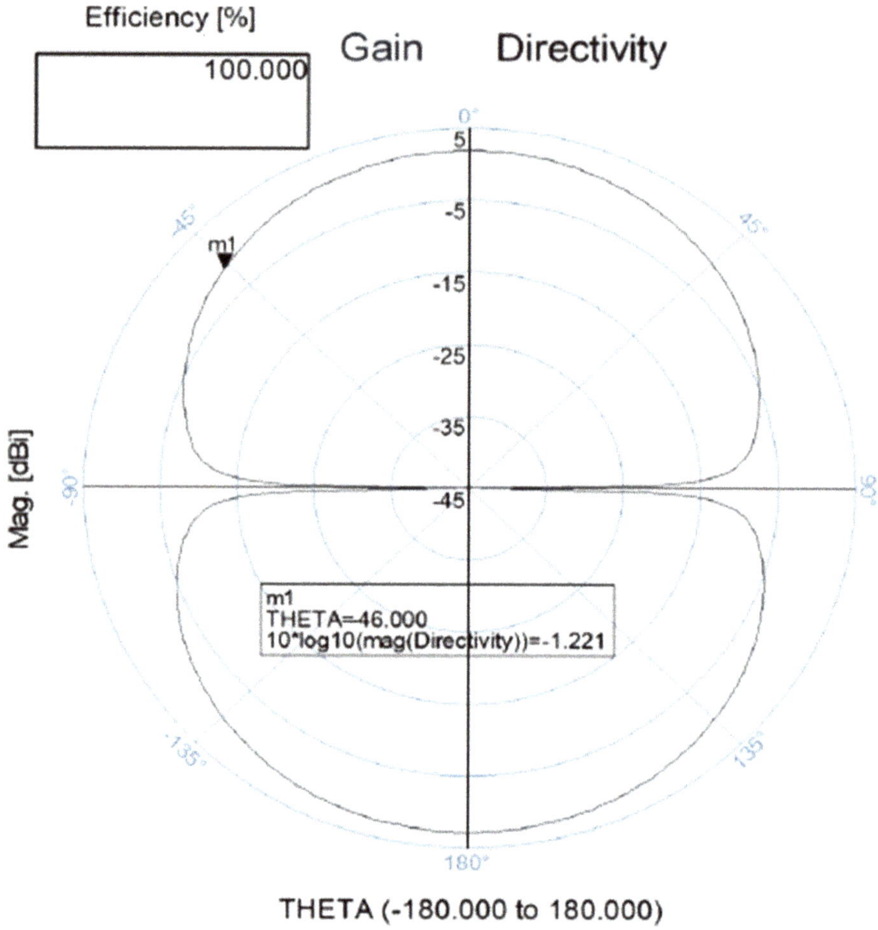

Figure 13.44. Radiation pattern of the wideband notch antenna with a fractal structure at 8 GHz.

An E PHEMT LNA, low noise amplifier, was connected to a notch antenna. The radiating element is connected to the LNA via an input matching network. An output matching network connects the amplifier port to the receiver. A DC bias network supplies the required voltage to the amplifiers. The amplifier specification is listed in table 12.1. The amplifier's complex S parameters are listed in table 12.2, and the noise parameters are listed in table 12.3. The active notch antenna's S_{21} parameters (gain) are presented in figure 13.46. The active antenna's gain is 22 ± 2.5 dB for frequencies ranging from 200 MHz to 900 MHz. The active antenna's gain is 12.5 ± 2.5 dB for frequencies ranging from 1 GHz to 3 GHz. The active notch antenna's noise figure is presented in figure 13.47. The active notch antenna's noise figure is 0.5 ± 0.3 dB for frequencies ranging from 300 MHz to 3.0 GHz. The active notch antenna's S_{22} parameters are lower than −5 dB for frequencies from 0.5 GHz to 3 GHz.

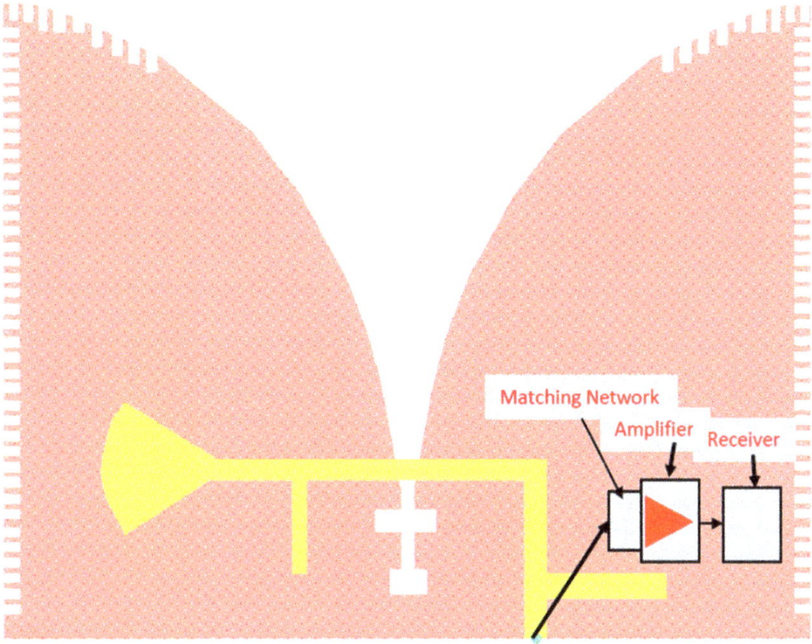

Figure 13.45. A wideband active notch antenna with a fractal structure.

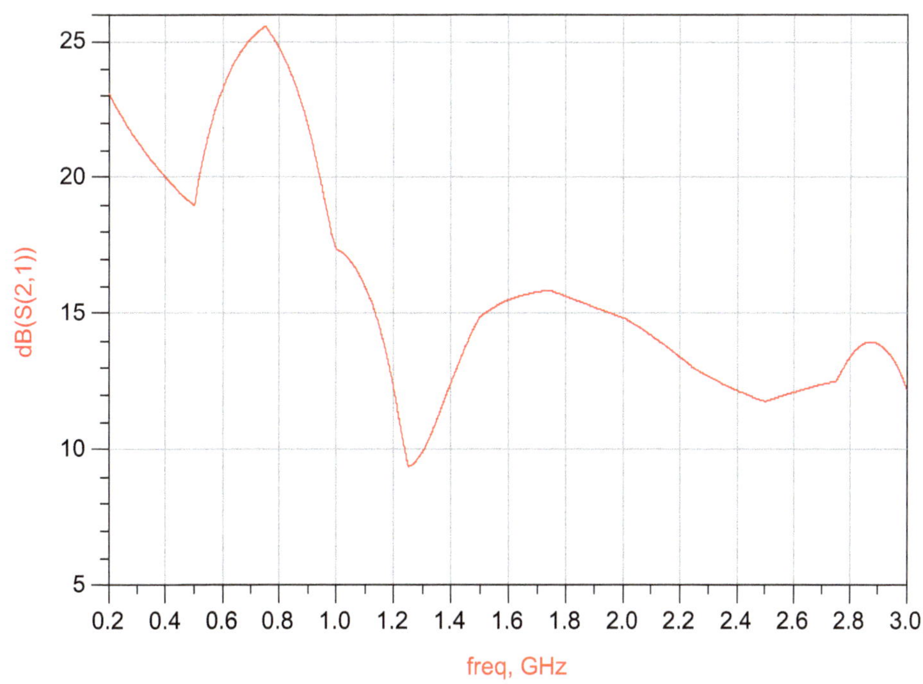

Figure 13.46. Active notch antenna's S_{21} parameters.

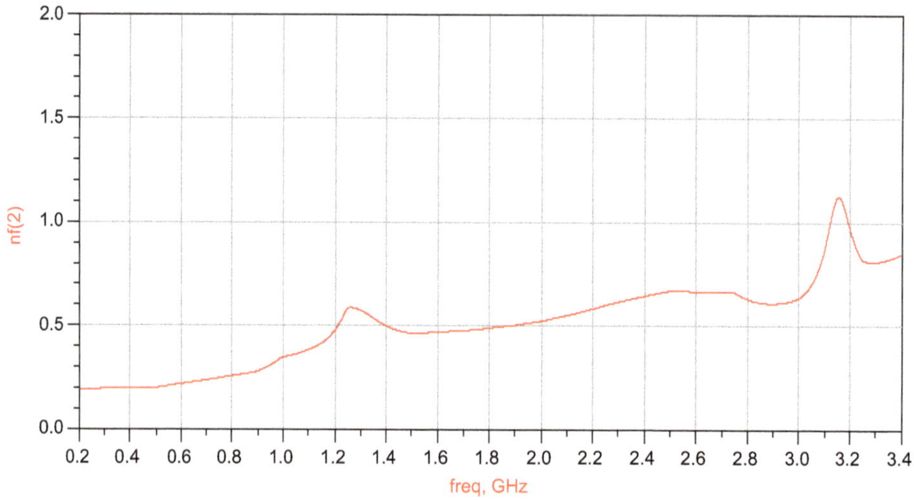

Figure 13.47. Active notch antenna's noise figure.

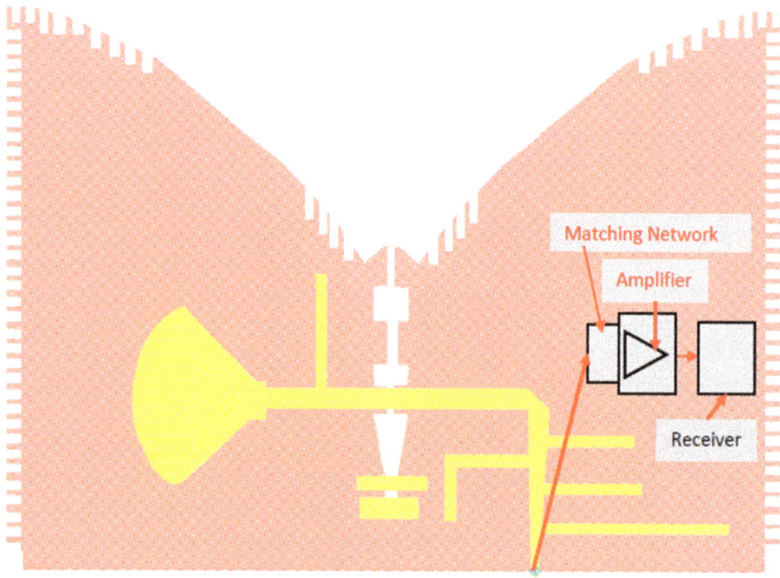

Figure 13.48. A wideband fractal active notch antenna with a fractal structure.

13.15 New compact ultra-wideband active notch antenna 0.4 GHz to 3 GHz

A wideband notch antenna with a fractal structure has been designed. The antenna is printed on RT-Duroid 5880 dielectric substrate with a dielectric constant of 2.2 and 1.2 mm thick. The notch antenna is shown in figure 13.48. The dimensions are

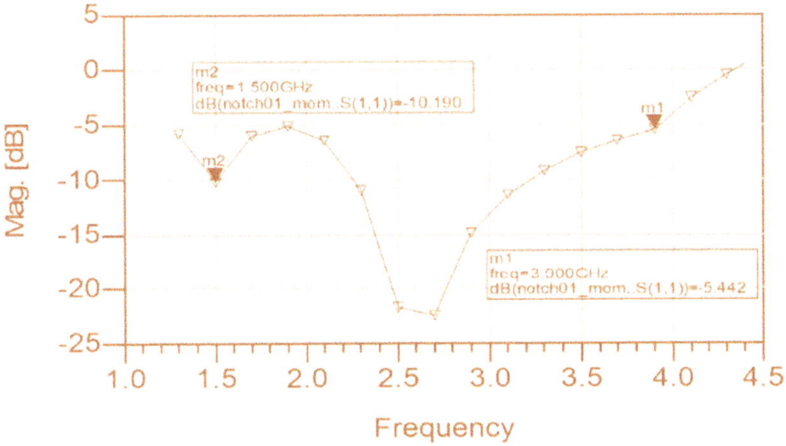

Figure 13.49. A fractal active notch antenna's S_{11} parameters.

52.2 × 36.8 mm. The antenna's center frequency is 1.7 GHz. The antenna's bandwidth is around 100% for S_{11} lower than −5 dB, as shown in figure 13.49. The notch antenna's VSWR is better than 3:1 for frequencies from 0.4 GHz to 3 GHz. The antenna's beamwidth is around 84°. An E PHEMT LNA, low noise amplifier, was connected to a notch antenna. The radiating element is connected to the LNA via an input matching network. An output matching network connects the amplifier port to the receiver. A DC bias network supplies the required voltage to the amplifiers. The amplifier specification is listed in table 12.1. The amplifier's complex S parameters are listed in table 12.2, and the noise parameters are listed in table 12.3. The active notch antenna's S_{21} parameters (gain) are presented in figure 13.50. The active antenna's gain is 20 ± 2.5 dB for frequencies ranging from 400 MHz to 1.3 GHz, and 12.5 ± 2.5 dB for frequencies ranging from 1.3 GHz to 3 GHz. The active notch antenna's noise figure is presented in figure 13.51. It is 0.5 ± 0.3 dB for frequencies ranging from 300 MHz to 3.0 GHz. The S_{22} parameters are lower than −5 dB for frequencies from 0.5 GHz to 3 GHz.

13.16 Conclusions

The chapter presents new compact ultra-wideband slot and notch antennas with frequencies ranging from 1 GHz to 18 GHz. The slot and notch antennas were analyzed by using 3D full-wave software. The antenna's bandwidth is from 50% to 100% with a VSWR better than 3:1. The antenna's gain is around 3 dBi with efficiency higher than 90%. The electrical parameters were computed in the vicinity of the human body. A compact new ultra-wideband notch antenna 1 GHz to 6 GHz and a wideband notch antenna 5.8 GHz to 18 GHz are presented. The space-filling technique and Hilbert curves were employed to design the fractal notch antennas. The fractal notch antennas were analyzed using 3D full-wave software. The

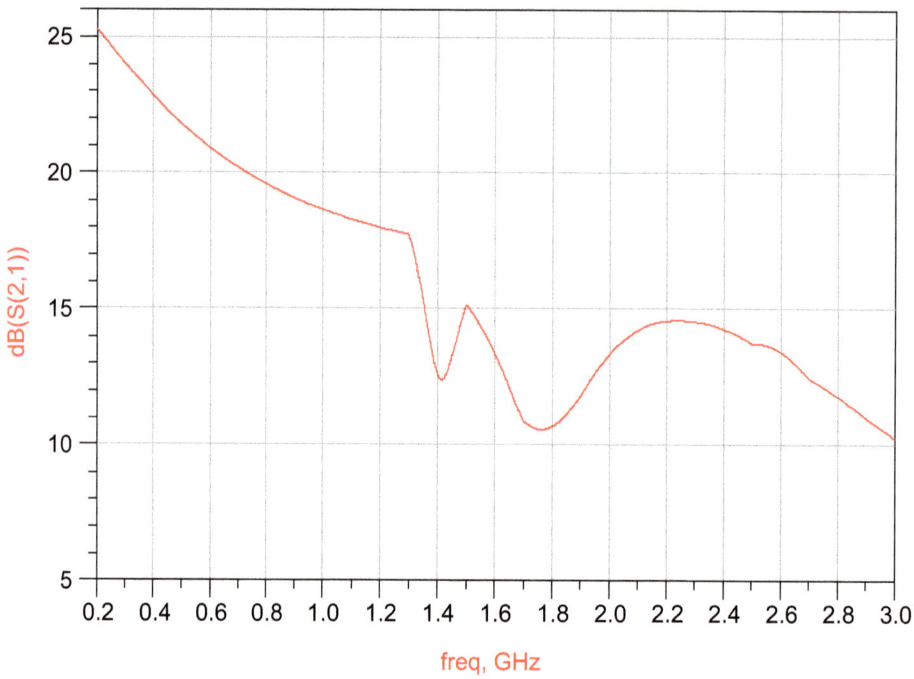

Figure 13.50. A fractal active notch antenna's S_{21} parameters.

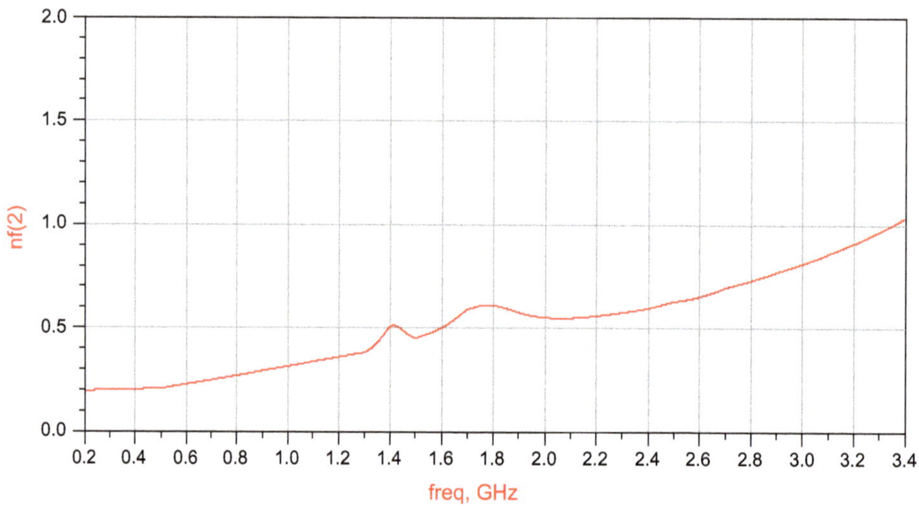

Figure 13.51. Active fractal notch antenna's noise figure.

antenna's bandwidth is around 100% with a VSWR better than 3:1, and the gain is around 3.5 dBi with efficiency higher than 90%. By using a fractal structure the notch antenna's length and width can be reduced by up to 50%.

This chapter also presents new compact ultra-wideband active slot and notch antennas with frequencies ranging from 1 GHz to 6 GHz.

References

[1] Balanis C A 1996 *Antenna Theory: Analysis and Design* 2nd edn (New York: Wiley)

[2] Godara L C (ed) 2002 *Handbook of Antennas in Wireless Communications* (Boca Raton, FL: CRC Press)

[3] Kraus J D and Marhefka R J 2002 *Antennas for all Applications* 3rd edn (New Delhi: McGraw Hill)

[4] James J R, Hall P S and Wood C 1981 *Microstrip Antenna Theory and Design* (London: Peter Peregrinus)

[5] Sabban A and Gupta K C 1991 Characterization of radiation loss from microstrip discontinuities using a multiport network modeling approach *IEEE Trans. Microwave Theory Tech.* **39** 705–12

[6] Sabban A 1991 Multiport network model for evaluating radiation loss and coupling among discontinuities in microstrip circuits *PhD Thesis* (University of Colorado at Boulder)

[7] Sabban A 1986 Microstrip antenna arrays *U.S. Patent US 1986/4,623,893*

[8] Sabban A 1983 A new wideband stacked microstrip antenna *IEEE Antenna and Propagation Symp. (Houston, TX, USA)*

[9] Sabban A 2015 *Low Visibility Antennas for Communication Systems* (New York: Taylor and Francis)

[10] Sabban A 1981 Wideband microstrip antenna arrays *IEEE Antenna and Propagation Symp. MELCOM (Tel Aviv)*

[11] Sabban A 2014 *RF Engineering, Microwave and Antennas* (Israel: Saar Publications)

[12] Fujimoto K and James J R (ed) *Mobile Antenna Systems Handbook* (Boston, MA: Artech House)

[13] Sabban A 2016 New wideband notch antennas for communication systems *Wirel. Eng. Technol. J.* **7** 75–82

[14] Sabban A 2012 Dual polarized dipole wearable antenna *US Patent number: 8203497* USA

[15] Sabban A 2012 Wideband tunable printed antennas for medical applications *IEEE Antenna and Propagation Symp. (Chicago, IL, USA)*

[16] Sabban A 2013 New wideband printed antennas for medical applications *IEEE Trans. Antennas Propag.* **61** 84–91

[17] Sabban A 2013 Comprehensive study of printed antennas on human body for medical applications *Int. J. Adv. Med. Sci.* **1** 1–10

Chapter 14

Aperture antennas for wireless communication systems

Reflector, horn antennas and antenna arrays are usually used as base station antennas for wireless communication systems. Reflector and horn antennas are defined as aperture antennas. The longest dimension of an aperture antenna is higher than several wavelengths. Reflector and horn antennas are used as transmitting antennas in wireless communication base stations.

14.1 The parabolic reflector antenna's configuration

The parabolic reflector antenna [1] consists of a radiating feed that is used to illuminate a reflector that is curved in the form of an accurate parabolic with diameter D as presented in figure 14.1. This shape enables a very narrow beam to be obtained. To provide the optimum illumination of the reflecting surface, the level of the parabola illumination should be greater by 10 dB in the center than that at the parabola edges.

Reflector geometry is presented in figure 14.2. The following relations given in equations (14.1)–(14.4) can be derived from the reflector geometry.

Parabolic reflector antenna gain

$$G \cong 10\log_{10}\left(\alpha\frac{(\pi D)^2}{\lambda^2}\right) \tag{14.1}$$

$$PQ = r'\cos\theta' \tag{14.2}$$

$$2f = r'(1 + \cos\theta') \tag{14.3}$$

doi:10.1088/978-0-7503-5224-6ch14

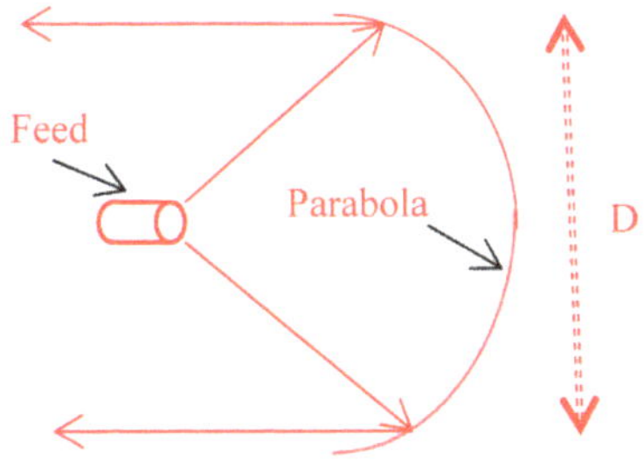

Figure 14.1. Parabolic antenna.

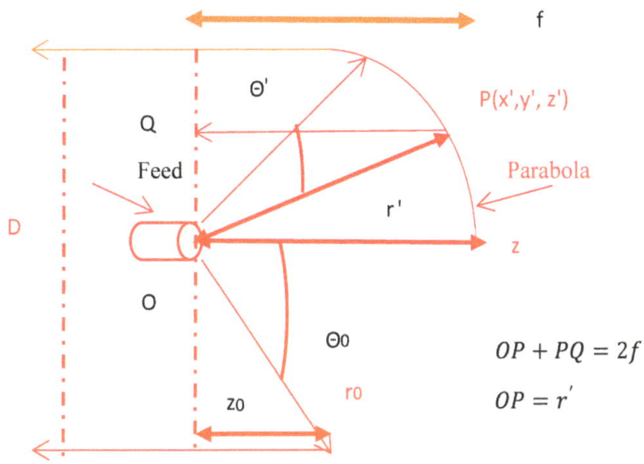

Figure 14.2. Reflector geometry.

$$2f = r' + r'\cos\theta' = \sqrt{(x')^2 + (y')^2 + (z')^2} + z' \tag{14.4}$$

The relation between the reflector diameter D and θ is given in equations (14.5)–(14.8).

$$\theta_0 = \tan^{-1}\frac{D/2}{z_0} \tag{14.5}$$

$$z_0 = f - \frac{(D/2)^2}{4f} \tag{14.6}$$

$$\theta_0 = \tan^{-1}\left|\frac{D/2}{z_0}\right| = \tan^{-1}\left|\frac{D/2}{f - \frac{(D/2)^2}{4f}}\right| = \tan^{-1}\left|\frac{f/2D}{\left(\frac{f}{D}\right)^2 - \frac{1}{16}}\right| \tag{14.7}$$

$$f = \frac{D}{4}\cot\left(\frac{\theta_0}{2}\right) \tag{14.8}$$

14.2 Reflector directivity

Reflector directivity is a function of the reflector geometry and feed radiation characteristics as given in equations (14.9) and (14.10).

$$D_0 = \frac{4\pi U_{max}}{P_{rad}} = \frac{16\pi^2}{\lambda^2} f^2 \left| \int_0^{\theta_0} \sqrt{G_F(\theta')} \tan\left(\frac{\theta'}{2}\right) d\theta' \right|^2 \tag{14.9}$$

$$D_0 = \frac{(\pi D)^2}{\lambda^2} \left[\cot^2\left(\frac{\theta_0}{2}\right) \left| \int_0^{\theta_0} \sqrt{G_F(\theta')} \tan\left(\frac{\theta'}{2}\right) d\theta' \right|^2 \right] \tag{14.10}$$

The reflector aperture efficiency is given in equation (14.11). The feed radiation pattern may be presented as in equation (14.12).

$$\epsilon_{ap} = \left[\cot^2\left(\frac{\theta_0}{2}\right) \left| \int_0^{\theta_0} \sqrt{G_F(\theta')} \tan\left(\frac{\theta'}{2}\right) d\theta' \right|^2 \right] \tag{14.11}$$

$$\begin{aligned} &G_F(\theta') = G_0^n \cos^n(\theta')\, for\ 0 \leqslant \theta' \leqslant \pi/2 \\ &G_F(\theta') = 0\, for\ \pi/2 < \theta' \leqslant \pi \\ &\int_0^{\pi/2} G_0^n \cos^n(\theta') \sin(\theta') d\theta' = 2 \end{aligned} \tag{14.12}$$

Where

$$G_0^n = 2(n+1)$$

Uniform illumination of the reflector aperture may be achieved if $G_F(\theta')$ is given by equation (14.13).

$$\begin{aligned} &G_F(\theta') = \sec^4\left(\frac{\theta'}{2}\right) for\ 0 \leqslant \theta' \leqslant \pi/2 \\ &G_F(\theta') = 0 \text{ for } \pi/2 < \theta' \leqslant \pi \end{aligned} \tag{14.13}$$

The reflector aperture efficiency is computed by multiplying all the antenna efficiencies due to spillover, blockage, taper, phase error, cross-polarization losses and random error over the reflector surface. $\epsilon_{ap} = \epsilon_s \epsilon_t \epsilon_b \epsilon_x \epsilon_p \epsilon_r$.

ϵ_s– Spillover efficiency, written in equation (14.14).

ϵ_t– Taper efficiency, written in equation (14.15).

ϵ_b– Blockage efficiency.
ϵ_p– Phase efficiency.
ϵ_x– Cross-polarization efficiency.
ϵ_r– Random error over the reflector surface efficiency.

$$\epsilon_s = \frac{\int_0^{\theta_0} G_F(\theta')\sin\theta' d\theta'}{\int_\pi^{\theta_0} G_F(\theta')\sin\theta' d\theta'} \tag{14.14}$$

$$\epsilon_t = 2\cot^2\left(\frac{\theta_0}{2}\right)\frac{\left|\int_0^{\theta_0} \sqrt{G_F(\theta')}\tan\left(\frac{\theta'}{2}\right)d\theta'\right|^2}{\int_0^{\theta_0} G_F(\theta')\sin\theta' d\theta'} \tag{14.15}$$

In the literature [1] we can find graphs that present the reflector antenna efficiencies as a function of the reflector antenna geometry and feed radiation pattern. However, equations (14.10)–(14.16) give us a good approximation of the reflector's directivity.

$$D_0 = \frac{(\pi D)^2 \epsilon_{ap}}{\lambda^2} \tag{14.16}$$

14.3 Cassegrain reflector

The parabolic reflector or dish antenna consists of a radiating element that may be a simple dipole or a waveguide horn antenna. This is placed at the center of the metallic parabolic reflecting surface as shown in figure 14.3. The energy from the radiating element is arranged so that it illuminates the sub-reflecting surface. The energy from the sub-reflector is arranged so that it illuminates the main reflecting surface. Once the energy is reflected it leaves the antenna's system in a narrow beam.

14.4 Horn antennas

Horn antennas are used as a feed element for radio astronomy, satellite tracking and communication reflector antennas, phased arrays' radiating elements, and in antenna calibration and measurements. Figure 14.4(a) presents an E plane sectoral horn, figure 14.4(b) presents an H plane sectoral horn, figure 14.4(c) presents a pyramidal horn, and figure 14.4(d) presents a conical horn.

14.4.1 *E* plane sectoral horn

Figure 14.5 presents an E plane sectoral horn. Horn antennas are fed by a waveguide. The excited mode is TE_{10}. Field expressions over the horn's aperture

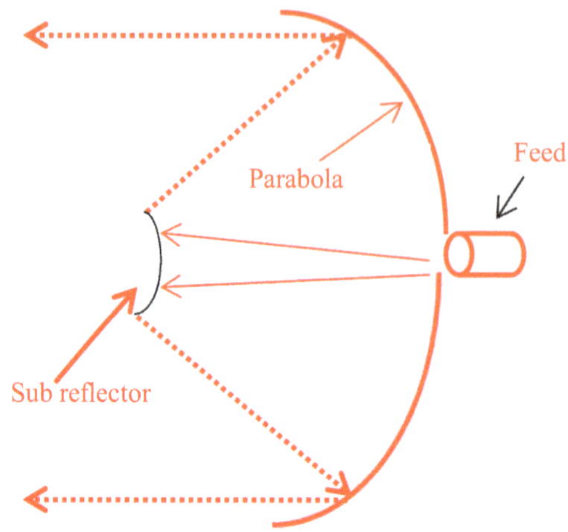

Figure 14.3. Cassegrain feed system.

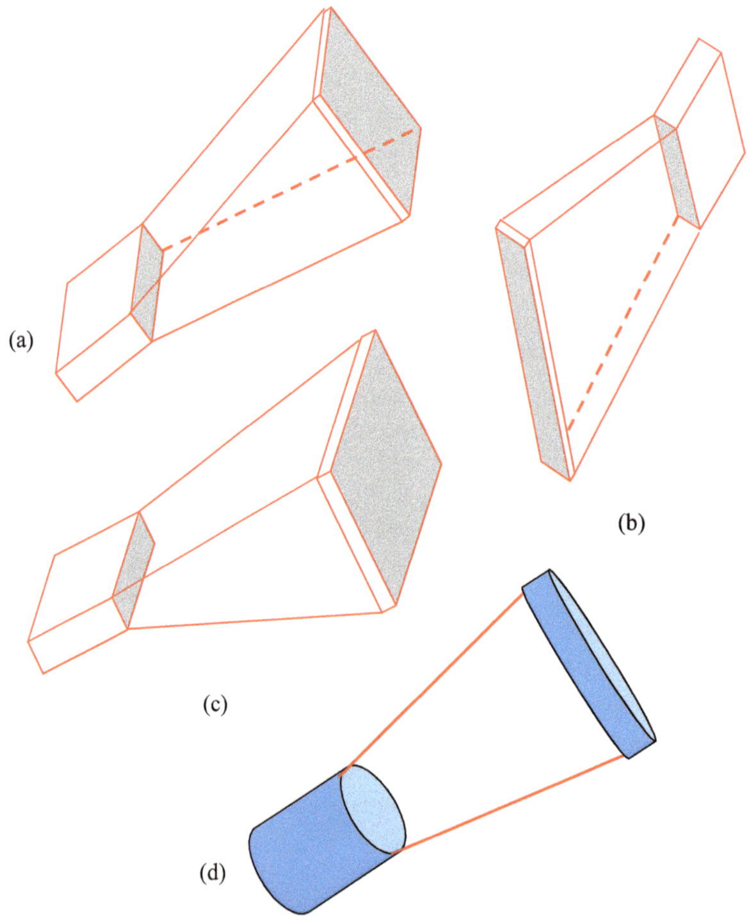

Figure 14.4. (a) *E* plane sectoral horn. (b) *H* plane sectoral horn. (c) Pyramidal horn. (d) Conical horn.

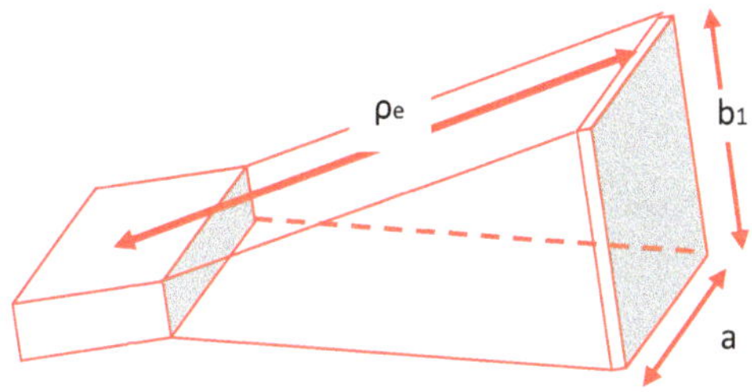

Figure 14.5. *E* plane sectoral horn.

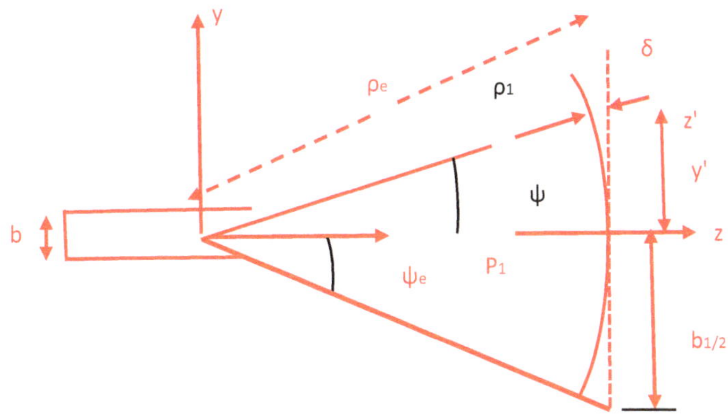

Figure 14.6. *E* plane sectoral horn geometry.

are similar to the fields of a TE_{10} mode in a rectangular waveguide with the aperture dimensions of, a, b_1. The fields in the antenna's aperture are given in equations (14.17)–(14.19).

$$E_y'(x', y') \approx E_1 \cos\left(\frac{\pi}{a}x'\right)e^{-j\left[ky'^2/(2\rho_1)\right]} \tag{14.17}$$

$$H_x'(x', y') \approx \frac{E_1}{\eta} \cos\left(\frac{\pi}{a}x'\right)e^{-j\left[ky'^2/(2\rho_1)\right]} \tag{14.18}$$

$$H_z'(x', y') \approx jE_1\left(\frac{\pi}{ka\eta}\right)\sin\left(\frac{\pi}{a}x'\right)e^{-j\left[ky'^2/(2\rho_1)\right]} \tag{14.19}$$

The horn length is ρ_1, as shown in figure 14.6. The extra distance along the aperture sides compared with the distance to the center is δ and is given by equations (14.20)–(14.23).

$$\delta = \rho_e - \sqrt{\rho_e - (b_1/2)^2} = \rho_e\left(1 - \sqrt{1-(b_1/2\rho_e)^2}\right) = b_1^2/8\rho_e \tag{14.20}$$

$$\frac{\delta}{\lambda} = S = b_1^2/8\lambda\rho_e \tag{14.21}$$

S represents the quadratic phase distribution as given in equation (14.21).

$$(\delta(y') + \rho_1)^2 = \rho_1^2 + (y')^2 \tag{14.22}$$

$$\delta(y') = -\rho_1 + \sqrt{\rho_1^2 + (y')^2} = -\rho_1 + \rho_1\sqrt{1+(\frac{y'}{\rho_1})^2} \tag{14.23}$$

The maximum phase deviation at the aperture $\varnothing_{max}$ is given by equation (14.24).

$$\varnothing_{max} = k\delta(y') : (y' = b_1/2) \;=\; kb_1^2/8\rho_1 \tag{14.24}$$

The total flare angle of the horn, $2\psi_e$, is given in equation (14.25).

$$2\,\psi_e = 2\tan^{-1}(b_1/2\rho_1) \tag{14.25}$$

Directivity of the E plane horn

The maximum radiation is given by equations (14.26)–(14.28).

$$U_{max} = \frac{r^2}{2\eta}\,|E|_{max}^2 \tag{14.26}$$

$$U_{max} = \frac{2ka^2\rho_1}{\pi^3\eta}\,|E|^2|F(t)|^2 \tag{14.27}$$

$$|F(t)|^2 = \left[C^2\left(\frac{b_1}{\sqrt{2\lambda\rho_1}}\right) + S^2\left(\frac{b_1}{\sqrt{2\lambda\rho_1}}\right)\right] \tag{14.28}$$

C and S are Fresnel integers and are given in table 14.1.

The total radiated power by the horn is given in equation (14.29).

$$P_{rad} = \frac{ab_1}{4\eta}\,|E|^2 \tag{14.29}$$

The directivity of the E plane horn D_E is given in equation (14.30).

$$D_E = \frac{4\pi U_{max}}{P_{rad}} = \frac{64a\rho_1}{\pi\lambda b_1}\left[C^2\left(\frac{b_1}{\sqrt{2\lambda\rho_1}}\right) + S^2\left(\frac{b_1}{\sqrt{2\lambda\rho_1}}\right)\right] \tag{14.30}$$

Figure 14.7 presents the H plane horn radiation pattern as a function of S.

Where,

$$S = b_1^2/8\lambda\rho_e.$$

Table 14.1. Fresnel integers.

x	$C_1(x)$	$S_1(x)$	$C(x)$	$S(x)$
0.0	0.626 66	0.626 66	0.0	0.0
0.1	0.526 66	0.626 32	0.100 00	0.000 52
0.2	0.426 69	0.623 99	0.199 92	0.004 19
0.3	0.3269	0.617 66	0.299 40	0.014 12
0.4	0.227 68	0.605 36	0.397 48	0.033 36
0.5	0.129 77	0.585 18	0.492 34	0.064 73
0.6	0.034 39	0.555 32	0.581 10	0.110 54
0.7	−0.056 72	0.514 27	0.659 65	0.172 14
0.8	−0.141 19	0.460 92	0.722 84	0.249 34
0.9	−0.216 06	0.394 81	0.764 82	0.339 78
1	−0.277 87	0.316 39	0.779 89	0.438 26
1.1	−0.322 85	0.227 28	0.763 81	0.536 50
1.2	−0.347 29	0.130 54	0.715 44	0.623 40
1.3	−0.348 03	0.030 81	0.638 55	0.686 33
1.4	−0.323 12	−0.065 73	0.543 10	0.713 53
1.5	−0.272 53	−0.151 58	0.445 26	0.697 51
1.6	−0.198 86	−0.218 61	0.365 46	0.638 89
1.7	−0.1079	−0.259 05	0.323 83	0.549 20
1.8	−0.008 71	−0.266 82	0.333 63	0.450 94
1.9	0.086 80	−0.239 18	0.394 47	0.373 35
2.0	0.165 20	−0.178 12	0.488 25	0.343 42
2.1	0.213 59	−0.091 41	0.581 56	0.374 27
2.2	0.222 42	0.007 43	0.636 29	0.455 70
2.3	0.188 33	0.100 54	0.626 56	0.553 15
2.4	0.116 50	0.168 79	0.554 96	0.619 69
2.5	0.021 35	0.196 14	0.457 42	0.619 18
2.6	−0.075 18	0.174 54	0.388 94	0.549 99
2.7	−0.148 16	0.107 89	0.392 49	0.452 92
2.8	−0.176 46	0.013 29	0.467 49	0.391 53
2.9	−0.150 21	−0.081 81	0.562 37	0.410 14
3.0	−0.076 21	−0.146 90	0.605 72	0.496 31
3.1	0.021 52	−0.158 83	0.561 60	0.581 81
3.2	0.107 91	−0.111 81	0.466 32	0.593 35
3.3	0.149 07	−0.022 60	0.405 70	0.519 29
3.4	0.126 91	0.073 01	0.438 49	0.429 65
3.5	0.049 65	0.133 35	0.532 57	0.415 25
3.6	−0.048 19	0.129 73	0.587 95	0.492 31
3.7	−0.119 29	0.062 58	0.541 95	0.574 98
3.8	−0.126 49	−0.034 83	0.448 10	0.565 62
3.9	−0.064 69	−0.110 30	0.422 33	0.475 21
4.0	0.032 19	−0.120 48	0.498 42	0.420 52
4.1	0.106 90	−0.058 15	0.573 69	0.475 80
4.2	0.112 28	0.038 85	0.541 72	0.563 20

4.3	0.043 74	0.107 51	0.449 44	0.554 00
4.4	−0.052 87	0.100 38	0.438 33	0.462 27
4.5	−0.108 84	0.021 49	0.526 02	0.434 27
4.6	−0.081 88	−0.071 26	0.567 24	0.516 19
4.7	0.008 10	−0.105 94	0.491 43	0.567 15
4.8	0.089 05	−0.053 81	0.433 80	0.496 75
4.9	0.092 77	0.042 24	0.500 16	0.435 07
5.0	0.015 19	0.098 74	0.563 63	0.499 19
5.1	−0.074 11	0.064 05	0.499 79	0.561 39
5.2	−0.091 25	−0.030 04	0.438 89	0.496 88
5.3	−0.018 92	−0.092 35	0.507 78	0.440 47
5.4	0.070 63	−0.059 76	0.557 23	0.514 03
5.5	0.084 08	0.034 40	0.478 43	0.553 69
5.6	0.006 41	0.089 00	0.451 71	0.470 04
5.7	−0.076 42	0.042 96	0.538 46	0.459 53
5.8	−0.069 14	−0.051 35	0.529 84	0.546 04
5.9	0.019 98	−0.082 31	0.448 59	0.516 33
6.0	0.082 45	−0.011 81	0.499 53	0.446 96
6.1	0.039 46	0.071 80	0.549 50	0.516 47
6.2	−0.053 63	0.060 18	0.467 61	0.539 82
6.3	−0.072 84	−0.031 44	0.476 00	0.455 55
6.4	0.008 33	−0.077 65	0.544 60	0.496 49
6.5	0.075 74	−0.013 26	0.481 61	0.545 38
6.6	0.031 83	0.068 72	0.468 99	0.463 07
6.7	−0.058 28	0.046 58	0.546 74	0.491 50
6.8	−0.057 34	−0.046 00	0.483 07	0.543 64
6.9	0.033 17	−0.064 40	0.473 22	0.462 44
7.0	0.068 32	0.020 77	0.545 47	0.444 70
7.1	−0.009 44	0.069 77	0.473 32	0.536 02
7.2	−0.064 43	0.000 41	0.488 74	0.457 25
7.3	−0.008 60	−0.067 93	0.539 27	0.518 44
7.4	0.065 82	−0.015 21	0.460 10	0.516 07
7.5	0.020 18	0.063 53	0.516 01	0.460 70
7.6	−0.061 37	0.023 67	0.515 64	0.538 85
7.7	−0.025 80	−0.059 58	0.462 78	0.482 02
7.8	0.058 28	−0.026 68	0.539 47	0.489 64
7.9	0.026 38	0.057 52	0.475 98	0.532 35
8.0	−0.057 30	0.024 94	0.499 80	0.460 21
8.1	−0.022 38	−0.057 52	0.522 75	0.532 04
8.2	0.058 03	−0.018 70	0.463 84	0.485 89
8.3	0.013 87	0.058 61	0.537 75	0.493 23
8.4	−0.058 99	0.007 89	0.470 92	0.524 29
8.5	−0.000 80	−0.058 81	0.514 17	0.465 34

(Continued)

Table 14.1. (*Continued*)

x	$C_1(x)$	$S_1(x)$	$C(x)$	$S(x)$
8.6	0.057 67	0.007 29	0.502 49	0.536 93
8.7	−0.016 16	0.055 15	0.482 74	0.467 74
8.8	−0.050 79	−0.025 45	0.527 97	0.522 94
8.9	0.034 61	−0.044 25	0.466 12	0.488 56
9.0	0.035 26	0.042 93	0.535 37	0.499 85
9.1	−0.049 51	0.023 81	0.466 61	0.510 42
9.2	−0.010 21	−0.053 38	0.529 14	0.481 35
9.3	0.053 54	0.004 85	0.476 28	0.524 67
9.4	−0.020 20	0.049 20	0.518 03	0.471 34
9.5	−0.039 95	−0.034 26	0.487 29	0.531 00
9.6	0.045 13	−0.025 99	0.508 13	0.467 86
9.7	0.008 37	0.050 86	0.495 44	0.532 50
9.8	−0.049 83	−0.010 94	0.501 92	0.467 58
9.9	0.029 16	−0.041 24	0.449 61	0.532 15
10.0	0.025 54	0.042 98	0.499 89	0.468 17
10.1	−0.049 27	0.004 78	0.499 61	0.531 51
10.2	0.017 38	−0.045 83	0.501 86	0.468 85
10.3	0.032 33	0.036 21	0.495 75	0.530 61
10.4	−0.046 81	0.010 94	0.507 51	0.470 33
10.5	0.013 60	−0.045 63	0.488 49	0.528 04

14.4.2 *H* plane sectoral horn

An *H* plane sectoral horn is shown in figure 14.8, and an *H* plane sectoral horn's geometry is shown in figure 14.9.

Field expressions over the horn's aperture are similar to the fields of the TE_{10} mode in a rectangular waveguide with the aperture dimensions of, a, b_1. The fields in the antenna's aperture are written in equations (14.30) and (14.31).

$$E_y'(x', y') \approx E_2 \cos\left(\frac{\pi}{a}x'\right)e^{-j\left[kx'^2/(2\rho_2)\right]} \tag{14.31}$$

$$H_x'(x', y') \approx \frac{E_2}{\eta} \cos\left(\frac{\pi}{a}x'\right)e^{-j\left[kx'^2/(2\rho_2)\right]} \tag{14.32}$$

The horn length is ρ_l. The extra distance along the aperture sides compared with the distance to the center is δ and is given by equations (14.33)–(14.36).

$$\delta = \rho_h - \sqrt{\rho_h - (a_1/2)^2} = \rho_h\left(1 - \sqrt{1-(a_1/2\rho_h)^2}\right) = a_1^2/8\rho_h \tag{14.33}$$

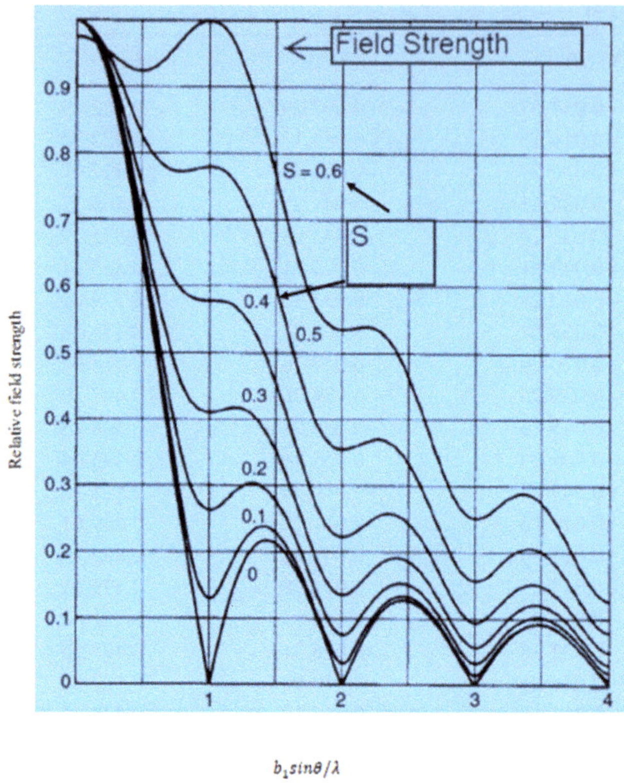

Figure 14.7. *H* plane horn radiation pattern as function of *S*, $S = b_1^2/8\lambda\rho_e$.

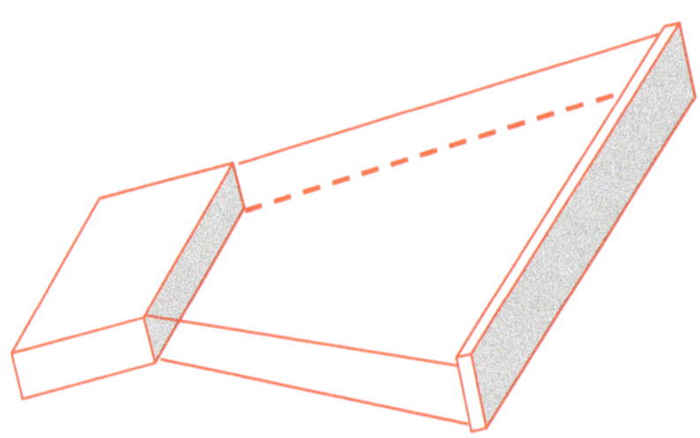

Figure 14.8. *H* plane sectoral horn.

$$\frac{\delta}{\lambda} = S = a_1^2/8\lambda\rho_h \tag{14.34}$$

S represents the quadratic phase distribution as written in equation (14.34).

$$(\delta(x') + \rho_2)^2 = \rho_2^2 + (x')^2 \tag{14.35}$$

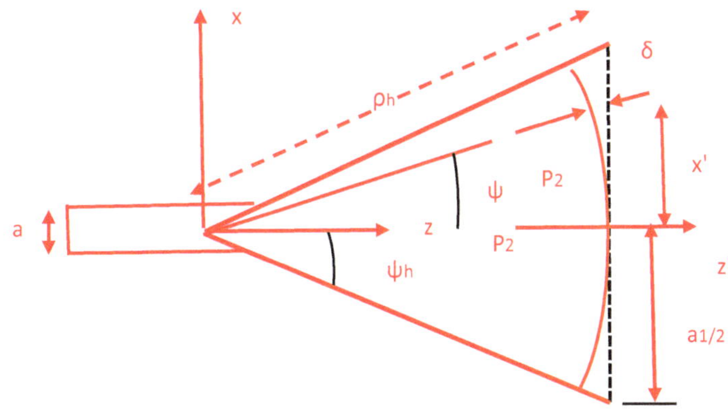

Figure 14.9. *H* plane sectoral horn's geometry.

$$\delta(x') = -\rho_2 + \sqrt{{\rho_2}^2 + (x')^2} = -\rho_2 + \rho_1\sqrt{1+\left(\frac{x'}{\rho_2}\right)^2} \tag{14.36}$$

The maximum phase deviation at the aperture $\varnothing_{max}$ is given by equation (14.37).

$$\varnothing_{max} = k\delta(x') \vdots (x' = a_1/2) = k{a_1}^2/8\rho_2 \tag{14.37}$$

The total flare angle of the horn, $2\psi_e$, is given in equation (14.38).

$$2\psi_h = 2\tan^{-1}(a_1/2\rho_2) \tag{14.38}$$

Directivity of the *E* plane horn

The maximum radiation is given by equations (14.39) and (14.41).

$$U_{max} = \frac{r^2}{2\eta}|E|^2_{max} \tag{14.39}$$

$$U_{max} = \frac{b^2\rho_2}{4\lambda\eta}|E|^2|F(t)|^2 \tag{14.40}$$

$$|F(t)|^2 = [(C(u) - C(v))^2 + (S(u) - S(v))^2] \tag{14.41}$$

Where

$$u = \frac{1}{\sqrt{2}}\left(\frac{\sqrt{\rho_2\lambda}}{a_1} - \frac{a_1}{\sqrt{\rho_2\lambda}}\right) \text{ and } v = \frac{1}{\sqrt{2}}\left(\frac{\sqrt{\rho_2\lambda}}{a_1} + \frac{a_1}{\sqrt{\rho_2\lambda}}\right)$$

C and *S* are Fresnel integers.

The total radiated power by the horn is given in equation (14.42).

$$P_{rad} = \frac{ab_1}{4\eta}|E|^2 \tag{14.42}$$

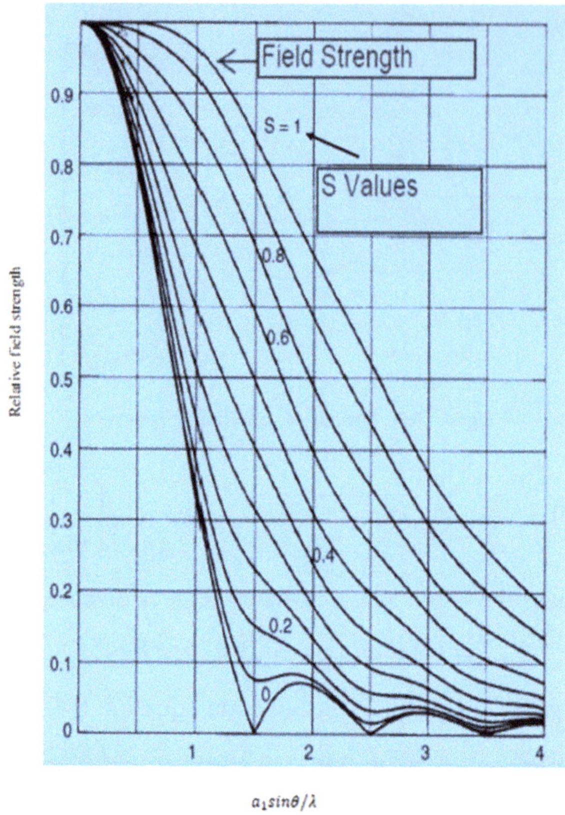

Figure 14.10. H plane horn radiation pattern as a function of S, $S = a_1^2/8\lambda\rho_h$.

The directivity of H plane horn D_H is given in equation (14.43).

$$D_H = \frac{4\pi U_{max}}{P_{rad}} = \frac{4b\pi\rho_2}{\lambda a_1}[(C(u) - C(v))^2 + (S(u) - S(v))^2] \tag{14.43}$$

H plane horn radiation pattern as a function of S, $S = a_1^2/8\lambda\rho_h$, is shown in figure 14.10.

14.4.3 Pyramidal horn antenna

The pyramidal horn antenna is a combination of the E and H horns as shown in figure 14.11. The pyramidal horn antenna is realizable only if $\rho_h = \rho_e$.

The directivity of the pyramidal horn antenna D_P is given in equation (14.44).

$$D_P = \frac{4\pi U_{max}}{P_{rad}} = \frac{\pi\lambda^2}{32ab}D_E D_H \tag{14.44}$$

Kraus [3] gives the following approximation, equations (14.45)–(14.48), for the pyramidal horn's beamwidth.

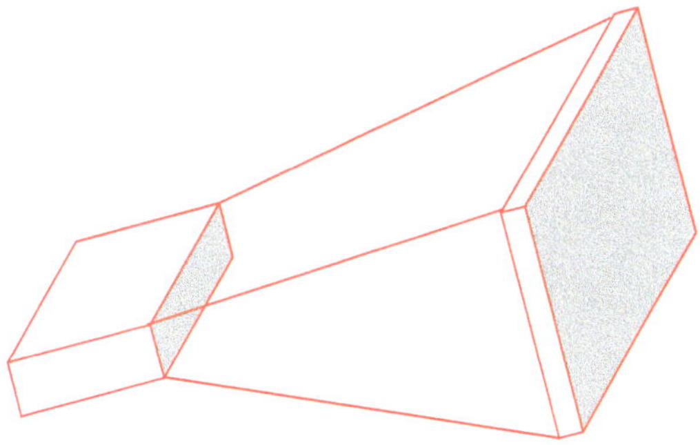

Figure 14.11. Pyramidal horn antenna.

$$\theta^{e}_{3dB} = \frac{56}{A_{e\lambda}} \tag{14.45}$$

$$\theta^{h}_{3dB} = \frac{67}{A_{h\lambda}} \tag{14.46}$$

$$\theta^{e}_{10dB} = \frac{100.8}{A_{e\lambda}} \tag{14.47}$$

$$\theta^{h}_{10dB} = \frac{120.6}{A_{h\lambda}} \tag{14.48}$$

The aperture dimensions of the wavelength in the E plane is $A_{e\lambda}$. The aperture dimensions of the wavelength in the H plane is $A_{h\lambda}$. The pyramidal horn's gain is given by equation (14.49).

$$G = 10\log_{10}4.5A_{h\lambda}A_{e\lambda}\mathrm{dBi} \tag{14.49}$$

The relative power at any angle, $P^{h}_{dB}(\theta)$, is given approximately by equation (14.50).

$$P^{h}_{dB}(\theta) = 10\left(\frac{\theta}{\theta^{h}_{10dB}}\right)^2 \tag{14.50}$$

14.5 Antenna arrays for wireless communication systems

14.5.1 Introduction

An array of antenna elements is a set of antennas used for transmitting or receiving electromagnetic waves. An array of antennas is a collection of N similar radiators with the same three-dimensional radiation pattern. All the radiating elements are fed with the same frequency and with a specified amplitude and phase relationship for the drive voltage of each element. The array functions as a single antenna, generally with a higher antenna gain than would be obtained from the individual elements. In antenna arrays electromagnetic wave interference is used to enhance the electromagnetic signal in one desired direction at the expense of other directions. It may also be used to null the radiation pattern in one particular direction. The theory of antenna arrays is presented in [1–11]. Several printed arrays and the gain limitation of printed arrays are presented in [4–10]. Analysis and computations of losses in microstrip lines are given in [4–10]. MM wave arrays are presented in [7–10].

14.5.2 Array radiation pattern

The polar radiation pattern of a single element is called the 'element pattern' (EP). The array pattern is the polar radiation pattern, which would result if the elements were replaced by isotropic radiators, with the same amplitude and phase of excitation as the actual elements, and spaced at points on a grid corresponding to the far-field phase centers of the radiators. If we assume that all the polar radiation patterns of the elements taken individually are identical (within a certain tolerance) and that the patterns are all aligned in the same direction in azimuth and elevation, then the total array antenna pattern is obtained by multiplying the array factor (AF) by the element pattern. The total array antenna pattern, ET, is ET = AF·EP.

The radiated field strength at a certain point in space in the far-field is calculated by adding the contributions of each element to the total radiated fields. The summation of the contribution of each element to the total radiated fields is called the array factor (AF). The field strengths fall off as $1/r$, where r is the distance from the antenna to the field point as shown in figure 14.12. We must take into account the amplitude and phase angle of the radiator excitation, and also the phase delay that is due to the time it takes the signal to get from the source to the field point. This phase delay is expressed as $2\pi r/\lambda$, where λ is the free space wavelength of the electromagnetic wave.

Contours of equal field strength may be interpreted as an amplitude polar radiation pattern. Contours of the squared modulus of the field strength may be interpreted as a power polar radiation pattern. Figure 14.12 shows a four element

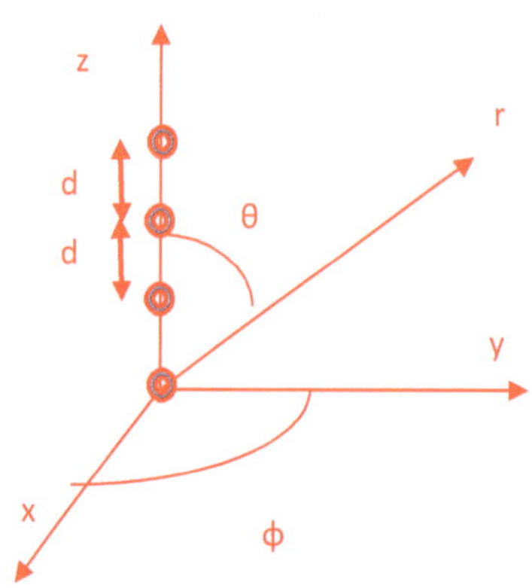

Figure 14.12. Coordinate system for external field calculations.

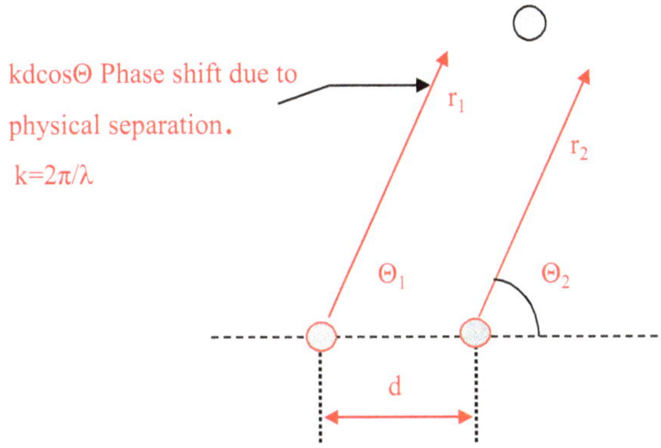

Figure 14.13. Two-element array.

array. The array factor of an N element array is given in equation (14.51). β represents the phase difference between the elements in the array.

Figure 14.13 shows a two-element array.

$$AF = 1 + e^{j\varphi} + e^{j2\varphi} + \ldots + e^{j\varphi(N-1)} = \sum_{n=1}^{N} e^{j\varphi(n-1)} \tag{14.51}$$

Where $\varphi = kd\cos\theta + \beta$. Where $k = 2\pi/\lambda$.

The series summation is given in equation (14.52).

$$AF = \frac{\sin\left(\frac{N\varphi}{2}\right)}{\sin\left(\frac{\varphi}{2}\right)} \tag{14.52}$$

The array factor will be zero when $\sin\left(\frac{N\varphi}{2}\right) = 0$. The array nulls will occur as given in equation (14.53).

$$\theta_n = \cos^{-1}\left[\frac{\lambda}{2\pi d}\left(-\beta \pm \frac{2n}{N}\pi\right)\right] n \neq N = 1, 2, 3\ldots \tag{14.53}$$

For $\varphi = \pm 2m\pi$ the array's maximum level is given in equation (14.54).

$$\theta_m = \cos^{-1}\left[\frac{\lambda}{2\pi d}(-\beta \pm 2m\pi)\right] m = 0,\ 1,\ 2,\ 3\ldots \tag{14.54}$$

The array's 3 dB beamwidth is given in equation (14.55).

$$\theta_{3dB} = \cos^{-1}\left[\frac{\lambda}{2\pi d}\left(-\beta \pm \frac{2.782}{N}\right)\right] \tag{14.55}$$

The peak value of the side lobe is given in equation (14.56).

$$\theta_{SL} = \cos^{-1}\left[\frac{\lambda}{2\pi d}\left(-\beta \pm \frac{3\pi}{N}\right)\right] \tag{14.56}$$

The side lobe level for $\beta = 0$ is -13.46 dB, as calculated in equation (14.57).

$$AF_{SL} = 20\log_{10}\left(\frac{2}{3\pi}\right) = -13.46\text{dB} \tag{14.57}$$

14.5.3 Broadside array

In a broadside array the main beam is perpendicular to the array. The array will radiate maximum energy perpendicular to the array if all elements in the array are fed with the same amplitude and phase level, $\beta = 0$.

The array factor will be zero when $\sin(\frac{N\varphi}{2})$. The array nulls will occur as given in equation (14.58).

$$\theta_n = \cos^{-1}\left(\pm\frac{n\lambda}{Nd}\right) \quad n \neq N = 1, 2, 3... \tag{14.58}$$

For $\varphi = \pm 2m\pi$ the array maximum level is given in equation (14.59).

$$\theta_m = \cos^{-1}\left(\frac{m\lambda}{d}\right) \quad m = 0, 1, 2, 3... \tag{14.59}$$

The array's 3 dB beamwidth is given in equation (14.60) when $\pi d/\lambda \ll 1$.

$$\theta_{3dB} = \cos^{-1}\left(\frac{1.391\lambda}{\pi Nd}\right) \quad \pi d/\lambda \ll 1 \tag{14.60}$$

The peak value of the side lobe is given in equation (14.61).

$$\theta_{SL} = \cos^{-1}\left[\frac{\lambda}{2d}\left(\pm\frac{(2s+1)}{N}\right)\right] \quad s = 0, 1, 2, 3... \tag{14.61}$$

14.5.4 End-fire array

In an end-fire array the main beam is in the direction of the array axis. The array will radiate maximum energy in the direction of the array axis. The array will radiate maximum energy in the direction $\theta = 0$ if $\beta = -kd$. The array will radiate maximum energy in the direction $\theta = 180$ if $\beta = kd$.

The array nulls will occur as given in equation (14.62).

$$\theta_n = \cos^{-1}\left(1-\frac{n\lambda}{Nd}\right) \quad n \neq N = 1, 2, 3... \tag{14.62}$$

For $\varphi = \pm 2m\pi$ the array maximum level is given in equation (14.63).

$$\theta_m = \cos^{-1}\left(1-\frac{m\lambda}{d}\right) \quad m = 0, 1, 2, 3... \tag{14.63}$$

The array's 3 dB beamwidth is given in equation (14.64) when $\pi d/\lambda \ll 1$.

$$\theta_{3dB} = \cos^{-1}\left(1-\frac{1.391\lambda}{\pi Nd}\right) \quad \pi d/\lambda \ll 1 \tag{14.64}$$

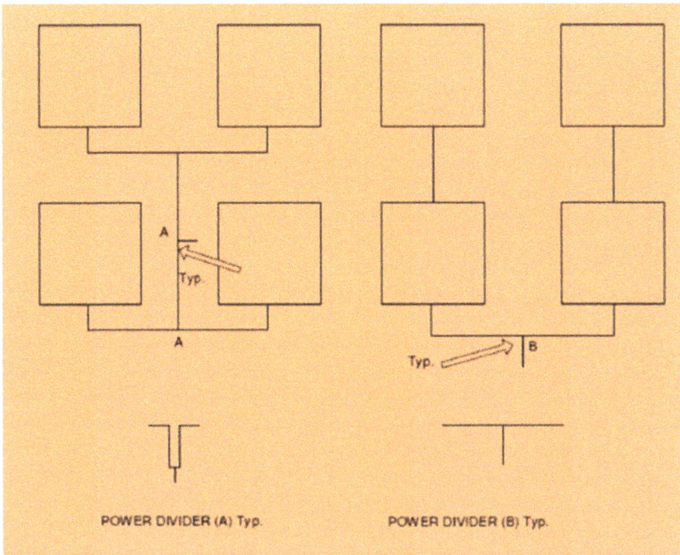

Figure 14.14. Configuration of a microstrip antenna array. (a) Parallel feed network. (b) Parallel series feed network.

The peak value of the side lobe is given in equation (14.65).

$$\theta_{SL} = \cos^{-1}\left[\frac{\lambda}{2d}\left(1-\frac{(2s+1)}{N}\right)\right] s = 0,\ 1,\ 2,\ 3\ldots \tag{14.65}$$

14.5.5 Printed arrays

In figure 14.14(a) a parallel feed network of a microstrip antenna array is shown. In figure 14.14(b) a parallel series feed network of a microstrip antenna array is presented.

Array directivity, D, may be written as: $D = D_0 = U_{max}/U_0 \sim AF_{max}^2 = N$.

A half power beamwidth may be written as: (HPBW) = 2 * (90 – arccos(1.39λ/πNd)).

In table 14.2 the array directivity and beamwidth as a function of the number of elements are listed. The radiation pattern of a 16 broadside element array is shown in figure 14.15. The array directivity is 12 dB.

14.5.6 Stacked microstrip antenna arrays

A stacked Ku band, 16 elements, microstrip antenna array was designed at 14.5 GHz as shown in figure 14.16. The resonator and feed network were printed on a substrate with a relative dielectric constant of 2.5 with thickness of 0.5 mm. The resonator is a circular microstrip resonator with diameter $a = 4.2$ mm. The radiating element was printed on a substrate with a relative dielectric constant of 2.2 with a thickness of 0.5 mm. The distance between the radiating elements is around 0.75λ. The array's dimensions are 125 × 40 × 1 mm. The antenna's bandwidth is 10% for a

Table 14.2. Array directivity as a function of the numbers of elements.

N-elements	D_0—dB	HPBW—θ_{3dB}
4	6	26°
8	9	10°
12	10.8	9°
16	12	7°
32	15	3.5°
64	18.1	1.75°

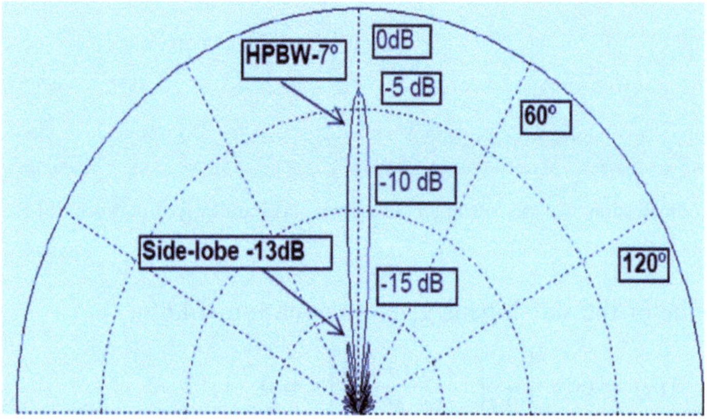

Figure 14.15. Radiation pattern of a 16 broadside element array.

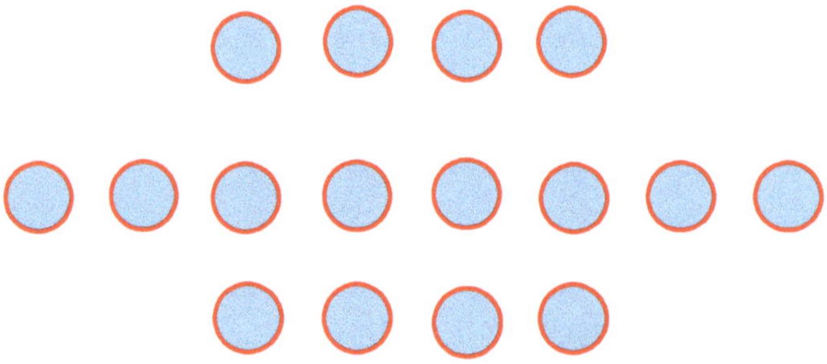

Figure 14.16. Stacked Ku band, 16 elements, microstrip antenna arrays.

VSWR better than 2:1. The antenna's beamwidth is around 10° × 25°, and the gain is around 18.5 dBi. Losses in the feed network are around 1 dB.

A stacked Ku band, 32 elements, microstrip antenna array was designed at 14.5 GHz as shown in figure 14.17. The resonator and feed network were printed on a substrate with a relative dielectric constant of 2.5 with a thickness of 0.5 mm. The resonator is a circular microstrip resonator with diameter $a = 4.2$ mm. The radiating

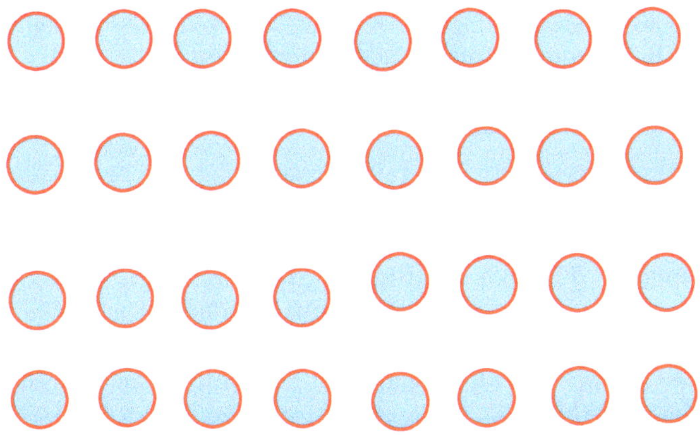

Figure 14.17. Stacked Ku band, 32 elements, microstrip antenna array.

Table 14.3. Measured results of a stacked microstrip antenna array.

Array	F (GHz)	Bandwidth %	Beamwidth °	Gain dBi	Dimensions mm
16	14.5	10	10 × 25	18.5	125 × 40 × 1
32	14.5	10	10 × 20	20.5	125 × 125 × 1
16	10	13.5	23 × 23	17.0	60 × 60 × 1.6
64	34	10	10 × 10	23.5	55 × 55 × 0.5

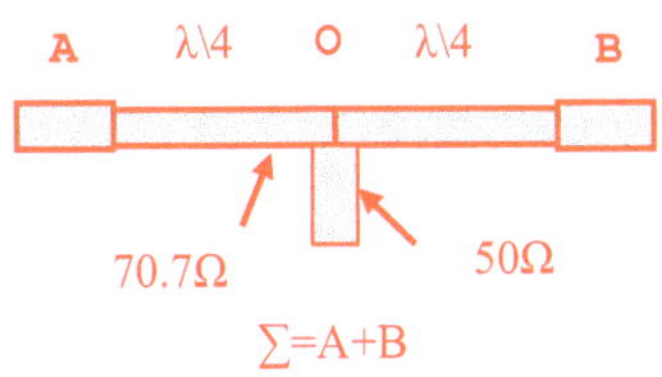

Figure 14.18. Power combiner/splitter.

element was printed on a substrate with a relative dielectric constant of 2.2 with thickness of 0.5 mm. The distance between the radiating elements is around 0.75λ. The array's dimensions are 125 × 125 × 1 mm. The antenna's bandwidth is 10% for a VSWR better than 2:1. The antenna's beamwidth is around 10° × 20°, and the measured gain is around 20.5 dBi. Losses in the feed network are around 1.5 dB. The measured results of several stacked microstrip antenna arrays are listed in table 14.3. The measured gain of the 64 elements array at 34 GHz is around 23.5 dBi.

A basic configuration of a power combiner/splitter is shown in figure 14.18. This configuration of a power combiner/splitter is used in the feed network of several printed arrays. The power combiner/splitter consists of two sections of a quarter-wavelength transformer. The distance from A to O is $\lambda/4$. The impedance at point O is 100 Ω. For an equal-split splitter the impedance of the quarter-wavelength transformer is 1.41 × Z_0, or 70.7 Ω for $Z_0 = 50$ Ω. For an input signal V, the outputs at ports A and B are equal in magnitude and in phase.

14.5.7 Ka band microstrip antenna arrays

Microstrip antenna arrays with integral feed networks may be broadly divided into arrays fed by parallel feeds and series fed arrays. Usually series fed arrays are more efficient than parallel fed arrays. However, parallel fed arrays have a well-controlled aperture distribution. Two Ka band microstrip antenna arrays that consist of 64 radiating elements have been designed on a 10 mil Duroid substrate with $\varepsilon_r = 2.2$. The first array uses a parallel feed network and the second uses a parallel series feed network as shown in figures 14.19(a) and (b). A comparison of the performance of the arrays is given in table 14.4. The results given in table 14.4 verify that the parallel series fed array is more efficient than the parallel fed array due to minimization of the number of discontinuities in the parallel series feed network.

The parallel series fed array has been modified by using a five-centimeter coaxial line to replace the same length of microstrip line. The results given in table 14.4 indicate that the efficiency of the parallel series fed array that incorporates a coaxial line in the feed network is around 67.6% due to the minimization of the microstrip line length. Two microstrip antenna arrays that consist of 256 radiating elements have been designed. The first array, Type A as shown in figure 14.20(a), uses a power divider that minimizes the number of microstrip discontinuities. The second array,

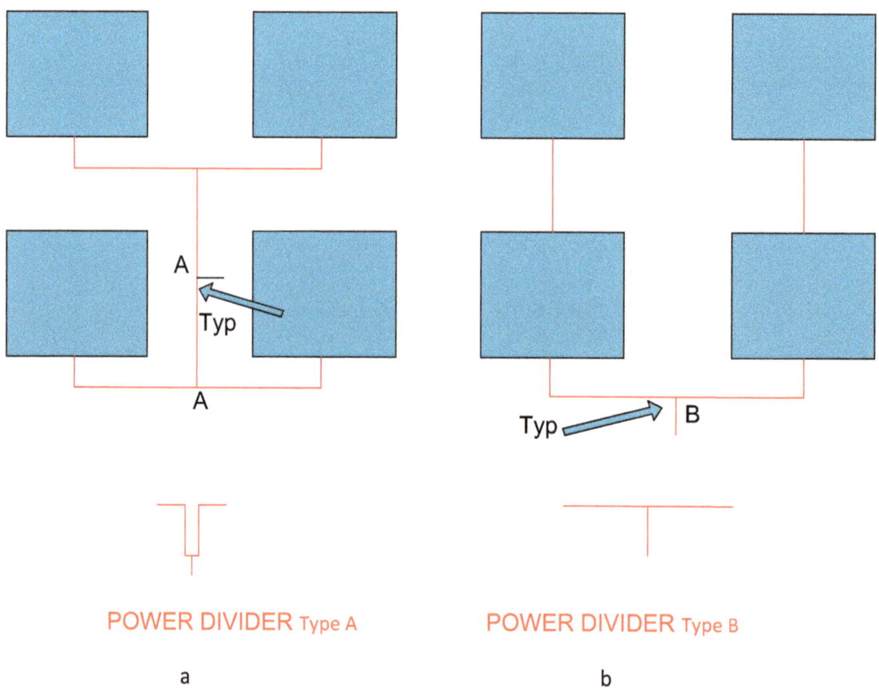

Figure 14.19. Configuration of 64 elements microstrip antenna arrays. (a) Parallel feed network. (b) Parallel series feed network.

Table 14.4. Performance of 64 elements microstrip antenna arrays.

Parameter	Corporate feed	Parallel feed	Corporate feed mix
Number of elements	64	64	64
Beamwidth °	8.5	8.5	8.5
Computed gain (dBi)	26.3	26.3	26.3
Microstrip line loss (dB)	1.1	1.2	0.5
Radiation losses (dB)	0.7	1.3	0.7
Mismatch loss (dB)	0.5	0.5	0.5
Expected gain	24.0 dBi	23.3 dBi	24.6 dBi
Efficiency %	58.9	50.7	67.6

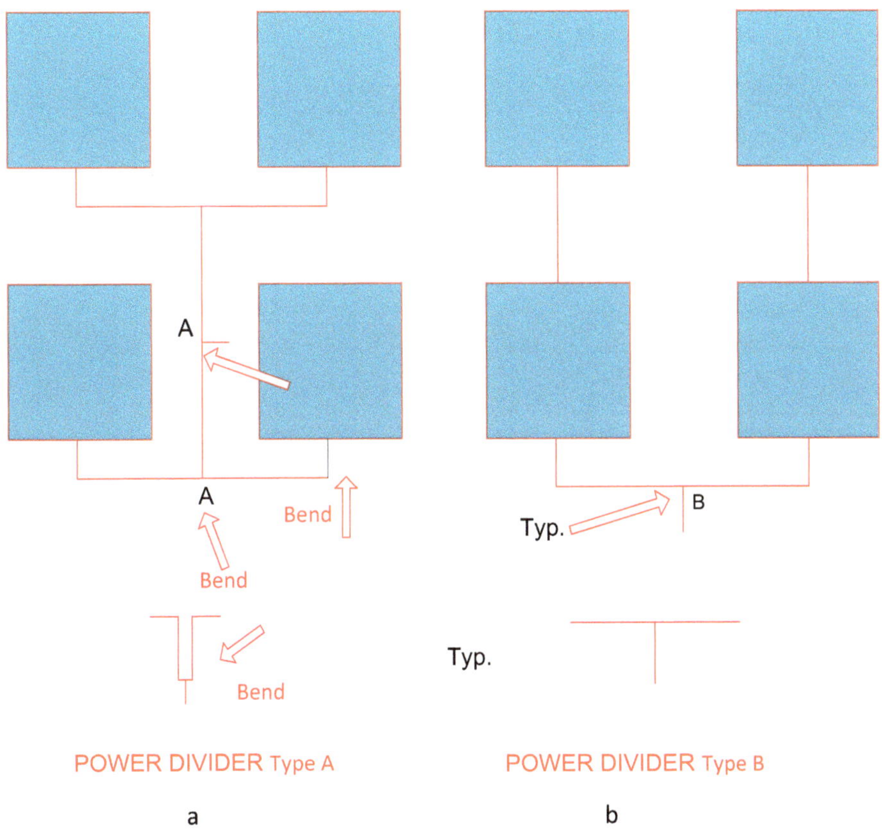

Figure 14.20. Configuration of 256 elements microstrip antenna arrays. (a) Parallel feed network. (b) Parallel series feed network.

Type B as shown in figure 14.20(b), incorporates more bend discontinuities in the feeding network. A comparison of the performance of the arrays is given in table 14.5. The Type A array with 256 radiating elements has been modified by using a 10 cm coaxial line to replace the same length of microstrip line. The performance comparison of the arrays, given in table 14.5, shows that the gain of the modified array has increased by 1.6 dB. The results given in table 14.5 verify that the Type A array is more efficient than the Type B array due to the minimization of the number of bend discontinuities in the Type A array feed network. The measured gain results are very close to the computed gain results and verify the loss computation.

14.6 Integrated outdoor unit for mm wave communication systems

In this section, an example of a design of a mm wave satellite communication system is presented. The block diagram of an integrated outdoor unit (ODU) for mm wave satellite communication applications is presented in figure 14.21. The ODU consists of a receiving and a transmitting channel.

Table 14.5. Performance of 256 elements microstrip antenna arrays.

Parameter	Type A	Type B	Type C
Number of elements	256	256	256
Beamwidth °	4.2	4.2	4.2
Computed gain (dBi)	32	32	32
Microstrip line loss (dB)	3.1	3.1	1.5
Radiation losses (dB)	1	1.9	1
Mismatch loss (dB)	0.5	0.5	0.5
Expected gain (dBi)	27.43	26.5	29.03
Efficiency %	34.9	28.2	50.47

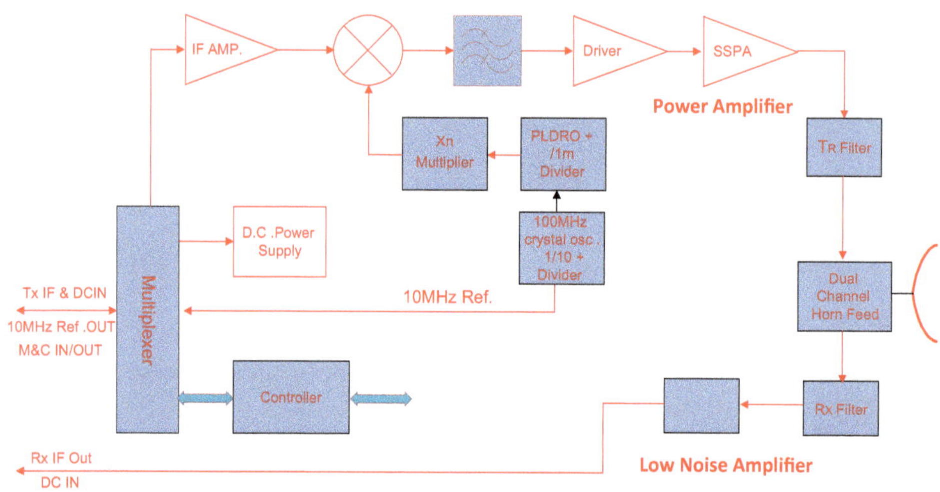

Figure 14.21. ODU basic block diagram.

14.6.1 Outdoor unit, ODU, description

The outdoor system is divided into three main parts. The transmitter, the 'ODU', the antenna assembly and the receiver front end, the 'LNB'. The antenna assembly consists of two main parts: the reflector antenna and mounting structure, and the dual channel feed horn. The ODU module's design is presented in this section. The design and theory of RF components have been presented in books and in articles, see [1–15]. At the low frequency I/O terminal four types of input and output signals are fed. Transmitting, Tx, input IF signal (in a frequency band of 2500–3000 GHz), DC input power supply (28 V ± 10%), output reference (10 MHz), and monitor and control at the 22 KHz IN/OUT signal.

The DC input is fed through the multiplexer to the power supply from which regulated voltages are supplied to different parts of the ODU. The output 10 MHz reference signal is obtained from a 100 MHz internal crystal oscillator by a 1:10 frequency divider and is routed to the I/O connector through the multiplexer. The monitor and control (22 KHz) signal is separated from the other signals by the multiplexer between the I/O connector and the controller. The control command, ON/OFF, is processed by the controller, and applied to the SSPA supply ON/OFF switch. The monitoring signals, originating from different parts of the ODU, are processed by the controller and installed in the 22 KHz serial link. The PLDRO Lock/Unlock monitoring signal is used also (via the controller) for switching the SSPA supply ON/OFF accordingly (automatic shut down). An additional alarm monitoring signal will be provided by switching the 10 MHz signal from the ON to OFF state (by the controller) if any of the alarm signals (PLDRO unlocked, SSPA supply in the OFF state or power supply alarm) appear at the controller inputs. The input IF signal (in the frequency band: 2500–3000 GHz) is separated from the other inputs by the multiplexer and is fed to the IF amplifier. The IF signal is amplified to an appropriate level (the IF amplifiers amplify the input IF signal from the minimum input level to the nominal level at the mixer input) and is fed to the mixer input. The frequency band of the input signal is converted to the transmitted frequency band by the mixer and the LO signal. The band pass filter, following the mixer output, attenuates the LO signal leaking (at 27 GHz), image signal and other spurious products. The filtered Tx signal is amplified by the driver and the power amplifier (PA) to the appropriate output power level. The Tx band pass following the power amplifier attenuates all the spurious signal levels below the relevant specified levels and also the output noise in the receiving (Rx) frequency band, below the thermal noise at the LNB (Rx link) input. The LO source consists of a PLDRO, multiplier and a band pass filter. The PLDRO is locked to a 100 MHz (internal) crystal oscillator. The output signal (at 9 or 13.5 GHz) is multiplied by the frequency multiplier to the specified frequency (27 GHz) and filtered to eliminate the spurious signals. All the above-mentioned components are installed in a mechanical enclosure which protects them from the outside environment and dissipates the heat, by convection, to the air outside the enclosure. Two connectors are mounted on the enclosure (Box): Input I/O (Type F connector) and Output (ISO PBR 320 for

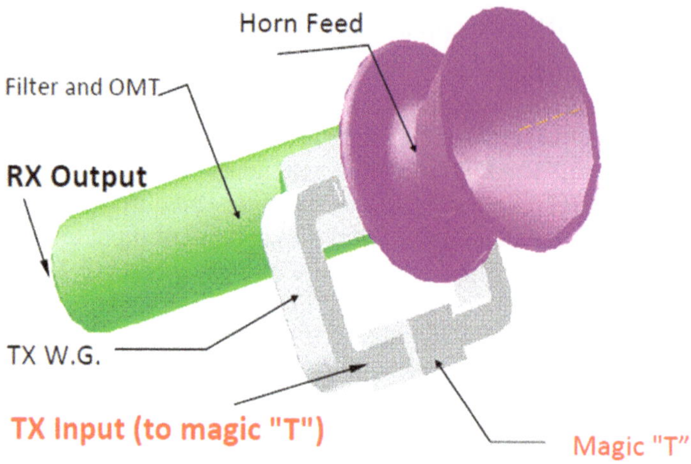

Figure 14.22. Outdoor unit drawing.

WR-28 waveguide). The ODU box will be mounted directly to the dual channel feed horn through the waveguide flange. The ODU, dual channel feed horn and the LNB are assembled on a special mounting cradle see figure 14.22. The cradle is mounted on the antenna boom and allows the rotation of the ODU, the LNB and the antenna feed around the feed axis for polarization adjustment. The ODU specifications is listed in table 14.6.

14.6.2 The low noise unit, LNB

The LNB is of universal standard with a circular C-120 flange at the input. The LNB is mounted directly on to the dual channel feed horn through the C-120 flange. This LNB is produced by several manufacturers and is available on the commercial market. The full description and specification of the LNB is listed in table 14.6 with the exception that the maximum noise figure should be 0.9 dB for both receiving bands. The LNB output may be connected through the ODU to enable the use of only one coaxial cable between the indoor unit, IDU, and the outdoor assembly. For that purpose, a special power supply must be added into the ODU enclosure, an appropriate protocol must be specified and the multiplexer must be redesigned to allow the filtering of different signals at the ODU's I/O terminal. The antenna assembly losses up to the antenna feed output are 0.8 dB, as presented and listed here.

Calculations of the antenna assembly losses

– Dual band feed horn's insertion loss	0.5 dB max.
– Transmitter filter and WG28 WG loss	0.2 dB max.
– Mismatch loss (WSWR < 1.35):	0.1 dB max.
	Total loss: 0.8 dB max.

Table 14.6. ODU specifications.

Description	Specification
Transmit frequency range	29.5–30.0 GHz
Receiving frequency range, in two bands	10.7–12.75 GHz
Low	10.7–11.7 GHz
High	11.7–12.75 GHz
Transmit: on one linear (Horizontal, *H*, or Vertical, *V*) polarization plane. Receive a dual linear orthogonal polarized signal by switching between *H* and *V* polarization	
Selecting the polarization orientation. Continuous adjustment of the polarization orientation: identical antenna tilt angle.	H or V±45° Ka and Ku signals
min. EIRP and 1 dB gain compression	
@max. antenna diameter 0.75 m	40 dBW
@max. antenna diameter 0.95 m	45 dBW
@max. antenna diameter 1.30 m	50 dBW
Off-axis EIRP density, max.	10 W m^{-2}
EIRP stability over temperature	±2 dB
G/*T* figure of merit	>14 dB K^{-1}
Description	Specification
Spurious emission below the total EIRP: in/out 29.5–30.0 GHz band	60 dB max.
Noise emission max. in 29.5–30.0 GHz band	
SSPA—ON	−75 dBW Hz^{-1} of EIRP
SSPA—OFF	−105 dBW Hz^{-1} of EIRP
Phase noise, max.: Freq. offset (For the transmitter) 10 Hz	−32 dBC Hz^{-1}
100 Hz	−62 dBC Hz^{-1}
1 kHz	−72 dBC Hz^{-1}
10 kHz	−82 dBC Hz^{-1}
> 100 kHz	−92 dBC Hz^{-1}
Spurious phase noise at AC line frequency	−32 dBC max.
Level of the sum of all phase noise components in frequency range from AC line freq. to 1 MHz.	−38 dBC max.
Amplitude variation (transmitter only), in any 200 kHz band	0.2 dB max.
2 MHz band	0.4 dB max.
40 MHz band	1.0 dB max.
500 MHz band	2.5 dB max.
Group delay variations (transmitter only) in any 2 MHz band	2 ns max. ptp
40 MHz band	4 ns max. ptp
Radiation pattern (transmitter only),	
1.8° < Φ < 7°	29–25 log Φ dBi max.

7° < Φ < 9.2°	8 dBi max.
9.2° < Φ < 48°	32–25 log ΦdBi max.
48° < Φ < 180°	0 dBi max.
Tx cross polar gain: 1.8° < Φ < 7°	19–25 log Φ dBi max.
7° < Φ < 9.2°	−2 dBi max.
Tx cross-pole isolation within 1/10 dB beam contour	−25/−22 dB max.
Antenna pointing, manual adjustment: elevation	10°–50°
Azimuth	0°–360°
Pointing accuracy (mechanical coarse and fine adjustment)	10% of the 3 dB beamwidth
TX power consumption: SIT I	20 W
SIT II	30 W
SIT III	50 W
Safety SSPA switch-OFF: avoiding interference to other users	Automatic shut down
Description	Specification
Monitoring functions	Lock alarm
SSPA ON/OFF status	
Control	Power supply alarm
SSPA ON/OFF indication on ODU	Presence det. alarm SSPA ON/OFF. Green/Red LED
Operation environment: temperature	−30° to +50°
Solar radiation	500 W m^{-2} max.
Humidity	0%–100% (condensing)
Rain/wind	40 mm h^{-1} max/45 km h^{-1} max.
Survival conditions: temperature	−40° to +60°
Solar radiation:	1000 W m^{-2}
Humidity:	0%–100% (condensing)
Precipitation: rain:	100 mm h^{-1} max.
Freezing rain:	12 mm h^{-1} max.
Snowfall:	50 mm h^{-1} max.
Static load:	25 mm of ice on all surfaces
Wind:	120 km h^{-1} max.
Storage and transportation:	
Temperature:	−40° to +70°
Shock and vibration	As required for commercial freights
IDU and ODU, DC power supply	28 V ± 10%
Tx IF IN (2.5–3 GHz)	−40 dBm
M&C: 10 MHz Ref. output power	22 kHz PWK
Freq. stability: connector: Tx output flange	−40 dBm ± 20 ppm
(Tx to antenna feed)	Type F—ISO PBR 320

The calculations of the minimum SSPA output power is listed here.

<u>**Calculations of the minimum SSPA output power**</u>

Reflector antenna	Antenna I	Antenna II	Antenna III
Antenna gain at Ka band	45.0 dBi	47.0 dBi	49.5 dBi
Tx to feed output loss	0.8 dB	0.8 dB	0.8 dB
Net gain	44.2 dBi	46.2 dBi	48.7 dBi
Specified EIRP	40.0 dBW	45.0 dBW	50.0 dBW
Minimum SSPA output	−4.2 dBW	−1.2 dBW	+1.3 dBW

14.6.3 Solid state power amplifier, SSPA, output power requirements

The ODU specifications are listed in table 14.6.

14.6.4 Isolation between receiving, Rx, and transmitting, Tx, channels

Since the Tx and the Rx signals may appear at the feed with the same polarization, the isolation between the transmitter, Tx, port and the receiver, Rx, port is mainly achieved by the Rx input filter. The main objectives of the Rx input filter is to prevent the gain degradation of the LNB and to eliminate a spurious response (mixing with the harmonics of the LOs in the LNB) and to reject signals at the image frequencies.

The isolation may be estimated for the following data:

Maximum Tx signal level at the Tx port	+33 dBm
Output 1 dB compression point of the LNB	+5 dBm
LNB Gain (max.)	60 dB

According to these assumptions the input 1 dB comp. point of the LNB is around 55 dBm and the isolation must be at least 88 dB. The LNA inside the LNB is tuned to 10.7–12.75 GHz. Following the LNA, a band pass filter is used to reject the image signal.

14.7 Solid state power amplifier, SSPA

14.7.1 Specifications

1. Input power: −15 dBm.
2. The output power level results from the system parameters are listed in table 14.7.

Reflector antenna design

For 50 dBW EIRP the required antenna gain should be 49.5 dBi. By using equations (14.1)–(14.16) we find that the antenna's diameter should be 1.35 m. The antenna's efficiency is around 50% and the reflector antenna's beamwidth is around 0.4°. The

Table 14.7. Output power requirements for three, SIT, situations.

	EIRP spec. dBW	Required PA output *(1) dBm (dBW)	PA to fed output loss	Antenna gain dBi	Antenna diameter m
SIT1	40	25.8 (−4.2)	0.8	45	0.6
SIT2	45	28.8 (−1.2)	0.8	47	0.75
SIT3	50	31.3 (+1.3)	0.8	49.5	1.35

* (1) Output power required at P1dB.

Table 14.8. Reflector antenna's electrical parameters.

	EIRP spec. dBW	Antenna gain dBi	Antenna diameter m	Beamwidth°
SIT1	40	45	0.8	0.65
SIT2	45	47	1	0.52
SIT3	50	49.5	1.35	0.4

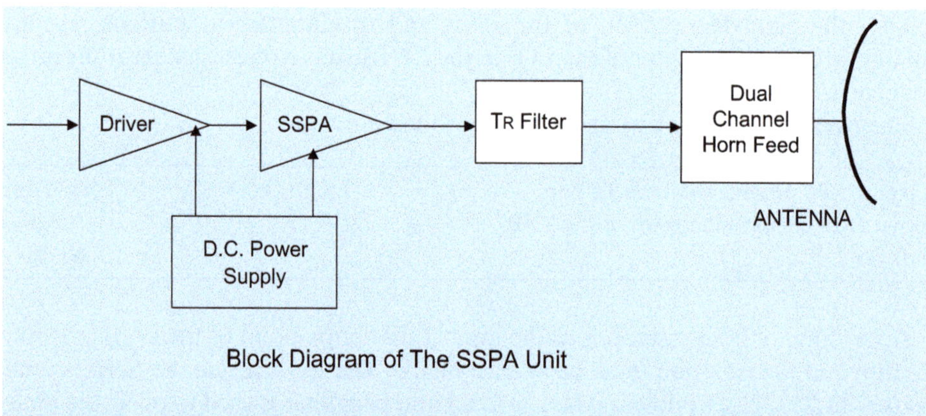

Figure 14.23. Block diagram of the SSPA unit.

focal point distance for $f/D = 0.3$ is around 0.4 m. The reflector antenna's electrical parameters are listed in table 14.8.

14.7.2 SSPA general description

The module is an integral MIC assembly that includes the up-converter, SSPA, output power detector and transition to the antenna as shown in figure 14.23. The key element for the realization of the ODU SSPA is the output stage basic MMIC power amplifier.

14.7.3 SSPA electrical design

1. Input power: −15 dBm.
2. The output power requirements result from the following system parameters: A SSPA block diagram is shown in the figure 14.23.

Power amplifier power and gain budget:

1. Input power: −15 dBm.
2. The output power requirements result from the following system parameters as given in table 14.9–14.11.

Power supply requirement are listed in table 14.12.

Table 14.9. Situation 1.

	P_{in}	Low PA	Medium PA	High PA
Power	−15 dBm	−5 dBm	13.5 dBm	26 dBm
Gain		10 dB	18.5 dB	12.5 dB
P1dB		8 dBm	19 dBm	27 dBm

Power supply: +5 V, 7 W.

Table 14.10. Situation 2.

	P_{in}	Low PA	Medium PA	High PA
Power	−15 dBm	−3 dBm	16 dBm	28.5 dBm
Gain		12 dB	19 dB	12.5 dB
P1dB		8 dBm	19 dBm	29 dBm

Power supply: +5 V, 10 W.

Table 14.11. Situation 3.

	P_{in}	Low PA	Medium PA	High PA
Power	−15 dBm	−1 dBm	18 dBm	31 dBm
Gain		14 dB	19 dB	13 dB
P1dB		8 dBm	19 dBm	31.5 dBm

Power supply: +5 V, 15 W.

Table 14.12. Supplies.

	Drain voltage	Gate voltage	DC power	Drain current amp.
SIT1	5 V	−5 V	7 W	1.4
SIT2	5 V	−5 V	10 W	2
SIT3	5 V	−5 V	15 W	3

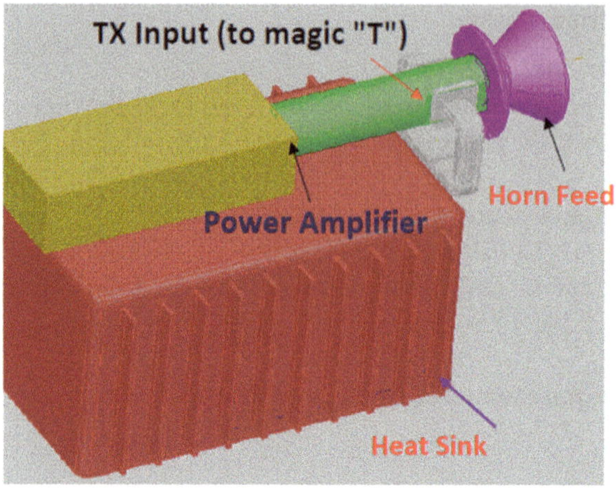

Figure 14.24. ODU package.

14.7.4 The ODU mechanical package

The ODU package combines all the antenna feed components and driving electronics. The RF chain consists of a horn feed, transducer, filter and the LNB as shown in figure 14.24. The ODU package is designed to ensure full compliance with the mechanical requirements as listed in table 14.6. The package houses the SSPA, up-converter, PLDRO, controller and a power supply. The ODU is mounted on the boom of the antenna using mechanics, which allow adjustment of the polarization angle, while maintaining an accurate position of the horn on the antenna's focal point. The ODU's package is designed for cost effectiveness and large volume production, while it protects its inner components from precipitation, dissipates the heat generated within and is coated and painted to absorb minimum solar radiation.

14.8 Solid state high power amplifiers, SSPAs, for mm wave communication system

This section describes the design and performance of new compact and low-cost Ka band power amplifiers. The main features of the power amplifiers are: 27 dBm to 35 dBm minimum output power for −16 dBm input power over the frequency range of 27.5–31 GHz. To reduce losses, MIC, MMIC and waveguide technologies are employed in the development and fabrication of this set of power amplifiers. Employing waveguide technology has minimized losses in the power combiner.

Three power amplifiers are described in this chapter. The first amplifier is a 0.5 W power amplifier, the second is a 1.5 W power amplifier, and the third amplifier is a 3.2 W power amplifier.

14.8.1 Introduction

An increasing demand for wide bandwidth in communication links makes the Ka band attractive for future commercial systems. This section describes the design and

Table 14.13. Power amplifier specifications.

Parameter	Specification
Frequency range	27.5–31 GHz
Input power	−16 to −20 dBm
Output power, minimum	27, 32,35 dBm minimum
Input VSWR	2:1
Output VSWR	2:1
Spurious level	−60 dBc
Supply voltage	±5 V
Connectors	K-connectors
Operating temperature	−30 °C to 60 °c
Storage temperature	−50 °C to 80 °C
Humidity	100%

performance of compact and low-cost Ka band power amplifiers. The main features of the power amplifiers are: 27–35 dBm minimum output power for −16 dBm input power over the frequency range 27.5–31 GHz.

14.8.2 Power amplifiers: specifications

The transmitting channel specifications were listed in table 14.6. The power amplifier specifications are listed in table 14.13.

14.8.3 Description of the 0.5 W, 1.5 W power amplifiers

A block diagram of the 0.5 W and 1.5 W power amplifier is shown in figure 14.25. The 0.5 W power amplifier consists of a low power MMIC amplifier, band pass filter, medium power MMIC amplifier and 0.5 W power amplifier. The 1.5 W power amplifier consists of the same modules as the 0.5 W power amplifier. The 0.5 W MMIC amplifier is connected to a four-way microstrip power divider, a high power module with four MMIC power amplifiers, a four-way waveguide power combiner and DC supply unit. Three stages of MMIC power amplifiers amplify the input signal from −16 dBm to 18 dBm. In the fourth stage, a 0.5 W MMIC power amplifier is connected to a four-way hybrid ring microstrip power divider printed on 10 mil Duroid. The output ports of the power divider are connected to four 0.5 W MMIC power amplifiers. The 0.5 W power amplifiers are combined via a four-way waveguide power combiner to yield a 32.5 dBm minimum output power level. The DC bias voltages of each MMIC amplifier has been experimentally optimized to achieve the required gain and output power level.

14.8.4 Gain and power budget for 0.5 W and 1.5 W amplifiers

The power amplifiers' gain and power budget is listed in table 14.14. The expected large signal gain of the 1.5 W amplifier is 49 dB. The expected output power is 33 dBm.

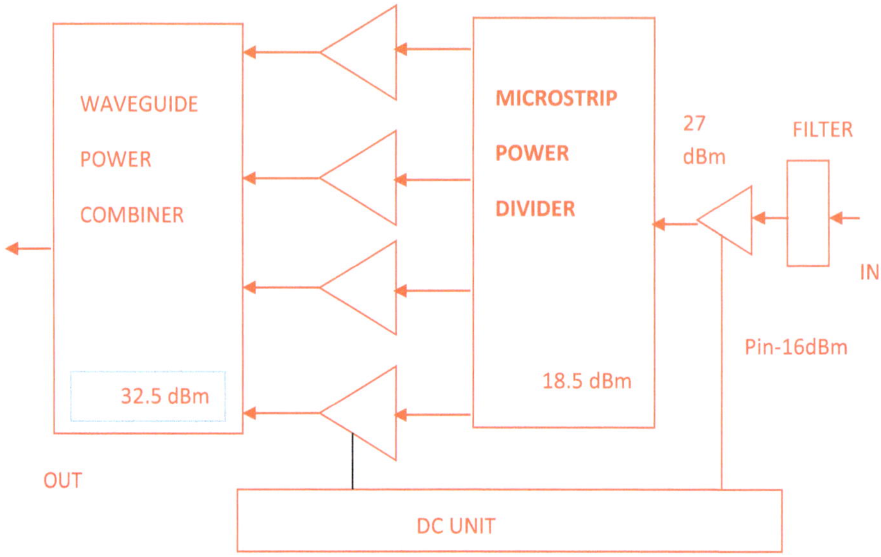

Figure 14.25. 1.5 W power amplifier block diagram.

Table 14.14. Transmitter gain and power budget.

Component	Gain/Loss (dB)	P_{out} (dBm) 0.5 W	P_{out} (dBm) 1.5 W
Input power	–	−16	−16
Amplifier	15	−1	−1
Filter	−1	−2	−2
Amplifier	20	18	18
Amplifier	9	27	27
4-way power divider	−7.5	–	19.5
Amplifier	8	–	27.5
4-way power combiner	5.5	–	33
Total	49	–	33

14.8.5 Description of the 3.2 W power amplifier

A block diagram of the 3.2 W power amplifier is shown in figure 14.26. The 3.2 W power amplifier consists of the same modules as the 0.5 W power amplifier. The 0.5 W MMIC amplifier is connected to a two-way microstrip power divider, a high power module with two 2 W MMIC power amplifiers, and a two-way microstrip power combiner. Three stages of MMIC power amplifiers amplify the input signal from −16 dBm to 18 dBm. In the fourth stage, a 0.5 W MMIC power amplifier is connected to a two-way microstrip power divider printed on 5 mil Alumina substrate. The output ports of the power divider are connected to two 2 W MMIC power amplifiers. The 2 W output power is combined via a two-way low loss power combiner to yield a

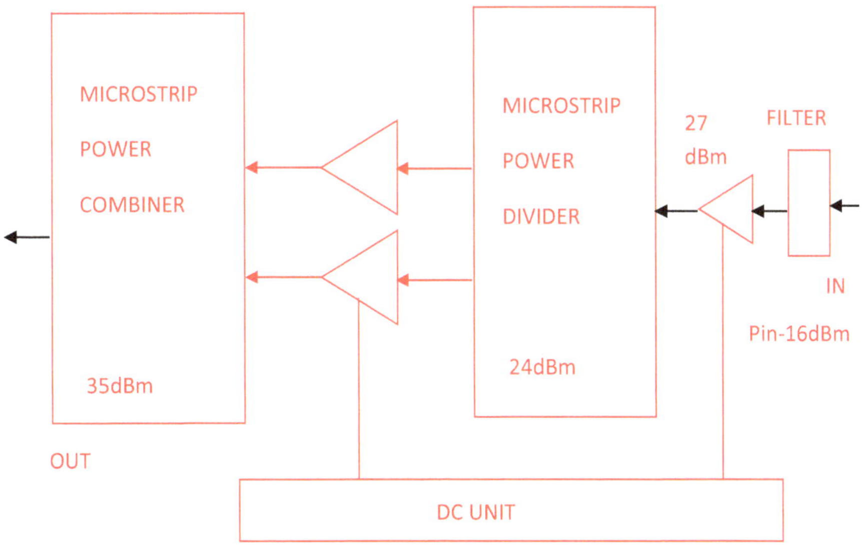

Figure 14.26. 3.2 W power amplifier block diagram.

Table 14.15. Measured test results of the modular and integrated power amplifier.

Parameter	Measured results. Modular 1.5 W AMP	Measured results. Integrated 1.5 W AMP
Frequency range (GHz)	27–31	27–31
Input power (dBm)	−16	−16
Output power (dBm)	31.5	31.5
Spurious level (dBc)	> −50	> −50
VSWR (Input/Output)	2:1	2:1

35 dBm minimum output power level. The DC bias voltages of each MMIC amplifier have been experimentally optimized to achieve the required gain and output power level.

14.8.6 Measured test results

Two different 1.5 W power amplifier modules were fabricated and tested. The first power amplifier is a modular unit and consists of seven modules. The unit consists of a band pass filter, low power amplifier, medium power amplifier, four-way microstrip power divider module, four 0.5 W power amplifiers, a four-way microstrip power combiner and a DC supply unit. The second power amplifier is an integrated power amplifier as described in section 14.7.5. The test results of the modular and integrated power amplifier are given in table 14.15. The measured test results of the four-way waveguide power combiner are listed in table 14.16.

Table 14.16. Measured test results of the four-way waveguide power combiner.

Parameter	Measured results
Frequency range	27–31 GHz
Insertion loss	0.5 dB
Amplitude balance	0.2 dB max.
Phase balance	$\pm 5°$ max.
VSWR (Input/Output)	1.5:1

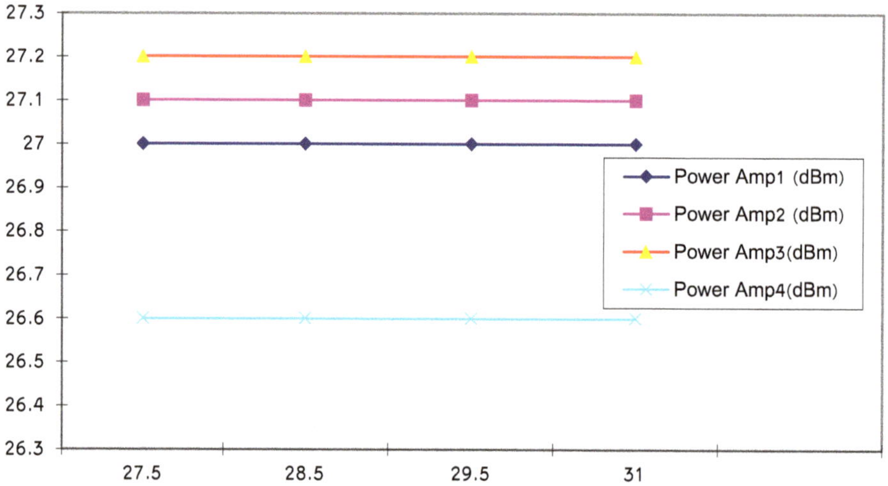

Figure 14.27. Output power balance of the 0.5 W power amplifier.

Figure 14.27 presents the output power balance of the 0.5 W power amplifier. Figure 14.28 presents the output power and gain of the 1.5 W power amplifier. A photo of the 1.5 W power amplifier is shown in figure 14.29. The four-way waveguide power combiner is covered by a metallic cover.

Figure 14.29 presents the low power and medium power MMIC amplifiers. A microstrip four-way power divider is used to supply the power to each of the four 0.5 W MMIC amplifiers. The output power of the 0.5 W MMIC amplifiers is combined by a four-way waveguide power combiner. A photo of the 1.5 W power amplifier with a four-way waveguide power combiner is shown in figure 14.30.

A photo of the 1.5 W power amplifier without the four-way waveguide power combiner is shown in figure 14.31. The first carrier in this photo is a side-coupled band pass filter. The second and third carrier contains medium power MMIC amplifiers. The fourth carrier contains a 0.5 W MMIC power amplifier. The output of the 0.5 W MMIC power amplifier is connected to a microstrip four-way power divider.

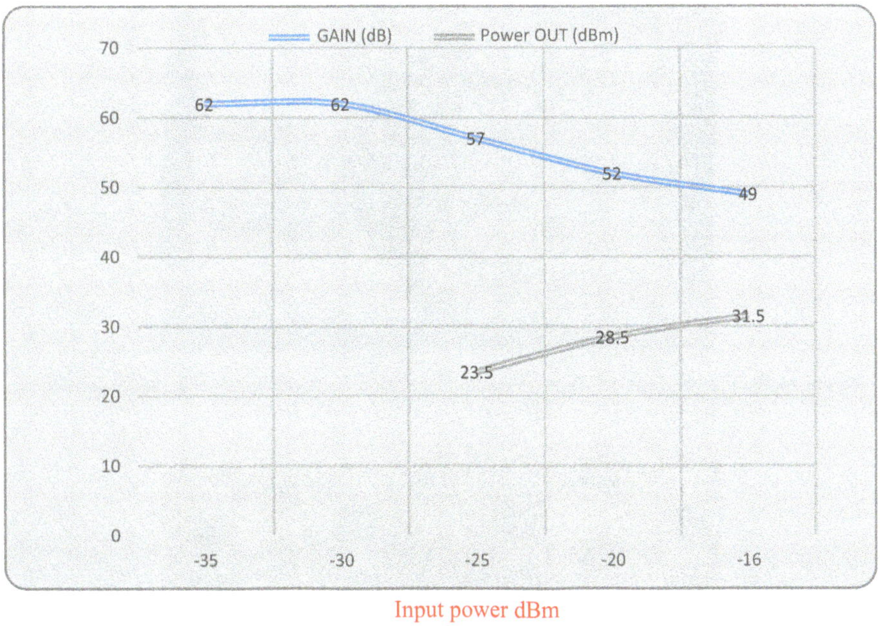

Figure 14.28. Output power and gain of the 1.5 W power amplifier.

Figure 14.29. Photo of the 1.5 W power amplifier.

14.9 Integrated Ku band automatic tracking system

Tracking communication systems use the mono-pulse tracking principle. The mono-pulse tracking principle is the most popular way to obtain accurate information about the target angular position in radar systems, see [1–12]. Tracking radars and scan radars systems use mono-pulse antennas in the accurate determination of angular deviation with respect to the antenna axis, off-bore-sight angle. Also, radio communication systems that need to track a narrow beam antenna to the transmitter use mono-pulse tracking systems. Satellite and tracking communication systems use the mono-pulse tracking principle to secure communication links with very narrow

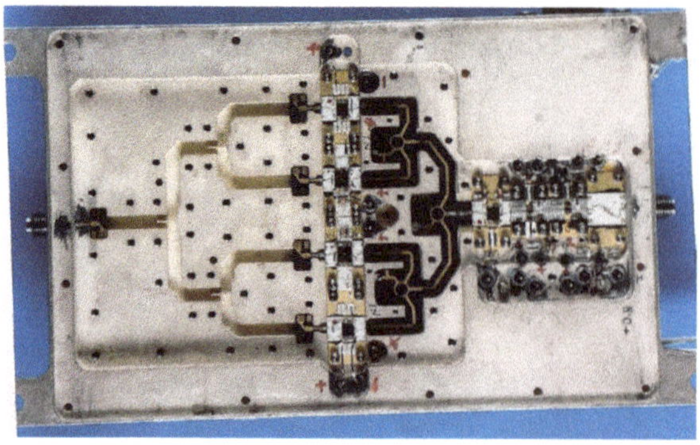

Figure 14.30. Photo of the 1.5 W power amplifier with a four-way waveguide power combiner.

Figure 14.31. Photo of the 1.5 W power amplifier without the four-way waveguide power combiner.

beam antennas. Identification military and commercial systems also use this principle to obtain narrow beam information links.

A Ku band, two axis, automatic tracking system may be implemented by using a four element mono-pulse antenna connected to a phase comparator network. The concept is widely used in the communication industry. The antenna may be a reflector antenna or an antenna array. The elements of the reflector feed may be placed at a distance of at least, $\lambda/2$, a half-wavelength apart. When the target is located along the antenna's bore sight each element is equidistant from the antenna's elements and all the signals are received in phase. However, when the target is off-bore-sight, due to the difference in path length (L), a phase difference is introduced between the signals that is used to calculate the error angle (θ). The tracking system consists of antennas, a mono-pulse comparator, up- and down-converter, mono-pulse processor at baseband frequencies, DC and control units, as shown in figure 14.32. The development and design of the Ku band automatic tracking system is based on MIC, waveguide and MMIC technology. The power divider and combiners in the power amplifier module are waveguide power combiners to minimize losses in the transmitter. Packed MMIC components are used to minimize the system's volume and weight.

PARAMETER	KU BAND	KU BAND	U BAND

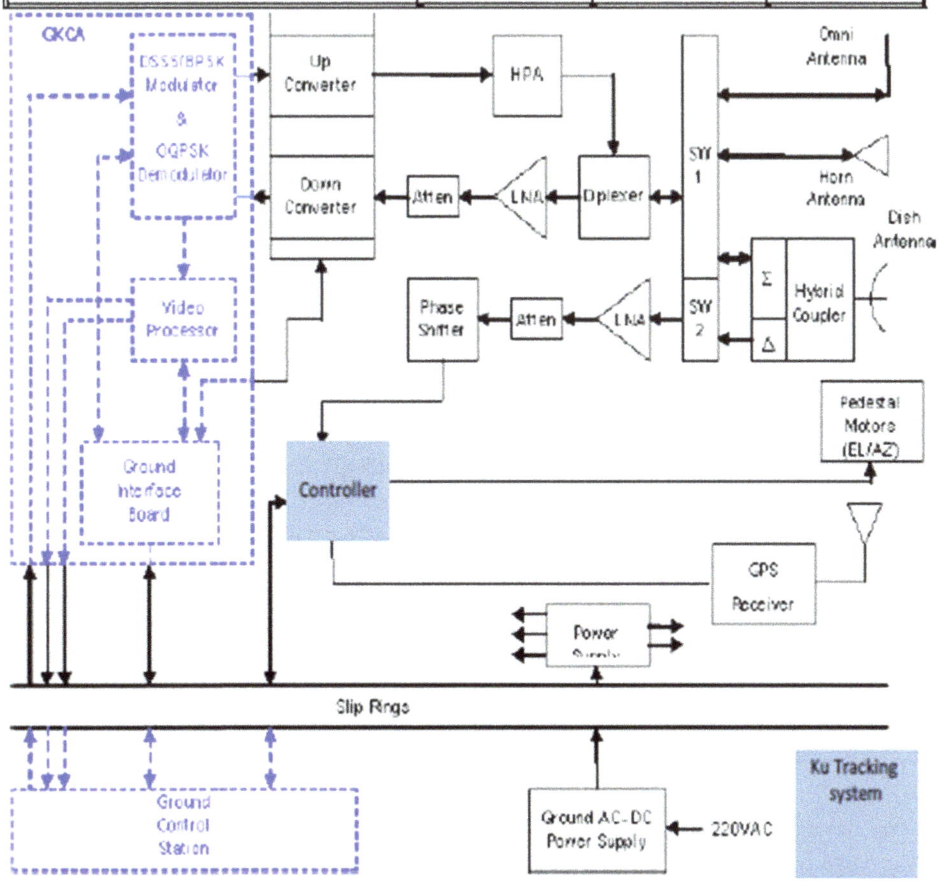

Figure 14.32. Block diagram of a Ku band automatic tracking system.

Table 14.17. Downink, DNL link, budget calculations.

Parameter	Ku band	Ku band	Ku band
Frequency	15.005 GHz	15.005 GHz	15.005 GHz
Distance	**50 km**	**100 km**	**200 km**
Rx noise figure	2.0 dB	2.0 dB	2.0 dB
IF bandwidth	8 MHz	8 MHz	8 MHz
Tx power	44 dBm	44 dBm	44 dBm
Tx components loss	5.5 dB	5.5 dB	5.5 dB
Tx antenna gain (vertical)	20 dBi	20 dBi	20 dBi
Tx pointing loss	1.5 dB	1.5 dB	1.5 dB
Tx random loss	0.5 dB	0.5 dB	0.5 dB
Tx EIRP	56.5 dBm	56.5 dBm	56.5 dBm
Free space loss	150 dB	156 dB	162 dB
Atmospheric absorption	2.24 dB	4.49 dB	8.98 dB
Precipitation absorption	0.0 dB	0.0 dB	0.0 dB
Total propagation loss	152.24 dB	160.49 dB	170.98 dB
Rx antenna gain	45 dBi	45 dBi	45 dBi
Rx polarization loss	0.5 dB	0.5 dB	0.5 dB
Rx pointing loss	1.0 dB	1.0 dB	1.0 dB
Rx components loss (MP)	5.0 dB	5.0 dB	5.0 dB
Effective carrier power at Rx input	**−57.24 dBm**	**−65.49 dBm**	**−75.98 dBm**
Rx noise threshold	103.28 dBm	103.28 dBm	103.28 dBm
Calculated *C*/*N* @ Rx input	46.04 dB	37.79 dB	27.3 dB
Required *C*/*N* @ Rx input	5.2 dB	5.2 dB	5.2 dB
Calculated fade margin	**40.84 dB**	**32.59 dB**	**22.1 dB**

14.9.1 Automatic tracking system: link budget calculations

The downlink budget tracking system's calculations are listed in table 14.17 for three distances 50 km, 100 km and 200 km. The transmitting antenna's gain is 20 dBi. The transmitter power is 44 dBm. The receiving antenna's gain is 45 dBi. The tracking system consists of three antennas: a reflector antenna, horn and omnidirectional antenna. The uplink tracking system budget calculations are listed in table 14.18 for three distances 50 km, 100 km and 200 km. The transmitting antenna's gain is 45 dBi. The transmitter power is 43 dBm, and the receiving antenna's gain is 20 dBi.

The downlink budget, dish to omnidirectional channels, tracking system calculations are listed in table 14.19 for three distances 70 km, 75 km and 80 km. The transmitting antenna's gain is 2.15 dBi. The transmitter power is 44 dBm, and the receiving dish antenna's gain is 45 dBi.

The downlink budget, horn to omnidirectional channels, tracking system calculations are listed in table 14.20 for three distances 2 km, 5 km and 10 km. The transmitting omnidirectional antenna's gain is 2.15 dBi. The transmitter power is 44 dBm, amd the receiving horn antenna's gain is 14 dBi.

Table 14.18. Uplink, UPL, budget calculations.

Parameter	Ku band	Ku band	Ku band
Frequency	14.759 GHz	14.759 GHz	14.759 GHz
Distance	**50 km**	**100 km**	**200 km**
Rx noise figure	3.0 dB	3.0 dB	3.0 dB
IF bandwidth	16 MHz	16 MHz	16 MHz
Tx power	43 dBm	43 dBm	43 dBm
Tx components loss	6.0 dB	6.0 dB	6.0 dB
Tx antenna gain	45 dBi	45 dBi	45 dBi
Tx pointing loss	1.5 dB	1.5 dB	1.5 dB
Tx random loss	0.5 dB	0.5 dB	0.5 dB
Tx EIRP	80 dBm	80 dBm	80 dBm
Free space loss	150 dB	156 dB	162 dB
Atmospheric absorption	2.14 dB	4.27 dB	8.55 dB
Precipitation absorption	0.0 dB	0.0 dB	0.0 dB
Total propagation loss	152.7 dB	160.77 dB	170.55 dB
Rx antenna gain (vertical)	20 dBi	20 dBi	20 dBi
Rx polarization loss	0.5 dB	0.5 dB	0.5 dB
Rx pointing loss	0.5 dB	0.5 dB	0.5 dB
Rx components loss	8.3 dB	8.3 dB	8.3 dB
Rx PG	16.9 dB	16.9 dB	16.9 dB
Effective carrier power at Rx input	**−44.54 dBm**	**−53.17 dBm**	**−62.95 dBm**
Rx noise threshold	99.28 dBm	99.28 dBm	99.28 dBm
Calculated C/N @ Rx input	54.74 dB	46.11 dB	36.33 dB
Required C/N @ Rx input	8.5 dB	8.5 dB	8.5 dB
Calculated fade margin	**46.24 dB**	**37.61 dB**	**27.83 dB**

The downlink budget, omnidirectional to omnidirectional, tracking system calculations are listed in table 14.21 for three distances 1 km, 2 km and 3 km. The transmitting omnidirectional antenna's gain is 2.15 dBi. The transmitter power is 44 dBm, and the receiving omnidirectional antenna's gain is 2.15 dBi.

14.9.2 Ku band tracking system antennas

14.9.2.1 Mono-pulse parabolic reflector antenna

The **parabolic reflector** antenna [1] consists of a radiating feed that is used to illuminate a reflector that is curved in the form of an accurate parabolic with diameter D as presented in figure 14.33. This shape enables a very narrow beam to be obtained. To provide the optimum illumination of the reflecting surface, the level of the parabola's illumination should be greater by 10 dB in the center than that at the parabola edges. The parabolic reflector antenna's gain may be calculated by using equation (14.66). α is the parabolic reflector antenna's efficiency.

Table 14.19. Downlink, DNL, budget calculations: dish–omni.

Parameter	Ku band	Ku band	Ku band
Frequency	15.005 GHz	15.005 GHz	15.005 GHz
Distance	**70 km**	**75 km**	**80 km**
Rx noise figure	2.0 dB	2.0 dB	2.0 dB
IF bandwidth	8 MHz	8 MHz	8 MHz
Tx power	44 dBm	44 dBm	44 dBm
Tx components loss	5.5 dB	5.5 dB	5.5 dB
Tx antenna gain (vertical)	2.15 dBi	2.15 dBi	2.15 dBi
Tx pointing loss	0.0 dB	0.0 dB	0.0 dB
Tx random loss	0.0 dB	0.0 dB	0.0 dB
Tx EIRP	40.65 dBm	40.65 dBm	40.65 dBm
Free space loss	152.9 dB	153.5 dB	154.0 dB
Atmospheric absorption	2.07 dB	2.22 dB	2.37 dB
Precipitation absorption	0.0 dB	0.0 dB	0.0 dB
Total propagation loss	154.97 dB	155.72 dB	156.37 dB
Rx antenna gain	45 dBi	45 dBi	45 dBi
Rx polarization loss	0.5 dB	0.5 dB	0.5 dB
Rx pointing loss	1.0 dB	1.0 dB	1.0 dB
Rx components loss (MP)	5.0 dB	5.0 dB	5.0 dB
Effective carrier power at Rx input	**−75.82 dBm**	**−76.57 dBm**	**−77.22 dBm**
Rx noise threshold	103.28 dBm	103.28 dBm	103.28 dBm
Calculated C/N @ Rx input	27.46 dB	26.71 dB	26.06 dB
Required C/N @ Rx input	5.2 dB	5.2 dB	5.2 dB
Calculated fade margin	**22.26 dB**	**21.51 dB**	**20.86 dB**

Parabolic reflector antenna gain—

$$G \cong 10\log_{10}\left(\alpha\frac{(\pi D)^2}{\lambda^2}\right) \tag{14.66}$$

Reflector antenna specifications

Frequency: 14.5–15.3 GHz.
Gain: 45 dBi.
Beamwidth: 0.8° to 0.9°.
Dish diameter: 1.6 m.

Mono-pulse comparator, rat-race coupler

A rat-race coupler is shown in figure 14.34. The rat-race circumference is 1.5 wavelengths. The distance from A to Δ port is $3\lambda/4$. The distance from A to $\sum$ port is $\lambda/4$. For an equal-split rat-race coupler, the impedance of the entire ring is fixed at

Table 14.20. Downlink, DNL link, link budget calculations: horn–omni.

Parameter	Ku band	Ku band	Ku band
Frequency	15.005 GHz	15.005 GHz	15.005 GHz
Distance	**3 km**	**5 km**	**10 km**
Rx noise figure	2.0 dB	2.0 dB	2.0 dB
IF bandwidth	8 MHz	8 MHz	8 MHz
Tx power	44 dBm	44 dBm	44 dBm
Tx components loss	5.5 dB	5.5 dB	5.5 dB
Tx antenna gain (vertical)	2.15 dBi	2.15 dBi	2.15 dBi
Tx pointing loss	0.0 dB	0.0 dB	0.0 dB
Tx random loss	0.0 dB	0.0 dB	0.0 dB
Tx EIRP	40.65 dBm	40.65 dBm	40.65 dBm
Free space loss	125.5 dB	129.9 dB	136.0 dB
Atmospheric absorption	0.13 dB	0.22 dB	0.45 dB
Precipitation absorption	0.0 dB	0.0 dB	0.0 dB
Total propagation loss	125.63 dB	130.12 dB	136.45 dB
Rx antenna gain (1)	14 dBi	14 dBi	14 dBi
Rx polarization loss	0.5 dB	0.5 dB	0.5 dB
Rx pointing loss	0.5 dB	0.5 dB	0.5 dB
Rx components loss (MP)	5.0 dB	5.0 dB	5.0 dB
Effective carrier power at Rx input	**−76.98 dBm**	**−81.47 dBm**	**−87.8 dBm**
Rx noise threshold	103.28 dBm	103.28 dBm	103.28 dBm
Calculated C/N @ Rx input	26.3 dB	21.81 dB	15.48 dB
Required C/N @ Rx input	5.2 dB	5.2 dB	5.2 dB
Calculated fade margin	**21.1 dB**	**16.61 dB**	**10.28 dB**

$1.41 \times Z_0$, or 70.7 Ω for $Z_0 = 50$ Ω. For an input signal V, the outputs at ports 2 and 4 are equal in magnitude, but 180 degrees out of phase.

Figure 14.35 presents the orientation between the mono-pulse antenna and the target. The distance between the two elements is d. A wave-front is incident at an angle θ. The phase difference between the two antennas is a phase $\Delta\Phi$. The angle θ may be calculated by using equation (14.67).

$$\theta = \sin^{-1}\left(\frac{\lambda\Delta\Phi}{2\pi d}\right) \tag{14.67}$$

Ku band mono-pulse reflector antenna feed

A mono-pulse double layer antenna was designed at 15 GHz. The mono-pulse double layer antenna consists of four circular patch antennas as shown in figure 14.36. The resonator and the feed network were printed on a substrate with a relative dielectric constant of 2.5 with a thickness of 0.8 mm. The resonator is a circular microstrip resonator with diameter $a = 4.2$ mm. The radiating element was printed on a substrate with a relative dielectric constant of 2.2 with a thickness of 0.8 mm.

Table 14.21. Downlink, DNL, budget calculations: omni–omni.

Parameter	Ku band	Ku band	Ku band
Frequency	15.005 GHz	15.005 GHz	15.005 GHz
Distance	**1 km**	**2 km**	**3 km**
Rx noise figure	2.0 dB	2.0 dB	2.0 dB
IF bandwidth	8 MHz	8 MHz	8 MHz
Tx power	44 dBm	44 dBm	44 dBm
Tx components loss	5.5 dB	5.5 dB	5.5 dB
Tx antenna gain (vertical)	2.15 dBi	2.15 dBi	2.15 dBi
Tx pointing loss	0.0 dB	0.0 dB	0.0 dB
Tx random loss	0.0 dB	0.0 dB	0.0 dB
Tx EIRP	40.65 dBm	40.65 dBm	40.65 dBm
Free space loss	115.964 dB	122.0 dB	125.5 dB
Atmospheric absorption	0.0449 dB	0.09 dB	0.13 dB
Precipitation absorption	0.0 dB	0.0 dB	0.0 dB
Total propagation loss	116.0 dB	122.09 dB	125.63 dB
Rx antenna gain	2.15 dBi	2.15 dBi	2.15 dBi
Rx polarization loss	0.5 dB	0.5 dB	0.5 dB
Rx pointing loss	0.0 dB	0.0 dB	0.0 dB
Rx components loss (MP)	5.0 dB	5.0 dB	5.0 dB
Effective carrier power at Rx input	**−78.7 dBm**	**−84.79 dBm**	**−88.33 dBm**
Rx noise threshold	103.28 dBm	103.28 dBm	103.28 dBm
Calculated *C/N* @ Rx input	24.58 dB	18.49 dB	14.95 dB
Required *C/N* @ Rx input	5.2 dB	5.2 dB	5.2 dB
Calculated fade margin	**19.38 dB**	**13.29 dB**	**9.75 dB**

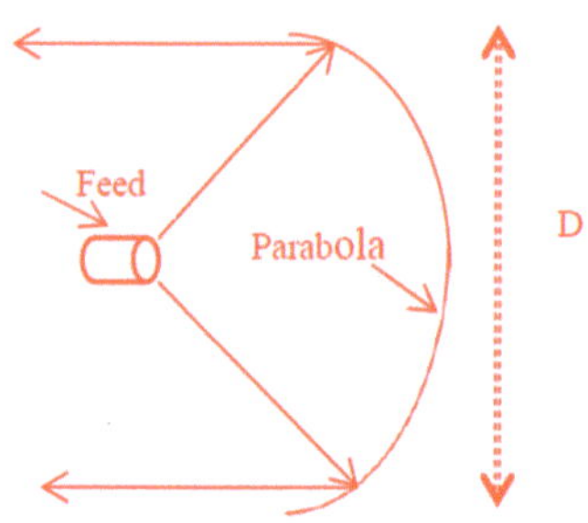

Figure 14.33. Parabolic antenna.

The radiating element is a circular microstrip patch with diameter $a = 4.5$ mm. The four circular patch antennas are connected to three 3 dB 180° rat-race couplers via the antenna feed-lines, as shown in figure 14.36. The comparator consists of three strip-line 3 dB 180° rat-race couplers printed on a substrate with a relative dielectric constant of 2.2 with a thickness of 0.8 mm. The comparator has four output ports: a sum port $\sum$, difference port Δ, elevation difference port Δ_{El} and azimuth difference port Δ_{Az}, as shown in figure 14.36. The antenna's bandwidth is 10% for a VSWR

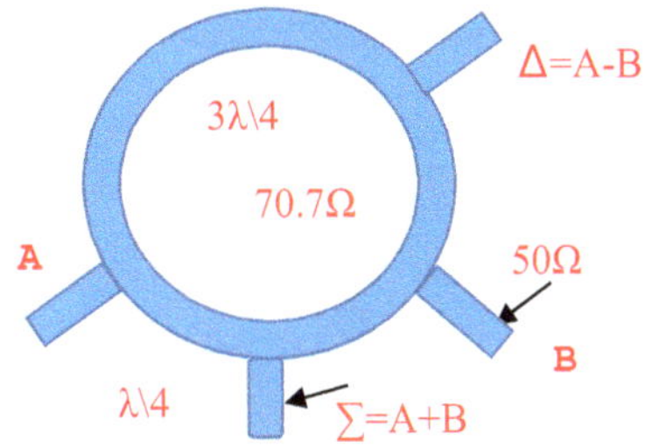

Figure 14.34. Rat-race coupler.

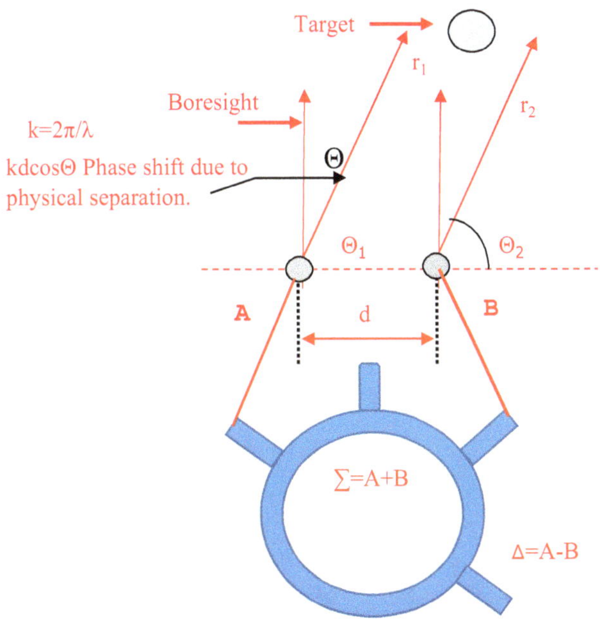

Figure 14.35. Orientation between a mono-pulse antenna and the target.

better than 2:1. The beamwidth is around 36°, and the measured antenna's gain is around 10 dBi.

Mono-pulse comparator specifications

Frequency: 14.5–15.3 GHz.
Insertion loss: 0.6 dB.
VSWR: 1.3:1.
The comparator losses are around 0.7 dB. The elevation error in the tracking system is calculated from the difference signal Δ_{EL} and the azimuth error from the difference signal Δ_{Az}. The resulting local minimum in the Δ port at the center of the bore sight is very deep, more than −20 dB, as shown in figure 14.37. A high angular accuracy in the tracking process is achieved by comparing the sum and

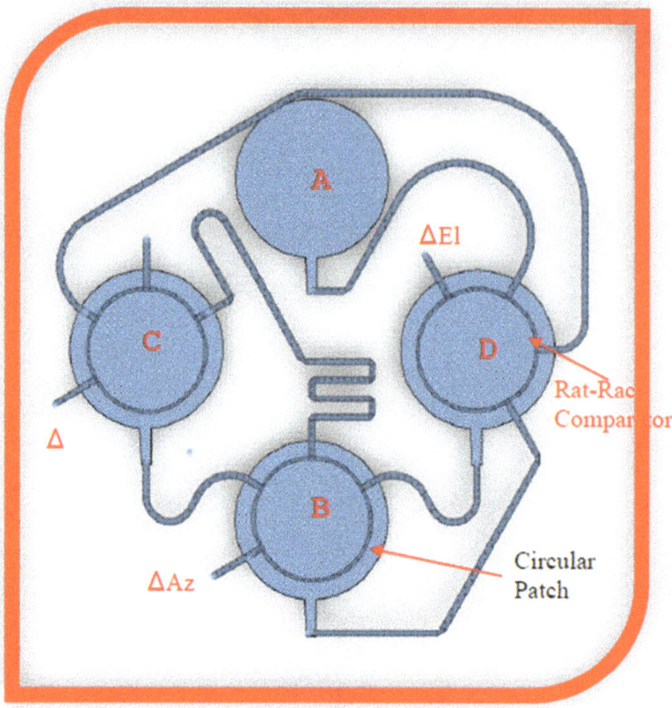

Figure 14.36. A microstrip stacked mono-pulse antenna and mono-pulse comparator.

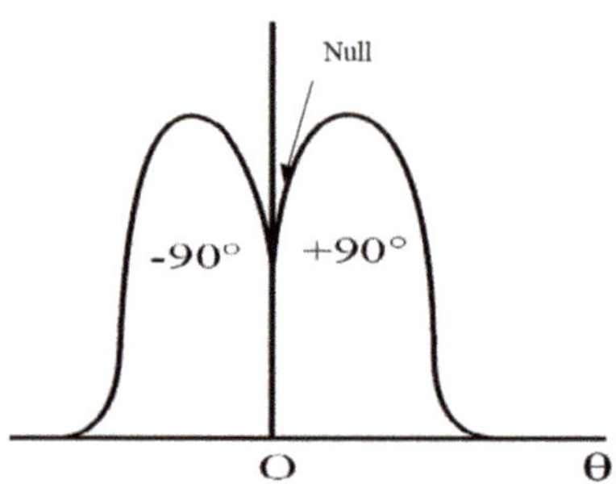

Figure 14.37. Mono-pulse antenna radiation pattern at the Δ port.

difference signals. A unique tracking algorithm is implemented inside the mono-pulse processor.

Reflector antenna performance
Frequency: 14.5–15.3 GHz.
Gain: 45 dBi.
Beamwidth: 0.9°.
Dish diameter: 1.6 m.

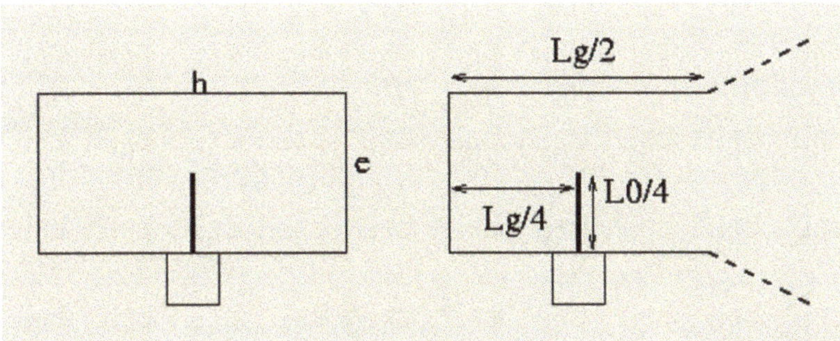

Figure 14.38. Ku band horn antenna feed.

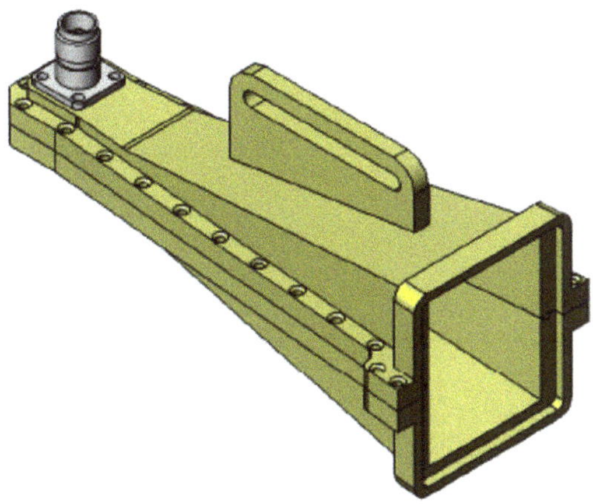

Figure 14.39. Ku band horn antenna assembly.

14.9.2.2 Horn antenna

Horn antenna specifications
Frequency: 14.5–15.3 GHz.
Gain: 14 dBi.
Dimensions: 160.5 × 67.6 × 42.5 mm.
The horn antenna is fed by a coaxial connector as shown in figure 14.38. The transition from coax to waveguide is presented in figure 14.38. The Ku band horn antenna assembly is presented in figure 14.39. A Ku band horn antenna fabrication drawing is shown in figure 14.40. The dimensions in inches of a Ku band waveguide are listed in table 14.22. The horn antenna's dimensions are 160.5 × 67.6 × 42.5 mm, and the gain is around 14 dBi.

14.9.2.3 Omnidirectional antenna

The omni antenna is a quarter-wavelength monopole antenna. The antenna's length is around 5 mm. The monopole ground plane diameter is 120 mm.

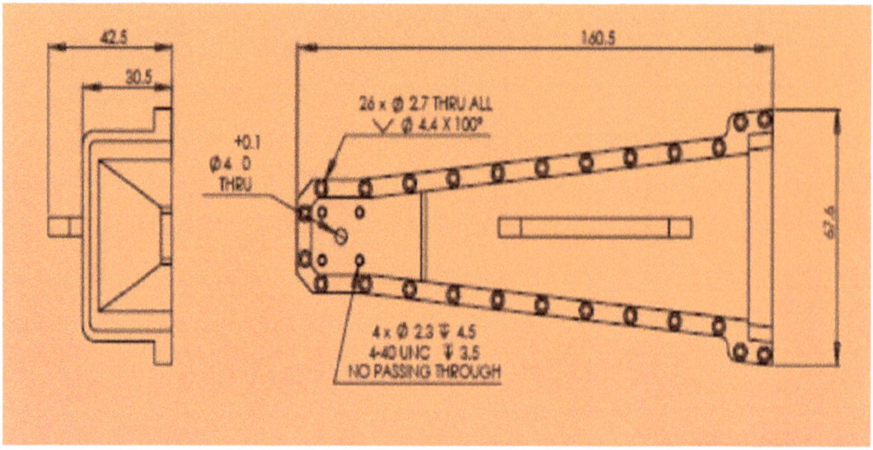

Figure 14.40. Ku band horn antenna fabrication drawing.

Table 14.22. Ku band waveguide dimensions in inches.

WR size	AMC model	Frequency GHz	Material type	Inside dimension	Outside dimension	Wall size	Cover flange
62	207	12.4–18.0	6061 AI	0.622 × 0.311	0.702 × 0.391	0.04	UG1665/U

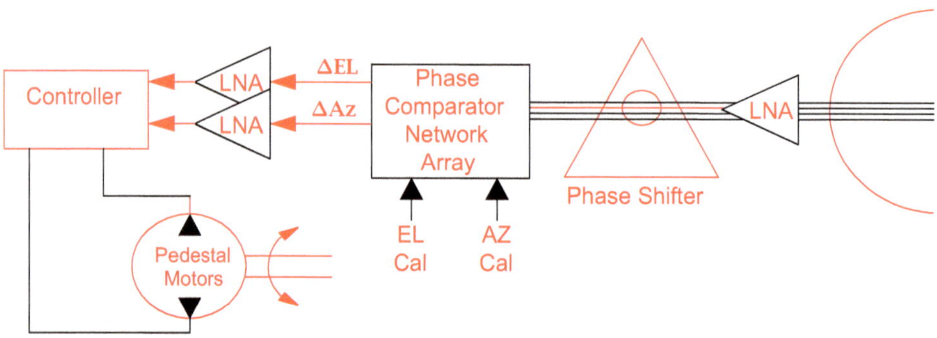

Figure 14.41. Mono-pulse processor.

Omnidirectional antenna specifications
Frequency: 14.5–15.3 GHz.
Gain: 2 dBi.
Dimensions: Ø 12 mm × 120 mm.

14.9.3 Mono-pulse processor

A block diagram of a tracking mono-pulse system is shown in figure 14.41.

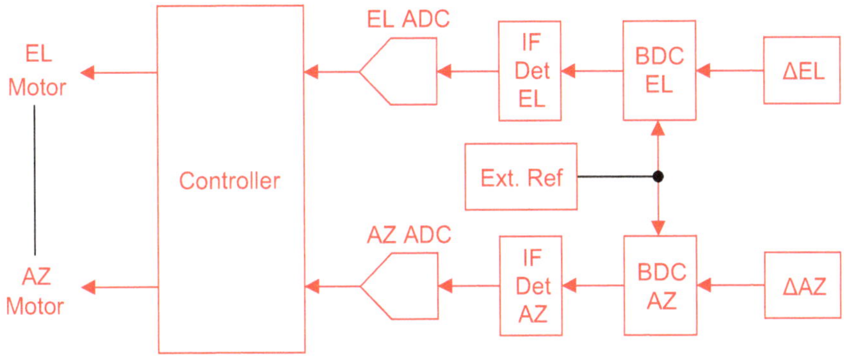

Figure 14.42. Tracking system controller.

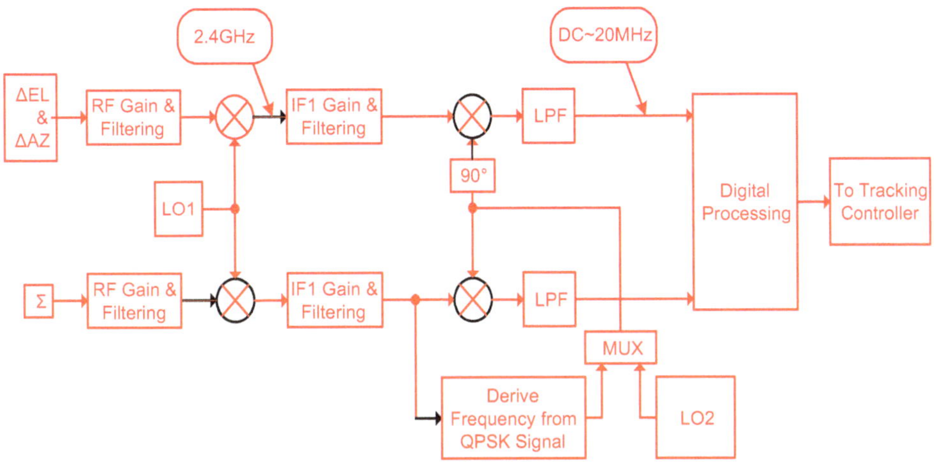

Figure 14.43. Mono-pulse tracking unit.

The tracking mono-pulse system consists of antennas, a low noise amplifier (LNA), phase comparator array and a controller. A block diagram of a tracking controller system is shown in figure 14.42. A detailed block diagram of a mono-pulse tracking unit is shown in figure 14.43. The sum, elevation difference and azimuth difference signals are amplified by an LNA and down-converted by a mixer to 2.4 GHz. The signals are amplified and down-converted to DC and up to 20 MHz. The down-converted signals are fed to a digital processing unit. The digital processing unit supplies the tracking data to the tracking controller shown in figure 14.42. A block diagram of a mono-pulse tracking unit is shown in figure 14.43. The mono-pulse tracking unit consists of sum and difference channels. Each channel consists of an amplifier, filter, IF down-converter mixer, IF amplifier and filter, down-converter mixer to baseband and filter. The baseband signal from the sum and difference channels is connected to a digital processing unit that provides the tracking commands to the tracking controller.

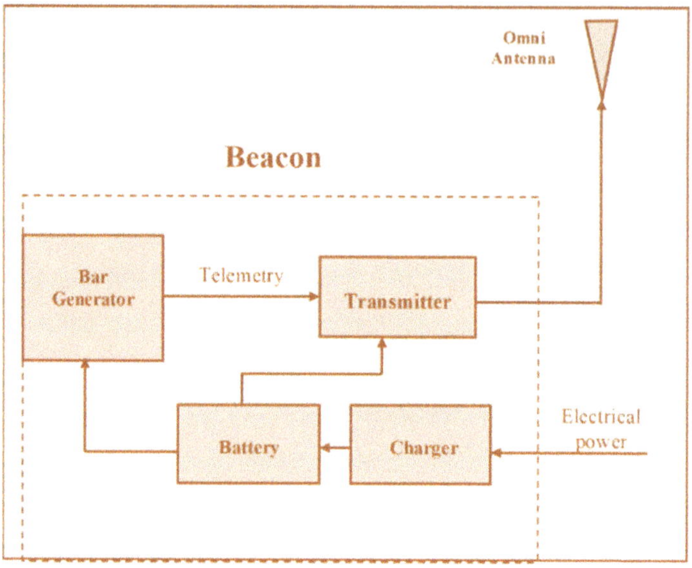

Figure 14.44. Omnidirectional link.

Pointing system specifications
Elevation range: 0° to 50°.
Azimuth range: Continuous in either direction, 360° continuous.
Accuracy: ± 0.1° azimuth, elevation
Position repeatability: ± 0.1°.
Position step resolution: Continuous.

Platform dynamics
Pitch: ±5° s^{-1}.
Tangential acceleration: ±0.5*g*.
Turning: 45° s^{-1} and 3° s^{-2}.
A block diagram of the omnidirectional link is shown in figure 14.44. The omnidirectional link consists of a transmitter, omnidirectional antenna and an electrical power supply unit.

14.9.4 High power amplifier

The design of RF amplifiers and components can be found in books and papers [13–16]. The desired output power of 47 dBm may be achieved by combining nine 6.5 W packed MMIC amplifiers modules. The HPA specifications are listed in table 14.23.

14.9.4.1 HPA design based on a 6.5 W Ku band power amplifier

Basic power amplifier module performance
6.5 W Ku band power amplifier.
Frequency range: 13–16 GHz.
38 dBm nominal P_{sat}.

Table 14.23. HPA specifications.

Parameter	Value	Tolerance
Tx frequency	14.5 GHz to 14.8 GHz	
1 dB compression point	47 dBm	Min. under all conditions
Tx gain	80 dBm	Min. under all conditions
Tx gain adjustment range	+6.0 dB to −20 dB	
Tx level flatness	±0.8 dB	Over any 36 MHz BW
Tx gain stability	±0.8 dB	Over all temperature and frequency
Tx linearity	−53 dBc	2 carriers @ 6 dB back-off
Input impedance	50 ohms	
Input connector	SMA	
Input VSWR	01:01.2	Nominal
Input NF	6 dB	Maximum
Output impedance	50 ohms	
Output connector	WR75	W/Gasket
Output VSWR	01:01.2	Nominal
Output port protection	Open/Short internally protected	
Visual indicators	Green LED: Power ON Red LED: summary alarm	
BIT indications	FWD VSWR REV VSWR HI VOLTAGE LOW VOLTAGE AGC SAT (low RF) HI TEMP	
Power	28 V DC/TDB Amp	Not to exceed
Temperature	Operational: −40 °C to + 55 °C Storage: −60 °C to + 75 °C	
ODU	IP65	
Vibration	1*g* random	Operational
Shock	10*g*	Operational
Weight	5 kg	

24 dB nominal gain.
14 dB nominal return loss.
0.25 μm PHEMT. MMIC technology.
10 lead flange package.
Bias conditions: 8 V @ 2.6 A I_{dq}.
Package dimension: 0.45 × 0.68 × 0.12 in.

HPA DC power
DC voltage 8 V.
Current: 35 A.

Total power: 300 W max.

A power amplifier block diagram with nine 6.5 W power modules is presented in figure 14.45. The −20 dBm input power is amplified by around 50 dB gain medium power module to 27 dBm. The output of the 27 dBm amplifier is connected to a printed nine-way power divider.

The output ports of the power divider are connected to nine 6.5 W modules with 23 dB gain. The output power of each 6.5 W module is around 38 dBm.

The nine 6.5 W modules are combined by a waveguide combiner. The output power of the HPA is 47 dBm. The transmitter gain and power budget are listed in table 14.24.

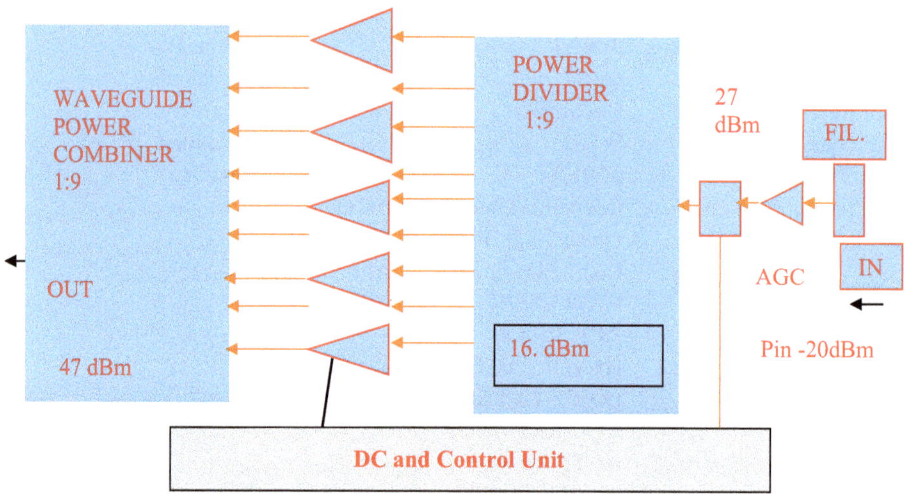

Figure 14.45. Power amplifier block diagram 9 × 6.5 W power modules.

Table 14.24. Transmitter gain and power budget.

Component	Gain/Loss (dB)	P_{out} (dBm) Amp.
Input power min.		−20
Amplifier	19–24	−1
Filter	−1	−2
Amplifier	20	18
Amplifier	9	27
9-way power divider	−10.5	16.5
Amplifier	23	38
1:9 Power combiner	10	47
Total with AGC	70–76	

A packed MMIC 6.5 W power module is shown in figure 14.46. A HPA DC power tree is shown in figure 14.47. The expected results of the 1:9 waveguide power combiner are listed in table 14.25.

14.9.4.2 HPA design with 41.5 dBm power modules

The desired output power of 47 dBm may be achieved by combining four 14 W MMIC power amplifier modules. The HPA specifications are listed in table 14.23.

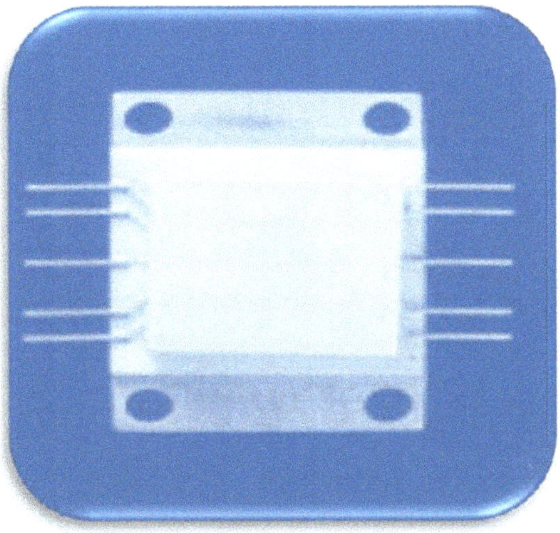

Figure 14.46. Packed MMIC 6.5 W power module.

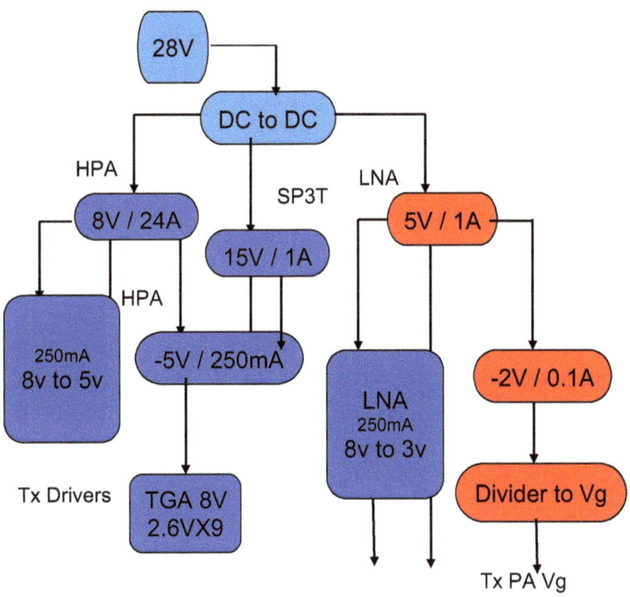

Figure 14.47. HPA DC power tree, 9 × 6.5 W modules.

Table 14.25. Expected results of the 1:9 waveguide power combiner.

Parameter	Measured results
Frequency range	14.5–15.3 GHz
Insertion loss	0.8 dB
Amplitude balance	0.4 dB max.
Phase balance	+5° max.
VSWR (Input/Output)	1.5:1

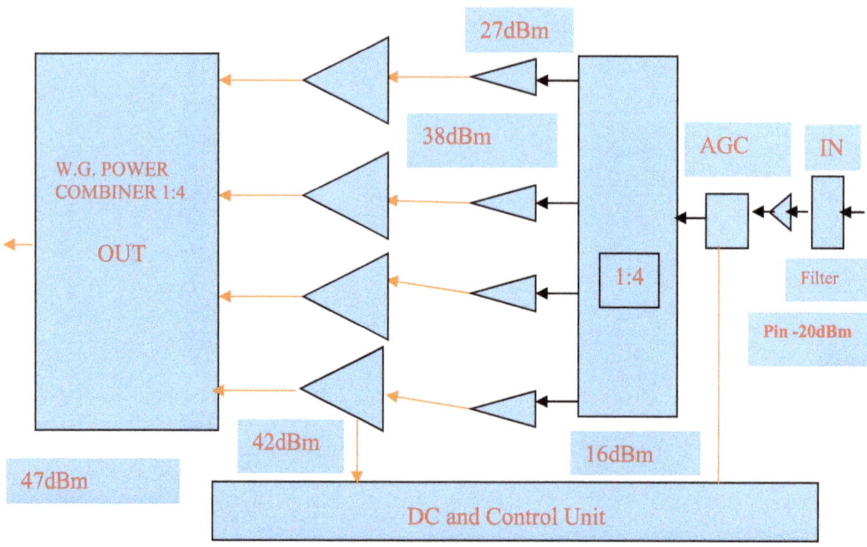

Figure 14.48. Power amplifier block diagram 4 × 14 W power modules.

Basic 42 dBm power amplifier module performance

f = 13.75–14.5 GHz.

Output power at 1 dB gain compression point: P1dB 41.5–42.0 dBm.

Power gain at 1 dB gain compression point: G1dB 5.0–6.0 dB.

Drain current IDS1: 5.5–6.0 A.

Power added efficiency had VDS = 9 V IDSQ = 4.4 A 28%.

3rd order intermodulation distortion: IM3 dBc −25.

A power amplifier block diagram with four 14 W power modules is presented in figure 14.48. The −20 dBm input power is amplified by around 50 dB gain medium power module to 27 dBm. The output of the 27 dBm amplifier is connected to a printed four-way power divider. The output ports of the power divider are connected to four 6.5 W modules with 23 dB gain. The output power of each 6.5 W module is around 38 dBm.

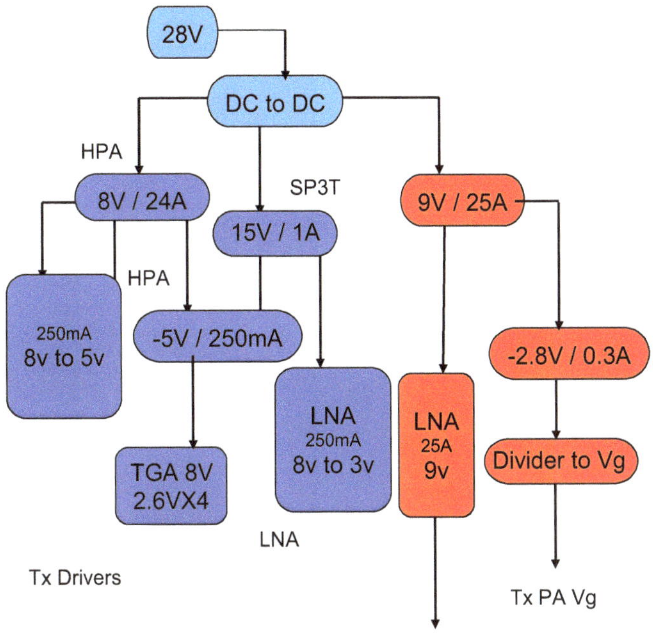

Figure 14.49. HPA DC power tree, 4 × 14 W modules.

The four 6.5 W modules are combined to a four 14 W power module. The four 14 W modules are combined by a waveguide combiner. The output power of the HPA is 47 dBm. The transmitter gain and power budget are listed in table 14.10. A HPA DC power tree of the 4 × 14 W modules is presented in figure 14.49.

HPA DC power
DC voltage: 9 V.
Current: 35 A.
Total power: 315 W max.
The transmitter gain and power budget are listed in table 14.26. The waveguide power divider and combiner of the power amplifier with 9 × 6.5 W power modules is shown in figure 14.50.
The waveguide power divider and combiner of the power amplifier with 4 × 14 W power modules is shown in figure 14.51.

14.9.5 Tracking system down-/up-converter

Design of a down- and up-converter unit is presented in this section.

14.9.5.1 Tracking system up-converter design

The up-converter consists of a RF amp with moderate NF to provide a low overall system NF, a high IP_3 double-balanced mixer with a 20 dBm local oscillator drive to

Table 14.26. Gain and power budget for HPA with a four-way power combiner.

Component	Gain/Loss (dB)	P_{out} (dBm) Amp.
Input power min.		−20
Amplifier	19–24	−1
Filter	−1	−2
Amplifier	20	18
Amplifier	9	27
4-way power divider	−7	20
Amplifier	23	38
Amplifier	5	43
4-way power combiner	5	48
Total with AGC	70–76	

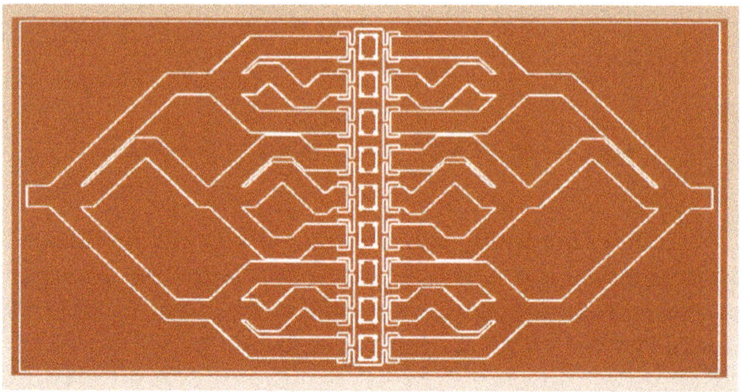

Figure 14.50. Power amplifier with 9 × 6.5 W power modules.

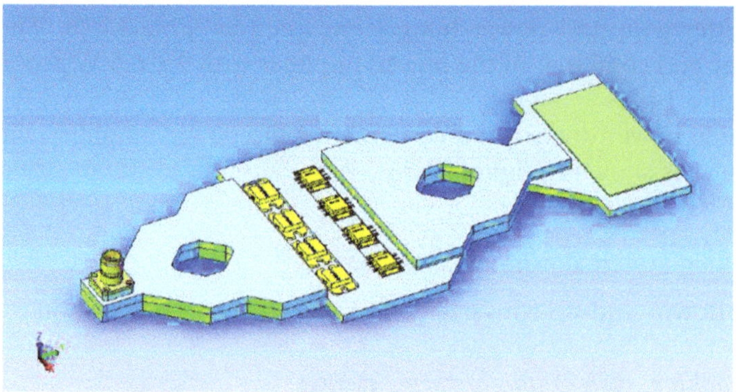

Figure 14.51. Power amplifier with 4 × 14 W power modules.

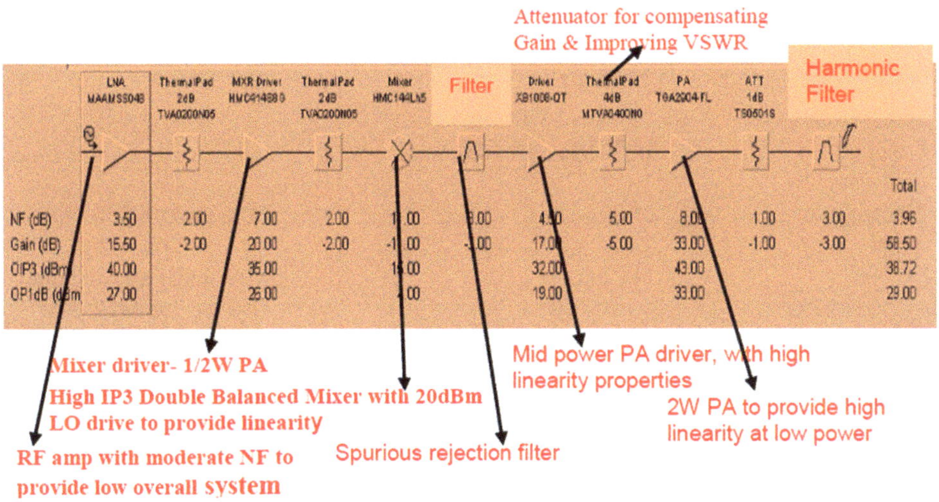

Figure 14.52. Block diagram of the up-converter.

provide linearity, medium power amplifiers, a 2 W power amplifier with high linearity and band pass filters. A block diagram of the up-converter is shown in figure 14.52.

The up-converter specifications are listed in table 14.27.

The block diagram of the up-converter proves that the up-converter's gain is around 58 dB and the noise figure is around 4 dB.

The up-converter's performance results are listed in table 14.28. The up-converter's gain is around 58.5 dB and the noise figure is around 4 dB. The power consumption is listed in table 14.29.

14.9.5.2 Tracking system down-converter design

The down-converter consists of a RF amplifier with moderate NF to provide a low overall system NF, a high IP_3 double-balanced image reject mixer with a 20 dBm local oscillator drive to provide linearity, a medium power amplifier with high linearity and band pass filters. A block diagram of the down-converter is shown in figure 14.53.

The down-converter specifications are listed in table 14.30.

The block diagram of the down-converter proves that the down-converter gain is around 61 dB and the noise figure is around 2 dB. The down-converter's performance results are listed in table 14.31. The up-converter's gain is around 61 dB and the noise figure is around 2 dB. The power consumption is listed in table 14.32.

14.9.6 Tracking system's interface

A block diagram of the tracking system's interface is shown in figure 14.54. A HPA interface block diagram is shown in figure 14.55, and a Ku band automatic tracking system assembly is shown in figure 14.56.

Table 14.27. Up-converter specifications.

Parameter	Value	Tolerance
IF input frequency	2.38 GHz 2.48 GHz	
RF output frequency	14.5 GHz 14.8 GHz	
1 dB compression point	21dBm	Min. under all conditions
BUC gain	54 dBm	Min. under all conditions
BUC level flatness	+/−0.8 dB	Over any 36 MHz BW
BUC gain stability	+/−0.8 dB	All temperature and frequency
BUC linearity	−53 dBc	2 carriers 300K apart @ 6 dB back-off
Input impedance	50 ohms	
Input connector	SMA	
Input VSWR	01:01.2	Nominal
Input NF	3 dB	Maximum
Output impedance	50 ohms	
Output connector	SMA	W/Gasket
Output VSWR	01:01.2	Nominal
Visual indicators	Green LED: Power ON Red LED: Summary alarm	
BIT indications	HI VOLTAGE LOW VOLTAGE AGC SAT (low RF) HI TEMP	
Power	28 V DC/TDB Amp	
Temperature	Operational: −40 °C to +55 °C Storage: −60 °C to +75 °C	
Vibration	1g random	Operational
Shock	10g	Operational
Weight	0.85 kg	

Table 14.28. Up-converter performance.

Parameter	Nominal	Delta
Gain dB	58.5	±4.5
NF dB	4	±0.8
OP1dBm	29	±1.1
ORR3 @ −41 dBm input (dBc)	42.5	—

Table 14.29. Up-converter power consumption.

Component	Voltage [V]	Current [mA]	Power [W]
IF LNA	5	180	0.9
Mixer driver	5	500	2.5
PA driver	5	100	0.5
PA driver	−0.5	10	0.005
PA	7	700	4.9
PA	−0.57	50	0.0285
	Total	1540	8.8335

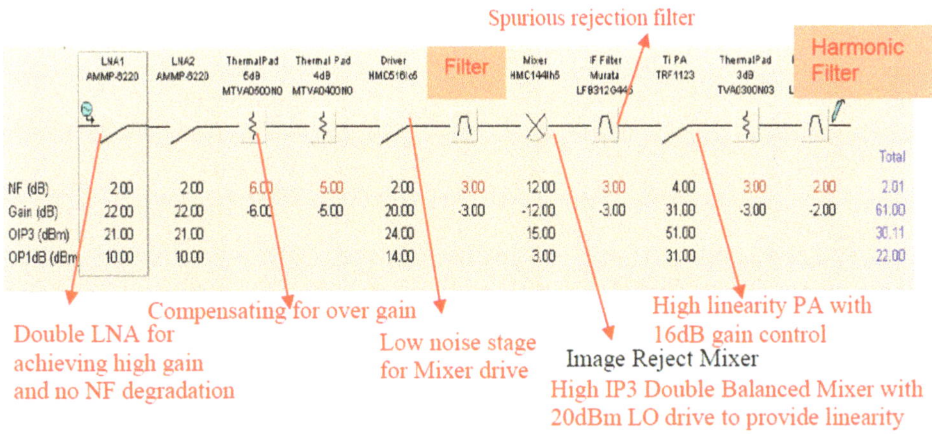

Figure 14.53. Block diagram of the down-converter.

Control

LO frequency set—SPI or RS-232 (both available):
SPI standard protocol.
RS-232 standard protocol.
Indications (via RS-232):
Over voltage.
Under voltage.
Lock (Synthesizer).

Tracking system's interface with HPA

Controls (via RS-232):
On/Off power.
Indications (via RS-232):
High temperature.
Over voltage.

Table 14.30. Down-converter specifications.

Parameter	Value	Tolerance
IF output frequency	2.38 GHz–2.48 GHz	
RF input frequency	15 GHz–15.3 GHz	
1 dB compression point	2 dBm	Min. under all conditions
BDC gain	58 dBm	Min. under all conditions
BDC level flatness	±0.8 dB	Over any 36 MHz BW
BDC gain stability	±0.8 dB	Over all temperature and frequency
BDC linearity	−53 dBc	2 carriers 300K apart @ 6 dB back-off
Input impedance	50 ohms	
Input connector	SMA	
Input VSWR	01:01.2	Nominal
Input NF	2.4 dB	Maximum
Output impedance	50 ohms	
Output connector	SMA	W/Gasket
Output VSWR	01:01.2	Nominal
Visual indicators	Green LED: Power ON red LED.	
BIT indications	HI VOLTAGE LOW VOLTAGE AGC SAT (low RF)	
Power	28 V DC/TDB Amp	Not to exceed
Temperature	Operational: −40 °C to +55 °C. Storage: −60 °C to +75 °C	
Vibration	1*g* random	Operational
Shock	10*g*	Operational
Weight	0.45 kg	
LO phase noise	SSB phase noise offset 10 Hz −35 dBc 100 Hz −65 dBc 1 KHz −77 dBc 10 KHz −85 dBc 100 KHz −95 dBc 1 MHz −110 dBc	
	10 MHz	1.2 ppm
Reference	Internal	TCXO

Table 14.31. Down-converter performance.

Parameter	Nominal	delta
Gain dB	61	±3.7
NF dB	2	±0.6
OP1dBm	22	±2
ORR3@−62 dBm input (dBc)	62.53	—

Table 14.32. Down-converter power consumption.

Component	Voltage [V]	Current [mA]	Power [W]
LNA1	3	55	0.165
LNA2	3	55	0.165
MXR driver	3	65	0.195
IF PA	7	600	4.2
IF PA	5	25	0.125
IF PA	−5	15	0.075
	Total	815	4.925

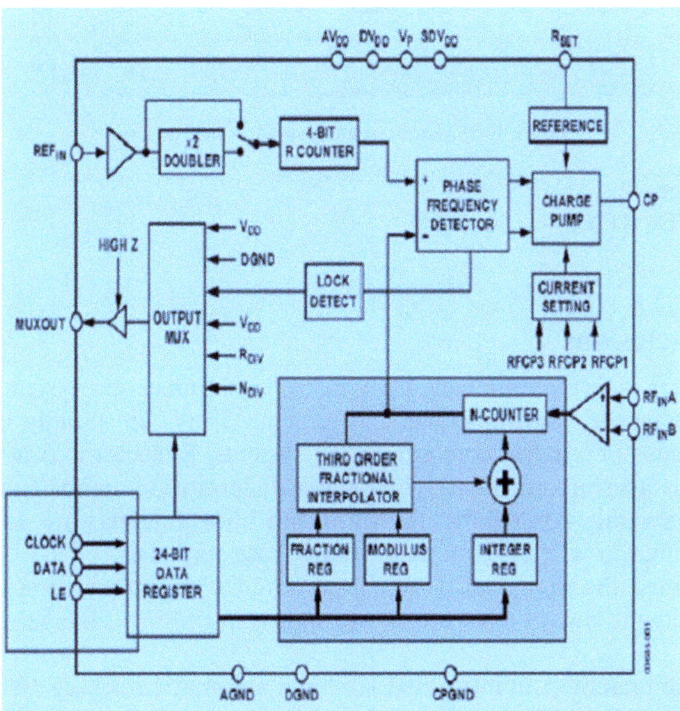

Figure 14.54. Block diagram of a tracking system's interface.

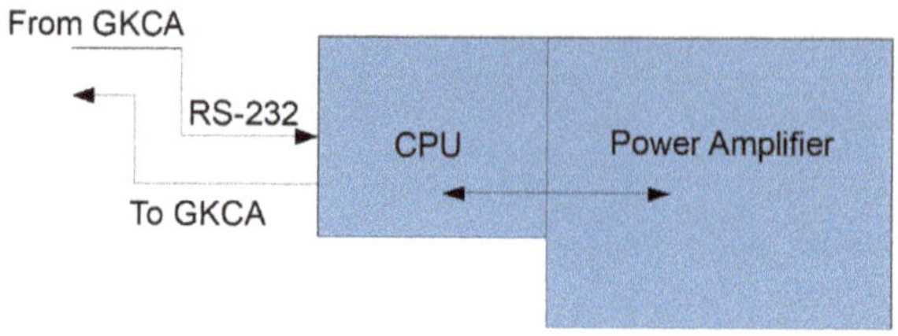

Figure 14.55. Block diagram of a HPA interface.

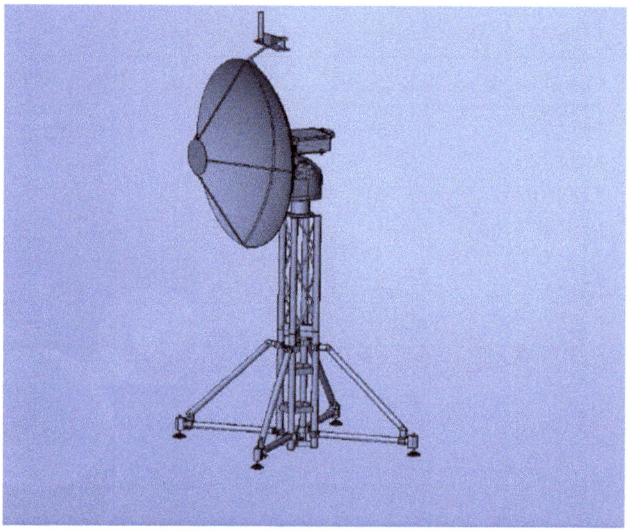

Figure 14.56. Ku band automatic tracking system.

Under voltage.
VSWR (FWD, REV).
AGC status.

14.10 Conclusions

In this chapter, aperture antennas for wireless communication systems have been presented. Reflector, horn antennas and antenna arrays are usually used as base station antennas of wireless communication systems. Reflector and horn antennas are defined as aperture antennas. The longest dimension of an aperture antenna is higher than several wavelengths. Reflector and horn antennas are used as transmitting antennas in wireless communications' base stations.

We presented the design of a mm wave satellite communication system. The design considerations of the receiving and transmitting channels have been discussed.

Finally, we presented an integrated Ku band automatic tracking system, and the design considerations of the modules.

References

[1] Balanis C A 1996 *Antenna Theory: Analysis and Design* 2nd edn (New York: Wiley)
[2] Godara L C (ed) 2002 *Handbook of Antennas in Wireless Communications* (Boca Raton, FL: CRC Press)
[3] Kraus J D and Marhefka R J 2002 *Antennas for all Applications* 3rd edn (New Delhi: McGraw Hill)
[4] James J R, Hall P S and Wood C 1981 *Microstrip Antenna Theory and Design* (London: Peter Peregrinus)
[5] Sabban A and Gupta K C 1991 Characterization of radiation loss from microstrip discontinuities using a multiport network modeling approach *IEEE Trans. Microw. Theory Tech.* **39** 705–12
[6] Sabban A 1991 Multiport network model for evaluating radiation loss and coupling among discontinuities in microstrip circuits *PhD thesis* University of Colorado at Boulder.
[7] Sabban A 2011 Microstrip antenna arrays *Microstrip Antennas* ed N Nasimuddin (Rijeka: InTech) pp 361–84
[8] Sabban A 1983 A new wideband stacked microstrip antenna *IEEE Antenna and Propagation Symp. (Houston, TX, USA)*
[9] Sabban A 2015 *Low Visibility Antennas for Communication Systems* (New York: Taylor and Francis)
[10] Sabban A 1981 Wideband microstrip antenna arrays *IEEE Antenna and Propagation Symp., MELCOM (Tel Aviv)*
[11] Sabban A 2014 *RF Engineering, Microwave and Antennas* (Israel: Saar Publications)
[12] Fujimoto K and James J R (ed) 1994 *Mobile Antenna Systems Handbook* (Boston, MA: Artech House)
[13] Rogers J and Plett C 2003 *Radio frequency Integrated Circuit Design* (Boston, MA: Artech House)
[14] Maluf N and Williams K 2004 *An Introduction to Microelectromechanical System Engineering* (Boston, MA: Artech House)
[15] Mass S A 1997 *Nonlinear Microwave and RF Circuits* (Boston, MA: Artech House)
[16] Sabban A 2016 *Wideband RF Technologies and Antenna in Microwave Frequencies* (New York: Wiley)

Chapter 15

Compact circular patch wearable metamaterials antennas for healthcare, IOT, and 5G systems

Novel sensors and antennas for healthcare devices, Internet of Things (IoT), and 5G devices are presented in this chapter. The major challenge in development of healthcare, IoT, 5G and communication systems is the evaluation of compact efficient antennas. Moreover, a low-cost wearable sensor may be evaluated by printing the microstrip antenna with the sensor feed network and the active devices on the same substrate. Design considerations, comparison between simulation and measured results of compact efficient antennas and sensors for wireless, 5G, energy harvesting, IoT, and medical systems are highlighted in this chapter. The electrical performance of the novel sensors and antennas on and near the user body were evaluated by employing electromagnetic software. Efficient passive and active metamaterial antennas and sensors were developed to improve the system electrical performance. The system dynamic range may be improved by using active printed antennas. Circular printed antennas' performance near the human body was evaluated by using electromagnetic 3D software. The circular metamaterial antennas are compact and flexible. The frequency range of the resonators, without circular split-ring resonators (CSRRs) is higher by 4% to 10% than the resonators with CSRRs. The gain of the circular polarized antennas without CSRRs is lower by 2–3 dB than the resonators with CSRRs. The gain of the new passive antennas with CSRRs is around 7–8.4 dBi. The bandwidth of the new circular polarized antennas with CSRRs is around 10% to 20%. The sensors voltage standing wave ratio (VSWR) is better than 3:1. The passive and active efficient metamaterials antennas improve the system performance.

doi:10.1088/978-0-7503-5224-6ch15

15.1 Introduction

Theory and design of printed antennas is discussed in [1]. The efficiency of compact printed antennas is low [2–4]. Printed metamaterials antennas and sensors are used in wireless communication systems and were discussed, evaluated and presented in several publications in this century [2–6]. Printed dipoles, planar inverted-F antenna (PIFA) and loop antennas, printed slots, microstrip antennas, and other compact antennas are employed in radars, IoT, 5G, and healthcare systems [2–6]. Several types of small efficient wideband wearable antennas are presented in [2–4]. Metamaterials are materials with periodic artificial structures. The metamaterial elements and structure define the electrical properties of the material. Metallic posts structures and periodic SRRs may be employed to produce structures with required permeability and dielectric constant as presented in [7–11]. Metamaterial technology may be employed to develop small efficient antennas for communication, wearable healthcare and IoT devices [12–16]. In [6] the development of a metamaterial microstrip antenna was presented. In [6] the antenna gain, and bandwidth are similar, to those of patch antennas. In [8] structures with negative dielectric permittivity are analyzed. In [9], a model and setup to simulate and measure the polarity of SRRs structures is presented. The model is used to compare measured results to computed results. In [12] a dual band transmission-line metamaterial antenna with two transmission-line arms, is presented. The antenna bandwidth is 3% with 60% efficiency and 2.6 dBi directivity. The antenna gain is around 0.8 dBi. Compact radiators such as printed loops and dipoles, patches, and FIPA antennas suffer from low efficiency [2–5, 16–28]. Compact efficient metamaterial antennas may be an important element in wearable medical systems. Wearable communication and medical systems are presented in [2–5]. Small efficient wearable metamaterials antennas for medical systems are presented in [15, 16]. Measurements of wearable antenna in the vicinity of the human body are presented in [17]. Active wearable antennas for wireless communication and medical systems are presented in [19].

A healthcare remote monitoring system in hospitals is presented in [29]. A wireless body area network with an adaptive thermal-aware protocol is discussed in [30]. In [31] a secure thermal-energy aware protocol is presented. In [32–40] wearable compact antennas for medical applications are presented. Wearable medical sensors can monitor and check patient daily health [34, 35]. Dual polarized dipole wearable antenna for medical applications are presented in [39]. Low-cost medical treatment may be achieved by online evaluation of a continuous measured medical information of a great number of patients.

Wearable antennas in today's world [2–5]

- Wearable antennas and sensors are used to monitor home medical devices to assist asthma, diabetes, and epilepsy patients.
- Wearable antennas and medical sensors may are used to monitor hospital activities such as patient care, patient health monitoring, and personal monitoring.
- Wearable antennas and sensors are used to operate IoT devices and systems.

Wearable antennas may be used in IoT devices. IoT is a system of interrelated computing networks, personal sensors and devices, commodities, mechanical and digital devices, peoples, healthcare sensors, animals, and wireless communication systems that have a unique identifier (UID) and can transfer data over a network without requiring human-to-human or human-to-computer assistance. IoT is an advanced 21st century emerging technology. It helps people every day live to get complete control over daily services and procedures and work smarter. It also helps to automate, home smart devices, companies, business and healthcare centers. IoT provides companies with real-time inspection and observation on how their organization functions online. IoT enables medical organizations and businesses to automate processes and reduce everyday expenses. It may improve company services, cut down on waste and improve product delivery. IoT helps to manufacture and deliver products efficiently and with less expense.

Metamaterials are employed to develop compact circular patch antennas with high efficiency for medical, IoT and communication systems. Detailed development of metamaterial antennas with CSRRs is presented in [2–5]. Circular patch metamaterial antennas have some advantages over square and rectangular patches. Such as symmetry, ease of generating circular polarization and propagation mode. The computed and measured bandwidth of the stacked circular patch wearable antenna with SRR for IoT and medical applications is around 15%, for S_{11} lower than −6 dB. The directivity and gain of the stacked circular patch with CSRR is around 8.4 dB and with 92% efficiency.

The progress in the design of compact antennas is shown in figure 15.1. Figure 15.1(a) highlights the gain of compact antennas. Figure 15.1(b) illustrates the bandwidth of compact antennas. The data presented in figure 15.1 are based on information given in [2–5].

15.2 Circular metamaterial patch with CSRR

A circular patch antenna with CSRR may be used as a sensor for IoT and medical applications. The CSRRs improve the antenna effective area and gain. The antenna presented in figure 15.1 consists of a circular patch, and a matching network. The antenna is printed on dielectric substrate 1.6 mm thick with dielectric constant of 2.2, and loss tangent of 0.001. Eighteen CSRRs are etched on the circular patch, this is the maximum number of CSRRs that can be etched on the patch. The radius of the antenna with CSRR, shown in figure 15.2(a), is 18 mm. The antenna resonates at 2.63 GHz. The resonant frequency of the dominant mode TM11 of the circular patch is given by equation (15.1) [1], c is the light velocity. Where a_e is the radius of the circular patch and it can be computed by using equation (15.2) [1]. ε_e is the effective dielectric constant. At 2.63 GHz the radius of the circular microstrip antenna should be 23.2 mm. The width of the CSRR strip is 0.15 mm. The outer radius of the CSRR is 2.6 mm, as presented in figure 15.2(b). The CSRR dimensions are optimized by using electromagnetic software. All the antennas presented in this chapter were analyzed by using electromagnetic software, see [41].

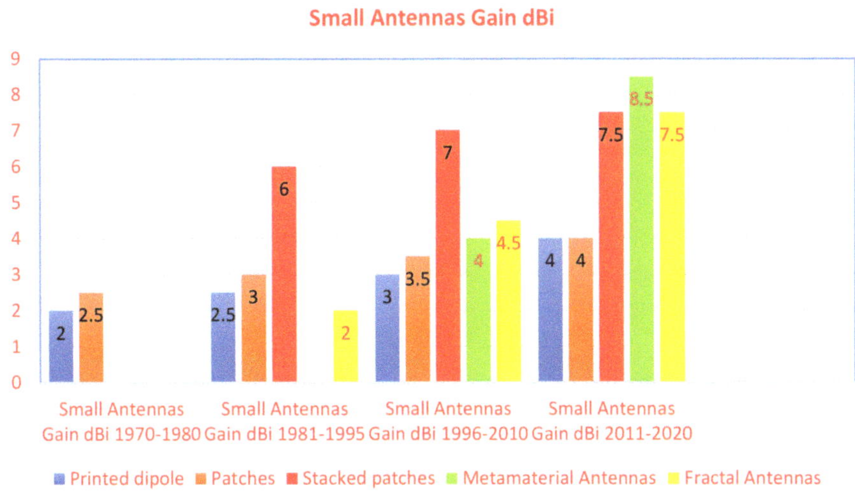

(a)

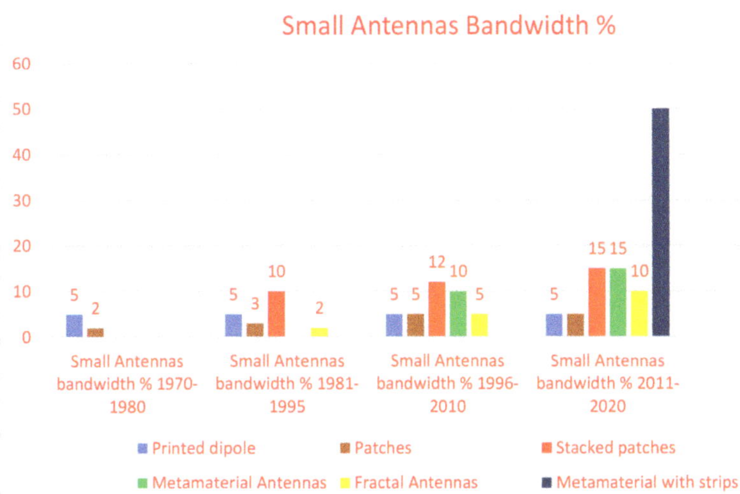

(b)

Figure 15.1. Advancement in development of compact antennas. (a) Antennas gain; (b) antennas bandwidth.

$$f = \frac{1.8412c}{2\pi a_e \sqrt{\varepsilon_e}} \tag{15.1}$$

$$a_e = \frac{1.8412c}{2\pi f \sqrt{\varepsilon_e}} \tag{15.2}$$

The diameter of the circular patch without CSRR is greater by 22% than the diameter of the antenna with CSRR. The CSRR's design detail was described

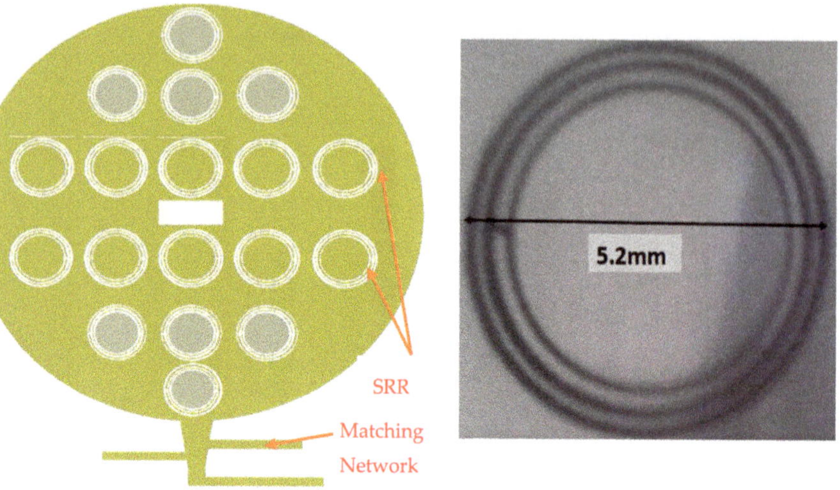

(a) (b)

Figure 15.2. (a) Circular patch with CSRR; (b) photo of a printed CSRR.

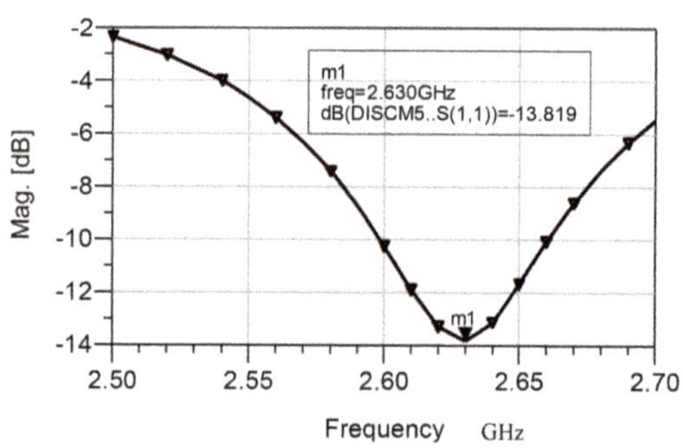

Figure 15.3. S_{11} of the compact wearable circular patch with CSRR.

in [2–5, 16]. The circular antenna bandwidth is around 5% for VSWR better than 2:1. The antenna bandwidth is around 7% for S_{11} lower than −7 dB as shown in figure 15.3. The antenna gain is around 7.6 dBi, as shown in figure 15.4. The antenna beam width is 80°. The circular patch efficiency is 83%.

The gain and directivity of the circular patch without CSRR is lower by 2–3 dB than the circular patch with CSRR.

15.3 Active receiving compact circular patch antennas

A receiving channel block diagram with an active antenna is shown in figure 15.5. A TAV541 PHEMT Low noise amplifier, LNA, is connected to the antenna, as shown in figures 15.5 and 15.6. The circular patch is etched on a substrate 1.6 mm thick, with dielectric constant of 2.2. The circular patch radius is 20 mm. The active

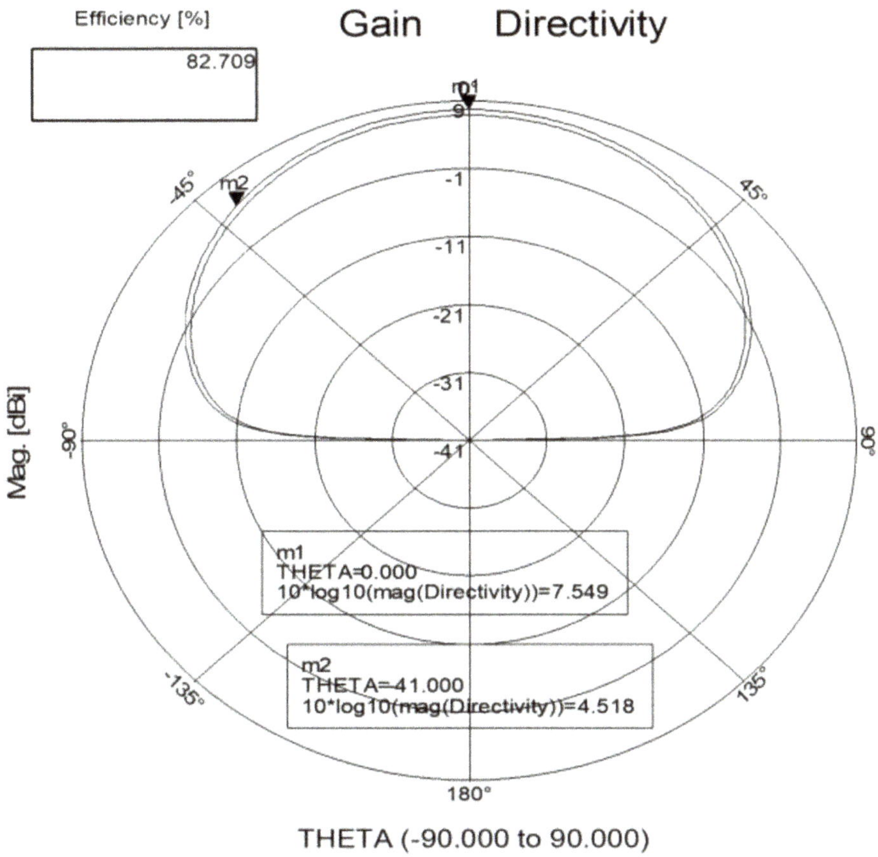

Figure 15.4. Radiation pattern of the circular patch with CSRR.

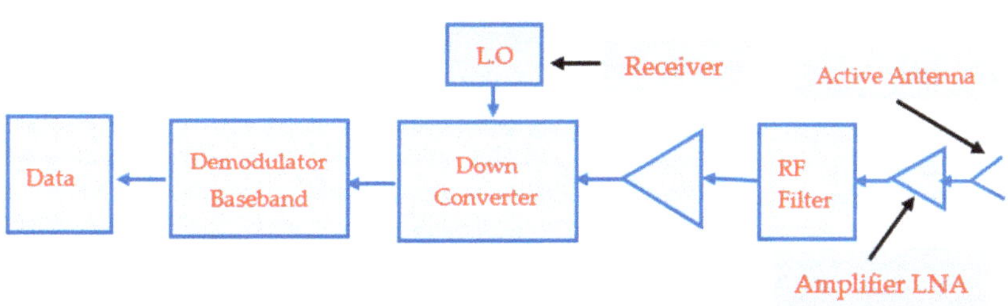

Figure 15.5. Receiving channel block diagram with wearable active antenna.

antenna structure is presented in figure 15.6. An input matching network, that is part of the antenna feed network, matches the circular patch to the low noise amplifier (LNA), as presented in figure 15.6. An output matching network matches the amplifier output to the receiver. The dimensions of the input and output matching network are around 22 × 28 mm and are etched on the same board. A DC bias circuit supplies the required voltages to the amplifiers. The amplifier requirements

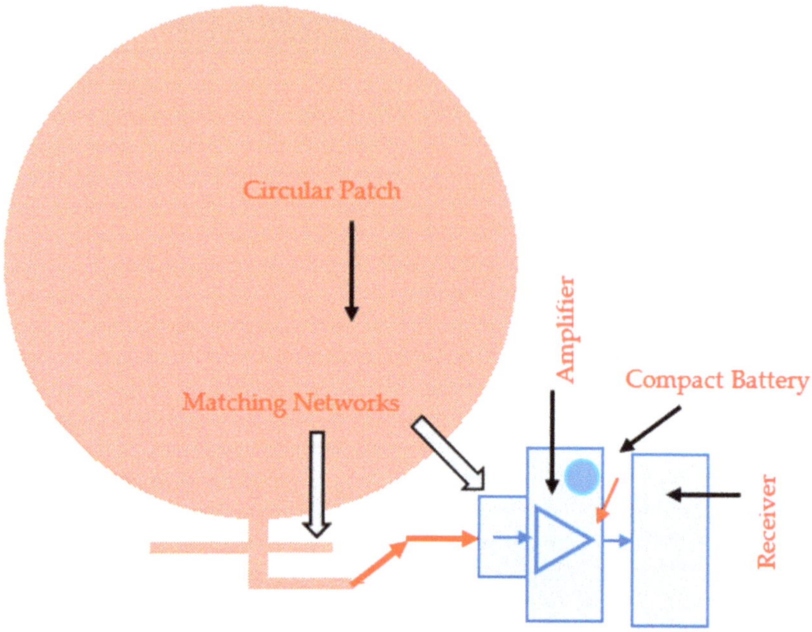

Figure 15.6. Circular wearable active receiving patch for medical and communication applications.

Table 15.1. S band LNA requirements.

Parameter	Specification	Remarks
Frequency	0.4–3 GHz	-
Gain at 1.9 GHz	18 dB	$V_{ds} = 3$ V
N.F at 1.9 GHz	0.5 dB	-
P1 dB at 1.9 GHz	19.0 dBm	-
OIP3 at 1.9 GHz	33.5 dBm	-
Max input power	17 dBm	-
V_{gs}	0.50 V	-
V_{ds}	3 V	$I_{ds} = 60$ mA
Supply voltage	±5 V	-
Package	Surface Mount	-
Operating temperature	−40 °C–80 °C	-

are listed in table 15.1. The amplifier complex S parameters are given in table 15.2. The amplifier noise parameters are given in table 15.2. The circular patch matched antenna S_{11} parameter on a human body is shown in figure 15.7. The circular patch VSWR without the matching network is around 5:1 in the frequencies from 2.4 to 3.2 GHz. However, VSWR of the antenna with the matching network is better than 2:1 in the frequencies from 2.4 to 3.2 GHz. The active antenna bandwidth is around

Table 15.2. *S* band LNA, *S* parameters and noise parameters.

F-GHz	S_{11}	$S_{11}{}^{\circ}$	S_{21}	$S_{21}{}^{\circ}$	S_{12}	$S_{12}{}^{\circ}$	S_{22}	$S_{22}{}^{\circ}$
1.04	0.74	−126.2	12.74	100.13	0.05	33.69	0.29	−94.96
1.21	0.71	−137.6	11.25	92.91	0.051	30.05	0.26	−104
1.53	0.687	−154.2	9.30	82.06	0.055	26.08	0.22	−119
1.75	0.67	−164.1	8.24	75.31	0.06	23.14	0.20	−128.4
2.02	0.67	−174.6	7.30	67.82	0.06	20.88	0.18	−138.8

Noise parameters

F-GHz	N. FMIN	N11X	N11Y	rn
1	0.16	0.3424	52.98	0.042
1.9	0.30	0.368	100.93	0.03
2	0.32	0.371	106.01	0.03
2.4	0.39	0.383	125.79	0.03
3	0.48	0.400	153.93	0.036
3.9	0.63	0.430	−167.3	0.06
5	0.81	0.465	−125.53	0.11

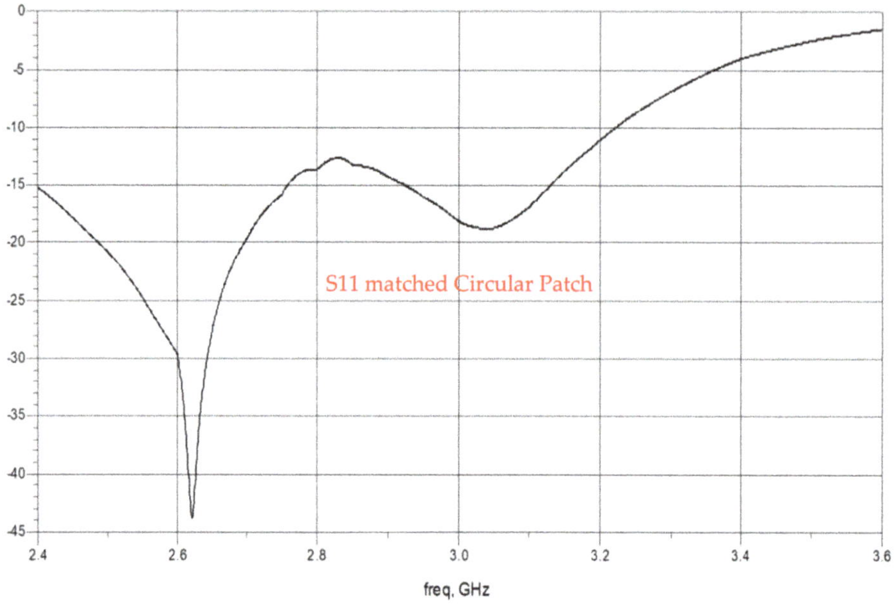

Figure 15.7. S_{11} of the receiving wearable circular patch.

25% for S_{11} lower than −9 dB. A Smith chart diagram of S_{11} with and without the output matching network is presented in figure 15.8(a). The antenna matching network was optimized, as presented in figure 15.8(b). The optimized matching network improved the antenna gain by 1.6 dB. The antenna bandwidth was improved to 44% for S_{11} lower than −6 dB. The active antenna gain on the human body is around 14 dB at 3 GHz and decreases to 11 dB at 3.6 GHz as shown in figure 15.9. The antenna noise figure is better than two for frequencies from 2 to 3.6 GHz. Measurement setups of RF devices were shown in [2]. There is a good match between simulated and measured electrical parameters.

15.4 Active transmitting wearable circular patch

A block diagram of a circular polarized transmitting link is shown in figure 15.10. An active transmitting circular patch is presented in figure 15.11. The antenna radius is 20 mm. The high-power amplifier (HPA) is an integral part of the antenna, see figure 15.11. The HPA, GaAs MESFET VNA25, and the antenna are assembled on the same substrate. The antenna is matched to the power amplifier by a printed matching circuit. The transmitter is matched to the amplifier by the input matching circuit. The matching circuits, dimensions are less than 23 × 29 mm. The amplifier requirements are given in table 15.3. In table 15.4 the HPA complex S parameters are given. The active transmitting circular patch VSWR is better than 2:1 for frequencies from 2 to 3 GHz as shown in figure 15.12. The patch matching network,

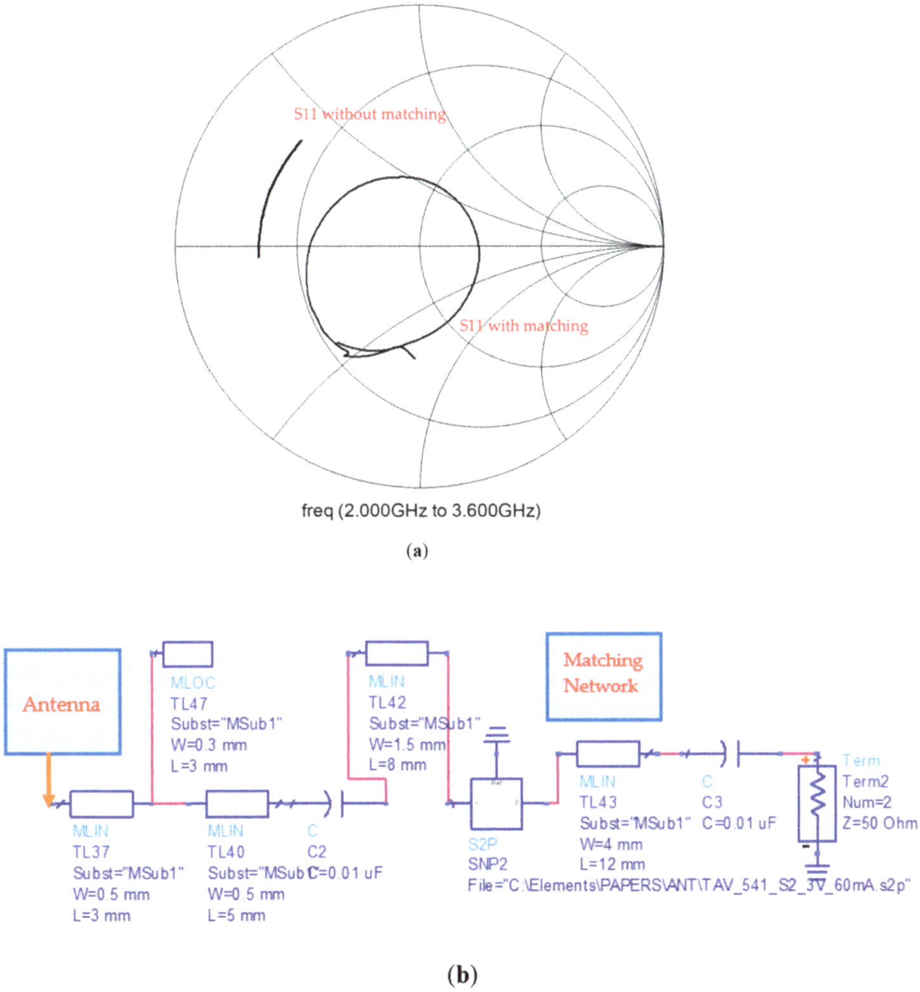

Figure 15.8. Circular active patch. (a) Smith chart, S_{11}; (b) matching network.

shown in figure 15.13(a), was optimized to improve the antenna bandwidth up to 50% for S_{11} lower than −6 dB. A Smith chart diagram of S_{11} with and without the output matching network is shown in figure 15.13(b). The optimized S_{21} parameter of the transmitting antenna, gain on the human body, is shown in figure 15.14. The active circular patch antenna gain, simulated and measured, is 11.0 ± 2 dB for frequencies from 1.9 to 3.3 GHz. The active transmitting circular patch output power is around 17.8 dBm.

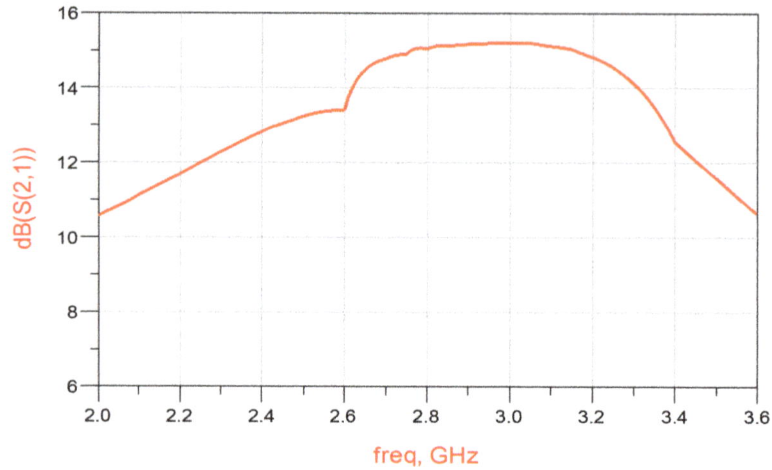

Figure 15.9. S_{21} of the receiving active circular patch.

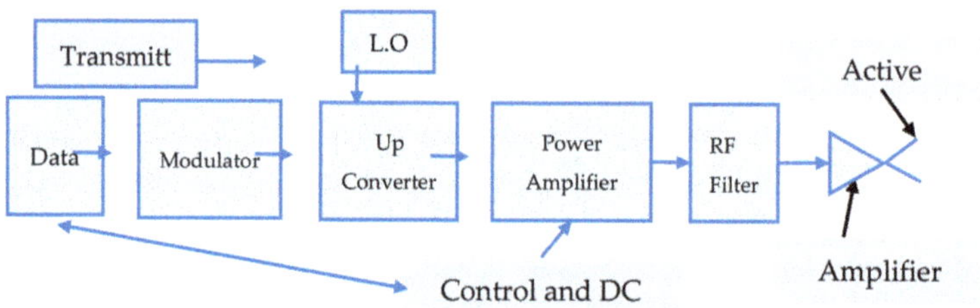

Figure 15.10. Transmitting channel block diagram with a compact wearable transmitting antenna.

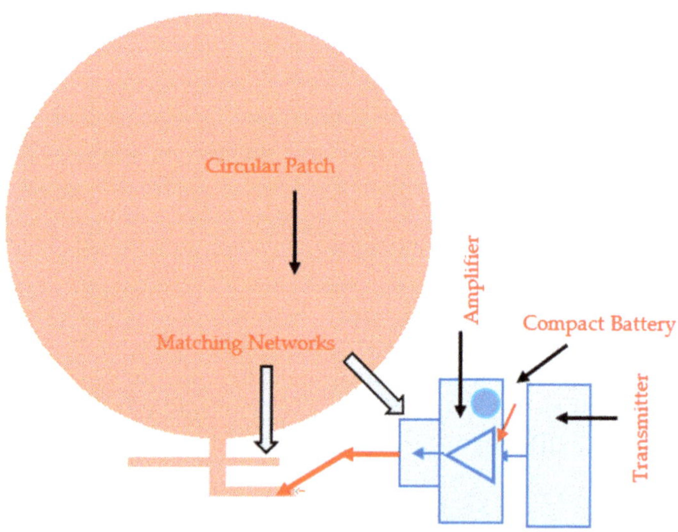

Figure 15.11. Transmitting active circular patch for medical and IoT applications.

Table 15.3.

Parameter	Specification	Remarks
Frequency range	1–4 GHz	-
Gain at 2.5 GHz	15 dB	$V_{ds} = 5$ V
N.F at 2 GHz	5.5 dB	-
P1 dB at 2 GHz	18.0 dBm	$I_{ds} = 85$ mA
OIP3 at 2 GHz	29 dBm	-
Input power	10 dBm	Maximum
V_{gs}	0.50 V	-
V_{ds}	4–5 V	-
I_{ds}	60–85 mA	-
Bias voltage	±5 V	-
Temperature range	−40 °C–85 °C	Operating

Table 15.4. *S* band HPA *S* parameters.

F-GHz	S_{11} dB	S_{11}°	S_{21} dB	S_{21}°	S_{12} dB	S_{12}°	S_{22} dB	S_{22}°
1.56	−12.3	137.6	17.5	134.9	67.8	−11.2	−18.9	92.4
1.6	−12.8	134.3	18.3	123.3	−44.2	−93.4	−18.9	113.7
1.8	−14.3	101.2	17.9	83	−43	−86.3	−22	69.5
2	−16.5	61.8	17.3	43.5	−40.4	94.6	−27	6.42
2.16	−18.5	22.1	16.8	12.9	−38	−105.5	−27.8	−70.2
2.4	−19.4	−53.9	15.7	−31.8	−36	−128	−22.2	−147.2
2.56	−17.7	99.7	15	−60	−34.6	−145.6	−19.3	−179.4
2.7	−15.7	131	14.3	−84.3	−33.8	−160.3	−17.5	158.1
2.86	−13.7	159	13.5	−111.1	−33	−177.7	−16	134.7
3	−12.2	179.1	12.7	−134.1	−32.4	167.4	15.2	116.3
3.18	−10.9	158.1	11.5	−164.9	−29.4	133.8	−16.1	176.4
3.4	-9.3	129.33	10.3	156.5	−28.9	107.7	−16.05	146.9

15.5 Active receiving compact stacked circular patch antenna

A stacked active circular patch is shown in figure 15.15. The PHEMT LNA, the same LNA as described in section 15.3, is assembled on the antenna board as shown in figure 15.15. The circular patch and the antenna matching network are assembled on a substrate 1.6 mm thick, with dielectric constant of 2.2. The circular resonator radius is 20 mm. The circular radiator is printed on a substrate 1.6 mm thick, with dielectric constant of 2.2. The circular radiator radius is 21 mm. The active antenna structure is presented in figure 15.15. An integral input matching circuit matches the circular patch to the LNA, as presented in figure 15.15. An integral output matching circuit matches the LNA output port to the system. A DC bias network supplies the

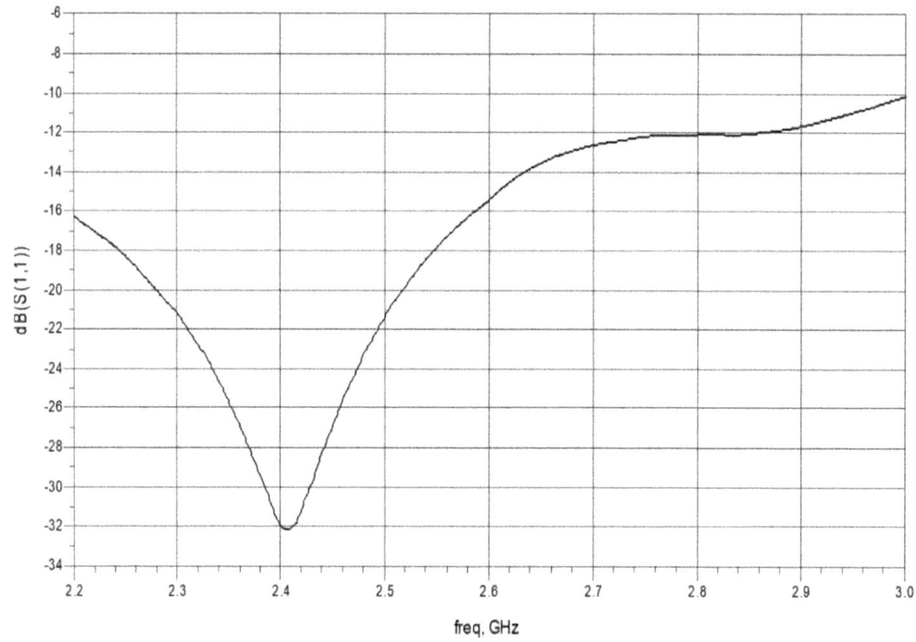

Figure 15.12. S_{11} of the transmitting S band active circular patch.

required bias voltages to the amplifiers. The circular patch S_{11} parameter on a human body is shown in figure 15.16. The active antenna bandwidth is around 33% for S_{11} lower than −9 dB.

The antenna gain is around 13.9 dB at 2.57 GHz and drops to 10.3 dB at 3.2 GHz. The antenna gain on the human body is presented in figure 15.17. The antenna gain is 12 ± 2 dB for frequencies from 2.4 to 3.2 GHz. The circular patch noise figure, shown in figure 15.18, is 0.5 ± 0.25 dB for frequencies from 2.4 to 3.4 GHz. The antenna feed circuit, shown in figure 15.19(a), was tuned and optimized to improve the antenna bandwidth up to 70% for S_{11} lower than −6 dB. Plots of S_{11} parameters on a Smith chart diagram, without and with the output matching circuit, are presented in figure 15.19(b). The optimized receiving antenna gain on the human body is shown in figure 15.20. The active circular patch antenna gain is 13.0 ± 3 dB for frequencies from 2 to 4 GHz. There is a good match between simulated and measured results.

15.6 Metamaterial wearable stacked circular patch antennas

A metamaterial wearable stacked circular patch is presented in figure 15.21. The circular patch and the feed circuit are printed on a substrate 1.6 mm thick, with dielectric constant of 2.2, see figure 15.21(a). The circular resonator radius is 18.2 mm.

The circular radiator is etched on substrate 1.6 mm thick, with dielectric constant of 2.2, shown in figure 15.21(b). The circular radiator radius is 18.6 mm.

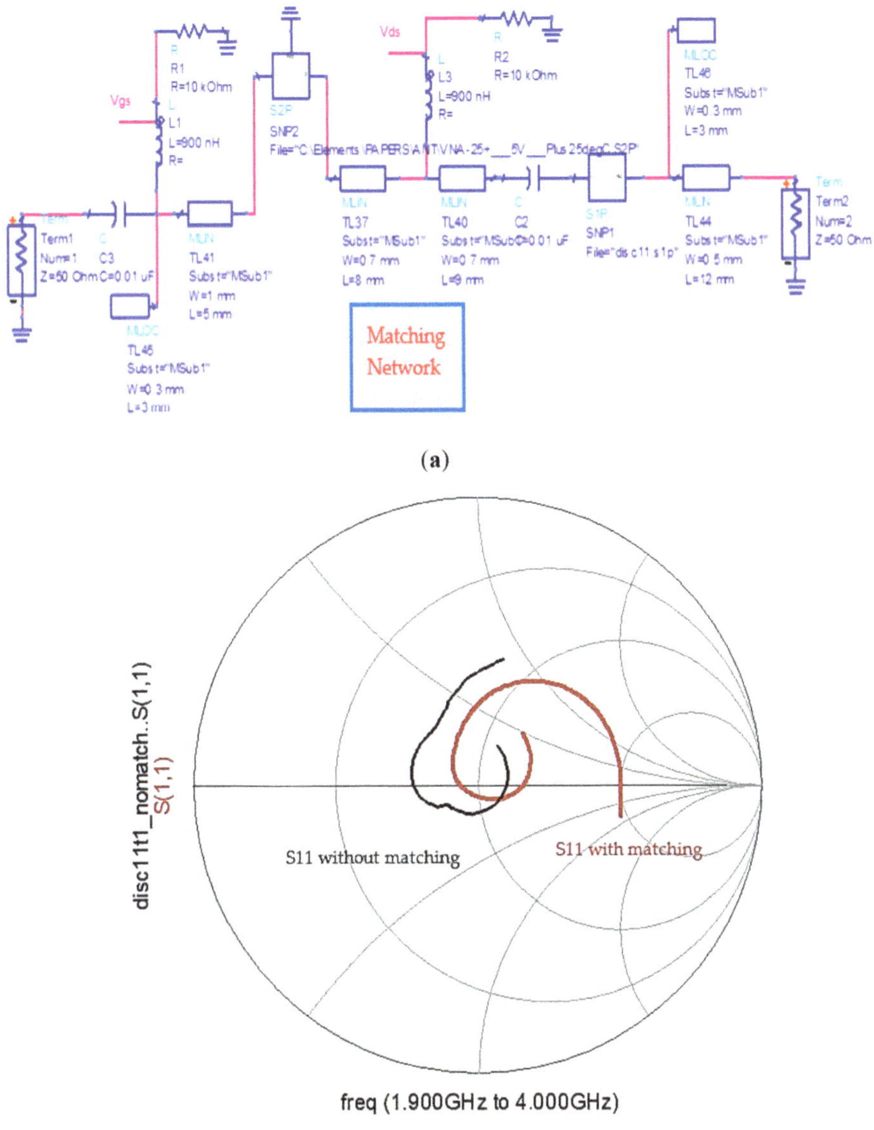

(b)

Figure 15.13. Transmitting circular active patch (a) matching network; (b) S_{11} on Smith chart of the active patch.

The radiator is a circular patch with eighteen CSRRs. The stacked circular patch is presented in figure 15.21(c). The spacing between the resonator and the radiator may be varied from 0 mm to 10 mm to get wider bandwidth from 6% to 12%. The radius of the circular patch without CSRR is greater by 25% than the radius of the circular patch with CSRR. The CSRR outer radius is 2.6 mm, shown in figure 15.21(d).

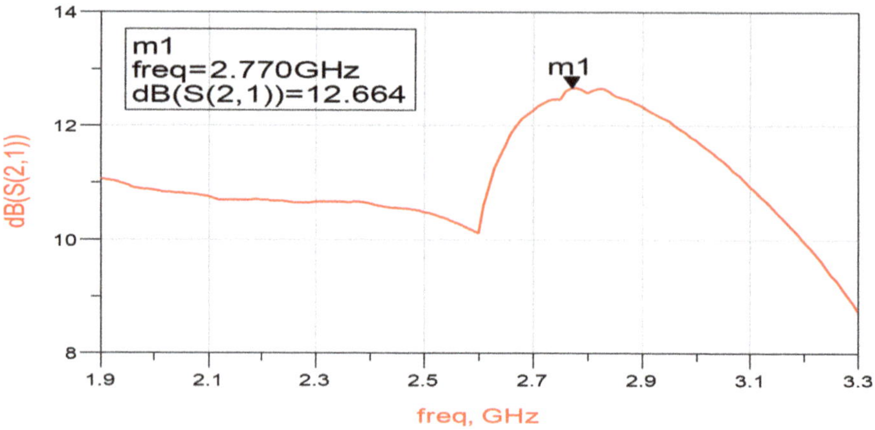

Figure 15.14. S_{21} of the active transmitting circular patch.

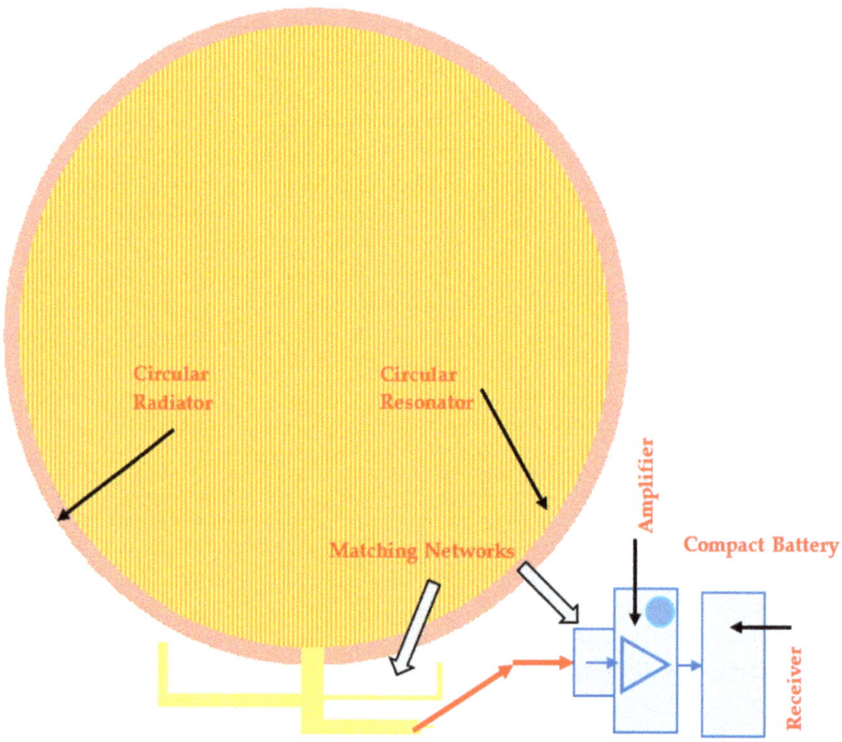

Figure 15.15. Active wideband receiving compact stacked circular patch.

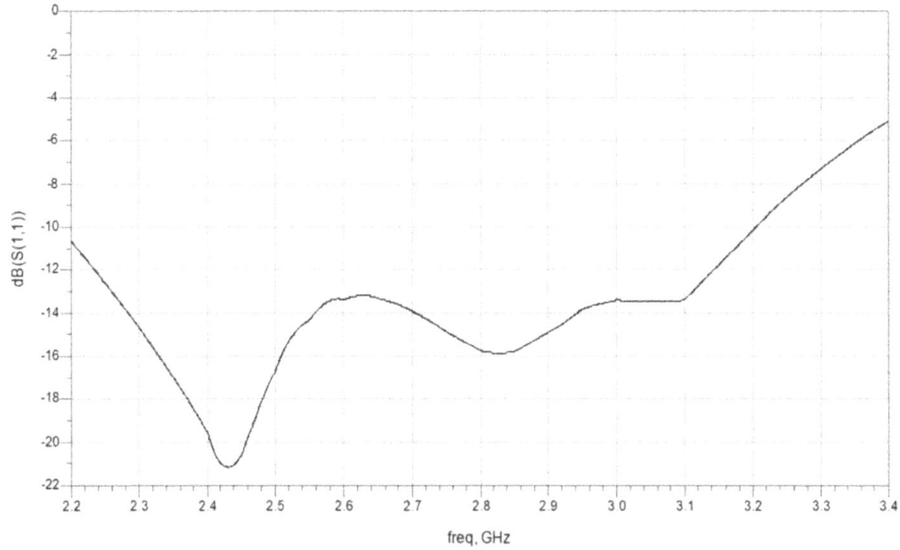

Figure 15.16. Stacked circular patch S_{11} parameter on a human body.

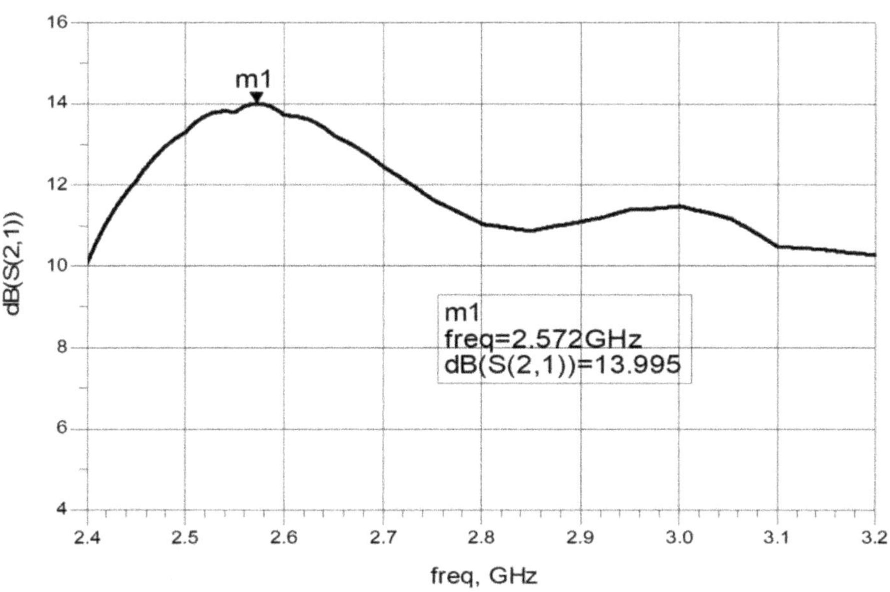

Figure 15.17. Gain, S_{21} parameter, of the stacked circular patch on a human body.

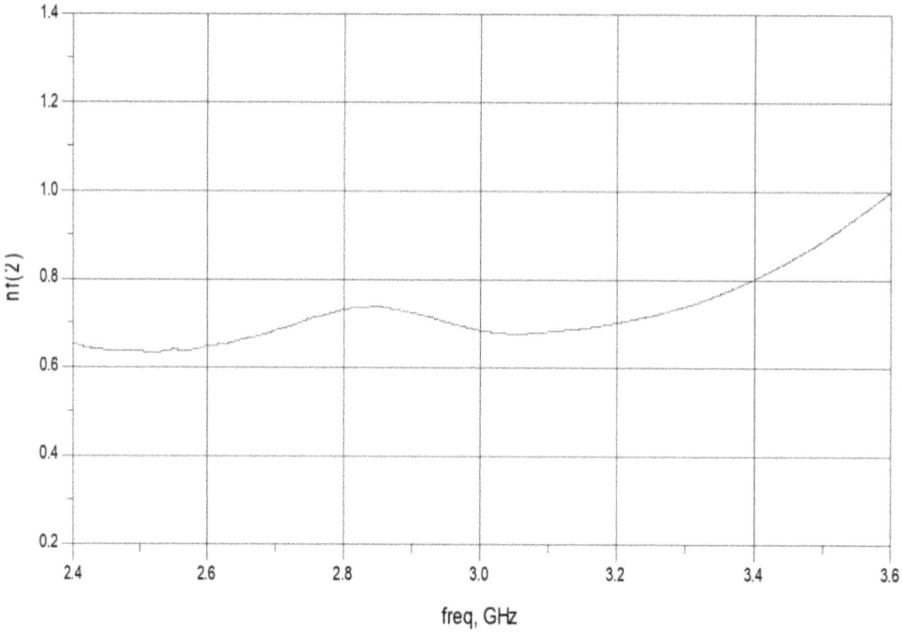

Figure 15.18. Noise figure of the stacked active circular patch.

The antenna S_{11} parameter is shown in figure 15.22. The antenna has 92% efficiency and 80° beam width. The antenna gain is around 8.4 dBi, as shown in figure 15.23(a). The 3D radiation pattern of the antenna with CSRR is shown in figure 15.23(b). The gain and directivity of the circular patch without CSRR is lower by 2 to 3 dB than the circular patch antenna with CSRR.

15.7 Applications of wearable antennas for healthcare and IoT systems

The antennas discussed in this chapter may be used in healthcare and IoT systems. In table 15.5 electrical characteristics of human body tissues are listed [16]. The sensors' electrical performance variation near a patient were evaluated by generating a model of the human body and the antenna, as shown in figure 15.24(a). The influence of the human body on the antenna performance is simulated by evaluating the antenna reflection coefficient on the human body. At fat tissues the dielectric constant is 5, and 45 at the stomach zone, and increases to 128 at the small intestine area. The variation of the electrical characteristics of the body tissues affects the electrical performance of the antenna. The sensor resonant frequency is shifted up to 8%, in different locations of the sensor on the patient's body. The circular polarized sensors may be located inside a belt, see figure 15.24(b). The belt electrical characteristics and thickness change the sensor electrical performance. The sensors electrical and mechanical parameters were tuned to achieve the best sensor electrical and mechanical parameters. The sensors' electrical performance were simulated and

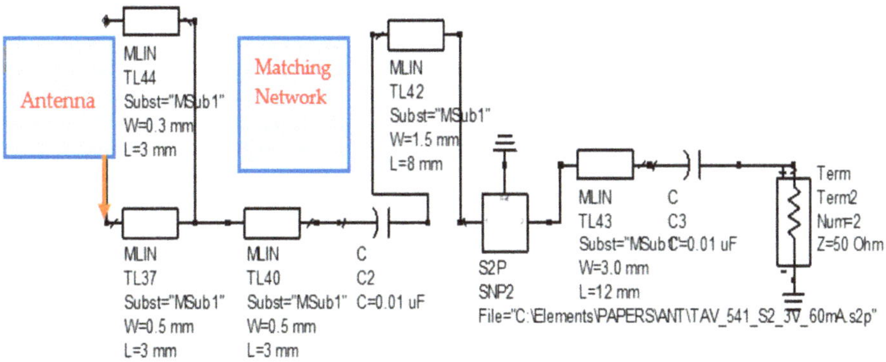

(a)

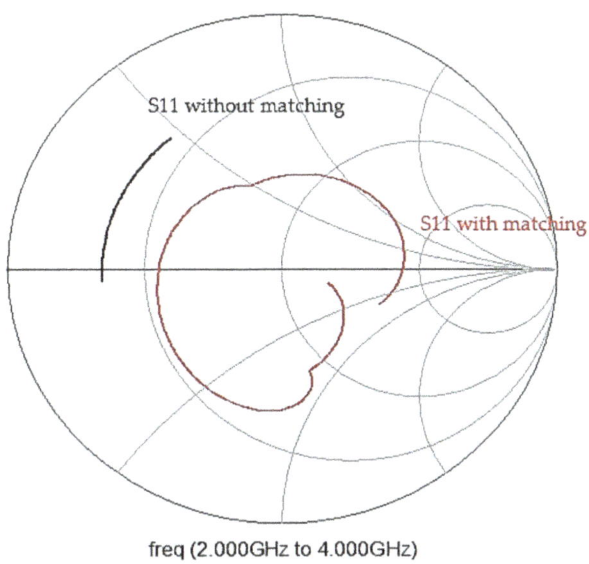

(b)

Figure 15.19. Receiving circular active patch (a) antenna matching network; (b) S_{11} of the antenna on a Smith chart.

measured for air spacing between the sensors and the human body up to 20 mm at different areas on the patient's body. Table 15.6 compares results between simulated and measured results of sensors without and with CSRR. Table 15.7 compares results between simulated and measured results of wearable antennas, based on information presented in [1, 2]. As presented in tables 15.6 and 15.7, there is a good match between simulated and measured results. Results listed in these tables verify that the gain of the antennas without CSRR is lower by around 2.5 dB than the

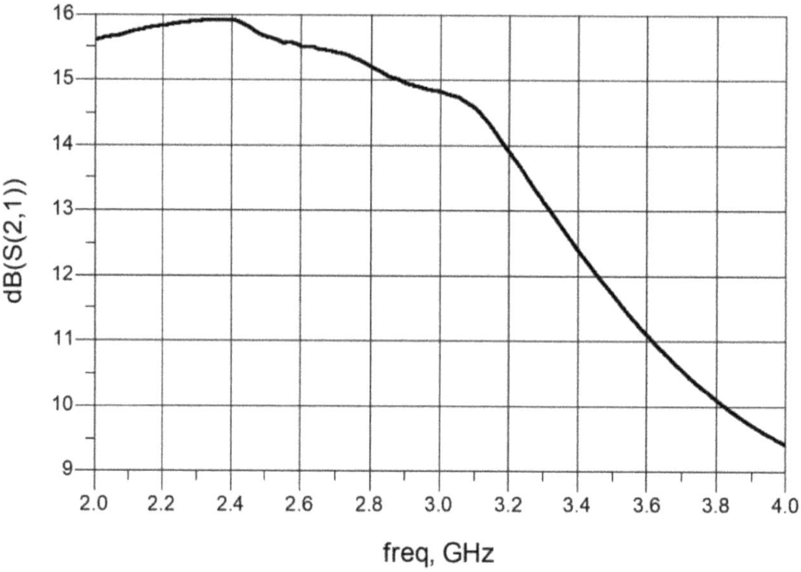

Figure 15.20. Gain, S_{21}, of the active stacked circular receiving patch.

antennas with CSRR. Electrical performance of dipoles, patches, loop, slot, and other antennas are given in [2–5].

Wearable healthcare systems and sensors can measure heartbeat, sweat rate, body temperature, blood pressure, perform gait analysis and other physiological parameters of the person wearing the healthcare device. Gait analysis is a useful tool both in clinical practice and biomechanical research. Gait analysis using wearable sensors provides quantitative and repeatable results over extended time periods with low cost and good portability, showing better prospects and making great progress in recent years. In sport, wearable sensors are used in several applications of gait analysis. Figure 15.25(a) presents a photo of the prototype of the active stacked antenna. Figure 15.25(b) presents the antenna modules and layout.

Figure 15.26 presents a healthcare monitoring network with wearable sensors for body area networks.

15.8 Conclusions

The active and passive antennas and sensors discussed in this article are compact, with a wide band, efficient, and low-cost. Wideband compact passive and active circular antennas may be used in wideband wearable wireless communication devices for commercial and medical applications. The antennas gain, bandwidth, matching network, efficiency, and noise figure are presented and summarized in this chapter. The directivity and gain of the antennas without CSSR are lower by 2–3 dB than the antennas with CSRR. The resonant frequency of the antennas without CSRR is lower by 5%–10% than the antennas with CSRR.

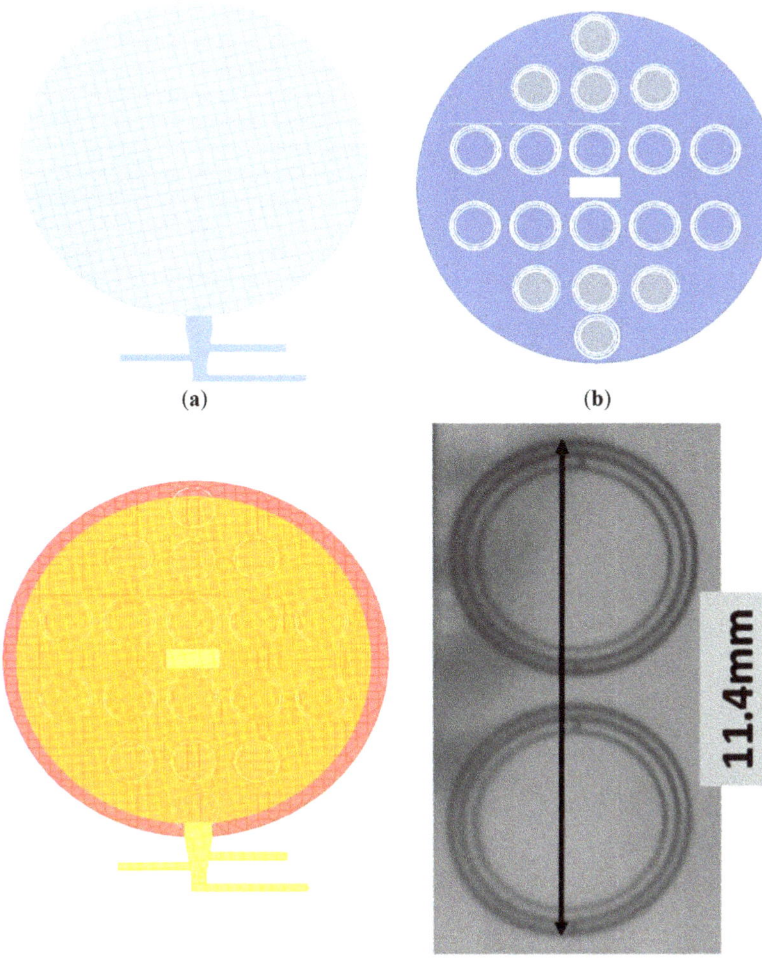

Figure 15.21. Wearable stacked circular patch with CSRR (a) resonator; (b) metamaterial radiator; (c) wearable stacked circular patch; and (d) CSRRs photo.

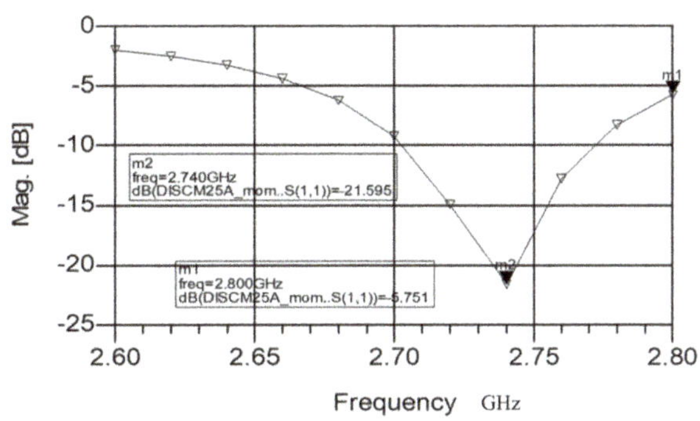

Figure 15.22. S_{11} of the stacked circular patch with CSRRs on a human body.

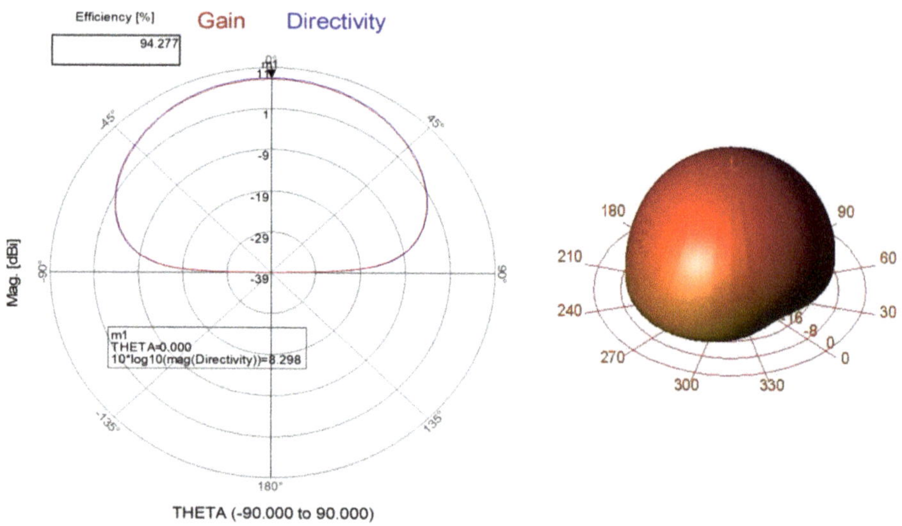

(a) (b)

Figure 15.23. (a) Radiation pattern of the stacked metamaterial circular patch at 2.7 GHz. (b) 3D radiation pattern of the stacked metamaterial circular patch.

Table 15.5. Electrical parameters of body tissues [16, 17].

Tissue	Parameter	440 MHz	600 MHz	1 GHz	1.25 GHz
Fat tissues	σ	0.047	0.05	0.054	0.06
	ε	5.00	5.00	4.72	4.55
Stomach tissues	σ	0.71	0.75	0.96	0.98
	ε	42.7	41.40	39.66	39.00
Blood	σ	1.76	1.78	1.91	1.99
	ε	57.2	56.5	55.40	55.00
Skin	σ	0.58	0.6	0.63	0.77
	ε	41.6	40.45	40.25	39.65
Lung tissues	σ	0.27	0.27	0.27	0.28
	ε	38.4	38.4	38.4	38.4
Kidney tissues	σ	0.90	0.90	0.90	0.91
	ε	117.45	117.45	117.45	117.45
Colon tissues	σ	1.00	1.05	1.30	1.45
	ε	63.5	61.9	60.00	59.40

Table 15.6. Comparison between printed antennas with and without CSRR.

Antenna	Frequency (GHz)	BW (%)	Simulated gain (dBi)	Measured gain (dBi)	Length. (mm)	Efficiency (%)
Circular patch with CSRR	2.62	9	7.9	7.8	36	86
Circular patch without CSRR	2.63	1	4.4	4.3	48	83
Printed dipole with CSRR	0.35	11	5.6	5.7	198	93
Dipole without CSRR	0.40	10	2.6	2.5	210	92
Stacked circular patch with CSRR	2.7	9	8.4	8.2	40	93
Stacked circular patch without CSRR	2.7	9	5.4	5.3	48	89

Table 15.7. Comparison of electrical characteristics of wearable antennas [1, 2].

Antenna	Frequency (GHz)	Bandwidth (%)	VSWR	Simulated gain (dBi)	Measured gain (dBi)
Printed dipole [2]	0.43	5–10	2:1	2–3	2–3
Dipole with CSRR	0.4	8–12	2:1	5–7	5–7
Loop [2]	0.43	5–10	4:1	0	−1–0
Circular patch with CSRR	2.62	9	2:1	7–7.5	7.2–7.8
Active circular receiving patch	2.5	25	2:1	11–13.5	11–14.0
Stacked circular patch with CSRR	2.7	9	2:1	7.8–8.5	7.8–8.4
Dipole (CSRR and strips) [2]	0.35	50	2.5:1	5–7.5	5–7.5
Patch [2]	2.2	1–3	2:1	2–3	2–3
Stacked patch [2]	2.2	10–15	2:1	4–5	4–5
Slot [2]	2.5	50	2:1	3	2–3
T shape slot [2]	2.5	60	2:1	3	2–3
Active slot [2]	2.5	40	3:1	12–20	12–20.5
Active T slot[2]	2.5	50	3:1	12–20	12–21
Active slot with CSRR [2]	2.5	50	2.5:1	10–16	11–15.8
Active stacked circular patch	2.5	25	2:1	12–14	11.5–15

Wireless wearable body area networks may be a good option for medical centers, public organizations, hospitals, patients and other users. Wearable sensors and technology provide an affordable service that may improve the everyday health and physiological response of users and patients. Wearable devices support the design of

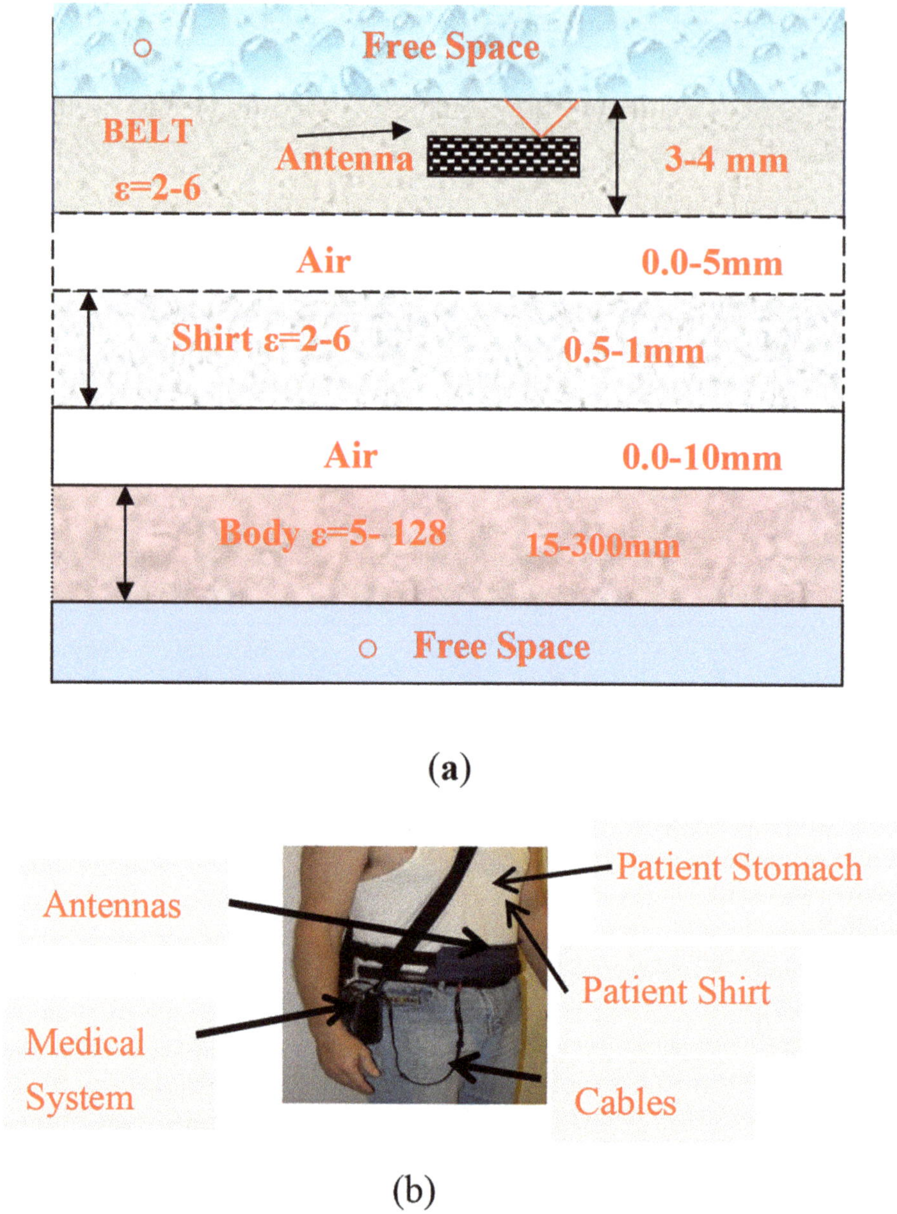

Figure 15.24. (a) Model of wearable antenna on human body; (b) wearable healthcare system.

personal medical networks with online response to treat and improve users and patients' daily life and health.

The output power of the active transmitting antennas is around 18.5 dBm. A description and comparison of the electrical performance of the wearable antennas without and with SRRs are given in this chapter.

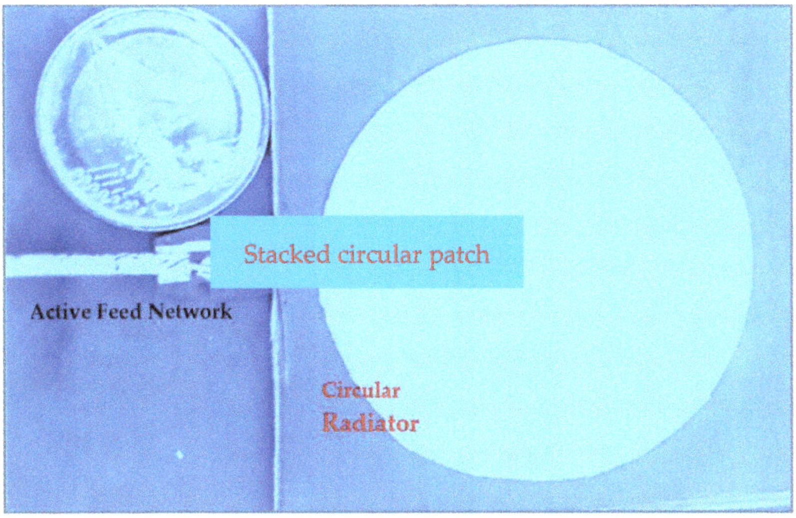

(a)

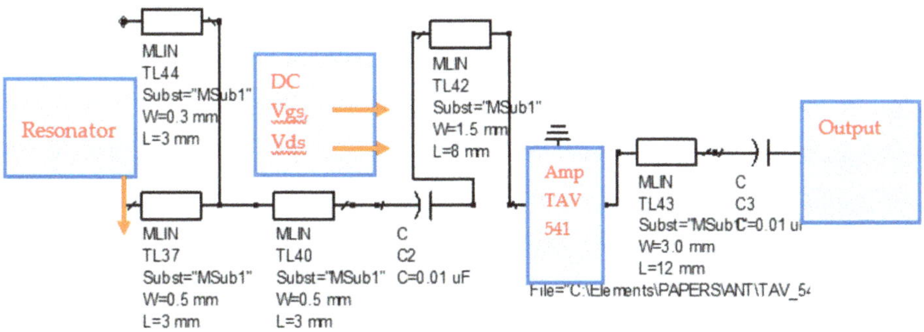

(b)

Figure 15.25. (a) A photo of the prototype of the active antenna; (b) antenna modules and block diagram.

Wideband compact metamaterial circular antennas with high gain and efficiency for RF and medical systems are presented in this chapter. The active circular patch antennas bandwidth is around 20%–40% for VSWR better than 2:1. The gain of the receiving circular patch antenna is around 12–14 dB.

In future work new circular polarized metamaterial antennas with high efficiency for communication, healthcare, IoT and wireless communication systems will be presented. In future work fractal antennas, without and with SRRs, with high gain and efficiency for healthcare and communication systems will be designed.

Parts of this chapter have been reproduced from reference [42].

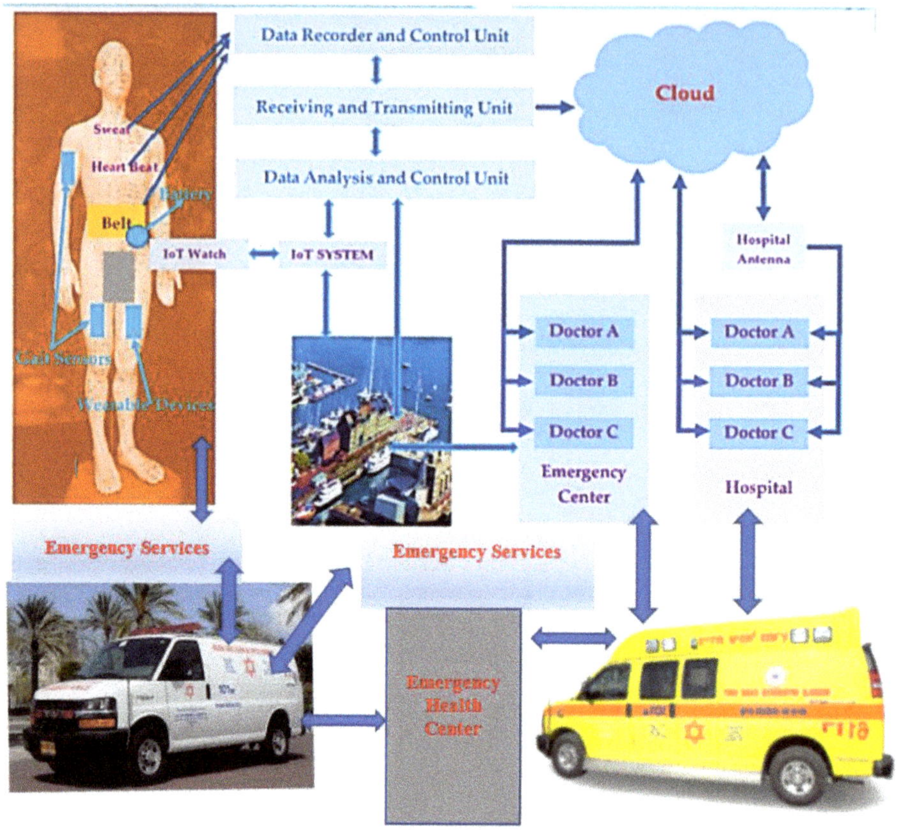

Figure 15.26. Healthcare monitoring medical network with wearable sensors.

References

[1] James J R, Hall P S and Wood C 1981 *Microstrip Antenna Theory and Design* (London: Institution of Engineering and Technology (IET))

[2] Sabban A 2017 *Novel Wearable Antennas for Communication and Medical Systems* (Boca Raton, FL: Taylor and Francis Group)

[3] Sabban A 2015 *Low Visibility Antennas for Communication Systems* (Boca Raton, FL: Taylor and Francis Group)

[4] Sabban A and Wideband R F 2016 *Technologies and Antennas in Microwave Frequencies* (Hoboken, NJ: Wiley)

[5] Sabban A 2013 New wideband printed antennas for medical applications *IEEE Trans. Antennas Propag.* **61** 84–91

[6] Pendry J B, Holden A J, Stewart W J and Youngs I 1996 Extremely low frequency plasmons in metallic mesostructures *Phys. Rev. Lett.* **76** 4773–6

[7] Pendry J B, Holden A, Robbins D and Stewart W 1999 Magnetism from conductors and enhanced nonlinear phenomena *IEEE Trans. Microw. Theory Tech.* **47** 2075–84

[8] Marques R, Mesa F, Martel J and Medina F 2003 Comparative analysis of edge and broadside coupled split ring resonators for metamaterial design—theory and experiment *IEEE Trans. Antennas Propag.* **51** 2572–581
[9] Marques R, Baena J D, Martel J, Medina F, Falcone F, Sorolla M and Martin F 2003 Novel small resonant electromagnetic particles for metamaterial and filter design *Proc. of the ICEAA'03* (Torino, Italy, 8–12 September 2003) 439–42
[10] Marques R, Martel J, Mesa F and Medina F 2002 Left-handed-media simulation and transmission of EM waves in subwavelength split-ring-resonator-loaded metallic waveguides *Phys. Rev. Lett.* **89** 183901
[11] Rajab K, Mittra R and Lanagan M 2005 Size reduction of microstrip antennas using metamaterials *Proc. of the 2005 IEEE APS Symp.* (Washington, DC, USA, 3–8 July 2005)
[12] Zhu J and Eleftheriades G 2008 A compact transmission-line metamaterial antenna with extended bandwidth *IEEE Antennas Wirel. Propag. Lett.* **8** 295–8
[13] Baena J D, Marques R, Martel J and Medina F 2003 Experimental results on metamaterial simulation using SRR-loaded waveguides *Proc. of the IEEE-AP/S Int. Symp. on Antennas and Propagation* (Columbus, OH, USA, 22–27 June 2003) 106–9
[14] Marques R, Martel J, Mesa F and Medina F 2002 A new 2D isotropic left-handed metamaterial design: theory and experiment *Microw. Opt. Technol. Lett.* **35** 405–8
[15] Shelby R A, Smith D R, Nemat-Nasser S C and Schultz S 2001 Microwave transmission through a two-dimensional, isotropic, left-handed metamaterial *Appl. Phys. Lett.* **78** 489–91
[16] Sabban A 2016 Small wearable meta materials antennas for medical systems *Appl. Comput. Electromagn. Soc. J.* **31** 434–3
[17] Sabban A 2016 Wearable antenna measurements in vicinity of human body *Wirel. Eng. Technol.* **7** 97–104
[18] Sabban A 2011 Microstrip antenna arrays *Microstrip Antennas* ed N Nasimuddin (London, UK: InTech) pp 361–84 http://intechopen.com/articles/show/title/microstrip-antenna-arrays
[19] Sabban A 2018 Active compact wearable body area networks for wireless communication, medical and IOT applications *MDPI ASI, Appl. Syst. Innov. J.* **1** 46
[20] Chirwa L, Hammond P, Roy S and Cumming D R S 2003 Electromagnetic radiation from ingested sources in the human intestine between 150 MHz and 1.2 GHz *IEEE Trans. Biomed. Eng.* **50** 484–92
[21] Werber D, Schwentner A and Biebl E M 2006 Investigation of RF transmission properties of human tissues *Adv. Radio Sci.* **4** 357–60
[22] Gupta B, Sankaralingam S and Dhar S 2010 Development of wearable and implantable antennas in the last decade: a review *Proc. of the 2010 10th Mediterranean Microwave Symp.* (Guzelyurt, Cyprus, 25–27 August 2010) 251–67
[23] Thalmann T, Popovic Z, Notaros B M and Mosig J R 2009 Investigation and design of a multi-band wearable antenna *Proc. of the 3rd European Conf. on Antennas and Propagation, EuCAP 2009* (Berlin, Germany, 23–27 March 2009) 462–5
[24] Salonen P, Rahmat-Samii Y and Kivikoski M 2004 Wearable antennas in the vicinity of human body *Proc. of the IEEE Antennas and Propagation Society Symp.* (Monterey, CA, 20–25 June 2004) 467–70
[25] Kellomäki T, Heikkinen J and Kivikoski M 2006 Wearable antennas for FM reception *Proc. of the 2006 First European Conf. on Antennas and Propagation* 6–10 (Nice, France, November 2006) 1–6

[26] Lee Y 2003 *Antenna Circuit Design for RFID Applications; Microchip (Application Note 710c)* (Chandler, AZ: Microchip Technology Inc)

[27] Sabban A 1986 Microstrip antenna and antenna arrays. *U.S. Patent* US 4,623,893, 18 November 1986

[28] Wheeler H A 1975 Small antennas *IEEE Trans. Antennas Propag.* **23** 462–9

[29] Jamil F, Ahmad S, Iqbal N and Kim D 2020 Towards a remote monitoring of patient vital signs based on IoT-based blockchain integrity management platforms in smart hospitals *Sensors* **20** 2195

[30] Jamil F, Iqbal M A, Amin R and Kim D 2019 Adaptive thermal-aware routing protocol for wireless body area network *Electronics* **8** 47

[31] Shahbazi Z and Byun Y-C 2020 Towards a secure thermal-energy aware routing protocol in wireless body area network based on blockchain technology *Sensors* **20** 3604

[32] Lin J and Itoh T 1994 Active integrated antennas *IEEE Trans. Microw. Theory Tech.* **42** 2186–94

[33] Mortazwi A, Itoh T and Harvey J 1998 *Active Antennas and Quasi-Optical Arrays* (New York: Wiley)

[34] Jacobsen S and Klemetsen Ø 2008 Improved detectability in medical microwave radio-thermometers as obtained by active antennas *IEEE Trans. Biomed. Eng.* **55** 2778–85

[35] Jacobsen S and Klemetsen O 2007 Active antennas in medical microwave radiometry *Electron. Lett.* **43** 606

[36] Ellingson S W, Simonetti J H and Patterson C D 2007 Design and evaluation of an active antenna for a 29–47 MHz radio telescope array *IEEE Trans. Antennas Propag.* **55** 826–31

[37] Segovia-Vargas D, Castro-Galan D, Munoz L E G and González-Posadas V 2008 Broadband active receiving patch with resistive equalization *IEEE Trans. Microw. Theory Tech.* **56** 56–64

[38] Rizzoli V, Costanzo A and Spadoni P 2008 Computer-aided design of ultra-wideband active antennas by means of a new figure of merit *IEEE Microw. Wirel. Components Lett.* **18** 290–2

[39] Sabban A 2012 Dual polarized dipole wearable antenna. *US Patent* 8203497, 19 June 2012

[40] Sabban A 1983 A new wideband stacked microstrip antenna *Proc. of the IEEE Antenna and Propagation Symp. (Houston, TX, USA, 23–26 May 1983)*

[41] ADS Momentum Software, Keysight (http://keysight.com/en/pc-1297113/advanced-design-system-ads?c.c.=IL&lc=eng)

[42] Sabban A 2020 New compact wearable metamaterials circular patch antennas for IoT, medical and 5G applications *Appl. Syst. Innov.* **3** 42

Chapter 16

Green electronic and communication technologies—going green

16.1 Introduction to green electronic technologies

It is clear in 2022 to world leaders and to the world communities that green technologies are the salvation of the world from an environment disaster. In the last decade the world suffered from fast weather changes. Moreover, the envirnoment suffers from severe climate changes, droughts, seawater acidification, rising of seawater levels, pollution of groundwater reserves, rise of environment temperature, and air pollution. A direct result of droughts and climate change is rapid spread of diseases and macro-parasites, and the extinction of several animal and bird species. These changes are almost irreversible. Countries, governments, green organizations, international organizations, and individuals should act immediately to save the planet.

High-tech communications industrial growth in the last century depleted and corrupted the Earth's resources. Electronics and computer waste, paper and plastics waste, cellular phones waste, and other daily garbage contain toxins, dangerous chemicals that pollute the air, soil and groundwater with dangerous toxins that subsequently are found in our drinking water supply. These toxic materials can be found in animal meat, fish, and food crops and livestock grown on polluted soil. This dangerous waste and air pollution affect human and animal health. The Earth's oceans, seas and lakes suffer from toxic and plastic pollution. Plastic waste damages oceans and sea creatures throughout the Earth. Plastic pollution kills birds, fish, including penguins, turtles, and other species. Large plastic pieces pose choking and strangulation hazards to fish and birds. Cattle, birds, and fish swallow tiny plastic particles and become contaminated. These birds and fish may be found in our food market. The contaminated birds and fish may be found on our plate and in our stomachs.

doi:10.1088/978-0-7503-5224-6ch16

Development of renewable green energy, green electronic, and waste management and recycling may significantly reduce air and water pollution.

16.2 Electronic and communication green technologies

In green electronics the use of hazardous and toxic materials such as lead, plastic materials and other toxic chemicals is not allowed or is limited. Electronics and communication companies are facing increasing pressure from legislators to avoid toxic and hazardous chemicals from their products. There is a strong environmental movement in the world to use green materials and in production of electronic and communication systems. There are now several initiatives in the United Nations organization with goals to re-use and to recycle electronic devices. Electronics manufacturers are faced with a challenge to develop green devices and to use green energy.

Methods to produce green efficient energy in the electronics industry are presented in this chapter. Energy harvesting technologies that can be used to produce, recycle and store green energy are presented in this chapter. Smartphones, computers, cellular phones, and other electronic devices contain toxics and hazardous materials that pollute the environment and consequently endanger human health. Electronics companies have to reduce the amount of electronic waste by developing products that minimize the amount of hazardous materials and to use components that can be recycled, and to develop products with longer lifecycle.

16.3 Renewable green energy for electronic and RF systems

Renewable green energy converts energy from natural resources, sun, light, water and wind, into electrical energy. Traditional nonrenewable energy resources include fuel, hydrogen, nuclear, coal, natural gas, and oil. Major types of renewable energy are solar energy, wind energy, water energy and biology fuel.

16.3.1 Solar energy

Plants use solar energy and sunlight in the photosynthesis process. Solar energy was available and used by human beings for thousands of years. Solar energy is divided into two main types: passive and active solar energy. Passive solar energy traps the Sun's energy without using any mechanical devices. Active solar energy uses mechanical devices to collect, store and distribute the energy. Solar cells directly convert energy from natural sunlight into electrical energy using the process of photovoltaics. Generating electricity from solar energy results in less consumption of fossil fuels, and reduction of pollution. A solar energy site is shown in figure 16.1.

16.3.1.1 More types of solar energy

- **Solar thermal energy** is the energy converted from the Sun's energy into heat.
- **Photovoltaic solar power** is the energy converted from the Sun's energy into electricity.
- **Concentrating solar power** is solar thermal energy which is used to generate solar power electricity.

Figure 16.1. Solar energy panels at a solar energy site in Israel.

A solar cell consists of a layer of silicon cells, a metal frame, a glass casing unit, and wiring to transfer electric current from the silicon. The conductive properties of silicon may be used to absorb and convert sunlight energy into electric power. sunlight causes electrons in a silicon cell to move and initiate a flow of electric current which is known as the '**photovoltaic effect**'. The first silicon photovoltaic cell was developed in 1954 by Bell Labs with around 4% efficiency. The efficiency of new solar panels is around 20% to 30%.

16.3.1.2 The photovoltaic process

The silicon photovoltaic solar cell absorbs the Sun's radiation. This radiation causes the silicon electrons to move and create an electric current flow. Metallic wires capture and feed this direct current to a solar inverter that converts the DC energy to AC, alternating current, electricity.

16.3.1.3 Advantage of solar energy

- There is no emission of carbon dioxide. Green and clean energy.
- No air and water pollution, environment friendly.
- Not degradable.
- Sun radiation is available almost every day. A reliable source of energy. The energy can be stored in batteries.
- Solar energy and light are free natural resources.
- Solar energy does not require expensive raw materials like fuel, coal, oil, or gas. Solar energy requires significantly lower operational labor than conventional power production. Raw materials should not be constantly extracted, refined, and transported to the power plant.
- Fuel, oil, coal, and gas used to produce conventional electricity are often transported cross-country or internationally. This transportation has additional costs, including monetary costs, and pollution costs of transport. These costs are avoided with solar energy.
- If a site has no connection to the state electricity facility it is called off-grid connection. On-grid means if the site remains connected to the state electricity facility it is called on-grid connection. In the off-grid case the facility, house, company, business, or a community is relying only on solar energy.

The ability to produce electricity off the grid is an important advantage of solar energy for isolated communities, facilities, and rural areas. Solar energy can be produced on or off the grid.

- In the case of solar energy there is no need to install long power lines that are significantly difficult and very expensive in some rural areas.
- The Sun is an unlimited free resource that can be sourced from several locations. One of the major advantages of solar energy is the ability to bypass the linkage between politics status and energy price, whereas in the fuel market there is a strong linkage between politics situations and energy price. Politics status manipulations doubled the price of fuel in the past fifty years.
- Price of solar energy is not affected by politics manipulations, war, and international relations.

16.3.1.4 Disadvantage of solar energy

- Solar energy depends on the weather and sunlight. Not available at night.
- Development and production of a solar site is expensive.
- Consumes a large area, see figures 16.1 and 16.2.
- Solar energy production is relatively inefficient. The efficiency of solar panels is less than 30%. A huge amount of surface area is required to produce adequate electricity.
- Solar electricity storage technology is expensive, bulky, and more appropriate to houses.

16.3.2 Renewable green wind energy

The wind kinetic energy may be converted to electricity to operate electric turbines and windmills. Windmills cannot be operated in an urban residential area. Offshore wind turbines provide steady, reliable clean energy in many countries. Figure 16.3(a) and (b) displays wind energy sites installation.

16.3.2.1 Advantages of green wind energy

- Green and clean energy with no air pollution. Turbines operation does not directly emit any CO_2.

Figure 16.2. Solar energy panels installation and site.

Figure 16.3. Green wind energy cites. (a) Wind energy turbines; (b) wind energy installation.

- Wind energy is environment friendly. No water pollution
- Not degradable.
- Wind energy is plentiful, readily available, and capturing its power does not pollute the world's valuable natural resources.
- Wind energy does not rely on constantly mining raw materials. Wind energy does not result in the destruction of forests and eco-systems.
- Wind is a free natural resource.
- Wind energy is a cheap energy that does not require expensive and ongoing raw materials like oil, coal, or gas.
- Wind energy requires lower operational labor than conventional power production. Raw materials should not be constantly extracted, refined, and transported to the power plant.
- The wind is an unlimited free commodity that can be sourced from several locations.
- One of the major advantages of wind energy is the ability to bypass the linkage between the politics situation and energy prices. Price of wind energy is less affected by politics manipulations, war, and international relations.
- Transportation costs and pollution costs are avoided with solar energy.

16.3.2.2 Disadvantages of wind energy

- Strong wind is not available everywhere. Dependent on the weather and wind level.
- Cannot be operated in an urban residential area.
- Large area is needed for wind energy sites.

- Land area is expensive. Development and production of a wind energy site is expensive.
- Wind turbines produce noise, from air passing over the blades and mechanical noise from the moving parts of the turbine. Very noisy.
- Wind turbines may scatter electromagnetic signals causing interference to communication systems.
- Birds get killed when they collide with the rotating blades of a turbine.

16.3.3 Hydroelectric green power

Water flow kinetic energy is converted to green electric energy. Waterfalls and water flow may be converted to electric energy. A water energy site is shown in figure 16.4(a)–(c).

16.3.3.1 Sources of hydroelectric

- Water flow due to rainfall.
- Waterfalls.
- Water stream, high water levels and low water levels.

Figure 16.4. Hydropower cites. (a) Waterfall; (b) water energy site; and (c) hydropower site.

The first hydroelectric plant was built at Niagara Falls in 1879. In 1881, streetlamps in the city were powered by hydropower. The energy available from the moving water depends on both the volume of the water flow and the change in the elevation of the waterfalls. The greater the water flow and the higher the waterfall, the more electricity that can be produced. Hydropower uses turbines and generators to convert the water kinetic energy into electricity. This energy is fed into the electrical grid to power industry, homes, hospitals, and stadiums. Hydroelectricity can be used to store energy in the form of potential energy between two areas with different heights with pumped-storage hydroelectricity. Water is pumped uphill into sites during periods of low demand. This energy can be released to generate energy when demand is high.

16.3.3.2 Advantages of hydropower energy

- Water is a free natural resource.
- Cheap energy compared to other energy sources.
- Help in flood control.

16.3.3.3 Disadvantages of hydropower energy

- Depends on the climate, weather and water stream.
- Cannot be operated in a urban residential area.
- Limited to locations of strong waterfalls and rivers.
- Consumes large space.
- Development and production of a hydropower energy site is expensive.

16.3.4 Green computing technologies

The huge growth of the computing industry has enhanced the environmental disaster on our planet. The huge growth of the computing industry increased energy consumption and electronic waste. This situation created an emergency need to develop green computers and to establish the field of green computing. The environmentally responsible use of computing devices, computer waste and their resources define green computing technologies. Green computing may be defined as the research and study of developing, engineering, manufacturing, employing, and disposing of computing devices and modules to minimize environmental hazards and pollution. Computer developers, manufacturers and vendors should be committed to develop green computing modules and devices by minimizing the use of hazard materials and improving the recycling process of computing and digital devices. Green information technology, green IT, and green computing technology are the same topic and have the same goals.

The Energy Star program was launched in 1992 by the Environmental Protection Agency (EPA) to enhance green computing technology and research.

Green computing technologies include the design and development of environmentally green manufacturing practices, efficient energy consumption by the computing industry, employing energy harvesting technologies, and enhancing waste disposal and recycling practices.

Major topics and initiatives of green computing

- **Energy consumption:** Reducing the energy consumption of computers and their modules and using them in an environmental friendly manner.
- **Green disposal:** Disposing computer waste and recycling unwanted computer modules, digital components and computing devices.
- **Green development and design:** Design and development of all computing devices with low energy consumption. Usage of energy harvesting technologies in the development and manufacturing of computing devices.
- **Green manufacturing:** Recycling computing devices, components and modules during the production of computers and their accessories. Reducing waste during the production process of computers and other computing accessories to minimize environmental pollution and climate change.

Green electronics, renewable green energy, green computing, recycling computer modules, and waste disposal management are the major subjects and initiatives in green technologies, and in green innovation in the computing industry.

16.3.5 Energy harvesting

In the last twenty years, the idea of employing free space energy in the forms of light, vibration, heat, RF energy, muscle motion and other types of natural energy, has become very attractive and useful. Several techniques to produce electrical energy from these sources of natural energy were presented in [1–3]. Energy harvesting devices can reduce the need to replace batteries every day and eliminate the usage of power cables. In order to collect as much free space energy as possible, it is important to harvest the electromagnetic energy from almost all communication devices and systems. Wideband dual polarized antennas can collect the available free space electromagnetic energy. The energy harvesting antenna must satisfy several essential specifications related to the communication and harvesting system. Highly efficient radiators are crucial because of the low-power densities of RF power in free space. The antennas should function at a specific frequency band and polarization. The antenna radiation pattern should have a wide beam width or an overall omnidirectional radiation pattern. Several types of antennas were used for harvesting energy applications [4–10]. In RF energy harvesting units and systems electromagnetic waves propagating in free space are collected, stored and used to charge electrical devices, batteries and/or other devices. There is a huge increase in the level of electromagnetic power in the free space. The expected amount of radio waves in the air in 2015 was around 5 Exa-bytes per month. The expected amount of radio waves in the air in 2018 was 13 Exa-bytes per month. In table 16.1 energy sources used in harvesting devices are listed.

In figure 16.5 the harvesting energy concept is shown. The RF energy harvesting module includes an antenna, a rectifying circuit and a rechargeable battery. The harvesting energy unit operates as a dual mode energy harvesting unit. The low noise amplifier is part of the receiver. The LNA DC bias voltages are supplied by the receiver DC unit.

Table 16.1. Energy sources used in harvesting devices.

Energy source	Applications	Efficiency (%)	Estimated harvested power
Light power	Outdoor/indoor	12–28	0.1 W cm^{-2}
Thermal energy	HumanIndustrial	~0.2 to ~5	0.08 mW cm^{-2}~1–11 mW cm^{-2}
Vibration energy	~Hz—human~kHz—machines	25 to 55	~5 μW cm^{-3}~1 mW cm^{-3}
RF power	0.8–3 GHz, WiFi, WLAN, LAN	~40 to 75	0.001 to 6 mW cm^{-2} 0.001 to 8 mW cm^{-2}

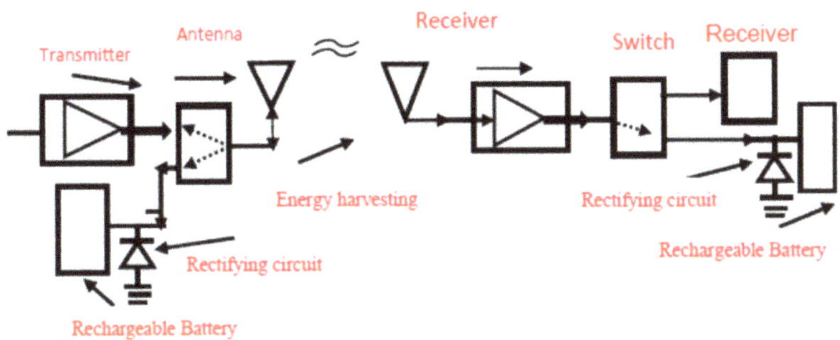

Figure 16.5. Dual mode energy harvesting concept for communication and medical systems.

16.4 Recycling in the electronics and computing industry

Recycling is the process of collecting and processing materials that would otherwise be thrown away as trash and turning them into new products. The huge growth in design and production of communication systems, computers and smartphones over the last century has resulted in most of the world population owning cellular phones, smartphones, smart watches, and other digital devices. As a result, the amount of unwanted smartphones, electronic components, computers and phones is increasing at a huge rate. With this huge amount of waste a new environmental disaster strikes our world. Electronic waste is filling up landfills rapidly and generates an environmental disaster. These electric devices contain hazardous chemicals and toxic materials that endanger the environment and the health of local residents. This electronic waste increases environmental air and water pollution. Green technologies are used to decrease environmental pollution. Recycling of computers waste, electronic waste, old batteries, plastic waste, paper waste, and bottles decrease environmental pollution. Recycling uses and saves original raw materials. Recycling minimizes waste and decreases air and water pollution. In figure 16.6(a) recycling process is shown. In figure 16.6(b) a recycling outdoor site is shown. In figure 16.7 a computer waste recycling site is presented.

Figure 16.6. Outdoor community recycling cite: (a) recycling process and (b) outdoor recycling cite.

Figure 16.7. Computer waste recycling cite.

16.4.1 Types of recycling materials

- Paper products recycling.
- Plastic products recycling.
- Metal products recycling.

16.4.2 Recycling advantages

- Conserves natural resources such as water, timber, and minerals.
- Minimizes the amount of waste sent to landfills and incinerators.
- Increases economic security by tapping a domestic source of materials.

- Prevents pollution by reducing the need to collect new raw materials.
- Helps to save electrical energy.
- Conserves valuable resources and supports production of electronic devices.

16.5 Innovations and challenges in green technologies

Engineers, developers and manufacturers in the electronics industry are now faced with a huge challenge of designing, developing, producing and installing electronic devices with great attention to the product's environmental impact throughout its entire lifecycle. Electronics devices create a lot of waste. Electronics manufacturers are encouraged to reduce the amount of electronic waste by designing products that reduce the amount of toxic materials and chemicals used. Electronic and computer manufacturers should recycle and re-use raw materials and parts as much as possible. Green electronic development and manufacturing include the development and manufacturing of products with green materials, energy harvesting unit, and longer lifecycles. This helps to minimize the amount of electronic and computer waste created by the electronic companies all over the planet. Green system development focuses on reducing the number of the device parts, green energy, green friendly materials and recycling devices and materials. Green technologies allow electronics developers and manufacturers to produce environmentally friendly reliable devices with a long lifecycle with high quality.

Future green electronic and computing technologies challenges

- Using renewable and harvesting energy in computing and electronic systems.
- Using recycled modules, components, and devices in production of electronic devices.
- Development and production of green computers, smartphones and electronic devices.
- Producing cheap efficient solar green energy.
- Recycling most of the computers and electronics waste.
- Development and production of green cars and airplanes.
- We should encourage people, young people, and kids to lead around the world initiatives to decrease air pollution, water pollution and climate changes.

References

[1] Paradiso J A and Starner T 2005 Energy scavenging for mobile and wireless electronics *IEEE Pervasive Comput.* **4** 18–27

[2] Sabban A 2019 Compact wearable meta materials antennas for energy harvesting systems, medical and IOT systems *MDPI Electron.* **8** 1340

[3] Valenta C R and Durgin G D 2014 Harvesting wireless power: survey of energy-harvester conversion efficiency in far-field, wireless power transfer systems *IEEE Microwave Mag.* **15** 108–20

[4] Nintanavongsa P, Muncuk U, Lewis D R and Chowdhury K R 2012 Design optimization and implementation for RF energy harvesting circuits *IEEE J. Emerg. Sel. Top. Curcuits Syst.* **2** 24–33

[5] Devi K K A, Sadasivam S, Din N M and Chakrabarthy C K 2011 Design of a 377 Ω patch antenna for ambient RF energy harvesting at downlink frequency of GSM 900 *Proc. of the 17th Asia Pacific Conf. on Communications (APCC'11)* (Sabah, Malaysia, October 2011) 492–5

[6] Rahim R A, Malek F, Anwar S F W, Hassan S L S, Junita M N and Hassan H F 2013 A harmonic suppression circularly polarized patch antenna for an RF ambient energy harvesting system *Proc. of the IEEE Conf. on Clean Energy and Technology (CEAT'13)* (Lankgkawi, Malaysia, November 2013) (IEEE) 33–7

[7] Krakauskas M, Sabaawi A M A and Tsimenidis C C 2014 Suspended patch microstrip antenna with cut rectangular slots for RF energy harvesting *Proc. of the 10th Loughborough Antennas and Propagation Conf. (LAPC'14)* (Loughborough, UK, November 2014) 304–7

[8] Sabban A 2015 *Low Visibility Antennas for Communication Systems* (Boca Raton, FL: Taylor & Francis Group)

[9] Sabban A 2016 *Wideband RF Technologies and Antenna in Microwave Frequencies* (New York: Wiley)

[10] Sabban A 2017 New wideband passive and active wearable antennas for energy harvesting applications *J. Sens. Technol.* **7** 53–70

IOP Publishing

Wearable Communication Systems and Antennas
(Second Edition)
Design, efficiency, and miniaturization techniques
Albert Sabban

Chapter 17

Analysis and design of wearable communication, medical and IOT systems

17.1 Introduction

This chapter describes the analysis and design process of RF devices, wearable systems, IoT devices and antennas. Computer aided design, CAD, software is discussed in this chapter. More information on electromagnetics, microwave engineering, wearable systems and antenna design process and measurements is presented in [1–6]. Wearable and printed antennas developed by using CAD commercial software are presented in [7–24]. CAD commercial software is presented in this chapter, see [24–27]. A detailed explanation about evaluation of losses from the feed network of wearable and printed antennas is presented in [28, 29].

17.2 Commercial electromagnetic software

In the last century some RF and communication commercial software was published, see [24–27]. The most popular commercial CAD tools that are used in the development of communication systems and antennas will be discussed in this section.

17.2.1 HFSS, high frequency structure simulator

ANSYS HFSS is full-wave electromagnetic (EM) simulation software for designing and developing high frequency devices such as microwave components, antennas, antenna arrays, RF modules, high-speed interconnects, filters, connectors, integrated circuits packages and printed circuit boards. HFSS is employed to analyze high frequency RF modules, high-speed communications systems, radar systems, wearable sensors, satellites, IoT devices, and other RF and digital devices. HFSS web, https://www.ansys.com/products/electronics/ansys-hfss.

doi:10.1088/978-0-7503-5224-6ch17

HFSS employs several solvers to give microwave and antennas engineers an accurate tool to solve all the electromagnetic problems. Through integration with ANSYS thermal, structural and fluid dynamics tools, HFSS provides a multi-physics analysis of microwave components, evaluating their electrical, thermal and structural reliability. The ANSYS HFSS simulation suite consists of a comprehensive set of solvers to address many electromagnetic designs. HFSS automatic adaptive mesh refinement allows engineers to focus on the design instead of spending time creating the best mesh.

RF Electromagnetic solvers

The highly accurate finite element method (FEM) is used by HFSS, MoM the large-scale method of moments, and SBR the ultra-large-scale asymptotic method of shooting and bouncing rays with advanced diffraction and creeping wave physics for enhanced accuracy (SBR+). The HFSS computation suite has the following solvers to solve electromagnetic problems:

HFSS Solvers

- Frequency domain solver.
- Time domain solver.
- Hybrid technologies.
- Integral equations.

HFSS SBR+

- Physical optics.
- Physical theory of diffraction.
- Uniform theory of diffraction.
- Shooting and bouncing ray.
- Creeping wave.

HFSS hybrid technologies: The FEM-IE hybrid technology is built upon FEM, IE MoM and the patented ANSYS domain decomposition method (DDM) to simulate and solve complex devices and systems. By applying the appropriate solver, local regions of high geometric detail and complex materials are addressed with finite element HFSS, while regions of large objects or installed platforms are addressed with 3D MoM HFSS-IE. The solution is delivered in a single setup through a single, scalable and fully coupled system matrix.

FEM (frequency domain) is a 3D, full-wave, frequency domain electromagnetic solver based on the FEM. Designers can compute *SYZ* parameters, generate component models to evaluate signal quality, calculate transmission path loss, analyze impedance mismatch, parasitic coupling and far-field radiation. Applications include MIC modules, printed and aperture antennas, cellular communication modules, MMIC circuits, high-speed digital and RF circuits, waveguides, RF connectors, filters and EMC designs.

Finite element transient (time domain): The finite element time domain solver is used to simulate transient electromagnetic field behavior and visualize fields and system responses in several applications such as TDR, time domain reflectometry, GPR pulsed ground-penetrating radar, lightning strikes, ESD electrostatic discharge and EM interference (EMI). It employs the same finite element meshing approach as the frequency domain solver, without the need to switch meshing technologies to switch simulation domains. The transient solver complements the frequency domain HFSS solver and allows designers to understand the EM characteristics on the same mesh in both time and frequency domains.

Integral equations (IE): The integral equations solver employs 3D MoM analysis for solving open radiation and scattering problems. It is ideal for radiation studies like antenna development and placement, and for scattering evaluation such as radar cross section. The solver can employ either multi-level fast multipole methods (MLFMM) or adaptive cross-approximation (ACA) to minimize memory requirements and computing time, allowing this software to be applied to very large structures.

ANSYS HFSS SBR+ SBR+ is an EM solver to empower SBR technology with simultaneous and consistent implementations of physical theory of diffraction (PTD), uniform theory of diffraction (UTD), and creeping wave for evaluating installed antenna performance on electrically large areas that may be thousands of wavelengths in size. SBR uses a ray tracing technique to model induced surface currents on the radiator platform or scattering geometry composed of metals and dielectrics. With the SBR+ solver, designers can get fast and accurate simulation of the antenna far-field radiation pattern, near-field distributions, coupling, *S*-parameters on electrically environmental, large and enormous platforms. Transmissions and reflections may be modeled in big structures like aircraft, satellites, cars, and antennas radomes. HFSS SBR+ provides good radar signature modeling, including ISAR images of electrically large targets.

Physical theory of diffraction: The PTD wedge correction feature is used for correcting PO currents along sharp edges of antenna platforms to refine EM field diffraction.

Uniform theory of diffraction: Designers may model UTD edge diffraction rays created by illuminated geometry edges and identified by PTD wedges. This is important when the major parts of the scattering geometry are otherwise shadowed from direct or multi-bounce illumination.

ANSYS RF option

The ANSYS RF option combined with HFSS generates an end-to-end high-performance RF computation flow. It consists of ANSYS EMIT, a unique multi-fidelity approach for predicting microwave system performance in complex EM environments with multiple sources of interference, and provides the diagnostic tools needed to quickly identify root-cause RFI cases. The RF option also includes ANSYS circuits, which include a harmonic balance circuit simulation, 2.5D planar MOM solver, and filter synthesis.

RF option solver main features:

- RF radio and component libraries.
- RF channel budget analysis.
- Built-in wireless propagation models.
- RF co-site and antenna coexistence analysis.
- Quick assessment and comparison of potential mitigation measures.
- Multi-fidelity behavioral radio models.

– Antenna-to-antenna coupling models.
– Automated diagnostics for rapid root-cause analysis.

Types of circuit analyses:
– DC analysis with multiple continuation options.
– Linear..
– Transient.
– Multi-tone harmonic balance analysis.

More details can be found in https://www.ansys.com/products/electronics/ansys-hfss.

17.2.2 Example of the process design using HFSS software

HFSS software was employed to design an antenna coupled to a micro bolometer at 100 GHz. Figure 17.1 presents electromagnetic analysis of a microstrip antenna coupled micro bolometer at 100 GHz using HFSS software. The power rate

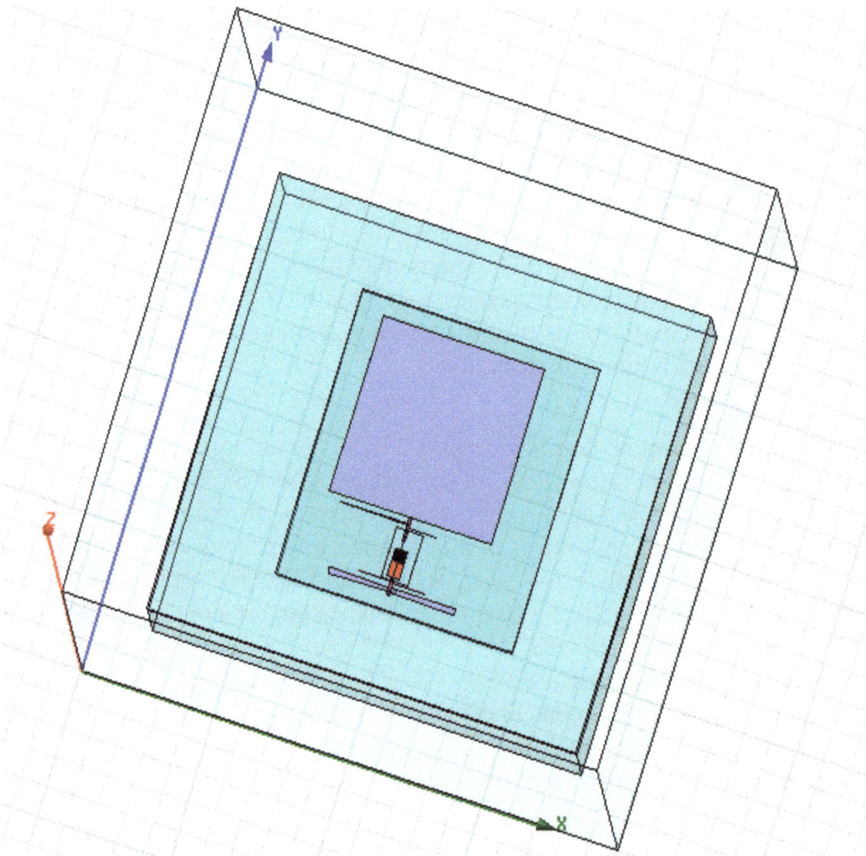

Figure 17.1. Analysis of a patch coupled micro bolometer using HFSS software.

dissipated on the resistor may be improved considerably by optimization of the antenna structure, the antenna feed network dimensions and resistor structure using HFSS optimization tools.

Design process of 100 GHz detection array using HFSS

- Optimization of the conductors types gold, platinum and titanium.
- Optimization of the conductor dimensions length and width.
- Optimization of the antenna type, dimensions, and shape.
- Optimization of the antenna feed and matching network.
- Optimization of the resistor shape and dimensions, length and width of the conductors.
- Optimization of the scarification layer thickness and electrical parameters.
- Optimization of the mutual coupling between the array elements.
- Optimization of the power rate dissipated on the resistor.

The detection array concept is based on a printed antenna coupled to a printed resistor. A direct antenna-coupling surface to a micromachined microbridge resistor is used for heating and sensing. The feed network determines the antenna efficiency. To reduce losses, the feed line and matching network dimensions were optimized and line dimensions are $60 \times 10 \times 1$ μm. The optimization process proved that the dissipated power rate on a titanium resistor is higher than the dissipated power rate on other conductors such as Pt. resistor. The dissipated power rate on a resistor connected to a titanium conductor is 37% of the input power. The optimization results proved that a patch antenna has several advantages over a dipole antenna as presented in table 17.1. The dimensions of the dipole are greater than the dimensions of the patch.

More information and figures about the MEMS detection array are given in section 4.7. Figure 17.2 presents mutual coupling analysis of a three-patches coupled micro bolometer using HFSS software. Figure 17.3 presents analysis of a dipole coupled micro bolometer at 100 GHz using HFSS software. Figure 17.4 presents a 3D radiation pattern of a 100 GHz patch using HFSS software.

17.3 Advance design system, ADS

ADS is an electronic design automation software system. It offers complete design integration to engineers that design RF devices such as wearable sensors, IoT devices, cellular and portable phones, RFID tags, wireless networks, radars, and satellite communications systems.

Table 17.1. Comparison between patch and dipole antennas.

Antenna	Dimensions (mm)	Bandwidth %	Gain (dB)	VSWR
Patch	0.6×0.6	8–10	4	2:1
Dipole	1.4×1.4	10	4	2:1

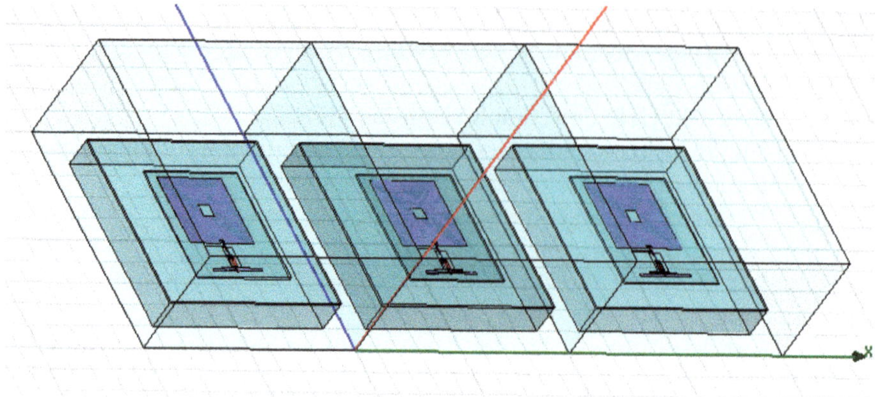

Figure 17.2. Mutual coupling analysis of a three-patches coupled micro bolometer using HFSS software.

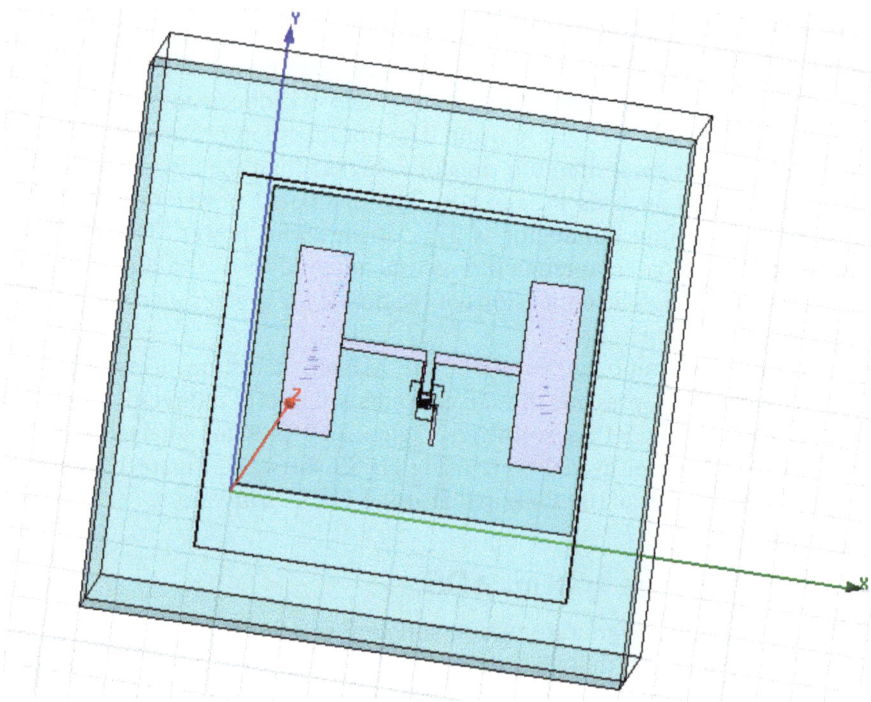

Figure 17.3. Analysis of a dipole coupled micro bolometer using HFSS software.

ADS supports RF engineers to develop RF devices, such as RF filters, power dividers, printed antennas, MIC modules, integrated MMICs for wireless communication, IOT, healthcare, and aerospace defense systems.

ADS offers a complete set of computation technologies such as frequency and time domain circuit simulation, electromagnetic simulation, system and modules analysis. ADS helps engineers to characterize and optimize all system modules such as harmonic balance simulation, circuit envelope simulation, transient convolution,

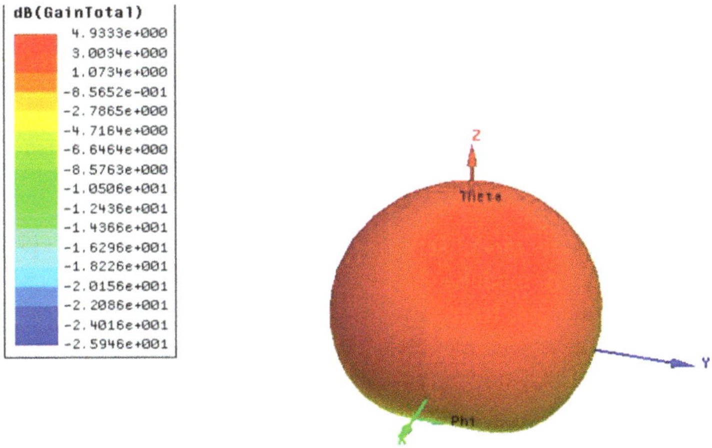

Figure 17.4. 3D radiation pattern of a 100 GHz patch using HFSS software

Ptolemy simulation, *X*-parameter, momentum and 3D EM simulation, including FEM and FDTD solvers.

ADS solvers features

- Project design environment to input schematics, launch simulations and manage design projects.
- Data display to manipulate and plot data.
- Linear simulator including *S*-parameters, noise, DC and small signal AC analysis.
- RF devices and system simulator
- Statistical simulator including design-of-experiments to get a robust designs.
- Optimizers to optimize design parameters to get the best performance and yield.
- Filter design guide to synthesize and analyzed lumped and distributed filters.
- Passive circuit design guide to synthesize and analyze matching networks and passive devices.
- Design guide developer studio to create customized design guides.
- Model libraries, RF system, RF passive and multilayer interconnects.
- Dynamic link to cadence for simulating composer RFIC schematic in ADS.
- Mentor dynamic access for bidirectional RF board design transfer.

ADS functionality

- **Design environment**—Graphical user interface for schematic entry and simulation setup within the ADS.
- **Data display**—Allows simulation results to be viewed either graphically, as they would appear on analyzers or other instruments, or as raw data.
- **Connection manager**—Reads data from, downloads data to, and controls selected instruments. The W2200BP includes the connection manager client only. The connection manager server must be downloaded and installed separately.

- **Linear simulator**—Frequency-domain circuit simulator that analyzes a large variety of RF and microwave linear devices.
- **Statistical design**—Optimization, sensitivity analysis, yield, yield optimization, design of experiments, and yield sensitivity histograms.
- **RF System simulator**—Model a complete RF system with accurate block level models that can later be replaced with device level models.

Circuit and components model sets

- **RF passive circuit models**—Models for many common RF parts such as capacitors, inductors, transformers, couplers, crystals and bond-wires.
- **Multilayer interconnect models**—Models for multiple coupled lines used in multilayer structures such as printed circuit boards, multichip devices, and IC packages.
- **RF system models**—Mixers, modulators, demodulators, gain blocks, and PLL components. Verification model extractors and verification models for RF system-level analysis.

Circuit and devices design guides

- **Passive circuit design guide**—Synthesis and analysis of commonly used microstrip components used by RF designers, such as Wilkinson dividers, quarter-wave matching networks, branch-line couplers, coupled line filters, and lumped-element matching elements.
- **Filter design guide**—Synthesis and analysis of commonly used passive filters.
- **Design guide developer studio**—Create an ADS design guide menu interface.

Momentum circuit designer—The momentum circuit designer is high-frequency physical design software that integrates standard and advanced layout editing features with momentum electromagnetic simulation technology and linear circuit simulation to improve physical circuit design. Momentum circuit designer includes ADS schematic capture, data display, momentum EM simulator, and layout editor, as well as layout and schematic design synchronization capabilities. An example of the design of a high pass LTCC filter process using ADS software is presented in section 4.9. Figure 17.5 presents the LTCC filter circuit analysis using ADS software. Figure 17.6 presents LTCC high pass filter electromagnetic analysis using ADS software. The length and width of the microstrip lines was optimized using ADS software to get the best electrical performance. Figure 17.7 presents S12 and S11 results of LTCC high pass filter at frequencies from 1 to 4 GHz.

FEM simulator

The FEM simulator has full-wave three dimensions electromagnetic simulation capabilities, based on the FEM method. The RF-Pro user interface that comes with this FEM element, makes setting up RF circuit co-simulation in ADS fast with no errors for the design of multi-technology RF modules that integrate RFIC, MMIC, package and PCB. It automates the extraction of nets and components for EM simulation without modifying the layout. The FEM simulator enables simulating 3D structures such as wire-bonds, connectors, and packaging with circuit and system

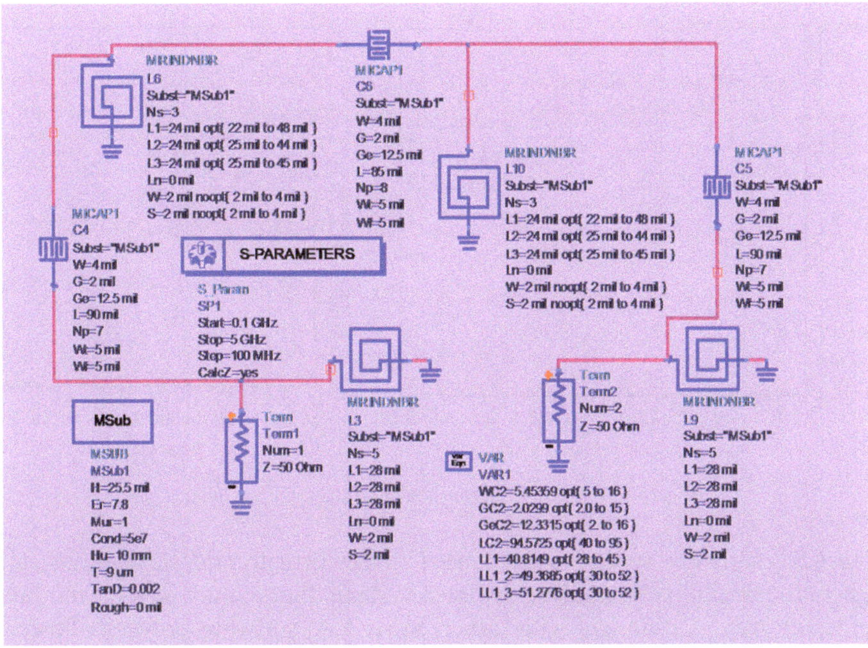

Figure 17.5. LTCC filter circuit analysis using ADS software.

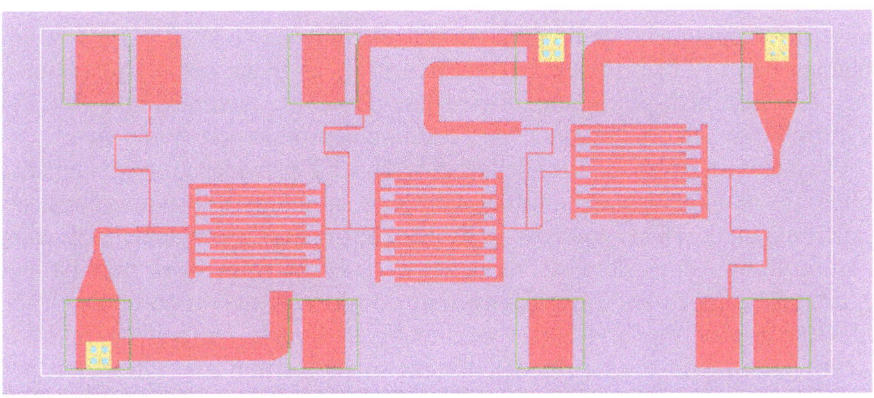

Figure 17.6. LTCC filter electromagnetic analysis using ADS software.

components. It is important for RF module designs where 3D interconnects and packaging must be simulated along with the circuit.

The frequency domain simulation can be used in electromagnetic professional software, in the 3D EM platform and in ADS.

17.4 CST electromagnetic software

CST is a high-performance **3D** electromagnetic **analysis** software for designing, evaluating, analyzing and optimizing RF components and systems.

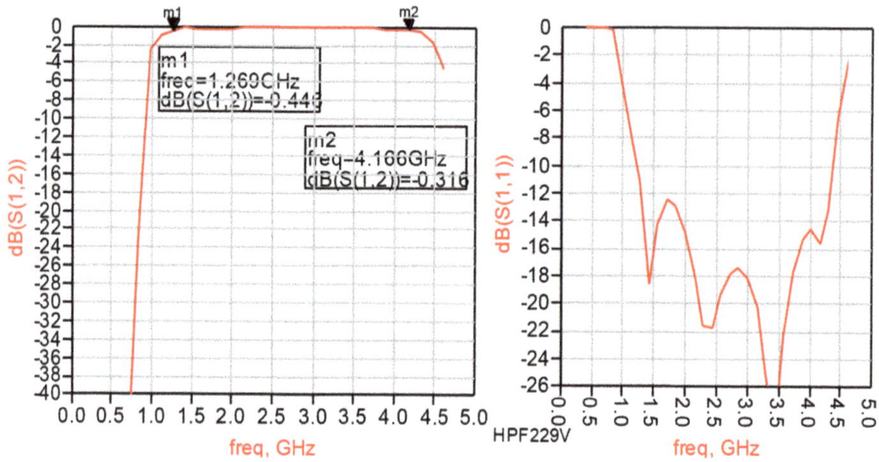

Figure 17.7. S12 and S11 results of LTCC high pass filter.

The CST software solvers can be used to perform hybrid simulations, giving designers the flexibility to analyze complex systems that include several modules in an efficient way. Co-design with other SIMULA software solvers allows EM simulation to be integrated into the design flow and drives the design process from the first design stages up to the final design process.

Common applications of EM analysis include the design of filters, antennas, couplers, electromagnetic compatibility, and interference (EMC/EMI), exposure of the human body to EM fields, electro-mechanical effects in motors and generators, and thermal effects in high-power amplifiers. More details are presented in CST Software web, https://www.3ds.com/products-services/simulia/products/cst-studio-suite/

With system assembly and modeling (SAM), CST provides an environment that simplifies the management of simulation projects, enabling the intuitive construction of electromagnetic (EM) systems and straightforward management of complex simulation flows using schematic modeling. The SAM framework may be used to analyze and optimize an entire device that consists of several components. SAM enables the use of the best solver technology for each component. SAM helps designers to compare the results of different solvers within one simulation project and perform post-processing automatically. SAM solver runs for hybrid and multi-physics simulations. For example, using the results of EM simulation to calculate thermal effects, then structural deformation, and then another EM simulation to analyze the RF device. This combination of different levels of simulation reduces the computational effort required to simulate and analyze a complex structure and device accurately.

The CST software design environment is a friendly user interface used by all the modules. It includes a schematic layout tool, a 3D modeling tool, a pre-processor for the electromagnetic solvers and post-processing tools. The ribbon-based interface uses tabs to display all the tools and options needed. Contextual tabs mean that the most relevant options for the task are always just a click away. The Project Wizard and the

Quick Start Guide provide guidance to new designers and offer access to a wide range of features. The 3D interactive modeling tool uses the ACIS 3D CAD kernel. This powerful tool enables complex models to be constructed within CST software.

CST Solvers

- Frequency domain solver.
- Integral equation solver.
- Transient solver.
- TLM solver.
- Eigen-mode solver.
- Resonant solver.
- Asymptotic solver.

CST Applications

- CST applications include static, stationary, low- and high-frequency problems, as well as devices with movement of charged particles.
- CST's products cover a wide range of RF devices. Typical applications are planar structures, couplers, connectors, filters, EMC and SAR problems, all kind of antennas, packages, LTCC structures, inductors, capacitors, sensors, waveguides, plasma sources, optical devices, recording units, actuators, motors and electromagnetic brakes.

17.5 Microwave office, AWR

Microwave office software, AWR, high-frequency design tool with its open design environment and advanced unified data model provides a flexible RF design tool. AWR seamlessly operates with a visual system simulator (VSS) system design and AXIEM and analyst EM simulation software tools in the NI AWR design environment platform to deliver a complete RF microwave circuit, system, and EM co-simulation environment.

AWR's **AXIEM** electromagnetic software is an EM simulation. The AXIEM product was developed specifically for three-dimensional planar applications such as RF printed devices and modules, MMIC, LTCC, and RFIC designs. More detail are given in, https://www.awr.com/serve/microwave-office-brochure-1.

The APLAC simulator provides multi-level analyses such as:

- Linear frequency domain analysis.
- Time domain analysis.
- Harmonic balance analysis.
- Phase noise computation.
- DC operation point computation.
- Linear and nonlinear noise including AC noise contributors, temperature.
- Yield predictions and optimization analysis.

AWR provides fast and accurate computation and offers robust circuit analysis and design insight, providing the linear and nonlinear time and frequency domain measurements required to properly evaluate and optimize high-frequency devices.

Simulation models, libraries of high frequency distributed transmission models, surface-mount vendor components, and process design kits from MMIC/RFIC companies enable accurate simulation of development prior to production, resulting in fewer and faster development iterations.

AWR design capabilities—The AWR user interface provides project management and design entry for high-frequency circuits, enabling engineers to easily build networks from a comprehensive library of RF components. The library supports parameterization for tuning and optimization processes, and hierarchical design with circuit, system, and EM simulation. AWR offers simulation controls and result graphs for standard microwave and RF measurements.

AWR automation features—Automation features expedite design tasks and manage network and measurement data, including labor-saving wizards to import PCB layout and/or open access schematic information from third-party tools.

AWR load-pull analysis—Readily develops amplifier input and output matching networks using complex swept load-pull data sets based on either measured or simulated data. Performance contours include available output power, gain, efficiency, intermodulation distortion levels, or any amplifier related performance features.

AWR synthesis and design assistance—Accelerates design starts with synthesis modules and designs assist wizards that generate topologies based on a set of user specified microwave performance requirements. Mixers, synthesized filters, impedance matching, and passive component networks are available for further tuning, optimization, EM simulation and verification.

Simulation APLAC—The harmonic balance, HB, simulator provides linear and nonlinear circuit analysis with powerful multi-rate HB, transient-assisted HB, and time variant analysis, supporting large-scale and highly nonlinear microwave circuits.

Planar EM—AXIEM provides the speed and accuracy to characterize and optimize passive structures, transmission lines, planar antennas, and large patch arrays.

3D EM helps to accelerate high-frequency product design from the design characterization stage through to full 3D EM verification. AWR 3D finite element solver provides fast and accurate EM analysis of interconnects such as bond-wires, vias, via fencing, and ball grids.

Automated circuit extraction—Using layout-based models for circuit extraction, ACE reduces the time required to model complex interconnects by automatically identifying transmission lines from the layout and partitioning them into existing models.

17.5.1 AWR for MMIC development

- Design capture with industry leading tuning.
- Linear and nonlinear frequency domain and time domain simulation.
- Layout with production ready GDSII export.

- Electromagnetic analysis from layout or schematic to most commercial EM solvers.
- Design rule check and layout versus schematic from a simulation schematic to ICED or Mentor.

System simulation and frequency planning using AWR
- Process design kits from a wide range of companies.
- Packaging design and module flow.

Automated circuit extraction technology, ACE
- ACE circuit extraction technology dramatically reduces from hours to seconds the time required to do the initial modeling of complex interconnects. The new technology enables designers to perform interconnect modeling at the earliest stages of the design flow, where problems can be identified and corrected before costly and time-consuming redesigns are required.

17.5.2 AWR for module design

- Full integration of APLAC's harmonic balance and time domain simulators.
- ACE technology delivers fast and accurate interconnect modeling.
- iNets technology offers automated interconnect construction for MMIC, module, and PCB development.
- EXTRACT flow enables EM and circuit extraction to be controlled by the schematic.
- AXIEM 3D planar EM solver technology delivers capacity, accuracy and speed.
- Multi-technology PDKs support circuit, system, schematic and layout co-design.

AWR software was employed to design Ka band switched filter bank. Figure 17.8 presents a nine section Ka band, band pass filter designed using AWR software.

AWR was employed to optimize the filter dimensions, see [6]. Figure 17.9 presents X-Ku band, band pass filter simulation results by using AWR software. Figure 17.10 presents Ka band, band pass filter simulation results by using AWR software.

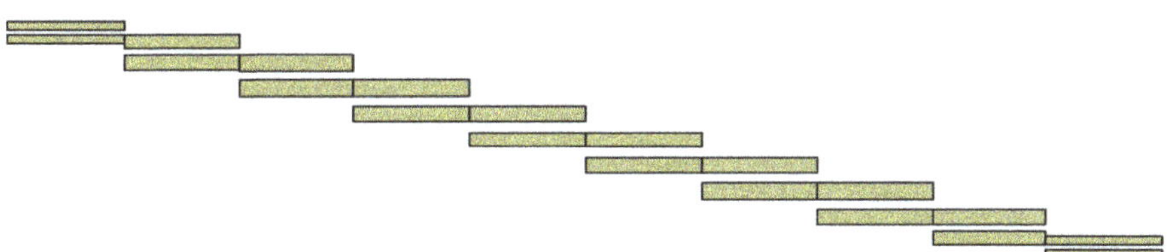

Figure 17.8. Nine section Ka band, band pass filter design using AWR software.

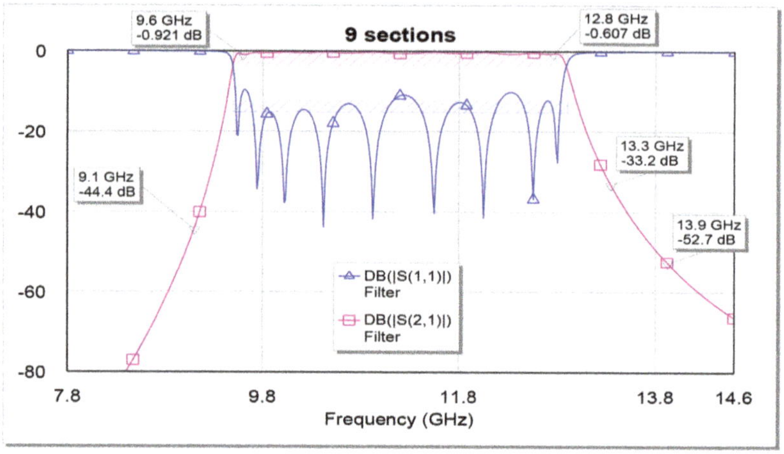

Figure 17.9. X-Ku band, band pass, filter simulation results by using AWR software.

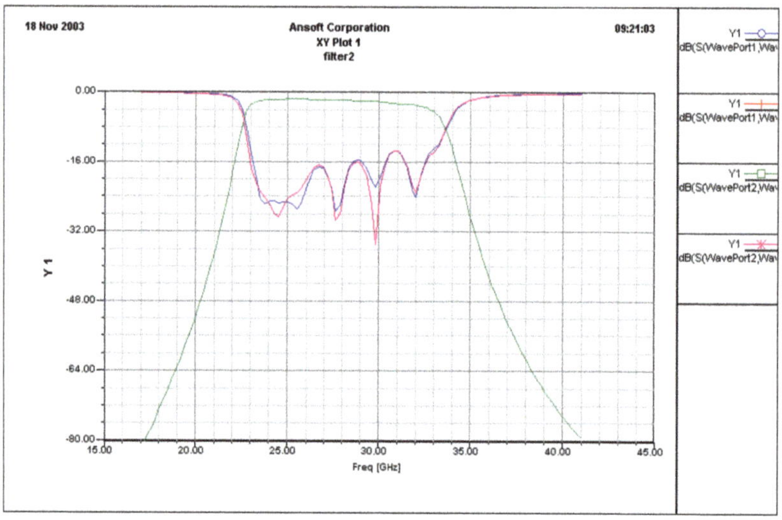

Figure 17.10. Ka band, band pass, filter simulation results by using AWR software.

17.6 Evaluation of losses in wearable sensors and antennas

Printed wearable sensors suffer usually from losses in the sensors and antennas feed network. Commercial software analyzes and computes the electrical parameters of the feed circuit and the radiating element. However, it does not provide us with information on how much power is lost in each transmission line and each microstrip discontinuity. Microstrip line losses and feed network losses increase considerably at mm-wave frequencies. Conductor, dielectric and radiation losses are the major components of loss in mm-wave microstrip devices and in printed antenna arrays. At frequencies ranging from 26 to 40 GHz, conductor losses are around 0.12 to 0.25 dB per wavelength, dielectric losses are around 0.05 dB per wavelength for a 50 Ω line on a 0.127 mm thick substrate with $\varepsilon r = 2.2$. The open nature of the microstrip structure suffers from radiation. For example radiation loss of a right-angle bend in a 50 Ω line on a 0.127 mm thick substrate with $\varepsilon r = 2.2$, is 0.12 dB at 30 GHz and 0.18 dB at 40 GHz. To minimize radiation loss, the number of discontinuities, such as steps, bends and T-junctions, should be minimized. Radiation from a curved microstrip line is much smaller compared to radiation from a right-angled bend. Moreover, in order to reduce radiation loss in the feed network, the width of the microstrip line is designed to be less than 0.12 λ on a 0.25 mm substrate with $\varepsilon r = 2.2$. Dielectric loss can be reduced by using substrates with a low dielectric loss.

Conductor loss may be minimized by designing the feed network length per wavelength as short as possible. By using a multilayer feed network design, the feed network length per wavelength is minimized considerably. Gold plating of the microstrip lines decrease conductor losses.

Microstrip feed network losses computations

Equations to calculate conductor loss and dielectric loss in microstrip lines are presented in [1]. Dielectric loss is incorporated in the multiport network analysis by considering a complex dielectric constant. Conductor losses are included in the analysis by defining an equivalent loss tangent δc, given by

$\delta c = \sqrt{2/\omega\mu\sigma}/h$, where σ is the strip conductivity, h is the substrate height and μ is the free space permeability.

17.7 Computation of radiation loss in wearable antennas feed network

The multiport network model, MNM, is used to compute radiation loss from wearable antennas feed lines by adding a number of open ports at the edges of the planar model for the discontinuity structure. The MNM model is employed to evaluate the voltage distributions at the open ports, voltages at the discontinuity edges are represented by equivalent magnetic current sources, as presented in figure 17.11 The amplitude M of the magnetic current elements is twice that of the edge voltage at that location and the phase of the magnetic current is equal to the

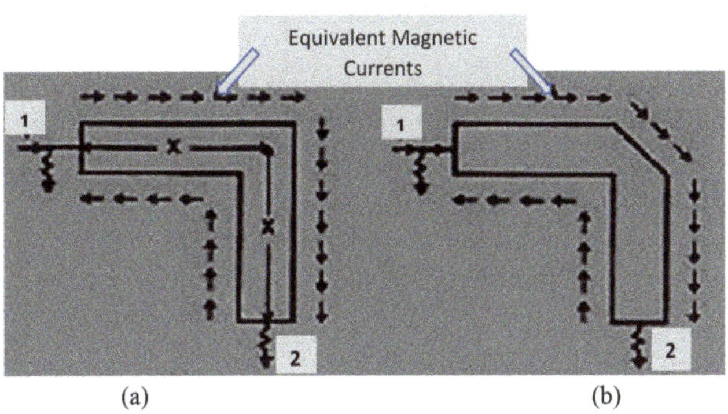

Figure 17.11. Equivalent magnetic current sources distribution. (a) Right-angle bend. (b) Chamfered bend.

phase of the corresponding voltage. The total radiation is computed by employing the superposition of the far-field radiated by each section. Referring to the coordinate system, shown in figure 17.12, the far-field pattern $F(\theta,\phi)$ may be written in terms of voltages at the various elements as given in equation (17.1). With the voltage at the ith element as $(V(i)\ e^{j\alpha(i)})$, we have

$$F(\theta, \phi) = \sum_{i=1}^{N} 2V(i)W(i)\exp\{k_0\gamma_0(i) + \alpha(i)\}F_i(\theta, \phi)$$

where

$$F_i(\theta, \phi) = \frac{\sin\left(\frac{k_0 W(i)}{2}\cos\theta\right)}{\frac{k_0 W(i)}{2}\cos\theta}\sin\theta \tag{17.1}$$

$$\gamma_0(i) = X_0(i)\sin\theta\cos\phi + Y_0(i)\cos\theta$$

The factor 2 in (17.1) accounts for the image of the magnetic current with respect to the ground plane. N is the number of ports, k_0 is the free space wave number, $X_0(i)$, $Y_0(i)$ specify the location of the ith magnetic current element, $W(i)$ is the width of the ith element. The radiated power is calculated by the integration of the Poynting vector over the half space and is given in equation (17.2).

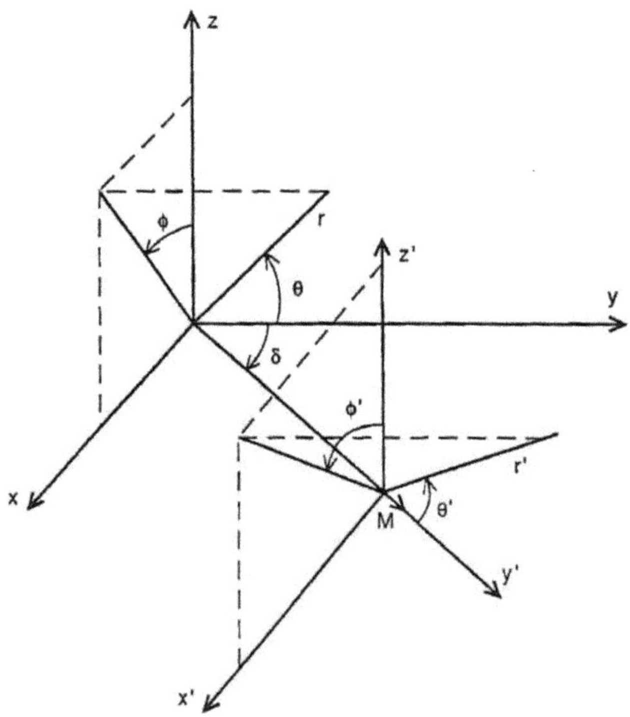

Figure 17.12. Coordinate system for computations of radiation losses.

$$P_r = \frac{1}{240\pi} \int_{\frac{-\pi}{2}}^{\frac{\pi}{2}} \int_{0}^{\pi} (|E_\theta|^2 + |E_\phi|^2) r^2 \sin\theta \, d\theta \, d\phi \tag{17.2}$$

The fields E_θ and E_ϕ are expressed in terms of $F(\theta,\phi)$ as given in equations (17.3) and (17.4).

$$E_\theta = \hat{a}_\theta(\frac{-jk_0}{4\pi r} F(\theta, \phi) F_\theta) \tag{17.3}$$

$$E_\phi = \hat{a}_\phi(\frac{-jk_0}{4\pi r} F(\theta, \phi) F_\phi) \tag{17.4}$$

where, F_ϕ and F_θ are given in equations (17.5) and (17.6).

$$F_\phi = \sin\phi' \sin\phi + \cos\delta \cos\phi \cos\phi' \tag{17.5}$$

$$F_\theta = -\sin\phi' \cos\theta \cos\phi + \cos\delta \cos\theta \sin\phi + \sin\delta \cos\phi' \sin\theta \tag{17.6}$$

$\cos\theta'$, and $\cos\phi'$ are given in equations (17.7) and (17.8).

$$\cos\theta' = \sin\theta \sin\phi \sin\delta + \cos\theta \cos\delta \tag{17.7}$$

$$\cos\phi' = \sin\theta \cos\phi / \sqrt{1 - \cos^2\theta'} \tag{17.8}$$

The radiated power may be expressed as a fraction of input power $10\log_{10}(P_r/P_i)$ dB.

The radiation loss may be expressed as given in equation (17.9).

$$\text{Radiation loss(dB)} = 10\log_{10}(1 - P_r/P_i) \tag{17.9}$$

Where P_i is the input power at port 1. P_i is computed from input current and the input impedance of the discontinuity terminated in matched loads at other ports. Detailed explanation about evaluation of losses from the feed network of wearable and printed antennas is presented in [28, 29]. In this chapter several results of radiation losses from microstrip discontinuities are presented.

Voltage (v) distribution at 10 GHz at the edges of a 50 Ω microstrip right-angled bend on GaAs substrate 0.127 mm thick with $\varepsilon r = 12.9$, is presented in figure 17.13. Voltage in volts distribution at 10 GHz at the edges of a 50 Ω microstrip chamfered bend on GaAs substrate 0.127 mm thick with $\varepsilon r = 12.9$, is presented in figure 17.14.

Results for the power radiated from right-angled bends for 50 Ω and 20 Ω lines etched on substrate 0.254 mm thick with 2.2 dielectric constant are presented in figure 17.15. Results for the power radiated from right-angled bends for 50 Ω line printed on substrate with 12.9 dielectric constant and 0.127 mm thick are presented in figure 17.15.

Where MNM is the multiport network model. PV refers to evaluation of radiation losses by using the Poynting vector approach, see [28, 29]. Radiation losses from right-angled bends are for bends with lower impedance printed on

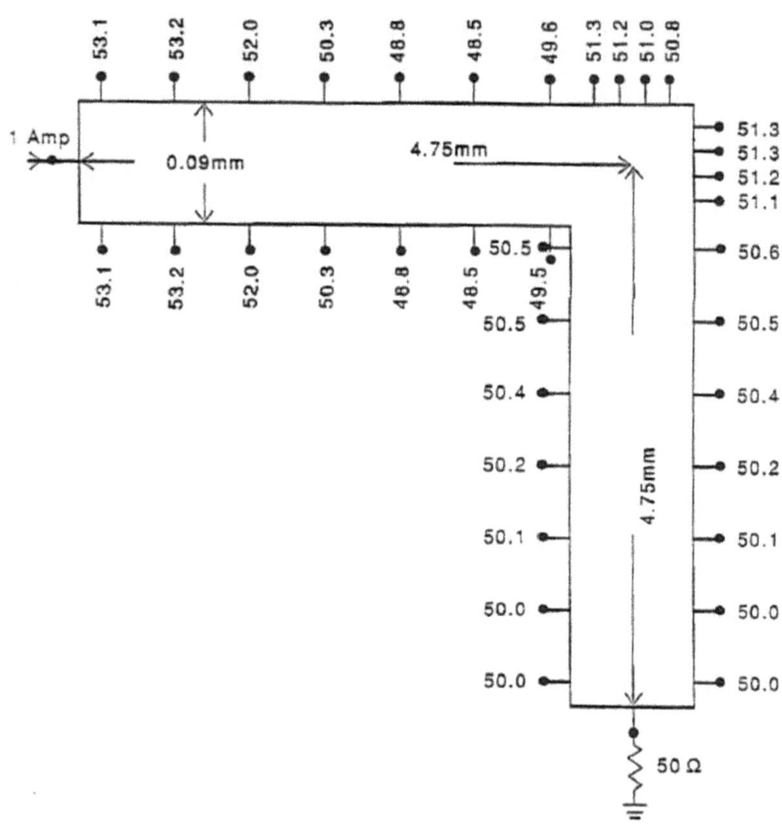

Figure 17.13. Voltage (v) distribution at the edges of a microstrip right-angled bend. $\varepsilon r = 12.9$; $h = 0.127$ mm; $Z0 = 50\ \Omega$; $f = 10$ GHz.

substrate with higher dielectric constant. The radiation loss for the 50 Ω right-angled bend printed on substrate with 12.9 dielectric constant and 0.127 mm thick is 0.062 dB at 40 GHz. The radiation loss for the 50 Ω right-angled bend printed on substrate with 2.2 dielectric constant and 0.254 mm thickness is 0.17 dB at 40 GHz. Results for the power radiated from a T-junction 50 Ω and 35.5 Ω lines etched on substrate 0.79 mm thick with 2.2 dielectric constant are shown in figure 17.16. Results for the power radiated from a 20 Ω open end etched on substrate 0.254 mm thick with 2.2 dielectric constant are shown in figure 17.17.

The radiation loss for the T-junction etched on substrate 0.79 mm thick with 2.2 dielectric constant are 0.18 dB at 12 GHz. Most of the radiated power from the T-junction originates from the junction area. The quarter wavelength 20 Ω open end resonates at 11 and 33 GHz and radiates around 90% of input power. At 11 and 33 GHz, the open end acts as a good quarter wavelength antenna. Results for the power radiated from a 50 Ω chamfered bend etched on substrate 0.254 mm thick with 2.2 dielectric constant are presented in figure 17.18. The radiated power from a chamfered bend is considerably lower than the radiated power from a right-angled

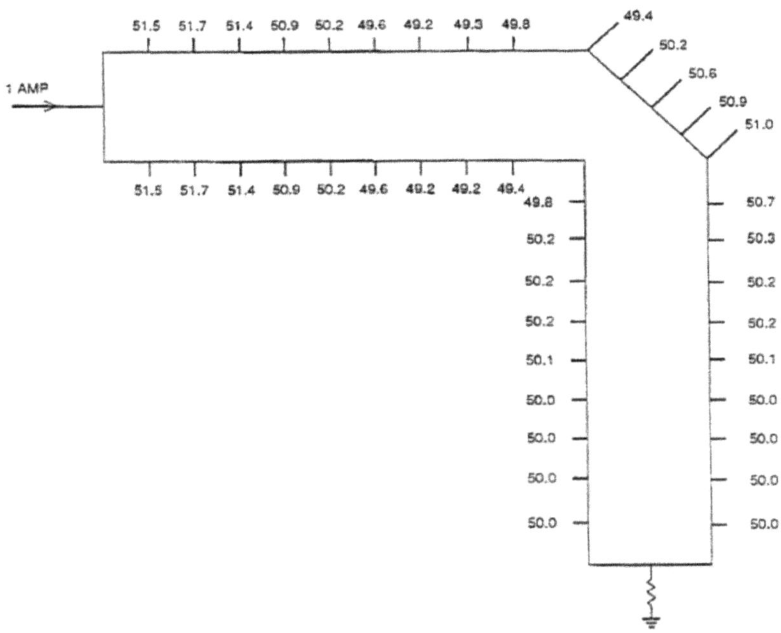

Figure 17.14. Voltage (v) distribution at the edges of a microstrip chamfered bend. $\varepsilon r = 12.9$; $h = 0.127$ mm; $Z0 = 50\ \Omega$; $f = 10$ GHz.

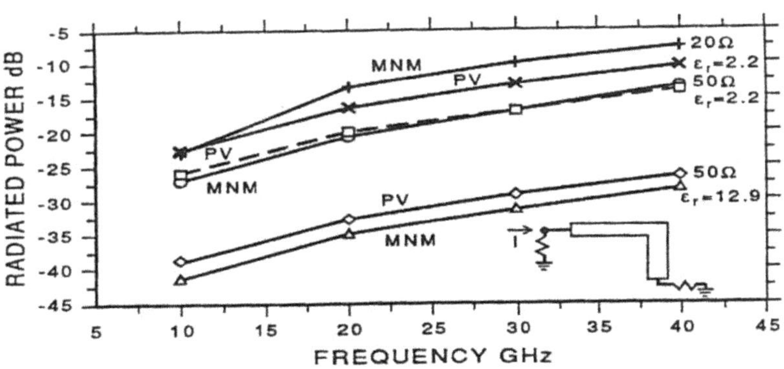

Figure 17.15. Radiation losses from right-angled bends.

bend. Radiation, conductor and dielectric losses cannot be ignored in the design of efficient wearable sensors and antennas.

17.8 Conclusions

In this chapter electromagnetics, microwave engineering, wearable systems and antennas analysis and design are presented. Electromagnetics commercial software CAD software are presented in this chapter. EM CAD software improves the design and evaluation of wearable devices for communication and medical applications.

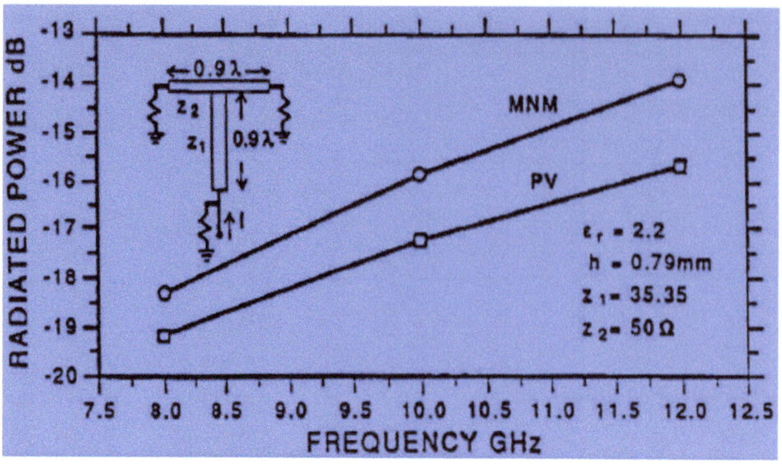

Figure 17.16. Radiation losses from a T-junction.

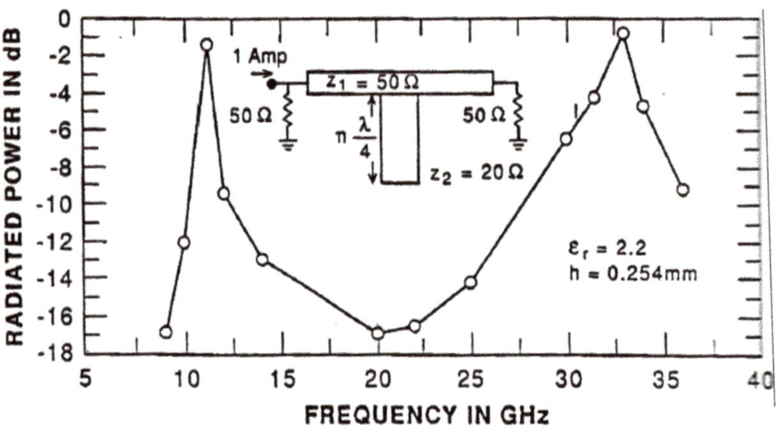

Figure 17.17. Radiation losses from a 20 Ω open end.

Printed wearable sensors suffer usually from losses in the sensors and antennas feed network. Commercial software analyzes and computes the electrical parameters of the feed network and the radiating element. However, it does not provide us with information on how much power is lost in each transmission line and each microstrip discontinuity. Losses in microstrip feed networks increase considerably at high frequencies. Microstrip line losses increase significantly at high frequencies. Conductor, dielectric and radiation losses determine the efficiency of mm-wave patch arrays. At frequencies from 30 to 40 GHz, conductor losses are around 0.14 to 0.22 dB per wavelength, dielectric losses are around 0.045 dB per wavelength for a 50 Ω line on a 0.127 mm substrate with $\varepsilon r = 2.2$. The open structure of the microstrip configuration suffers from radiation. The radiation loss of a right-angle bend in a 50 Ω line on a 0.127 mm substrate with $\varepsilon r = 2.2$, is 0.11 dB at 30 GHz and 0.18 dB at

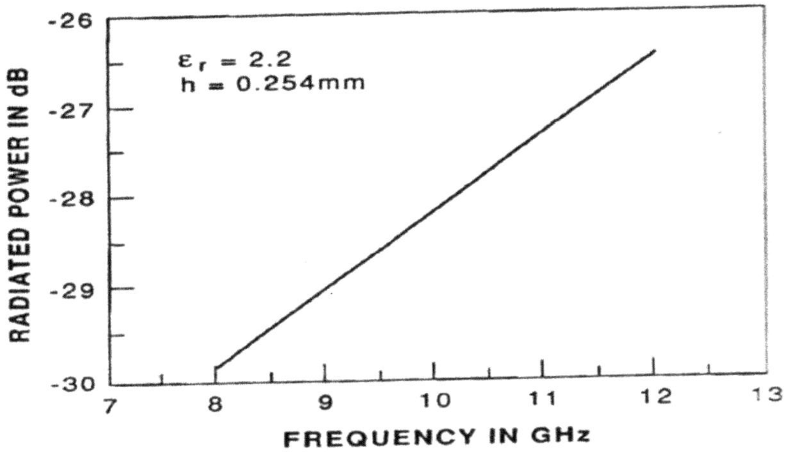

Figure 17.18. Radiation losses from a chamfered bend.

40 GHz. To minimize radiation loss, the number of discontinuities, such as bends and T-Junctions, should be minimized. The radiated power from a right-angled bend is considerably higher than the radiated power from a chamfered bend. Dielectric loss can be reduced by printing the circuit on substrates with a low dielectric loss.

References

[1] Balanis C A 1996 *Antenna Theory: Analysis and Design* 2nd edn (New York: Wiley)

[2] Godara L C (ed) 2002 *Handbook of Antennas in Wireless Communications* (Boca Raton, FL: CRC Press LLC)

[3] Kraus J D and Marhefka R J 2002 *Antennas for all Applications* 3rd edn (New York: McGraw Hill)

[4] Sabban A 2014 *RF Engineering, Microwave and Antennas* (Tel Aviv Israel: Saar Publication)

[5] Sabban A 2015 *Low Visibility Antennas for Communication Systems* (Boca Raton, FL: Taylor and Francis Group)

[6] Sabban A 2016 *Wideband RF Technologies and Antenna in Microwave Frequencies* (New York: Wiley)

[7] Sabban A 2016 Small wearable meta materials antennas for medical systems *Appl. Comput. Electromagn. Soc. J.* **31** 434–43

[8] Sabban A 2015 New compact wearable meta-material antennas *Global J. Res. Anal.* **IV** 268–71

[9] Sabban A 2015 New wideband meta materials printed antennas for medical applications *J. Adv. Med. Sci.* **3** 1–10 invited paper

[10] Sabban A 2013 New wideband printed antennas for medical applications *IEEE Trans. Antennas Propag.* **61** 84–91

[11] Sabban A 2013 Comprehensive study of printed antennas on human body for medical applications *J. Adv. Med. Sci.* **1** 1–10

[12] Sabban A 2013 Wearable antennas *Advancements in Microstrip and Printed Antennas* ed A Kishk (Rijeka: Intech)

[13] Sabban A 2011 *Microstrip Antenna Arrays, Microstrip Antennas* ed Nasimuddin Nasimuddin (Rijeka: InTech)

[14] Sabban A 2015 Dually polarized tunable printed antennas for medical applications *IEEE European Antennas and Propagation Conf. (EUCAP)* (Lisbon Portugal, April 2015)

[15] Sabban A 2014 New microstrip meta materials antennas *IEEE Antennas and Propagation* (Memphis, TN, July 2014)

[16] Sabban A 2013 Wearable antennas for medical applications *IEEE BodyNet* (Boston, MA, October 2013) 1–7

[17] Sabban A 2013 New meta materials antennas *IEEE Antennas and Propagation* (July 2013, Orlando, FL)

[18] Sabban A 2013 Meta materials antennas *New Tech Magazine* (Tel Aviv, Israel, June 2013) 16–9

[19] Sabban A 2012 Wideband tunable printed antennas for medical applications *IEEE APS/URSI Conf.* (Chicago IL. USA) July 2012 1–2

[20] Sabban A 2012 MM wave microstrip antenna arrays *New Tech. Mag.* 16–21

[21] Sabban A July 2011 New compact wideband printed antennas for medical applications *IEEE APS/URSI Conf.* (Spokane WA, July 2011) 251–4

[22] Sabban A 2010 Interaction between new printed antennas and human body in medical applications *Asia Pacific Symp.* (Japan, December 2010) 187–90

[23] Sabban A 2009 Wideband printed antennas for medical applications *APMC 2009 Conf. (Singapore, December 2009)* 393–6

[24] ANSYS HFSS https://ansys.com/products/electronics/ansys-hfss)

[25] Keysight ADS https://keysight.com/en/pc-1297113/advanced-design-system-ads?nid=-34346.0&c.c.=IL&lc=eng)

[26] CST Software https://3ds.com/products-services/simulia/products/cst-studio-suite/)

[27] Microwave office AWR https://awr.com/serve/microwave-office-brochure-1)

[28] Sabban A and Gupta K C 1991 Characterization of radiation loss from microstrip discontinuities using a multiport network modeling approach *IEEE Trans. MTT* **39** 705–12

[29] Sabban A 1991 Multiport network model for evaluating radiation loss and coupling among discontinuities in microstrip circuits *PhD Thesis* University of Colorado, Boulder

IOP Publishing

Wearable Communication Systems and Antennas (Second Edition)

Design, efficiency, and miniaturization techniques

Albert Sabban

Chapter 18

Measurements of wearable systems and antennas

18.1 Introduction

This chapter presents RF systems, electromagnetics, wearable sensors and antenna measurements. Measurement of wearable sensors and RF healthcare systems near the human body are presented in sections 18.2–18.5. RF measurement theory is discussed in sections 18.6 and 18.7, see [1–5]. *S* parameter measurements are an important step in development of RF devices, RF systems and antennas. Setups for RF systems and antennas measurements are described in this chapter. Measurements of input and output power of wearable systems will be also described. Intermodulation measurements, IP_2 and IP_3, will be also discussed.

It is more convenient to measure antennas in the receiving mode. Antenna radiation pattern is identical for receiving and transmitting modes if the tested antenna is reciprocal. Active antennas are not reciprocal. Radiation patterns of antennas are usually measured in the far-field. However, far-field radiation pattern measurements suffer from several disadvantages. First, a long free space area is needed. Antenna radiation pattern is affected from reflections from the ground and from walls. These reflections add errors to the measured radiation pattern results. Antenna radiation pattern measurements facilities are expensive. It is difficult and almost impossible to measure the antenna on the antenna's operating environment, such as the human body, an airplane or satellite. Some of these drawbacks can be solved by near-field and indoor measurements. Near-field measurements are presented in [1]. Small microwave and communication companies do not own antenna measurement facilities. There are several companies around the world that provide antenna measurement services, near-field and far-field measurements. One day, near-field measurements may cost around 4500 USD, and far-field measurements may cost around 2500 USD.

doi:10.1088/978-0-7503-5224-6ch18

18.2 Representation of wearable systems by N ports model

RF and wearable communication systems can be represented as multiport networks with N ports as presented in figure 18.1. We can assume that only one mode propagates in each port. The propagating fields in each port represent incident and reflected electromagnetic waves. The fields may be represented by equivalent voltages and currents as written in equations (18.1) and (18.2) [1–5].

$$\begin{aligned} V_n^- &= Z_n I_n^- \\ V_n^+ &= Z_n I_n^+ \end{aligned} \tag{18.1}$$

$$\begin{aligned} I_n^- &= Y_n V_n^- \\ I_n^+ &= Y_n V_n^+ \end{aligned} \tag{18.2}$$

The voltages and currents in each port are written in equation (18.3).

$$\begin{aligned} I_n &= I_n^+ - I_n^- \\ V_n &= V_n^+ + V_n^- \end{aligned} \tag{18.3}$$

The Z matrix represents the relations between the voltages and currents at each port as written in equation (18.4). The Y matrix represents the relations between the currents and voltages at each port as written in equation (18.5). The Z matrix is the inverse of the Y matrix.

$$[V] = \begin{bmatrix} V_1 \\ V_2 \\ V_N \end{bmatrix} = \begin{bmatrix} Z_{11} & Z_{12} & Z_{1N} \\ Z_{21} & Z_{22} & Z_{2N} \\ Z_{N1} & Z_{N2} & Z_{NN} \end{bmatrix} \begin{bmatrix} I_1 \\ I_2 \\ I_N \end{bmatrix} = [Z][I] \tag{18.4}$$

$$[I] = \begin{bmatrix} I_1 \\ I_2 \\ I_N \end{bmatrix} = \begin{bmatrix} Y_{11} & Y_{12} & Y_{1N} \\ Y_{21} & Y_{22} & Y_{2N} \\ Y_{N1} & Y_{N2} & Y_{NN} \end{bmatrix} \begin{bmatrix} V_1 \\ V_2 \\ V_N \end{bmatrix} = [Y][V] \tag{18.5}$$

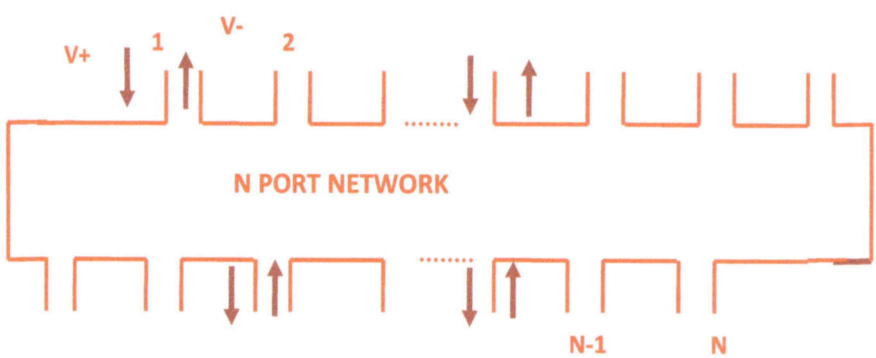

Figure 18.1. Multiport networks with N ports.

18.3 Scattering matrix

In RF networks we can measure power, voltage standing wave ratio (VSWR), and the location of the minimum field strength. However, we cannot measure voltages and currents in RF systems. The power reflection coefficient is the ratio between the reflected power and the input power. The voltage reflection coefficient is the square root of the power reflection coefficient. The scattering matrix is a mathematical presentation that describes how electromagnetic energy propagates through a multiport network. The *S* matrix presents accurately the properties of complex networks. The *S* matrix presents the relation between the forward voltage and the reflected voltage as given in equation (18.6). *S* parameters are defined for a given frequency and ports impedance and vary as a function of frequency for any non-ideal network. *S* parameters present the response of an N-port network to the input voltage signals at each port. The first number in the subscript refers to the responding port, while the second number refers to the incident port. So S_{32} means the response at port 3 due to a voltage at port 2. The S_{nn} elements represent reflection coefficients.

$$[V^-] = \begin{bmatrix} V_1^- \\ V_2^- \\ V_2^+ \end{bmatrix} = \begin{bmatrix} S_{11} & S_{12} & S_{1N} \\ S_{21} & S_{22} & S_{2N} \\ S_{N1} & S_{N2} & S_{NN} \end{bmatrix} \begin{bmatrix} V_1^+ \\ V_2^+ \\ V_2^+ \end{bmatrix} = [S][V^+] \tag{18.6}$$

The S_{nm} element represents transmission coefficients as presented in equation (18.7), where a_i represents the forward voltage in the *i* port.

$$\begin{aligned} S_{nn} &= \frac{V_n^-}{V_n^+} \mid a_i = 0 \quad i \neq n \\ S_{nm} &= \frac{V_n^-}{V_m^+} \mid a_i = 0 \quad i \neq m \end{aligned} \tag{18.7}$$

In equation (18.8) the forward and reflected voltages are represented by normalizing the *S* matrix. *S* parameters vary with frequency and are defined for a given frequency. In a reciprocal microwave network $S_{nm} = S_{mn}$ and $[S]^t = [S]$.

$$\begin{aligned} I &= I^+ - I^- = V^+ - V^- \\ V &= V^+ + V^- \\ V^+ &= \frac{1}{2}(V + I) \\ V^- &= \frac{1}{2}(V - I) \end{aligned} \tag{18.8}$$

The relation between the *Z* and *S* matrix is derived by using equations (18.8) and (18.9) and is written in equations (18.10) and (18.11).

$$\begin{aligned} I_n = I_n^+ - I_n^- &= V_n^+ - V_n^- \\ V_n &= V_n^+ + V_n^- \end{aligned} \tag{18.9}$$

$$
\begin{aligned}
[V] &= [V^+]+[V^-] = [Z][I] = [Z][V^+] - [Z][V^-] \\
([Z] + [U])[V^-] &= ([Z] - [U])[V^+] \\
[V^-] &= ([Z] + [U])^{-1}([Z] - [U])[V^+] \\
[V^-] &= [S][V^+] \\
[S] &= ([Z] + [U])^{-1}([Z] - [U])
\end{aligned}
\tag{18.10}
$$

$$
\begin{aligned}
[V^+] &= \frac{1}{2}([V] + [I]) = \frac{1}{2}([Z] + [U])[I] \\
[V^-] &= \frac{1}{2}([V] - [I]) = \frac{1}{2}([Z] - [U])[I] \\
\frac{1}{2}[I] &= ([Z] + [U])^{-1}[V^+] \\
[V^-] &= ([Z] - [U])([Z] + [U])^{-1}[V^+] \\
[S] &= ([Z] - [U])([Z] + [U])^{-1}
\end{aligned}
\tag{18.11}
$$

S parameters may be measured by using a network analyzer as presented in figure 18.2(a). Advanced network analyzers can have two to sixteen ports. Measured S_{11} results by using a network analyzer are presented in figure 18.2(b).

18.4 *S* parameter measurements for RF devices

S parameter measurement of an antenna is usually a one port measurement. The first step in the measurement process is to calibrate the network analyzer to the desired frequency range. A one port, S_{1P}, calibration process consists of three steps.

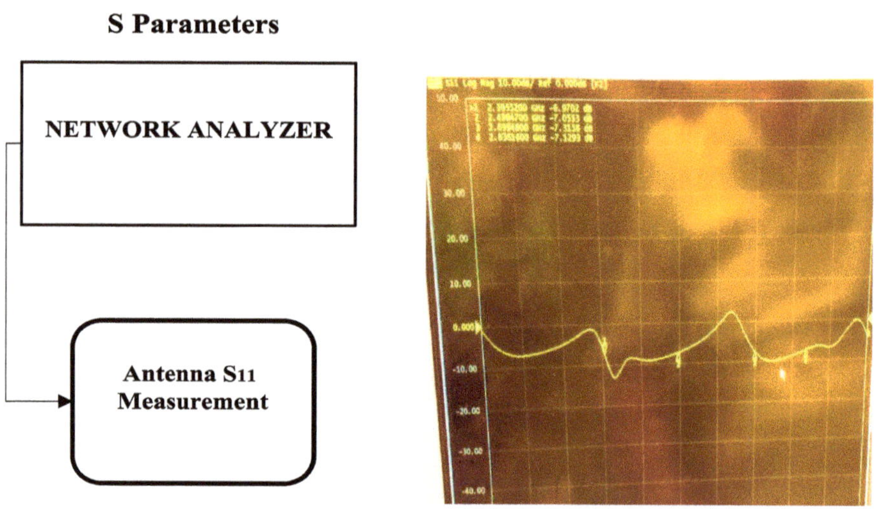

a. b.

Figure 18.2. (a) Measurements of $S11$ parameter for RF devices. (b) Photo of measured S_{11}.

- Short calibration. Connect a short connector to the input port and calibrate.
- Open calibration. Connect an open connector to the input port and calibrate.
- Load calibration. Connect a load connector to the input port and calibrate.

Connect the antenna to the network analyzer and measure the S_{11} parameters. Save and plot the S_{11} results. The measured result of the antenna's $S11$ parameters is shown in figure 18.2(b). A setup for S parameters measurement is shown in figure 18.3(a). The measurement results of the S_{11} parameters are shown in figure 18.3(b). A two port S parameter measurement setup is shown in figure 18.4. A two port, S_{2P}, calibration process consists of four steps.

- Short calibration. Connect a short connector to the input port and calibrate.
- Open calibration. Connect an open connector to the input port and calibrate.
- Load calibration. Connect load connector to the input port and calibrate.
- Through calibration. Connect port 1 to port 2 by the calibrated cable.

Measure the S parameters: S_{11}, S_{22}, S_{12}, and S_{21} for all channels. S_{21} parameter presents the RF head gain. Gain flatness and phase balance in a system with N channels is measured by comparing the S_{21} magnitude and the phase measured of all the channels. A setup to measure the RF head gain and flatness is presented in figure 18.4(a). A two port network analyzer is shown in figure 18.4(b). Typical measured S parameter results are listed in table 18.1. S parameters can be converted to dB by using equation (18.12).

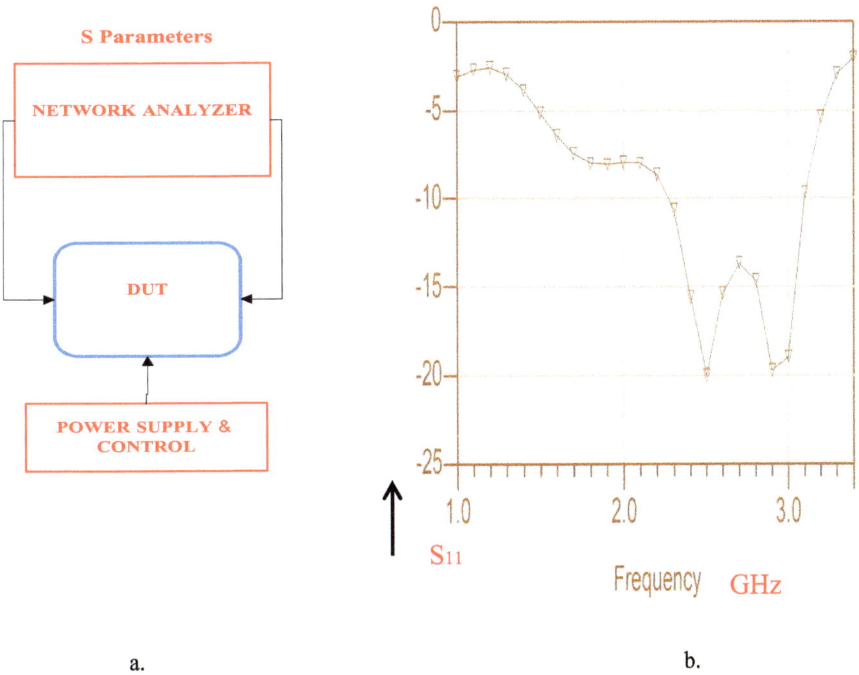

Figure 18.3. (a) Two port S parameter measurements. (b) S_{11} parameter results.

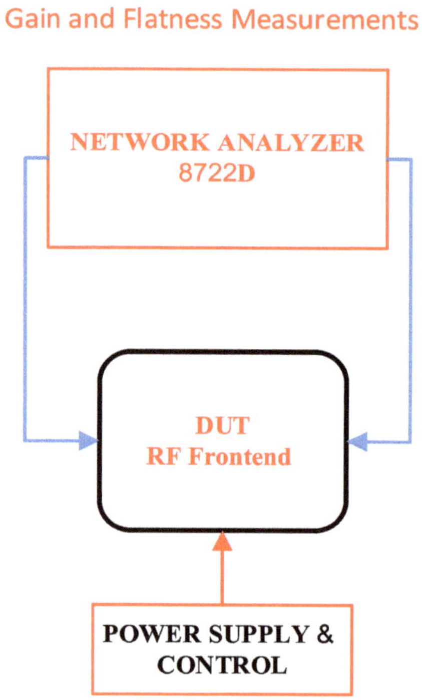

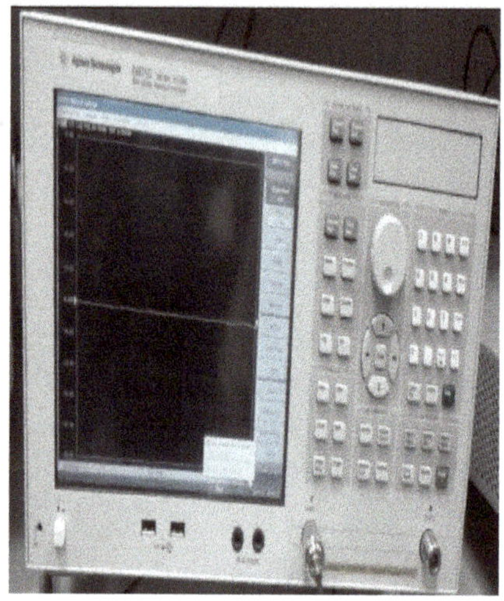

a. b.

Figure 18.4. (a) Measurements of gain and gain flatness. (b) Network analyser photo.

Table 18.1. Typical RF head *S* parameter measured results.

Channel	S_{11} (dB)	S_{22} (dB)	S_{12} (dB)	S_{21} (dB)
1	−12	−11	−22	31
2	−10.5	−11	−23	29
3	−11	−10	−20	29
n-1	−10	−9	−20	29
n	−9	−10.5	−19	28

$$S_{ij}(\text{dB}) = 20 \times \log\left[S_{ij}(\text{magnitude})\right] \tag{18.12}$$

Types of *S* parameters measurements

Small signal *S* parameters measurements—Passive networks are linear at any power level. In small signal *S* parameter measurements the signals have only linear effects on the network. **Large signal *S* parameters measurements**—The *S* matrix will vary with input signal strength. *S* parameters are measured for different power values.

18.5 RF transmission measurements

In figure 18.5(a) a block diagram of a RF transmission measurements setup is presented. The RF transmission measurements setup includes a sweep generator, the, DuT, device under test, receiving and transmitting radiators and a spectrum analyzer. In figure 18.5(b) the measured transmission results using a spectrum analyzer are presented. To improve the measurement accuracy, absorbing material can be located near electromagnetic reflection sources such as metals, walls and the room floor.

The Friis equation allows us to compute the received power as written in equations (18.13) and (18.14). The receiving antenna in the measurement setup can be an antenna with a known gain, G_R, and r is the distance between the antennas.

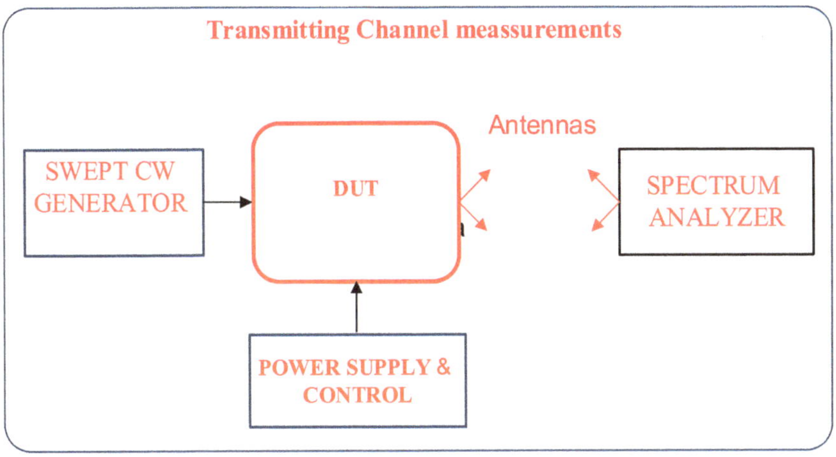

a.

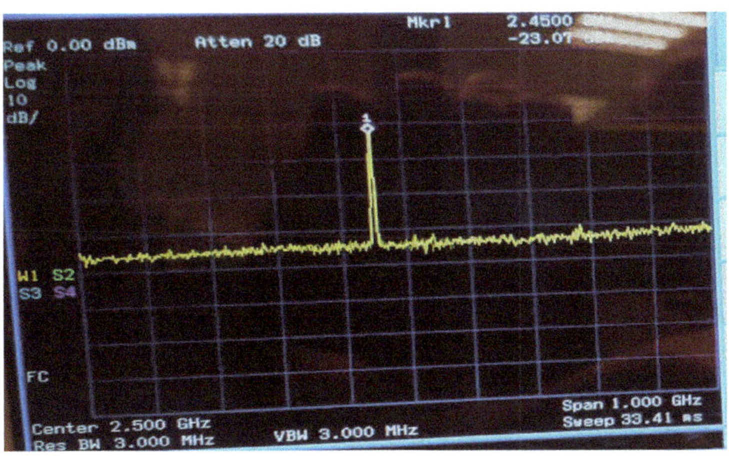

b.

Figure 18.5. (a) Block diagram of RF transmission measurements setup. (b) Measured S_{21} transmission results.

$$
\begin{aligned}
P_R &= P_T G_T G_R \left(\frac{\lambda}{4\pi r}\right)^2 \\
\text{For} - G_T &= G_R = G \\
G &= \sqrt{\frac{P_R}{P_T}}\left(\frac{4\pi r}{\lambda}\right)
\end{aligned}
\tag{18.13}
$$

$$
\begin{aligned}
P_R &= P_T G_T G_R \left(\frac{\lambda}{4\pi r}\right)^2 \\
\text{For} - G_T &\neq G_R \\
G_T &= \frac{1}{G_R}\frac{P_R}{P_T}\left(\frac{4\pi r}{\lambda}\right)^2
\end{aligned}
\tag{18.14}
$$

An example of the transmission measurement results can be presented as listed in table 18.2.

18.6 Output power and linearity measurements

In figure 18.6 a block diagram of measurements setup of output power and linearity of RF devices is presented. Measurements setup of output power and linearity of RF devices includes a sweep generator, device under test and a power meter or spectrum analyzer. In the process of output power and linearity measurements the input power is increased by 1 dB steps and we measure the device under test output power value and linearity.

18.7 Power input protection measurements of RF devices

In figure 18.7 a block diagram of measurements setup of power input protection of RF devices is presented. The output power and linearity measurements setup includes a sweep generator, power amplifier, device under test, attenuator and a spectrum, power meter or spectrum analyzer. In the process of input power protection measurements the input power is increased by 1 dB steps from 0 dBm

Table 18.2. An example of the transmission S_{21} tested antennas results.

Transmission measured S_{21} results of the tested antennas in dB				
F (GHz)	S_{21} (dB) Antenna 1	S_{21} (dB) Antenna 2	S_{21} (dB) Antenna 3	Remarks
1.1	−15	−14	−15	
1.15	−14	−13	−14	
1.2	−13	−12	−13	
1.25	−12	−11.5	−12.5	
1.30	−13	−12.5	−13	
1.35	−13.5	−13.5	−14	
1.4	−14	−14.5	−15	

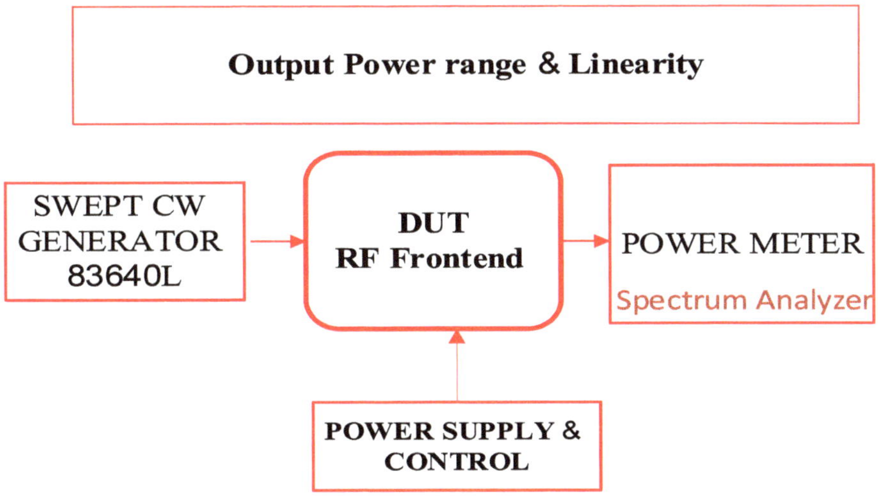

Figure 18.6. Output power and linearity measurements setup of RF devices.

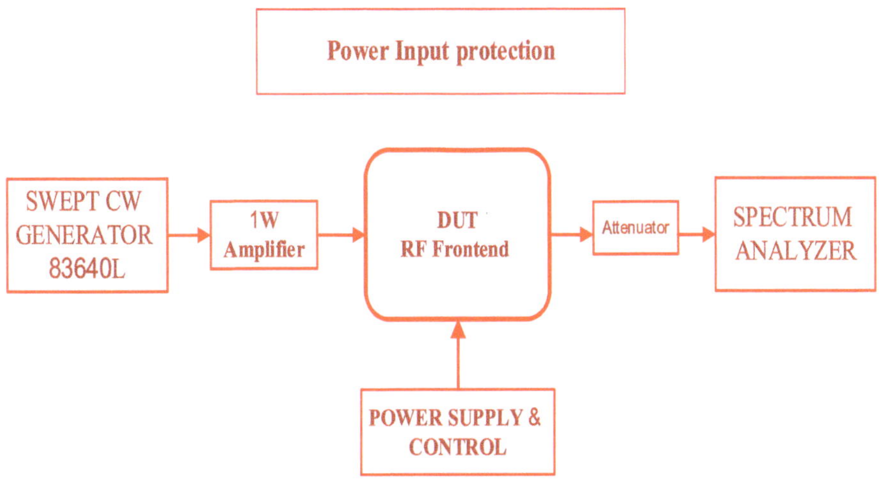

Figure 18.7. Power input protection measurement.

and we measure the output power level and observe that the device under test, DuT, functions with no damage.

18.8 Non-harmonic spurious measurements of RF devices

Figure 18.8(a) presents a block diagram of a non-harmonic spurious measurements setup of RF devices. Figure 18.8(b) presents a spectrum analyzer.

A non-harmonic spurious measurements setup includes a sweep generator, the device under test and a spectrum analyzer. In measurements of non-harmonic spurious, we increase the input power in 1 dB steps, up to the RF device 1 dB compression point, and measure the device output spurious values.

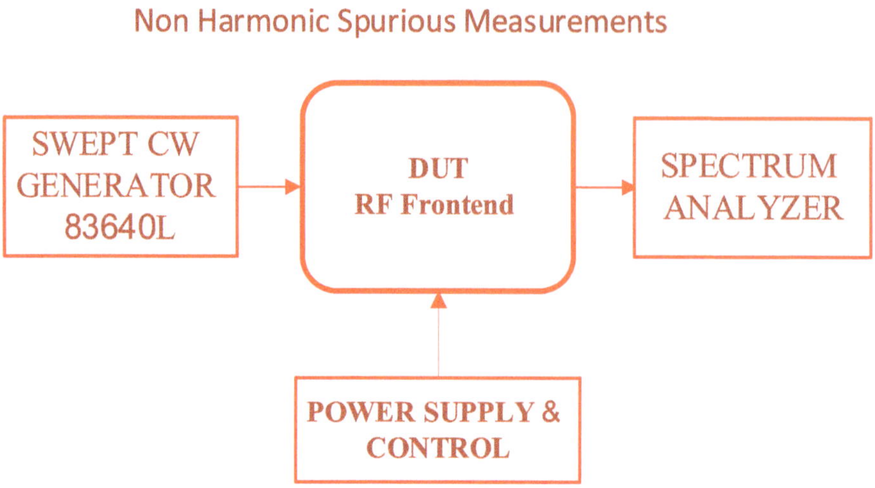

a.

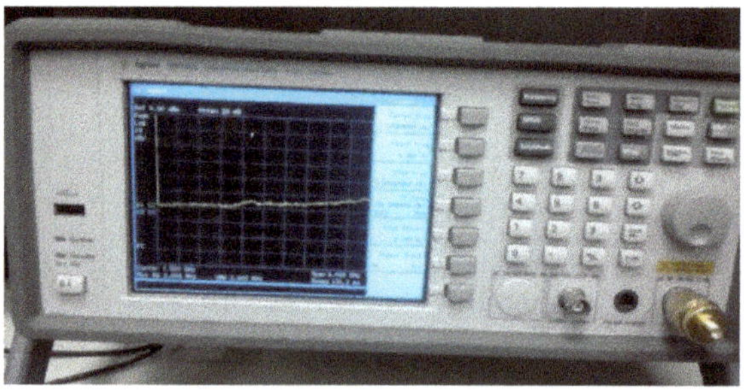

b.

Figure 18.8. (a) Block diagram of non-harmonic spurious measurements. (b) Spectrum analyser.

18.9 Switching time measurements of RF devices

Figure 18.9 presents the block diagram of the switching time measurements setup. The switching time measurements setup includes a sweep generator, the DuT, a detector, pulse generator and oscilloscope. Switching time may be measured by using the oscilloscope. In switching time measurements, we transmit an RF signal through the device. We inject a pulse through the switch control port. The pulse envelope may be detected on the oscilloscope screen.

18.10 IP_2 measurements

Figure 18.10 presents the setup for IP_2 and IP_3 measurements of RF devices. Figure 18.11 presents second order intermodulation results. IP_2 may be computed by equation (18.15).

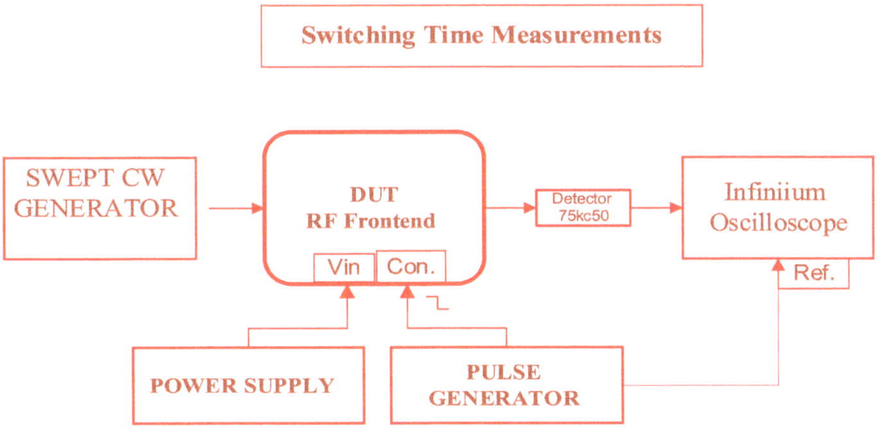

Figure 18.9. Switching time measurement setup of RF devices.

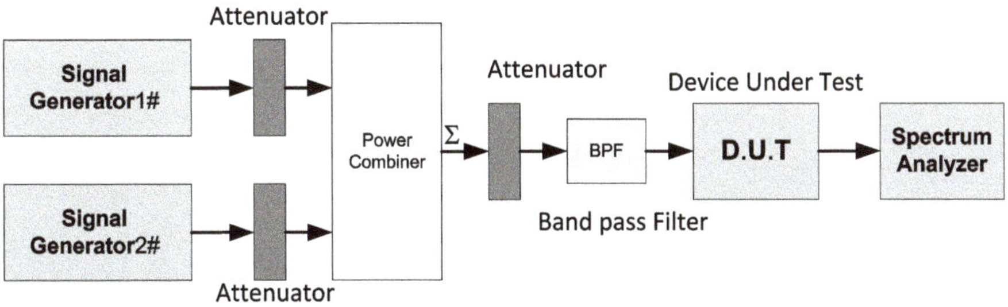

Figure 18.10. Setup for IP_2 and IP_3 measurements.

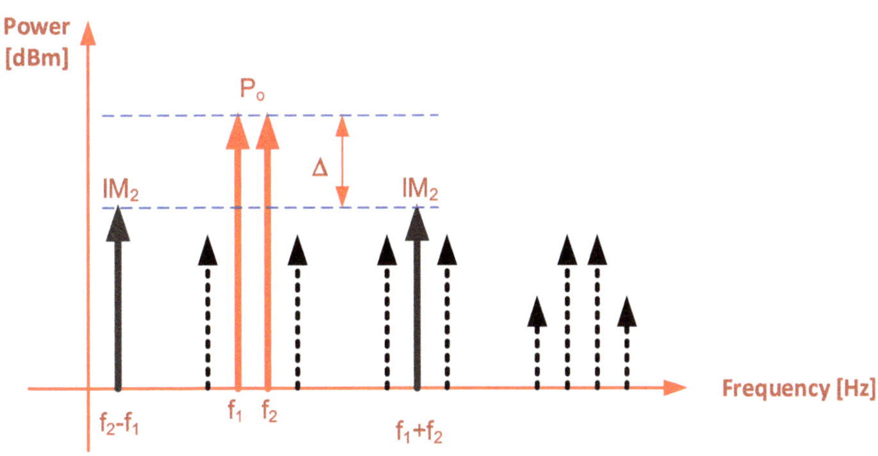

Figure 18.11. RF device second order intermodulation results.

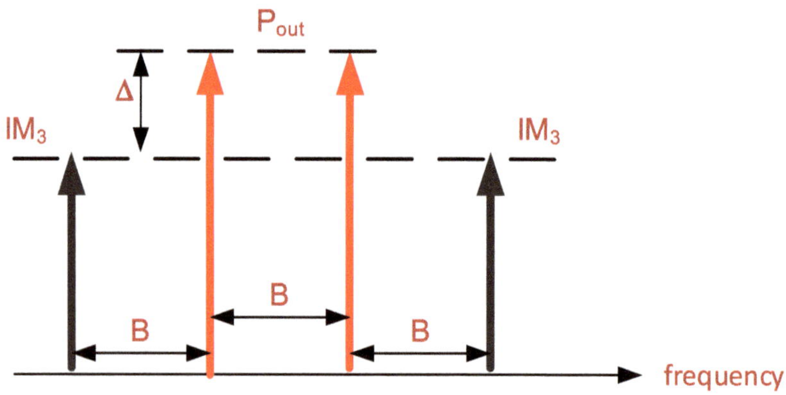

Figure 18.12. Third-order intermodulation results of RF devices.

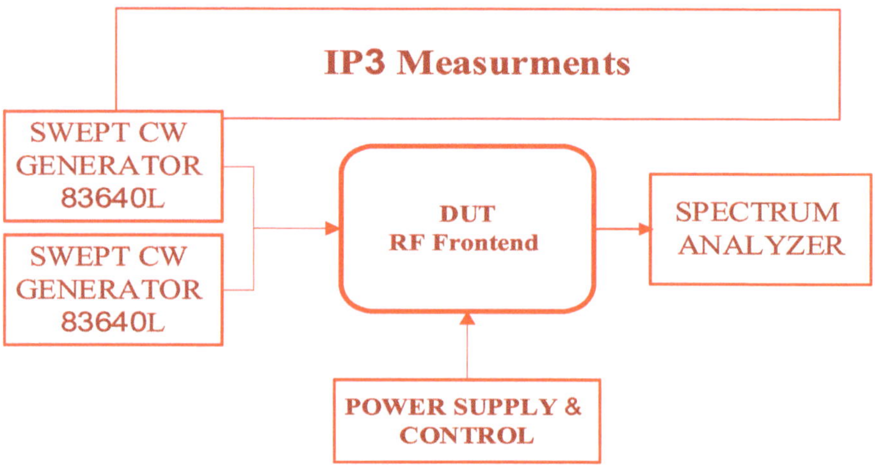

Figure 18.13. Two tone measurements.

$$IP_{2[\mathrm{dBm}]} = P_{\mathrm{out}[\mathrm{dBm}]} + \Delta_{[\mathrm{dB}]} \tag{18.15}$$

Third order intermodulation results are shown in figure 18.12. IP_3 may be computed by equation (18.16).

$$IP_{3[\mathrm{dBm}]} = P_{\mathrm{out}[\mathrm{dBm}]} + \frac{\Delta_{[\mathrm{dB}]}}{2} \tag{18.16}$$

18.11 IP_3 measurements

Figure 18.13 presents a block diagram of the output IP_3 measurements setup. The IP_3 setup includes two sweep generators, the DuT and a spectrum analyzer. In the IP_3 test we inject two signals to the DuT and measure the intermodulation signals.

Table 18.3. IP_3 measurements.

P_{in} (dBm)	19.999 (GHz) IM_1	$F_1 = 20$ (GHz)	$F_2 = 20.001$ (GHz)	20.002 (GHz) IM_2	IP_3 OUT	IP_3 INPUT
B	C (dBm)	D (dBm)	E (dBm)	F (dBm)	(dBm)	(dBm)
−17	−10.8	10	10	−13.8	20.4	−5.1
P_{in} (dBm)	29.999(GHz) IM_1	$F_1 = 30$ (GHz)	$F_2 = 30.001$ (GHz)	30.002 (GHz) IM2	IP_3	OUT
IP_3 INPUT						
−14.50	−17	10	10	−18.8	23.5	−0.1
P_{in} (dBm)	39.998 (GHz) IM_1	$F_1 = 39.999$ (GHz)	$F_2 = 40$ (GHz)	40.001 (GHz) IM2	IP_3	OUT
IP_3 INPUT						
−14.5	−12	10	10	−11.2	21	−3.9

The IP_3 test results of a receiving channel are listed in table 18.3. The input and output measured IP_3 may be computed by using equations (18.16–18.18).

$$IP3_{out} = P_{F1} + ((P_{F1} + P_{F2})/2) - P_{IM1})/2 \tag{18.17}$$

$$IP3_{IN} = \frac{(P_{F1} + P_{F2})}{2} + \frac{\left(\frac{(P_{F1}+P_{F2})}{2} - P_{IM2}\right)}{2} - \left(\frac{(P_{F1} + P_{F2}}{2} - P_{in}\right) \tag{18.18}$$

For example, the first signal is at 20 GHz with 10 dBm power level. The second signal is at 20.001 GHz with 10 dB power level. The first intermodulation signal is at 19.999 GHz with −10.8 dBm power level. The second intermodulation signal is at 20.001 GHz with −13.8 dBm power level. The power level of IP_3 at the input is −5.1 dBm, and is at the output 20.4 dBm.

18.12 Noise figure measurements

Figure 18.14 presents the block diagram of a noise figure measurements setup. The noise figure measurements setup includes a noise source, the DuT, an amplifier and a spectrum analyzer. The noise setup is calibrated without the DuT. We measure the difference in the noise figure, delta (Δ) value, when the noise source is on to the measured noise figure when the noise source is off. The excess noise ratio (ENR) is listed on the noise source for a given frequency. The measured NF is computed by using equation (18.19).

$$NF = 10 \quad \log(10^{0.1*\mathrm{ENR}})/((10^{0.1*\Delta}) - 1) \tag{18.19}$$

The noise figure measurements are listed in table 18.4.

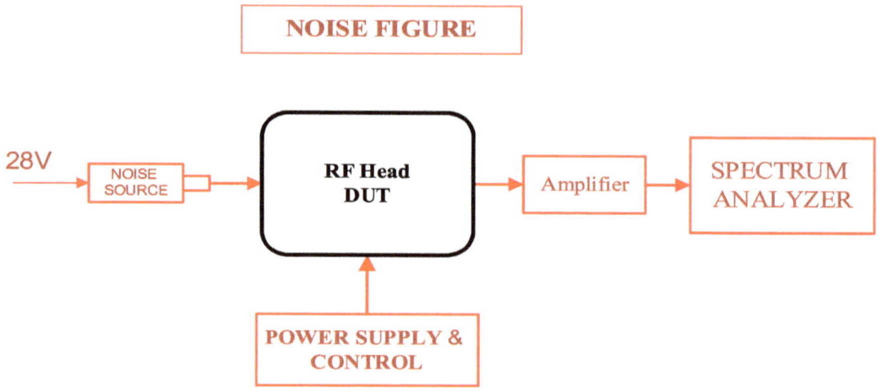

Figure 18.14. RF devices Noise figure measurements block diagram.

Table 18.4. Noise figure tested results of the RF device.

Parameter	Measurement 1	Measurement 2	Measurement 3
ENR	23.62	23.62	24
Delta [dB]	14	15	14
	9.8	8.75	10.16

18.13 Antennas' electrical performance measurements

Antennas' impedance and matching to the system are measured by measuring S_{11} parameters as functions of frequency. Antennas' radiation pattern, gain, directivity, beam width, polarization, and efficiency are measured in far-field or near-field ranges. During the testing of the antenna performance we check if the antenna meets the antenna and system specifications. S parameters measurements are presented in section 18.4.

18.13.1 Radiation pattern measurements

Antennas radiate electromagnetic fields that vary as a function of the angular direction in space. Radiation pattern measurements measure the radiated field at a constant distance from the antenna at various angles. The antenna's radiation pattern may be defined as the locus of the points that emit the same power level per unit surface. The radiated power per unit surface is proportional to the square of the electromagnetic electric fields. Accurate measurements of the antenna's electrical parameters such as gain, radiation pattern, side lobe level, and efficiency can be done in a far-field antenna range. In a far-field antenna range the antenna is placed in the far-field distance from the transmitting antenna. In measurement of large antennas, such as big reflectors, the far-field distance may be a thousand meters; in this case near-field techniques are used. In near-field measurements the fields are measured on a surface close to the antenna (around 3–10 wavelength). Fourier transform is used to transfer near-fields to far-fields. *R*, the far-field distance, the Fraunhofer distance, is written in equation (18.20).

$$R = 2D^2/\lambda \tag{18.20}$$

D is the larger antenna dimension and λ is the antenna wavelength. Radiation pattern graphs can be drawn by using Cartesian coordinates as presented in figure 18.15 [1–23]. In figure 18.16 a polar radiation pattern plot is presented. The Cartesian and polar plot are used to measure the antenna beam width, which is the angle at the −3 dB points around the maximum radiated field. In figure 18.17 a 3D radiation pattern of a loop antenna is presented.

Antennas main beam—An antenna main beam is the region around the direction of maximum radiation.

Beam width—The antenna angular range in which half of the maximum power is radiated is defined as the antenna beam width. This angular range of the main lobe is defined as the antenna 3 dB beam width. At these points the field strength falls by −3 dB from the maximum field value.

Side lobes' level—Radiation in undesired directions, away from the main beam, is called side lobes. The maximum value, in dB, of the side lobes away from the main beam characterize the antenna's radiation pattern.

18.13.2 Directivity and antenna effective area

Antenna directivity is the ratio between the maximum radiated power in a certain direction compared to the average radiated power to all directions over a sphere, as presented in equation (18.21) [1–5].

$$D = \frac{P(\theta, \phi)_{\text{maximal}}}{P(\theta, \phi)_{\text{average}}} = 4\pi\frac{P(\theta, \phi)_{\text{maximal}}}{P_{rad}} \tag{18.21}$$

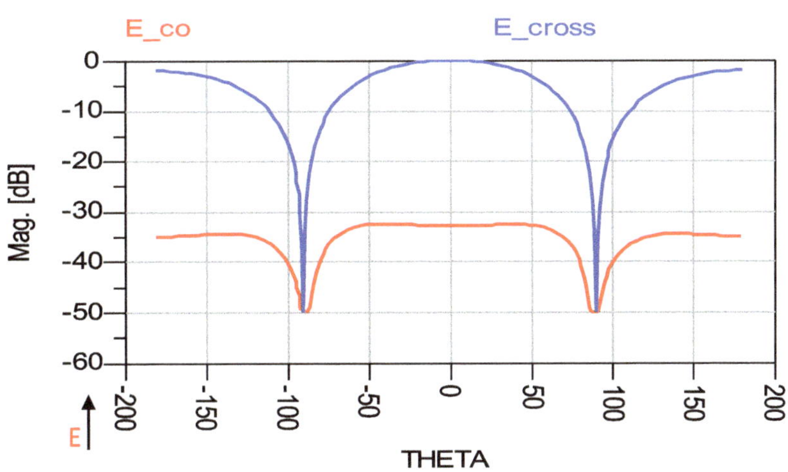

Figure 18.15. Radiation pattern of printed loop antenna on a human body.

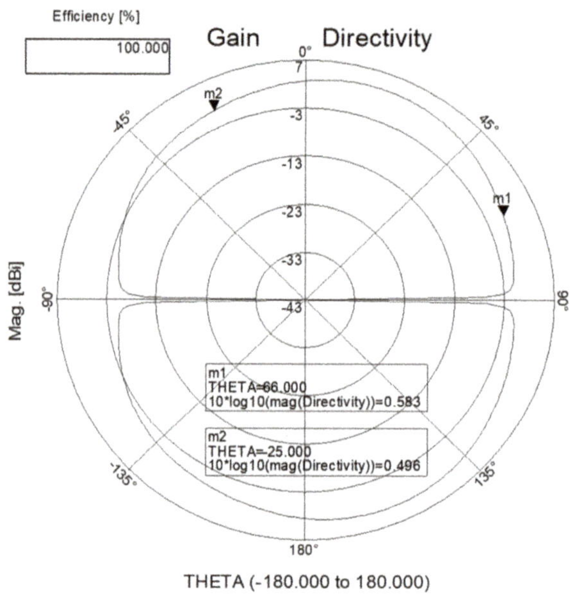

Figure 18.16. Radiation pattern of a wide band *S* band wearable printed slot antenna.

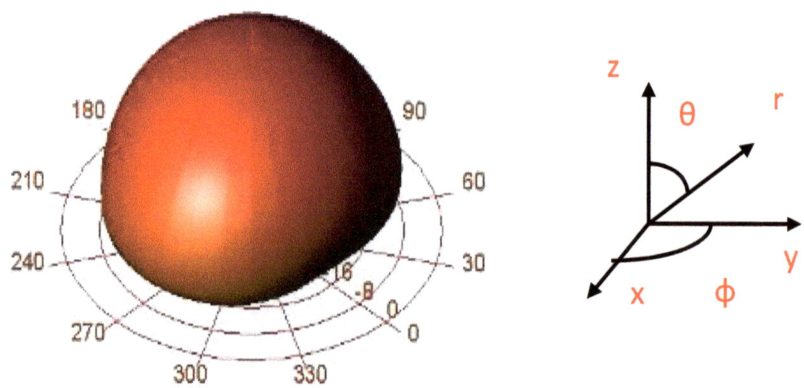

Figure 18.17. Loop antenna with ground, 3D radiation pattern on a human body.

where, $P(\theta, \phi)_{\text{average}} = \frac{1}{4\pi}\int\int P\left(\theta, \phi\right)\sin\theta d\theta d\phi = \frac{P_{\text{rad}}}{4\pi}$.

Antenna directivity, *D*, can be calculate, approximately, by using equation (18.22). The beam width in radian in the *AZ* plane and *EL* plane are used to calculate the antenna's directivity as presented in equation (18.22).

$$D \sim \frac{4\pi}{\theta E \times \theta H} \tag{18.22}$$

θE – Beam width in radian in *elevation* plane; θH – Beam width in *azimuth* plane.

Antenna's effective area (A_{eff})—The antenna's effective area that contributes to the antenna's directivity is given in equation (18.23).

$$A_{\mathrm{eff}} = \frac{D\lambda^2}{4\pi} \sim \frac{\lambda^2}{\theta E \times \theta H} \tag{18.23}$$

18.13.3 Radiation efficiency (α)

Radiation efficiency is the ratio between the antenna gain to the antenna directivity, $\alpha = G/D$. Radiation efficiency is the ratio of maximum power radiated to the total input power. Antenna efficiency takes into account losses and is the ratio between the total radiated power and the radiated power of an ideal lossless antenna. Efficiency is equal to the radiation resistance divided by total resistance of the feed point. Efficiency is also defined as the ratio of the radiated power to the total input power as written in equation (18.24).

$$\alpha = \frac{P_r}{P_r + P_l} \tag{18.24}$$

In figure 18.18 the radiation pattern of a wire loop antenna in free space is presented.

18.13.4 Typical antenna radiation pattern

In figure 18.19 a typical antenna radiation pattern is presented. The antenna's −3 dB beam is measured between the points where the maximum electrical field decays to $0.707E$. Half of the radiated power is concentrated in the antenna's main beam.

In figure 18.20 an elevated free space range is presented. In figure 18.21 an anechoic chamber is presented.

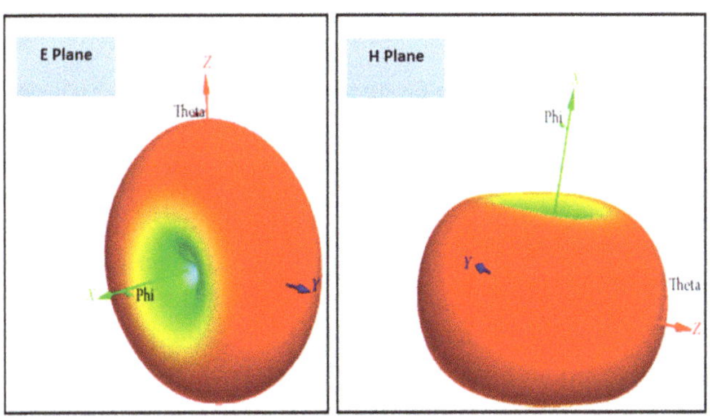

Figure 18.18. Radiation pattern of loop antenna in free space, *E* and *H* planes.

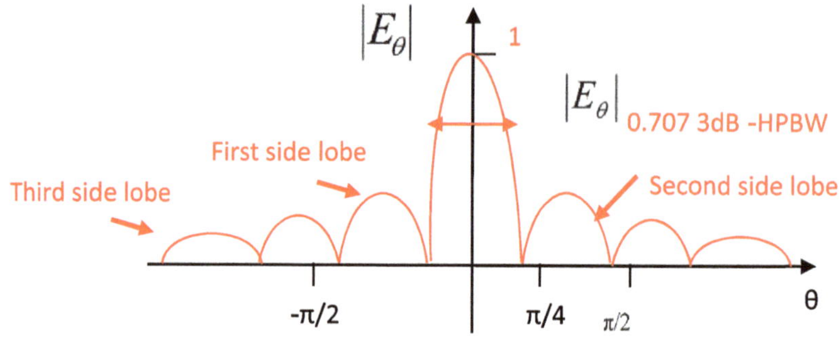

Figure 18.19. Radiation pattern of an antenna, main beam and side lobes.

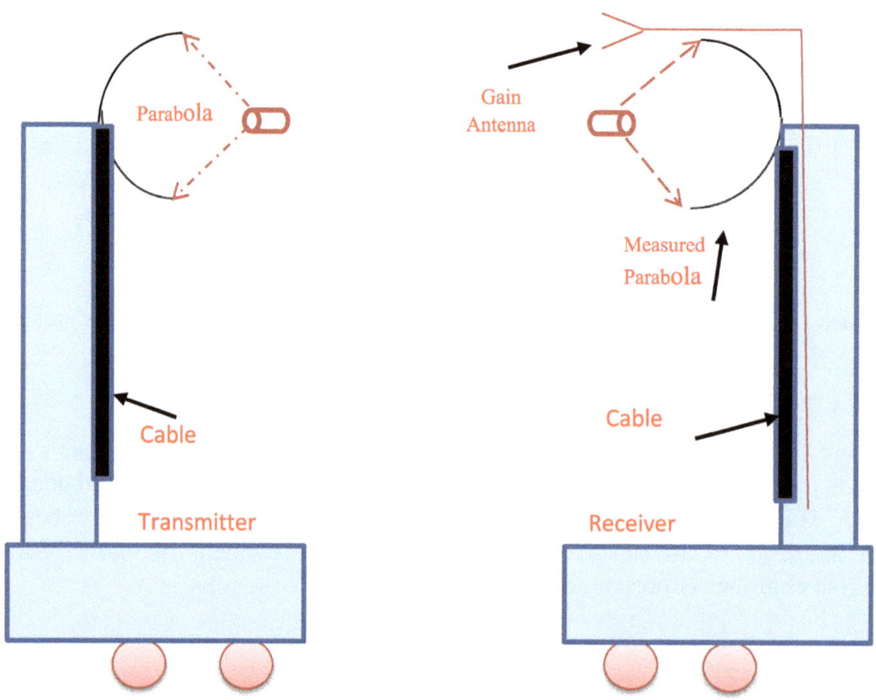

Figure 18.20. Antenna range for far-field radiation pattern measurements.

18.13.5 Gain measurements

Antenna gain (G)—Gain is defined as the ratio between the maximum radiated power to the power that would be radiated by an isotropic radiator. Figure 18.20 presents an antenna's far-field range for radiation pattern measurements. Antenna gain is measured by comparing the field intensity received by the antenna under test to the field intensity received by a standard horn as shown in figure 18.20. The gain as function of frequency of the standard horn is given in the antenna datasheets supplied by the standard horn manufacturer. Figure 18.21 presents an anechoic

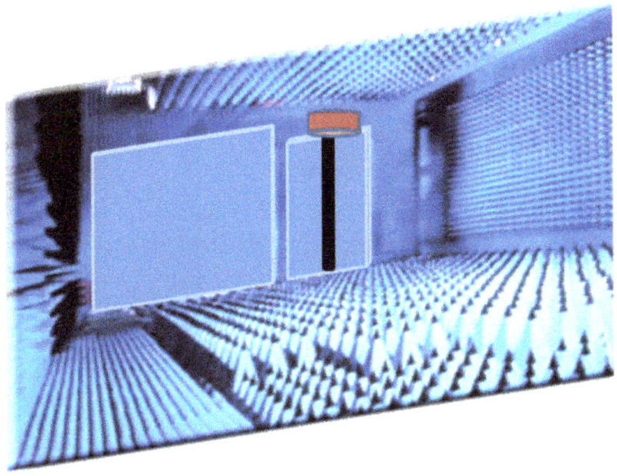

Figure 18.21. Anechoic chamber for indoor antenna measurements.

chamber used for indoor antenna measurements. The chamber's metallic walls are covered with absorbing materials.

Figure 18.21 presents an anechoic chamber used for indoor antenna measurements.

18.14 Antenna range setup

In figure 18.20 antenna range setup is presented. An antenna range setup consists of the following instruments:

- Transmitting system that consists of a wide band signal generator and transmitting antenna.
- The antenna under test.
- A receiving system.
- Positioning system.
- Recorder and plotter.
- Data processing system and a computer, antenna measurement software.

The signal generator should be with good spectral purity, controlled power level, with good frequency stability, and with a controlled frequency value. The receiving system includes the antenna under test, attenuators, a detector and amplifiers. Companies such as Agilent, Tektronix, Antritsu and others manufacture antenna measurement setups.

18.15 Conclusions

Microwave systems, communication systems, wearable systems, and antennas measurements are presented in this chapter. S parameters of wearable communication systems are measured by using a network analyzer setup.

Wearable sensors and antennas' electrical performance near or on the human body may be tested by employing a phantom. In the antenna development process the effect of the antenna's location on the human body should be modelled and simulated.

Setups for wearable systems and antenna measurements were presented in this chapter. Small antennas' radiation pattern, beam width, gain and efficiency can be measured in a far-field antenna range. However, the radiation pattern of large reflector antennas can be measured in a near-field antenna range.

Parts of this chapter have been reproduced from reference [6].

References

[1] Balanis C A 1996 *Antenna Theory: Analysis and Design* 2nd edn (New York: Wiley)

[2] Godara L C 2002 *Handbook of Antennas in Wireless Communications* (Boca Raton, FL: CRC Press)

[3] Kraus J D and Marhefka R J 2002 *Antennas for all Applications* 3rd edn (New Delhi: McGraw Hill)

[4] Sabban A 2014 *RF Engineering, Microwave and Antennas* (Tel Aviv: Saar Publication)

[5] Sabban A 2015 *Low Visibility Antennas for Communication Systems* (New York: Taylor and Francis)

[6] Sabban A 2016 *Wideband RF Technologies and Antenna in Microwave Frequencies* (New York: Wiley)

[7] Sabban A 2016 Small wearable meta materials antennas for medical systems *Appl. Comput. Electromagn. Soc. J.* **31** 434–43

[8] Sabban A 2015 New compact wearable meta-material antennas *Glob. J. Res. Anal.* **IV** 268–71

[9] Sabban A 2015 New wideband meta materials printed antennas for medical applications *J. Adv. Med. Sci.* **3** 1–10

[10] Sabban A 2013 New wideband printed antennas for medical applications *IEEE Trans. Antennas Propag.* **61** 84–91

[11] Sabban A 2013 Comprehensive study of printed antennas on human body for medical applications *J. Adv. Med. Sci.* **1** 1–10

[12] Sabban A 2013 Wearable antennas *Advancements in Microstrip and Printed Antennas* ed A Kishk (Rijeka: InTech)

[13] Sabban A 2011 Microstrip antenna arrays *Microstrip Antennas* ed N Nasimuddin (Rijeka: InTech)

[14] Sabban A 2015 Dually polarized tunable printed antennas for medical applications *IEEE European Antennas and Propagation Conf., EUCAP 2015 (Lisbon)*

[15] Sabban A 2014 New microstrip meta materials antennas *IEEE Antennas and Propagation* (Memphis, TN)

[16] Sabban A 2013 Wearable antennas for medical applications *IEEE BodyNet 2013* (Boston, MA) 1–7

[17] Sabban A 2013 New meta materials antennas *IEEE Antennas and Propagation* (Orlando, FL)

[18] Sabban A 2013 Meta materials antennas *New Tech Mag.* 16–9

[19] Sabban A 2012 Wideband tunable printed antennas for medical applications *IEEE APS/URSI Conf.* (Chicago, IL) 1–2

[20] Sabban A 2012 MM wave microstrip antenna arrays *New Tech Mag.* 16–21

[21] Sabban A 2011 New compact wideband printed antennas for medical applications *IEEE APS/URSI Conf.* (Spokane, WA) 251–4

[22] Sabban A 2010 Interaction between new printed antennas and human body in medical applications *Asia Pacific Symp.* (Japan) 187–90

[23] Sabban A 2009 Wideband printed antennas for medical applications *APMC 2009 Conf.* (Singapore) 393–6

IOP Publishing

Wearable Communication Systems and Antennas (Second Edition)

Design, efficiency, and miniaturization techniques

Albert Sabban

Chapter 19

Ethics in wearable healthcare and communication systems

Several ethical dilemmas must be considered when we develop wearable healthcare, IoT, computers, and wireless communication systems such as patient rights, patient privacy, spam mail, intellectual property, employee exploitation and justice.

19.1 Introduction to ethics theory and practice

For thousands of years ethics has played a major role in our world. The Bible presents many ethical issues. Ethic in Greek is ethos. Epicurus (341–270 BCE) and Aristotle (384–322 BCE) presented several ethical concepts and dilemmas. Ethics is a philosophical discipline concerning moral dilemmas. Ethics explores and studies the principles for deciding what is right and what is wrong in everyday life, in business, and in relations between peoples. Ethics is the analysis of the processes by which we decide what is right and what is wrong. Ethics is above obeying the law. Ethics tries to give answer to moral dilemmas.

Terms in an ethics dictionary

Moral—System of public norms or rules, written or not, about how people should behave.

Ethics—Ethics is the analysis of the processes by which we decide what is right and what is wrong. Ethics is the philosophy on morals.

Permissible—Dilemmas that are neither right nor wrong to do. Ethically 'neutral'.

Impermissible—Ethical dilemmas that are wrong to do. It is highly recommended not to do them.

doi:10.1088/978-0-7503-5224-6ch19

Supererogatory—Actions that are right to do, but it is not wrong not to do them.

Ethics types

- Nature of right or good, nature and justification of ethical issues are defined as meta-ethics.
- Standards and principles are defined as normative ethics.
- Actual application of ethical principles to some situations is called applied ethics.

19.2 The basics of ethics theory

Several theories and procedures have been presented to analyze and solve ethical problems, see [1–6]. The basic fundamentals of the theory and definitions of ethics were written around 3000 years ago, in the Bible and in ancient religious books. Ethical dilemmas were investigated by scholars in ancient Greece and in the Middle East.

Consequentialist theories

The utilitarian approach

Epicurus (341–270 BCE) said that the best life is life that produces the least pain and distress. Jeremy Bentham (1748–1832) stated that actions could be described as good or bad depending upon the amount and degree of pleasure or pain they would produce.

The egoistic approach is the ethics of self-interest. A person behaves according to their self-interest ignoring the interests of the public and others.

The common good approach

The common good approach is based on Plato (427–347 BCE) and Aristotle (384–322 BCE) ideas that our actions should contribute to ethical communal life. Moreover, Jean-Jacques Rousseau (1712–78) said that the best society should be guided by the 'general will' of the people, which would then produce what is best for the people in the community. The common good approach is based that the general good will of peoples will produce what is best for the community.

Ethical scepticism—Ethical standards are not universal but are relative to a person's culture and time. Global ethics states that a person must be sensitive and aware of different cultures. This is the relativist viewpoint which is called ethical scepticism.

Non-consequentialist theories

The duty-based approach is sometimes called deontological ethics. This approach is associated with Immanuel Kant (1724–1804). He stated that doing what is right is not about the consequences of our actions but about having the proper intention in performing the action.

The rights approach—Determines that the best ethical action is the act that protects the ethical rights of those who are affected by the action. It emphasizes the idea that all humans have a right to dignity, as well as animals and nonhumans.

The fairness or justice approach

The Law Code of Hammurabi in Ancient Mesopotamia (c 1750 BCE) held that all free men should be treated alike, just as all slaves should be treated alike. John Rawls (1921–2002) said that just, ethical principles are those that would be chosen by free and rational people in an initial situation of equality.

The divine command approach

The divine command approach sees that what is right is the same as that commanded by God. Ethical standards are the creation of God's will. Following God's will is seen as ethical.

Agent centered theories

The virtue approach—Ethics should be a major part of a person's life and not only with the discrete actions that a person may perform in any given situation. The virtue approach argues that ethical actions should be consistent with ideal human virtues.

The feminist approach—The feminist approach emphasizes the importance of the experiences of women and other marginalized groups to ethical discussion. The most important contributions of this approach are the principle of care as a legitimately primary ethical concern.

19.3 Medical ethics

Medical ethics is a field of applied medical and healthcare ethics. Medical ethics is the study of moral values and judgments as they apply to medicine and healthcare. As a scholarly discipline, medical ethics encompasses its practical application in clinical settings as well as work on its history, philosophy, theology, sociology, and anthropology. Medical ethics has a long history since ancient Egypt, 2700 BCE, the Code of Hammurabi in Babylon, 1750 BCE, and the Hippocratic oath, 460–370 BCE. In the UK, Percival's Code in 1803 was the basis of the American Medical Association (AMA) code issued in 1847.

Important medical ethics codes 1898 to 2000

- **Walter Reed (United States)** introduced **in 1898** written 'contracts' that allow healthy human people to be used in medical experiments.
- **Berlin code or Prussian code in 1900**. Medical experiments cannot be done when a person does not agree to give informed consent. Medical experiments cannot be done in the absence of unambiguous consent, or when information is not properly explained to a person.
- Helsinki Declaration on application to medical research, 1964. lately revised in 2000.

Medical ethics is based primarily on issues arising out of the practice of healthcare and medicine. Ethics is an essential part of medical practice and research. Some ethical principles are basic to the physician–patient relationship, but application in specific situations is often problematic due to disagreement about what is the right way to act. Medical ethics is based on the process of deciding what is the best way to act in a given situation. The best decision taken is based on the real-time information known about at a given time. For example, a physician cannot treat or operate on a patient without the acceptance of the patient or their family.

Biomedical ethics, bioethics

Bioethics is a broad subject, concerned with the moral issues raised by research and development in the biological sciences. Bioethics is an approach of understanding and examining what is 'right' and what is 'wrong' in biomedical research and practice, for example, injecting experimental drugs into animals, or how to operate on and treat animals in research labs. Bioethics does not require the acceptance of certain traditional values that are fundamental to medical ethics. Bioethics is a branch of applied ethics that studies the philosophical, social, and legal issues arising in medicine and the biological sciences. Bioethics is mainly concerned with human life and wellbeing, though it sometimes also treats ethical questions relating to the nonhuman biological environment. Replacing human organs and their functions raises several ethical dilemmas.

19.4 Ethical problems

Ethical problems due to the physician–patient relationship

- Several ethical problems are related to the relationship between physician and patient. For example, in a physician's office there are two patients, a well-known rich businessman and an old man. The physician must decide which person to serve first. The rich man gave a nice present to the physician. This case may be categorized as a problem caused by the fact of having to choose between what is right to do or whether they owe an obligation to a rich person.
- The physician's brother, a rich businessman and a foreign citizen are on a waiting list for surgery. The physician must decide which person to serve first.

Ethical problems due to lack of information

- The physician or biomedical researcher decides what to do without collecting all the information.
- The physician gives a patient the wrong medicine because he did not have time to perform all the medical tests.

Ethical problems due to malfunction and dishonesty

- The physician does not prescribe the best medicines to the patient because the medicines he uses are manufactured in a company that his brother owns.
- A biomedical researcher is recommended to use a medical system manufactured by a company that gave them a research grant. However, another company developed a better medical system.
- A medicine company sells a drug without declaring what side effects the drug may cause.

How to make ethical decisions

- Awareness, check if there is a moral issue.
- Check the facts.
- Check the moral issues.
- Examine what rules or values apply in each case.
- Check your duties and obligations.
- Who must decide and act?

- What are your ability, responsibilities and duties?
- Analyze the consequences for each situation in real time.
- Analyze and check all the options to solve the ethical dilemmas.

How do we decide what are ethical dilemmas?

Rational approaches

- Use well-known, and standard rules to make the right decision.
- The greatest good for the greatest number of people.
- The right action is the one that produces the best outcomes.
- Measure the action utility.
- The right action should be based on principles, values and codes.

Non-rational approaches

- Obedience to the boss and authorities. Following the rules or instructions of those in authority, whether or not you agree with them.
- Imitation of strong persons. Following the behaviour of the role model and leaders.
- Intuition and instincts, feeling in the mind.
- Feeling or desiring a subjective approach.
- There is no need to repeat a systematic process each time a moral issue arises like one that has been dealt with previously. Habit and tradition.

Solving process of ethical dilemmas

1. Determine whether the issue at hand is an ethical dilemma. Problem identification (technical facts, moral parameters, legal constraints, relevant human values).
2. Consult authoritative sources to see how experienced medical staff generally deal with such ethical dilemmas.
3. Consider alternative solutions. Develop alternative solutions (identify relevant ethical principles for each alternative; recognize ethical assumptions for each alternative; determine additional emerging ethical problems).
4. Discuss your proposed solution with those whom it will affect. Select one alternative solution (justify the proposed solution, check the selected solution against ethical theories). Present the proposed solution to those whom it will affect.
5. Apply the decision and act on it. Consider objections to the selected solution (objections arising from factual errors, fault reasons, conflicting values).
6. Evaluate the decision and be prepared to act differently in future.

Solving ethical dilemmas in healthcare

Factors that help to solve ethical dilemmas in healthcare.

1. Ethical principles.
2. Ethical rules.
3. Ethical codes.
4. Ethical theories.

5. Justice, respect and fairness.
6. Fidelity to the patients and the community.
7. The written law.
8. Doing no harm to patients.

19.5 Ethics in organizations and companies

In the last 30 years, healthcare organizations, biomedical companies, high-tech companies and professional organizations have become international companies. People from different countries and cultures work in the same organizations and business environments. There are significant differences in the moral and ethical values between different cultures. The organization can be managed efficiently only if the moral and ethical values are the same for all employees. Several ethical tips for organizations are listed below.

Ethical recommendations for organizations

- Develop a code of ethics. A list of ethical principles and beliefs that state how employees of an organization must behave.
- Implement the code in the company's culture top-down.
- The company management must support the implementation of the ethical code.
- Appoint an ethical committee.
- Treat ethics as a process.
- The company ethical committee must meet once a week and discuss ethical cases that have occurred.
- Create open lines of communication with the company management and ethical committee.
- The company management must set examples of good behavior.
- Educate employees to behave ethically. Ethical meetings, workshops, reports and publication of ethical dilemmas.
- Rewards and punishments for ethical behaviors.

Ethical codes

In the last 20 years, ethical codes have been a must in healthcare organizations, biomedical companies, high-tech companies and in professional organizations. Government authorities and international organizations only offer research grants to companies that have an ethical code. The ethical code must be written by company staff and not by company management. A company's ethical code must represent the company's moral code and state what the staff can and cannot do. The company's management must support the implementation of the ethical code in the company. The company's ethical committee must meet once a week and discuss ethical cases that have occurred. The company's ethical committee must publish a report that presents ethical cases that have happened in the company. The committee must represent all employees in the company.

Basic principles of a company ethical code

- Respect for people.
- Ethical standards such as justice, dignity, honesty, equality and professionalism.

- Obligation to the ethical code by company management and all employees.
- What the staff can and cannot do.
- The ethical code must protect the employee's rights and privacy.
- The ethical must reflect the company environment.

Benefits of managing ethic codes in organizations and companies

- Improves the culture of the organization.
- Creating a moral environment in the organization.
- Improves employee productivity, morale, teamwork and production.
- Establishes values for quality management, strategic planning and diversity management.
- Improves significantly public image.
- It is an insurance policy and the right thing to do in the long term.

Ethics problems in organizations that the ethical code of the company must solve

If a company's ethical code is relevant to the company and implemented by the company management, the ethical problems listed below should decline and be prevented.

- Employees lying to the management and supervisors.
- Management lying to employees, customers, vendors or the public.
- Misuse of organizational assets.
- Lying in reports and falsifying records.
- Sexual harassment.
- Stealing of company assets.
- Accepting or giving bribes.
- Withholding information from employees, customers, vendors or the public.

19.6 Ethical dilemmas in science research and development

Research topics come from areas that are important to people, and which affect society. The effect of research results and outcomes on society must be considered. For example, research and development of communication systems affect the environment and society. Communication systems radiate electromagnetic radiation that may have an effect, and therefore the rate of the radiation should be monitored. Research and development of nuclear bombs affects the environment and society. Research sites must be situated far from towns and villages, and the rate of nuclear radiation should be monitored. Federal governments and other funding organizations use research funding to affect the topics researchers explore. The research results must not be related to or affected by the funding agencies.

Ethical problems to consider in research and science

- Check the moral issues involved in the research.
- Fraudulent activity by scientists is a primary concern. Using research funds for personal expenses. Publishing results that are not accurate. Publishing results that are based on the recommendation of the funding agency or company.
- Publishing partial results.

- Credit should be given to all research participants.
- Taking care that research participants are not harmed physically or psychologically.
- Perform the research according to international regulations and ethical codes. For example, the American Educational Research Association, AERA, code. For a link to the AERA Code of Ethics, see http://www.aera.net/Portals/38/docs/About_AERA/CodeOfEthics(1).pdf.

Ethical problems in biomedical research with participants

- **Participants** should be given a real and correct explanation, with details, of the goals and procedures of the research.
- **Participants** must get a detail description of any potential risks, discomforts and side effects expected during and after the research.
- **Participants** should be informed about possible benefits to be obtained by participating in the research.
- **Researchers** should disclose any alternative procedures and information that might be advantageous to the patient.
- **Researchers** should answer any questions participants may have during the research.
- **Participants** should be told that they are free to withdraw and discontinue participation in research at any time.

To encourage participants to be tested during the research, researchers must take care of full anonymity of the research.

- **Anonymity**: After identifying the sampling frame, the researcher should forget the names or any other unique identifiers of the participants. They should take care that personal data will not be published in the media. They should inform them of the journal outlet they have planned.
- **Confidentiality:** If one cannot maintain anonymity then promise **confidentiality**. The researcher should guarantee that noone will be individually identifiable in any way. All research tables, reports, and publications will only discuss findings with no personal information.
- **Informed consent**: Be honest and fair with the research participants. The researcher should tell the participant everything they want to know about their research.

Medical and biomedical ethical codes

The major topics of the following ethical codes are presented. AERA code of ethics The Belmont Report from 1979, and the Nuremberg Code

Topics of AERA code of ethics

The code of ethics of the American Educational Research Association was approved by the AERA Council in February 2011. The code sets forth the ethical principles and standards that govern the professional work of education researchers. The current code is intended to provide guidance that informs and is helpful to education researchers in their research, teaching, service, and related professional work.

- Professional competence education researchers strive to maintain the highest levels of competence in their work.
- Education researchers with integrity are honest, fair, respectful of others in their professional activities in research, teaching, practice and service.
- Respect for people's rights, dignity and diversity. Education researchers respect the rights, dignity, and worth of all people and take care to do no harm in the conduct of their work. In their research, they have a special obligation to protect the rights, welfare, and dignity of the research participants.
- Ethical standards.

Education researchers adhere to the highest possible standards that are reasonable and responsible in their research, teaching, practice, and service activities.

The Belmont report 1979

The Belmont ethical code's main topics are:

- Voluntary consent of the research participants. A researcher must protect participants with diminished abilities, children and prisoners.
- Avoidance of accidental death or disability to the research participants. The researcher must ensure that no harm comes to the participants and minimize risks.
- Avoidance of unnecessary suffering to the research participants. Fairness in procedures in selecting participants.

Nuremberg Code 1947

The Nuremberg (ethical) Code is a list of rules established by a military tribunal on Nazi war crimes during World War II. The main topics are:

- Voluntary consent of the research participants.
- Avoidance of accidental death or disability of the research participants.
- Avoidance of unnecessary suffering to the research participants.
- Termination of research if harm is likely.
- Experiments should be conducted by highly qualified people.

19.7 Ethical dilemmas for using computers and the internet

Computers and the internet have no borders. In the internet era the world became a global village. Anyone can access information about almost any person in the world. We may declare that privacy has vanished in our universe. Anyone can send an email with a message to almost any person in the world. Hackers can access private and business computers and steal personal, commercial, military and medical data. Fake companies sell fake drugs via the internet.

List of ethical dilemmas for using computers and the internet

- Privacy violation. Anyone can access information about almost any person in the world.
- Virus attack. Unknown people spread viruses via email messages or websites.
- Virtual objects and people commit violent and sexual crimes via the internet.

- Mailing and placing of fake information via email, Facebook, Instagram and other internet social media.
- Theft of information and intellectual property via the internet.
- Hacking. Hackers access private and business computers and steal personal, commercial, military and medical data.
- Commercial crimes. Selling products by fake companies. Selling fake drugs via the internet.
- Hackers create fake identities and use other people's personal identities.
- A person may have several identities in the web.
- Unwanted intrusions into a person's daily life. Sending fake mail and information.
- Pirate downloading of software, music, books and movies without the owner's permission.
- Exploitation of computer resources. Students and staff may use a university's computer resources for their private needs. For example, students may use their student account to run their own digital business, may open a popular website and download music. University staff members may use the university's server or computer system to download or view or store private content that is against university policies.
- Plagiarism: copying or duplicating school assignments, books, information and articles. For example, copying home assignments. Copying and publishing research results from the internet.

Ethical recommended behaviors and the internet

- Apply ethical and moral values when using the internet and computers.
- Do not access files without the permission of the files' owners.
- Respect copyright laws and policies.
- Do not use computers to harm other users.
- Do not use computers to steal another's identity and information.
- Do not copy copyrighted music, software and movies without the author's permission.
- Respect the privacy of computer users, just as you want to protect your privacy.
- Ask permission from another user to use their computer and resources.
- Do not use rude and impolite language while chatting, sending emails, blogging and using social networks.
- We need to respect others' views and should not criticize people on the internet.

19.8 How to prevent and minimize ethical crimes in the digital media

Report to the authorities about illegal communication in the web.

- Report to the authorities about illegal communication and activities.
- Do not write user ID and passwords on paper or anywhere else.
- Complain to the Internet Crime authority, IC3. They accept online internet crime complaints from either the actual victim or from a third party.

A link to the IC3 website: https://www.ic3.gov/default.aspx.

How to protect a computer from spam email

- Do not open spam emails. Delete spam mails as soon as possible.
- Never respond to spam as this will confirm to the sender that it is a 'live' email address.
- Have a primary and secondary email address. One for people you know and one for all other purposes.
- Avoid giving out your email address unless you know how it will be used.
- Never purchase anything advertised via an unrecognized email.

Preventing auction fraud

- Before you bid, contact the seller with any questions you have.
- Review the seller's feedback.
- Be cautious when dealing with individuals outside of your own country.
- Ensure you understand refund, return, and warranty policies.
- Determine the shipping charges before you buy.
- Be wary if the seller only accepts wire transfers or cash.
- If an escrow service is used, ensure it is legitimate.
- Consider insuring your item.
- Be cautious of unsolicited offers.

Preventing credit card payment fraud on the internet

- Ensure that a site is secure and reputable before providing your credit card number online.
- If purchasing merchandise, ensure it is from a reputable source.
- Promptly reconcile credit card statements to avoid unauthorized charges.
- Do your research to ensure the legitimacy of the individual or company.
- Beware of providing credit card information when requested through emails.

Preventing identity theft on the internet

- Do your homework to ensure the business or website is legitimate.
- Be cautious of scams requiring you to provide your personal information.
- Ensure that websites are secure before submitting your credit card number.
- Attempt to obtain a physical address, rather than a PO box or maildrop.
- Never throw away credit card or bank statements in a usable form.
- Be aware of missed bills, which could indicate that your account has been taken over.
- Never give your credit card number over the phone unless you make the call.
- Monitor your credit statements monthly for any fraudulent activity.
- Report unauthorized transactions to your bank or credit card company.
- Review a copy of your credit report at least once a month.

Preventing extortion on the internet

- Ensure security is installed at every possible entry point.
- Ensure you are utilizing the most up-to-date patches for your software.

- Security needs to be multi-layered so that numerous obstacles will be in the way of an intruder.
- Identify all machines connected to the internet and assess the defense that's engaged.
- Identify whether your servers are utilizing any ports that have been known to represent insecurities.

Preventing phishing on the internet

- Avoid filling out forms in email messages that ask for personal information.
- Be suspicious of any unexpected email requesting personal information.
- Always compare the link in the email to the link that you are directed to.
- Log on to the official website, instead of 'linking' to it from an unrecognized email.
- Contact the actual business that sent the email to verify if the email is genuine.

19.9 Conclusions

For thousands of years ethics has played a major role in our world. Ethics is the study of the basis or principles for deciding what is right and what is wrong. Ethics is the analysis of the processes by which we decide what is right and what is wrong. Obeying the law is not ethics.

Ethical dilemmas arise every minute in our life in medicine, the internet, computing, biomedical research and business. Ethical problems can be related to the physician–patient relationship.

Bioethics is a broad ethics topic that deals with the moral issues raised by developments in the biological sciences. Bioethics is a way of understanding and exploring what is 'right' and what is 'wrong' in biomedical research and practice.

In the last 20 years, ethical codes have been a must in healthcare organizations, biomedical companies, high-tech companies and in professional organizations. A company's ethical code must represent the company's moral code and state what the staff can and cannot do.

The code of ethics of the American Educational Research Association sets the ethical principles and standards that govern the professional work of education researchers.

Computers and the internet have no borders. In the internet era the world became a global village. Anyone can access information about almost any person in the world. We may declare that privacy has vanished in our universe. Hackers can access private and business computers and steal personal, commercial, military and medical data. Fake companies sell fake drugs via the internet.

International laws and regulations must protect internet services and customers from internet crime.

References

[1] Thiroux J P and Krasemann K W 2011 *Ethics: Theory and Practice* 11th edn (London: Pearson)

[2] Lewis V 2016 *Bioethics: Principles, Issues, and Cases* 3rd edn (Oxford: Oxford University Press)

[3] Sabban A 2010 *Ethic and Design Lectures* (Karmiel: ORT Braude College)
[4] Sabban A 2012 *Product Design Lectures* (Karmiel: ORT Braude College)
[5] Link to AERA Code of Ethics http://aera.net/Portals/38/docs/About_AERA/CodeOfEthics (1).pdf)
[6] IC3 website https://ic3.gov/default.aspx)

www.ingramcontent.com/pod-product-compliance
Ingram Content Group UK Ltd.
Pitfield, Milton Keynes, MK11 3LW, UK
UKHW051910100926
8556IPUK00005B/841

9 780750 352208